FUZZY LOGIC AND SOFT COMPUTING

Bernadette Bouchon-Meunier is a principal researcher at LAFORIA-IBP, CNRS, Université Paris VI. She is Editor-in-Chief of the *International Journal of Uncertainty, Fuzziness and Knowledge-Based Systems.*

Ronald R. Yager is Director of the Machine Intelligence Institute at Iona College. He has published numerous articles and books in the area of fuzzy logic.

Lotfi A. Zadeh is Director of the Berkeley Initiative in Soft Computing (BISC) at the University of California at Berkeley. Professor Zadeh is the inventor of the field of fuzzy logic and the field of soft computing.

ADVANCES IN FUZZY SYSTEMS – APPLICATIONS AND THEORY

Honorary Editor: Lotfi A. Zadeh (*Univ. of California, Berkeley*)
Series Editors: Kaoru Hirota (*Tokyo Inst. of Tech.*),
George J. Klir (*SUNY at Binghamton*),
Elie Sanchez (*Neurinfo*),
Pei-Zhuang Wang (*Nat. Univ. of Singapore*),
Ronald R. Yager (*Iona College*)

Vol. 1: Between Mind and Computer: Fuzzy Science and Engineering
(*Eds. P.-Z. Wang and K.-F. Loe*)

Vol. 3: Comparative Approaches to Medical Reasoning
(*Eds. M. E. Cohen and D. L. Hudson*)

Vol. 4: Fuzzy Logic and Soft Computing
(*Eds. B. Bouchon-Meunier, R. R. Yager and L. A. Zadeh*)

Forthcoming volumes:

Vol. 2: Industrial Applications of Fuzzy Technology in the World
(*Eds. K. Hirota and M. Sugeno*)

Vol. 5: Fuzzy Sets, Fuzzy Logic, Applications
(*G. Bojadziev and M. Bojadziev*)

Vol. 6: Fuzzy Sets, Fuzzy Logic, and Fuzzy Systems: Selected Papers by Lotfi A. Zadeh
(*Eds. G. J. Klir and B. Yuan*)

Vol. 7: Genetic Algorithms and Fuzzy Logic Systems: Soft Computing Perspectives
(*Eds. E. Sanchez, T. Shibata and L. A. Zadeh*)

Advances in Fuzzy Systems – Applications and Theory Vol. 4

FUZZY LOGIC AND SOFT COMPUTING

Editors

Bernadette Bouchon-Meunier
LAFORIA-IBP, CNRS, Université Paris VI

Ronald R. Yager
Iona College

Lotfi A. Zadeh
University of California, Berkeley

World Scientific
Singapore • New Jersey • London • Hong Kong

Published by

World Scientific Publishing Co. Pte. Ltd.
P O Box 128, Farrer Road, Singapore 9128
USA office: Suite 1B, 1060 Main Street, River Edge, NJ 07661
UK office: 57 Shelton Street, Covent Garden, London WC2H 9HE

Library of Congress Cataloging-in-Publication Data

Fuzzy logic and soft computing / editors, Bernadette Bouchon-Meunier,
Ronald R. Yager, Lotfi A. Zadeh.
p. cm. -- (Advances in fuzzy systems; vol. 4)
Includes bibliographical references.
ISBN 9810223455
1. Expert systems (Computer science) 2. Fuzzy sets. 3. Neural
networks (Computer science) I. Bouchon-Meunier, B. (Bernadette),
1948– II. Yager, Ronald R., 1941– . III. Zadeh, Lotfi Asker.
IV. Series.
QA76.76.E95F883 1995
006.3--dc20 95-22063
CIP

Printed in Singapore.

PREFACE

Soft computing constitutes a collection of disciplines which include fuzzy logic, neural networks, genetic algorithms and probabilistic reasoning. It is fast emerging as a tool to help computer-based intelligent systems mimic the ability of the human mind to employ modes of reasoning that are approximate rather than exact. The basic thesis of soft computing is that precision and certainty carry a cost and that intelligent systems should exploit, wherever possible, the tolerance for imprecision and uncertainty. Considerable success has been achieved in the application of this principle, especially with the use of fuzzy logic, in the development of a large number of intelligent control systems. These types of systems have appeared in applications as diverse as large scale subway controllers and as small as video cameras. We are now at a juncture where the ideas implicit in soft computing will begin to have significant impact in many other domains of application, this is especially true of information-related applications such as database and information retrieval. Another area where we shall see great use being made of fuzzy logic technology is in the construction of "things that think", for example a coat that can adapt its thermal properties to different weather conditions.

This volume focuses on the current state of soft computing, especially the fuzzy logic component. It comprises seven sections. The first section consists of three articles on fuzzy logic and genetic algorithms. The second section focuses on the issue of learning in soft computing. The third section, on hybrid and fuzzy systems, concerns itself with the use of fuzzy logic in a number of different paradigms employed in soft computing. The fourth section is devoted to decision and aggregation techniques. The fifth section concentrates on the use of fuzzy technologies in database systems. The sixth section has ten articles on foundational issues in fuzzy set theory. The final section describes a number of applications in which fuzzy logic plays a major role.

B. Bouchon-Meunier

R. R. Yager

L. A. Zadeh

CONTENTS

APPLICATIONS OF FUZZY SETS

SPECIAL INVITED PAPER

FUZZY LOGIC AND GENETIC ALGORITHMS

EVOLUTIONARY ALGORITHMS FOR FUZZY LOGIC: A BRIEF OVERVIEW

Thomas BÄCK, Frank KURSAWE

University of Dortmund, Department of Computer Science, LS XI
D–44221 Dortmund, Germany
{baeck,kursawe}@LS11.informatik.uni-dortmund.de

ABSTRACT

Evolutionary algorithms are direct, global optimization algorithms gleaned from the model of organic evolution. The most important representatives, *genetic algorithms* and *evolution strategies*, are briefly introduced and compared in this paper, and their major differences are clarified. Furthermore, the paper summarizes the application possibilities of evolutionary algorithms in the design of *fuzzy logic controllers*. The optimization of fuzzy membership functions turns out to be a promising and successful application domain for evolutionary algorithms, while the automatic learning of fuzzy control rules by means of *fuzzy classifier systems* is still in an early stage of research.

1 Evolutionary Algorithms

Evolutionary algorithms are a class of direct, probabilistic search and optimization methods based on the model of organic evolution (where they also borrow most of the terminology from). The algorithms exploit the collective learning process within a *population* of *individuals*, and each of the individuals represents a search point in the space of potential solutions to a given problem. The start population (which is often randomly initialized) evolves towards increasingly better regions of the search space by means of randomized processes of *selection*, *mutation*, and *recombination*. The selection operator favours individuals of higher fitness (quality in terms of the objective function $f : M \to I\!R$ which characterizes the optimization problem) to reproduce more often than individuals of lower fitness. Recombination allows for the exchange of information (partial solutions) between individuals, and mutation introduces innovation into the population.

Using this high level of abstraction, we can formulate a general basic algorithm which subsumes the existing evolutionary algorithms. In the following, t denotes a *generation* (iteration) counter and $P(t) \in I^{\mu}$ is a population of μ individuals at generation t. I denotes the space of individuals and is not necessarily identical to the optimization problem's search space M because individuals may carry additional information. $P'(t) \in I^{\lambda}$ and $P''(t) \in I^{\lambda}$ are used to indicate intermediate populations of size λ ($\mu = \lambda$ is possible). Furthermore, we use $Q \in \{P(t), \emptyset\}$ to denote a set of individuals which might be taken into account by selection in addition to the intermediate population $P''(t)$. The resulting evolutionary algorithm consists of a simple loop of recombination, mutation, fitness evaluation, and selection which is iterated until a specific termination criterion is fulfilled:

ALGORITHM 1 (Basic evolutionary algorithm)

$t := 0;$
initialize $P(t) \in I^{\mu};$
evaluate $P(t);$
while not *terminate*$(P(t))$ **do**
  recombine: $P'(t) := r(P(t));$
  mutate: $P''(t) := m(P'(t));$
  evaluate $P''(t) \in I^{\lambda};$
  select: $P(t+1) := s(P''(t) \cup Q);$
  $t := t + 1;$
od
return(*best individual in* $P(t)$);

The currently most important and widely known concrete representatives of this general outline are *genetic algorithms* (GAs) [10, 12, 8], *evolution strategies* (ESs) [21, 22, 23], and *evolutionary programming* (EP) [7, 5, 6]. In the following, we restrict the presentation to a short overview of GAs and ESs (modern variants of EP are quite similar to ESs) to clarify the most important characteristics and differences of both approaches. The interested reader is referred to the original literature or the more detailed overview given in [2].

1.1 Genetic Algorithms

In traditional genetic algorithms, individuals are always represented by binary vectors $\vec{a} \in \mathbb{B}^l$ ($\mathbb{B} = \{0, 1\}$) of fixed length l. The reasons for this choice are both of historical and theoretical nature (the theory of genetic algorithms is based on the analysis of so-called *schemata* — similarity templates representing hyperspaces of $\mathbb{B}^l$ — and schema processing properties of the algorithm; see [8] for more details). In case of optimization problems which are not formalizable as functions $f : \mathbb{B}^l \to \mathbb{R}$, a coding mechanism is utilized to represent the search space of the optimization problem by binary vectors (finding such a code is often a difficult task, such that some recent applications of "genetic algorithms" are based on a combination of direct representations of candidate solutions and problem-specific genetic operators; see e.g. [18]).

Genetic algorithms put a strong emphasis on the recombination (*crossover*) operator as the main search operator. In its simplest form, crossover exchanges all bits to the right of a randomly chosen position between two individuals [10]. This one-point crossover can naturally be extended by sampling more than one breakpoint and alternately exchanging each second of the resulting segments [12]. In the extreme case of *uniform crossover*, a random decision whether to exchange it or not is made for each bit position of the individuals [26]. Besides the number of crossover points, the operator is characterized by the *crossover probability* p_c which denotes the probability per individual to undergo recombination (often, $p_c \approx 0.6$ is chosen)

The role of mutation is normally interpreted to be only of marginal importance in GAs (a "background" operator [10]). It works by occasionally inverting single bits of individuals with an extremely small probability p_m (e.g., $p_m \approx 0.001$ [12]). Recent investigations, however, clarify that the importance of mutation was so far underestimated and a more recommendable setting is given by $p_m = 1/l$ [19, 1].

Selection in genetic algorithms is a probabilistic operator which works by copying individuals from $P''(t)$ ($Q = \emptyset$) into the new parent population $P(t+1)$. Each individuals' selection probability (the probability to be copied) is given by the proportion of its fitness from the total population fitness (*proportional selection*):

$$p(\vec{a}_i) = \frac{f(\vec{a}_i)}{\sum_{j=1}^{\mu} f(\vec{a}_j)} .$$

Notice that this definition assumes positive fitness values and a maximization task; otherwise, so-called *scaling mechanisms* have to be used in combination with proportional selection [8].

Genetic algorithms always maintain a constant population size (i.e., $\mu = \lambda$) which is of the order of 50–100 individuals. Normally, the start population is randomly initialized (with probability 0.5 for a one respectively a zero bit) and the algorithm is terminated after a predefined number of generations has passed.

1.2 Evolution Strategies

Initially developed for experimental optimization purposes [21], evolution strategies are nowadays important computer algorithms for continuous parameter optimization problems $f : I\!R^n \rightarrow I\!R$ [22, 23]. In contrast to genetic algorithms, candidate solutions are directly represented by real-valued vectors $\vec{x} \in I\!R^n$, and individuals $\vec{a} = (\vec{x}, \vec{\sigma})$ consist not only of the vector $\vec{x}$, but also incorporate an additional, n-dimensional vector $\vec{\sigma} \in I\!R^n_+$ of positive standard deviations σ_i. These *strategy parameters* σ_i are utilized by the mutation operator to modify the corresponding object variables x_i ($i \in \{1, \ldots, n\}$).

Mutation works for each of the object variables x_i by adding normally distributed random numbers with expectation zero and variance σ_i^2 (indicated by the notation $N(0, \sigma_i^2)$). The standard deviations σ_i are neither constant nor explicitly controlled, but they also undergo a logarithmic-normally distributed variation mechanism:

$$\begin{aligned} \sigma_i' &= \sigma_i \cdot \exp(\tau' \cdot N(0,1) + \tau \cdot N_i(0,1)) \ , \\ x_i' &= x_i + \sigma_i' \cdot N_i(0,1) \ . \end{aligned}$$

The mutation of σ_i is based on a *global* factor $\tau' \cdot N(0,1)$ (the random number is sampled only once for the complete individual) and a *local* factor $\tau \cdot N_i(0,1)$ (the random number is sampled anew for each component). Schwefel recommends the settings $\tau' \sim 1/\sqrt{2n}$ and $\tau \sim 1/\sqrt{2\sqrt{n}}$ for the "learning rates" τ' and τ [22].

It is important to see that selection works implicitly on the strategy parameters $\vec{\sigma}$ by exploiting the link between advantageous changes of object variables (i.e., a large improvement of fitness) and useful standard deviations (i.e., appropriate *internal models* of the objective function topology). This mechanism of *self-adaptation* of strategy parameters allows for an adaptation of these parameters without the need for finding an appropriate exogenous control mechanism [24, 9]. Besides the standard deviations up to $n \cdot (n-1)/2$ covariances of the generalized n-dimensional normal distribution may also be taken into account for self-adaptation, which introduces linear *correlated mutations* to the algorithm. This mechanism is able to accelerate the search in case of a complicated local topology [23].

While mutation is the most important search operator in evolution strategies, recombination on strategy parameters and object variables is necessary for the self-adaptation process and often helpful for the progress of the search. Normally, an *intermediate recombination* operator is recommended for strategy parameters, i.e. a strategy parameter of an offspring individual results from taking the average of the corresponding strategy parameters of both parents. In case of the object variables, a *discrete recombination* operator where each x_i is randomly copied from either the first or the second parent (in analogy with uniform crossover in genetic algorithms) is normally preferred. The most appropriate choice, however, clearly depends on the particular objective function topology.

A further, secondary function of recombination in evolution strategies consists in changing the population size from μ to λ individuals (a setting of $\mu = 15$, $\lambda = 100$) is quite normal). This works by repeating the application of recombination on the level of individuals λ times (in contrast to genetic algorithms, recombination in evolution strategies is always applied, i.e., no parameter comparable with the crossover rate exists in ESs).

Finally, the selection operator in evolution strategies is completely deterministic and works by choosing the μ best individuals out of $P''(t)$ ($Q = \emptyset$, (μ,λ)-selection) respectively out of $P''(t) \cup P(t)$ ($Q = P(t)$, $(\mu+\lambda)$-selection) to become parents of the next generation. The (μ,λ)-selection is preferred in modern implementations of the ES, because it supports the self-adaptation mechanism (by providing the possibility to become extinct for inappropriate strategy parameter settings) and allows for an application of evolution strategies in case of time-varying or perturbed objective functions.

Since self-adaptation of strategy parameters is certainly the most distinguishing (and complicated) feature of ESs, we summarize the criteria which were identified by Schwefel to be critical for a successful self-adaptation mechanism [25]: A (μ,λ)-strategy should be used, with a not too small value of μ (i.e., the selective pressure should not be too strong), e.g., a (15,100)-strategy, and the recombination operator should be applied also on strategy parameters (especially intermediate recombination).

1.3 A Summary of Differences

Although, at first glance, the representation of individuals seems to be the distinguishing property of genetic algorithms and evolution strategies, the self-adaptation concept of strategy parameters — which is completely missing in genetic algorithms — is of much more importance. The process of tuning strategy parameters "by hand" for a particular application problem, which is often a time-consuming problem in applying a genetic algorithm, is not required for evolution strategies. Concerning the genetic operators, the roles of mutation in evolution strategies respectively recombination in genetic algorithms reflect en emphasis on different operators which might be explained by a phenotypical level of modeling in case of ESs and a genotypical level in case of GAs. Finally, the selection operators are characterized by deterministic, rank-based survival of the top group in evolution strategies versus a probabilistic mechanism with nonzero chances of survival even for the worst individuals in case of genetic algorithms. Currently, much of the theoretical basis to understand the pros and cons of these different principles of representations and operators is still missing and subject to active research.

2 Evolutionary Algorithms for Fuzzy Logic

The application possibilities for evolutionary algorithms in the field of fuzzy logic are documented by a number of recent research publications, which can roughly be divide into two groups:

- Optimization of the membership functions of fuzzy sets.
- Automatic learning of fuzzy rules.

In the following, we will briefly discuss the general idea for both topics and refer the interested reader to the literature cited for more details on the applications and fuzzy logic in general (see [15, 16]).

2.1 Optimization of Membership Functions

Fuzzy membership functions provide the characterization of fuzzy sets by establishing a connection between linguistic terms (such as "slow", "medium", "fast" for a speed variable) and precise numerical values of variables in a physical system. A fuzzy membership function approximates the confidence with which a numerical value is described by a linguistic term. A typical example of triangular fuzzy membership functions for a speed variable is given in figure 1.

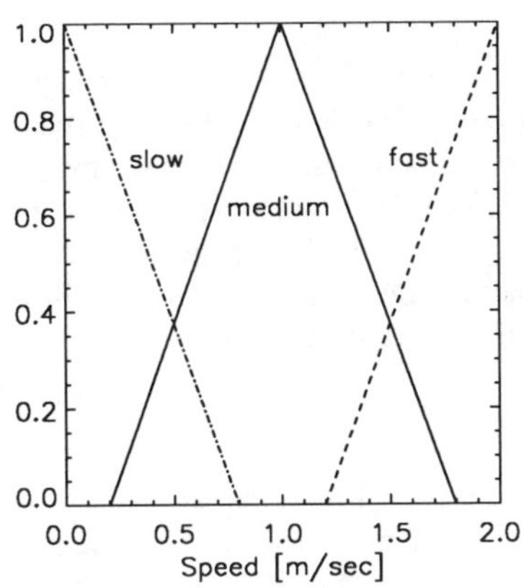

Figure 1: Membership functions for a physical variable "speed".

Notice that membership functions not necessarily have to be (isosceles) triangles as in this example. For instance, Gaussian membership functions

$$u(x) \quad = \quad \exp\left(\frac{-(x-c)^2}{2\sigma^2}\right)$$

provide an important alternative (e.g., see [13, 15, 29]). A further possibility for the shape of membership functions is to choose trapezoidals [14].

The correct choice of the membership functions, however, is by no means trivial but plays a crucial role in the success of an application. Several example applications demonstrate that evolutionary algorithms are capable of optimizing the membership functions of fuzzy logic controllers. The basic idea is to represent the complete set of membership functions by an individual and to evolve shape and location of the triangles (respectively the Gaussian curves). Each triangle may be described by its anchor points on the abscissa axis, and the Gaussian membership functions are characterized by c and σ.

Karr describes an application to the cart-pole balancing system and uses a genetic algorithm to evolve the membership functions of a fuzzy controller [13]. In order to evaluate the fitness of a controller, the system is simulated for a fixed simulation time, repeating the simulation four times for different initial conditions. The resulting, optimized fuzzy logic controller turns out to perform by far better than the controller based on membership functions designed by a human expert. Moreover, Karr also demonstrates that the genetic algorithm approach can be used successfully when the membership functions have to be altered in real time due to a variation of the cart mass (the expert-designed controller failed totally in this case) [13]. These promising results were recently confirmed by an application of the method for the online control of a laboratory pH system with drastically changing system characteristics [14].

Of course, rather than using a binary encoding of continuous parameters which characterize membership functions and a genetic algorithm, one might prefer an evolution strategy to optimize the membership functions. Wienholdt reports good results from an application of ESs to optimize *radial basis function* (RBFs — Gaussian membership functions) networks for time series prediction [29].

We conclude this section by referring to a problem which may arise from unconstrained variations of the membership function shape by the optimization algorithm: The *completeness* property, which requires that a fuzzy logic controller always be able to infer a control action for every state of the process (see [15]) might be violated if the anchor points of membership functions are shifted such that the possible range of values is no longer completely covered. In order to solve this problem, one might consider to introduce special constraints to the evolutionary algorithms' objective function.

2.2 Fuzzy Classifier Systems

Besides learning the membership functions, an even more challenging problem consists in the automatic learning of fuzzy *control rules*, i.e., the linguistic statements which are normally derived from expert knowledge [15]. This idea comes close to so-called *classifier systems* [4], rule-based systems which use a genetic algorithm as a rule-generation mechanism, such that the classifier system is capable of inductive learning [11]. Valenzuela-Rendón presented an extension of classifier systems which allows for the learning of fuzzy rules by incorporating fuzzyfication and defuzzyfication components as well as a fuzzy rule matching mechanism [27, 28]. Although the first application examples (the imitation of static linear respectively quadratic one-input one-output systems) are very simple, the approach may indicate a path towards automatic learning of fuzzy control rules. A promising further step into this direction was recently presented by Parodi and Bonelli, who extended Valenzuela-Rendón's approach by a mechanism which allows learning of fuzzy rules, membership functions, and output weights representing the relative importance of the fuzzy rules at the same time [20].

3 Conclusions

Evolutionary algorithms clearly represent a successful approach towards the optimization of fuzzy membership functions, and we expect this field of applications to grow remarkably in the near future. Of course, the method is only applicable if the quality function of the resulting fuzzy controller can be evaluated numerically without taking too much time, because evolutionary algorithms typically require the evaluation of a large number of individuals. On the other hand, the global search characteristics of these algorithms yield membership functions of surprisingly high quality in comparison with those defined by human experts.

Research concerning the automatic learning of fuzzy control rules by means of evolutionary algorithms respectively classifier systems is surely in its initial stage and much work remains to be done. Nevertheless, we are sure that this line of research should also be followed, because an automatic tool for this time-consuming task is highly desirable.

The development and optimization of fuzzy controllers are important topics for further research where evolutionary algorithms will surely prove to be very helpful.

References

[1] Th. Bäck (1992) The interaction of mutation rate, selection, and self-adaptation within a genetic algorithm, In: Männer and Manderick [17], p. 85–94.

[2] Th. Bäck and H.-P. Schwefel (1993) An overview of evolutionary algorithms for parameter optimization, *Evolutionary Computation*, 1(1):1–23.

[3] R. K. Belew and L. B. Booker (Eds.) (1991) *Proceedings of the 4th International Conference on Genetic Algorithms*, Morgan Kaufmann Publishers, San Mateo, CA.

[4] L. B. Booker, D. E. Goldberg, and J. H. Holland (1989) Classifier systems and genetic algorithms, In: J. G. Carbonell (Ed.), *Machine Learning: Paradigms and Methods*, The MIT Press / Elsevier, p. 235–282.

[5] D. B. Fogel (1991) *System Identification through Simulated Evolution: A Machine Learning Approach to Modeling*, Ginn Press, Needham Heights.

[6] D. B. Fogel (1992) *Evolving Artificial Intelligence*, PhD thesis, University of California, San Diego, CA.

[7] L. J. Fogel, A. J. Owens, and M. J. Walsh (1966) *Artificial Intelligence through Simulated Evolution*, Wiley, New York.

[8] D. E. Goldberg (1989) *Genetic algorithms in search, optimization and machine learning*, Addison Wesley, Reading, MA.

[9] F. Hoffmeister and Th. Bäck (1992) Genetic self–learning, In: F. J. Varela and P. Bourgine (Eds.), *Proceedings of the First European Conference on Artificial Life*, The MIT Press, Cambridge, MA, p. 227–235.

[10] J. H. Holland (1975) *Adaptation in natural and artificial systems*, The University of Michigan Press, Ann Arbor, MI.

[11] J. H. Holland, K. J. Holyoak, R. E. Nisbett, and P. R. Thagard (1986) *Induction: Processes of Inference, Learning, and Discovery*, MIT Press.

[12] K. A. De Jong (1975) *An analysis of the behaviour of a class of genetic adaptive systems*, PhD thesis, University of Michigan. Diss. Abstr. Int. 36(10), 5140B, University Microfilms No. 76–9381.

[13] C. L. Karr (1991) Design of an adaptive fuzzy logic controller using a genetic algorithm, In: Belew and Booker [3], p. 450–457.

[14] C. L. Karr and E. J. Gentry (1993) Fuzzy control of pH using genetic algorithms, *IEEE Transactions on Fuzzy Systems*, 1(1):46–53.

[15] C. C. Lee (1990) Fuzzy logic in control systems: Fuzzy logic controller — Part I, *IEEE Transactions on Systems, Man, and Cybernetics*, 20(2):404–418.

[16] C. C. Lee (1990) Fuzzy logic in control systems: Fuzzy logic controller — Part II, *IEEE Transactions on Systems, Man, and Cybernetics*, 20(2):419–435.

[17] R. Männer and B. Manderick (Eds.) (1992) *Parallel Problem Solving from Nature 2*. Elsevier, Amsterdam.

[18] Z. Michalewicz (1992) *Genetic Algorithms + Data Structures = Evolution Programs*, Artificial Intelligence, Springer, Berlin.

[19] H. Mühlenbein (1992) How genetic algorithms really work: I. mutation and hillclimbing, In: Männer and Manderick [17], p. 15–25.

[20] A. Parodi and P. Bonelli (1993) A new approach to fuzzy classifier systems, In: S. Forrest (Ed.), *Proceedings of the 5th International Conference on Genetic Algorithms*, Morgan Kaufmann Publishers, San Mateo, CA, p. 223–230.

[21] I. Rechenberg (1973) *Evolutionsstrategie: Optimierung technischer Systeme nach Prinzipien der biologischen Evolution*, Frommann–Holzboog, Stuttgart.

[22] H.-P. Schwefel (1977) *Numerische Optimierung von Computer-Modellen mittels der Evolutionsstrategie*, volume 26 of *Interdisciplinary Systems Research*, Birkhäuser, Basel.

[23] H.-P. Schwefel (1981) *Numerical Optimization of Computer Models*, Wiley, Chichester.

[24] H.-P. Schwefel (1987) Collective phenomena in evolutionary systems, In: *Preprints of the 31st Annual Meeting of the International Society for General System Research, Budapest*, volume 2, p. 1025–1033.

[25] H.-P. Schwefel (1990) Systems analysis, systems design, and evolutionary strategies, *Syst. Anal. Model. Simul.*, 7(11/12):853–864.

[26] G. Syswerda (1989) Uniform crossover in genetic algorithms, In: J. D. Schaffer (Ed.), *Proceedings of the 3rd International Conference on Genetic Algorithms*, Morgan Kaufmann Publishers, San Mateo, CA, p. 2–9.

[27] M. Valenzuela-Rendón (1991) The fuzzy classifier system: A classifier system for continuously varying variables, In Belew and Booker [3], p. 346–353.

[28] M. Valenzuela-Rendón (1991) The fuzzy classifier system: Motivations and first results, In: H.-P. Schwefel and R. Männer (Eds.), *Parallel Problem Solving from Nature — Proceedings 1st Workshop PPSN I*, volume 496 of *Lecture Notes in Computer Science*, Springer, Berlin, p. 338–342.

[29] W. Wienholt (1993) Optimizing the structure of radial basis function networks by optimizing fuzzy inference systems with evolution strategy, Internal Report IR-INI 93-07, Institut für Neuroinformatik, Ruhr-Universität Bochum.

GENERATING FUZZY RULES FROM EXAMPLES USING GENETIC ALGORITHMS[1]

Francisco HERRERA, Manuel LOZANO, Jose Luis VERDEGAY

Dept. of Computer Science and Artificial Intelligence
University of Granada,
18071 - Granada, Spain
e-mail: herrera,lozano,verdegay@robinson.ugr.es

ABSTRACT

The problem of generating desirable fuzzy rules is very important in the development of fuzzy systems. The purpose of this paper is to present a generation method of fuzzy control rules by learning from examples using genetic algorithms. We propose a real coded genetic algorithm for learning fuzzy rules, and an iterative process to obtain a set of rules that covers the examples set with a covering value previously defined.

Keywords: Fuzzy rules, learning, genetic algorithms.

1. Introduction

Fuzzy rules based systems have been shown to be an important tool to model complex systems, where due to the complexity or the imprecision, classical tools are unsuccessful.

The knowledge base of a fuzzy controller consists of a collection of rules describing the control actions. The performance of fuzzy control depends greatly on whether the control rules are reasonable or not.

There are different modes to derive them:

- Based on Expert Experience and Control Engineering Knowledge.
- Based on Operator's Control Actions.
- Based on the Fuzzy Model of a Process.
- Based on Learning or Self organizing Control.

The construction of fuzzy rules has been mainly based on the operator's control experience or actions. Unfortunately, adquiring rules from experts is not an easy task, and on the other hand, it is very difficult for a knowledge engineer to extract rules from static databases.

Recently there is an extensive literature of learning from examples in a fuzzy environment and various methods have been proposed. These methods include learning methods employing Descent Method, [1], or Neural Network, [10, 21], or fuzzy discretization and clustering techniques [14, 22], or mountain clustering method, [19], or frequencies in the identification of a system, [5], or Belief or Uncertainty measures, [3, 4, 15], or genetic algorithms [13], as well as tools to identify the structure of a fuzzy model [16, 17, 18].

Assuming that we can manage linguistic information from the experience of the human controller, and numerical information, in a learning process we can distinguish the following components:

1. A generation method of desirable fuzzy rules able to include the complete knowledge of the set of examples.

[1]This research has been supported under project PB92-0933.

2. A Combination and Simplify method.
    2.1. Resolving conflicts among the rules.
    2.2. Finding the final set of fuzzy rules able to approximate the input-output behavior of a real system.
3. A tuning method of the final set of rules.

We will focus this paper on the description of a method for generating desirable fuzzy rules from examples using genetic algorithms. We propose a generation process of fuzzy rules by mean of a special real coded genetic algorithm (RCGA), where a chromosome represents a fuzzy rule, and is evaluated by means of a frequency method. The RCGA finds good rules, and with an iterative process we obtain a set of rules covering the set of examples.

In order to that, we set out the paper as follows. Next section describes the fuzzy rule structure that we use. Section 3 introduces the genetic algorithms. Section 4 presents the generation process of fuzzy rules and section 5 presents a numerical example.

2. Fuzzy Rule Structure

We consider rules with a free structure, without an initial referential set of fuzzy sets.

The rules have the form

$$R_i : \text{IF } x_1 \text{ is } A_{i1} \text{ and } ... \text{ and } x_n \text{ is } A_{in} \text{ THEN } y_1 \text{ is } B_{i1} \text{ and } ... \text{ and } y_m \text{ is } B_{im}$$

where $x_1, ..., x_n$ and $y_1, ..., y_m$ are the process state variables and the control variables respectively; and $A_{i1}, ..., A_{in}$, $B_{i1}, ..., B_{im}$ are fuzzy sets in the universes of discourse $U_1, ..., U_n$, $V_1, ..., V_m$.

These fuzzy sets are characterized by their membership functions

$$A_{ij}(B_{ih}) : U_j(V_h) \longrightarrow [0,1], \ j = 1, ..., n, \ h = 1, .., m.$$

We consider normalized trapezoidal membership functions with parametric representation achieved by means of the 4-tuple $(a_{ij}^1, a_{ij}^2, a_{ij}^3, a_{ij}^4)$, $(b_{ih}^1, b_{ih}^2, b_{ih}^3, b_{ih}^4)$.

3. Genetic Algorithms

Genetic Algorithms (GA) are search algorithms that use operations found in natural genetics to guide the trek through a search space. GA use a direct analogy of natural behaviour. They work with a population of individuals, each representing a possible solution to a given problem. Each individual is assigned a fitness score according to how good a solution to the problem it is. GA are theoretically and empirically proven to provide robust search in complex spaces, giving a valid approach to problems requiring efficient and effective search [6].

Any GA starts with a population of randomly generated solutions, chromosomes, and advances toward better solutions by applying genetic operators, modeled on the genetic processes occurring in nature. In these algorithms we maintain a population of solutions for a given problem; this population undergoes evolution in a form of natural selection. In each generation, relatively good solutions reproduce to give offsprings that replace the relatively bad solution which die. An evaluation or fitness function plays the role of the environment to distinguish between good and bad solutions. Although there are many possible variants of the basic GA, the fundamental underlying mechanism operates on a population of chromosomes or individuals (representing possible solutions to the problem) and consists of three operations:

(1) evaluation of individual fitness,

(2) formation of a gene pool, and

(3) recombination using two basic genetic operators: crossover and mutation.

The figure 1 shows the structure of a simple GA.

Procedure Genetic Algorithm
begin (1)
  $t = 0$;
  inicialize $P(t)$;
  evaluate $P(t)$;
  While (**Not** *termination-condition*) **do**
  begin (2)
    $t = t + 1$;
    select $P(t)$ *from* $P(t-1)$;
    recombine $P(t)$;
    evaluate $P(t)$;
  end (2)
end (1)

Figure 1: Structure of a GA

The crossover operator combines the features of two parent structures to form two similar offsprings. It is applied at a random position with a probability of performance, the crossover probability, P_c. The mutation operator arbitrarily alters one or more components of a selected structure so as to increase the structural variability of the population. Each position of each solution vector in the population undergoes a random change according to a probability defined by a mutation rate, the mutation probability, P_m.

It is generally accepted that any GA to solve a problem must take into account the five following components:

1. *A genetic representation of solutions to the problem,*
2. *a way to create an initial population of solutions,*
3. *an evaluation function which gives the fitness of each individual,*
4. *genetic operators that alter the genetic composition of children during reproduction, and*
5. *values for the parameters that the GA uses (population size, probabilities of applying genetic operators, etc.).*

The basic principles of GA were first laid down rigorously by Holland [7], and are well described in many texts as [6,12].

4. Fuzzy Rule Generating Process

Let us suppose we know a set of p training examples $E_p = \{e_1, ..., e_p\}$ of the system consisting of the values that the variables take during an experiment in which the system is controlled by an expert:

"in the time $t = k$, the value of the variable vectors X and Y are ex^k and ey^k respectively"

If we wish to generate a set of rules describing the behavior of a system, then a condition over the set of generated rules, R, is necessary to establish. This is the requirement of covering all possible situation-action pairs, $e_k \in E_p$, that is, the *completeness property.* This may be formalized for some constant $\tau \in [0,1]$, requires the non-zero union of fuzzy sets $A_i(\cdot)$, $B_i(\cdot)$, $i = 1, ..., T$, $T = |R|$, and is formulated by the following expressions:

$$C_R(e_k) = \bigcup_{i=1..T} R_i(e_k) > \tau \quad k = 1, ..., p$$

$$R_i(e_k) = *(A_i(ex^k), B_i(ey^k))$$

$$A_i(ex^k) = *(A_{i1}(ex_1^k), ..., A_{in}(ex_n^k))$$

$$B_i(ey^k) = *(B_{i1}(ey_1^k), ..., B_{im}(ey_n^k))$$

where $*$ is a t-norm, and $R_i(e_k)$ is a *compatibility degree* between the rule R_i and the example e_k .

Given a set of rules R, we define the *covering value* of an example e_k with a base of rules R as

$$CV_R(e_k) = \sum_{i=1}^{T} R_i(e_k)$$

and we require in our method the condition

$$CV_R(e_k) \geq \epsilon \quad k = 1, ..., p$$

with $\epsilon = 1$ as an initial proposal.

We propose a rule generation process by means of a RCGA and a covering method of the set of examples.

- Every chromosome of the RCGA represents a fuzzy rule. The RCGA obtains a good rule with the following features:

  a) a great frequency value,
  b) a big set of positive examples with compatibility degree greater than τ,
  c) a small negative examples set,
  d) an adequate rate of symmetrical membership functions, and
  e) an adequate rate of membership functions wide.

- The covering method will be an iterative process that runs the RCGA over the set of examples, choices the best chromosome (rule), assigns to every example the relative covering value, and removes the examples with a covering value greater than ϵ from the set of examples.

First we describe the components of the RCGA, and fartherly we present the covering method.

4.1. Real Coded Genetic Algorithms

In order to describe the RCGA we present the five components indicated in section 3.

4.1.1. Representation

In the population of the RCGA a candidate solution C_r, $r = 1, ..., M$, represents a fuzzy rule

IF x_1 is A_{r1} and ... and x_n is A_{rn} THEN y_1 is B_{r1} and ... and y_m is B_{rm}

where the real values a_{rj}^1, a_{rj}^2, a_{rj}^3, a_{rj}^4, b_{rh}^1, b_{rh}^2, b_{rh}^3, b_{rh}^4 characterize the membership functions of A_{rj} and B_{rh} $j = 1, ..., n$, $h = 1, ..., m$, respectively.

Thus, C_r codes the vector values:

$$(a_{r1}^1, a_{r1}^2, a_{r1}^3, a_{r1}^4, ..., a_{rn}^1, a_{rn}^2, a_{rn}^3, a_{rn}^4, b_{r1}^1, b_{r1}^2, b_{r1}^3, b_{r1}^4, ..., b_{rm}^1, b_{rm}^2, b_{rm}^3, b_{rm}^4)$$

As was justified in [8,9], we propose to approach this problem with real coded genes together special genetic operators developed for them. Then a rule would be a chromosome vector coded as a vector of floating point numbers.

Finally, we represent a population of M chromosomes (rules) by C, and it is set up as follows:

$$C = (C_1, ..., C_M).$$

Now, the fundamental underlying mechanisms are developed.

4.1.2. Formation of an initial population or gene pool

The domain of every input variable X_j is a close real interval $U_j = [a_j, b_j]$. Similarly, the domain of the output variable Y_h is the real interval $V_h = [c_h, d_h]$.

The initial gene pool is created partially from $E_t \subseteq E_p$ (t chromosomes) and the remaining ($M - t$ chromosomes) initiated randomly as follows:

- Let $t = \min\{|E_p|, M/2\}$, then we choice randomly t examples from E_p, and for every one we determine the chromosome (rule) belonging to the initial gene pool. Suppose the example $e_k \in E_t$ and the component $ex_j^k \in [a_j, b_j]$, $\triangle ex_j^k = \min\{ex_j^k - a_j, b_j - ex_j^k\}$, let $\delta(ex_j^k)$ be a

  random value in the range $[0, \triangle ex_j^k]$, then we form the membership function by the 4-tuple

  $$(ex_j^k - \delta(ex_j^k), ex_j^k, ex_j^k, ex_j^k + \delta(ex_j^k)).$$

  The procedure is the same for the remaining components of e_k.
- The remaining $M - t$ chromosomes of the initial population are chosen randomly, every gene in its respective interval,

  $$C_r = (c_{r1}, ..., c_{rl}),$$

  $l = 4(n + m)$, with the requirement

  $$c_{4s+1} \leq c_{4s+2} \leq c_{4s+3} \leq c_{4s+4},\ s = 0, ..., n + m - 1.$$

4.1.3. Evaluation of individual fitness

We define the fitness function according to the five following features:

- Great *frequency value*,
- Great *average covering degree over positive examples*,
- Small *negative examples set*,
- Adequate *rate of membership functions wide*,
- Adequate *rate of symmetrical membership functions*.

a) *Frequency of a fuzzy rule*
The frequency of a fuzzy rule, R_i, through the set of examples E_p is defined as ([5]):

$$\Psi_{E_p}(R_i) = \frac{\sum_{k=1}^{p} R_i(e_k)}{p}$$

with $R_i(e_k)$ the compatibility degree.

b) *Set of positive examples*
The set of positive examples to R_i with compatibility degree greater than τ is defined as

$$E^+(R_i) = \{e_k \in E_p / R_i(e_k) \geq \tau\}$$

then $n_{R_i}^+ = |E^+(R_i)|$ defines the number of positive examples, and

$$G_{R_i} = \sum_{e_k \in E^+(R_i)} R_i(e_k)/n_{R_i}^+$$

the *average covering degree* on $E^+(R_i)$.

c) *Set of negative examples*
The set of negative examples to R_i is defined as:

$$E^-(R_i) = \{e_k \in E_p / R_i(e_k) = 0 \text{ and } A_i(ex^k) > 0\}$$

with $n_{R_i}^- = |E^-(R_i)|$ the number of negative examples.

An example is considered negative for a rule when it matches better with some other rule having the same antecedent but different consequent.

The *penalty function on the negative examples set* will be:

$$g_n(R_i^-) = \begin{cases} 1 & \text{if } n_{R_i}^- \leq 5 \\ \frac{1}{n_{R_i}^- - 5 + exp(1)} & \text{otherwise} \end{cases}$$

where we permit until 5 negative examples per rule without penalization.

In all cases we consider the negative examples over the complete set of training examples.

d) *Rule variable wides and rule modal variable wides*
The rule variable wides (RW) and rule modal variable wides (RMW) are:

$$RW_i = \frac{\sum_{j=1}^{h} \frac{WVR_{ij}}{DW_j}}{h} \quad \text{and} \quad RMW_i = \frac{\sum_{j=1}^{h} \frac{WVMR_{ij}}{WVR_{ij}}}{h}$$

where

$$WVR_{ij} = c_{ij}^4 - c_{ij}^1, \; WVMR_{ij} = c_{ij}^3 - c_{ij}^2$$

being $(c_{ij}^1, c_{ij}^2, c_{ij}^3, c_{ij}^4)$, $j = 1, ..., h = n + m$, the 4-tuple associated to every membership function, and DW_j is the *domain interval wide per variable.*

We define the *membership wide rate, MWR* as a functional of the two above parameters, defined by the following expression:

$$MWR(R_i) = g_1(RW_i) \cdot g_2(RMW_i)$$

where the functions g_i represent the required relations with respect to the wide.

* If we prefer small rate we can consider the function

$$g_i(x) = e^{1-ax}.$$

* If we prefer a constant relation among the domain wide and the variable wide or between the two variable interval wide we can consider the function:

$$g_i(x) = e^{-|1-ax|}$$

with a the constant relation.

e) *Rate of Symmetry, RS*

The rate of symmetry is defined in order to require a symmetry to the fuzzy numbers, and prevents the bad covering of extreme points. It is defined as follows:

$$RS(R_i) = \frac{1}{d^i}$$

where

$$d^i = Max_{j=1...n+m}\{d_i^j\},\ d_i^j = Max\left\{\frac{d_{i1}^j}{d_{i2}^j}, \frac{d_{i2}^j}{d_{i1}^j}\right\},\ \text{and}\ d_{i1}^j = c_{ij}^2 - c_{ij}^1,\ d_{i2}^j = c_{ij}^4 - c_{ij}^3.$$

Clearly, $RS(R_i) \leq 1$, and if the rule has a symmetry semantic, then $RS(R_i) = 1$.

An *evaluation function* to the rule R_i, and therefore a fitness function to the associated chromosome C_i can be defined as follows:

$$F(R_i) = \Psi_{E_p}(R_i) \cdot G_{R_i} \cdot g_n(R_i^-) \cdot MWR(R_i) \cdot RS(R_i).$$

with the objective of maximizing the fitness function.

4.1.4. Genetic Operators

During the reproduction phase of the GA we use the two classical genetic operators, mutation and crossover.

We consider the use of the non-uniform mutation proposed by Z. Michalewicz, [12]. If $C_v^t = (c_1, ..., c_k, ..., c_H)$ is a chromosome and the element c_k was selected for this mutation (the domain of c_k is $[c_{kl}, c_{kr}]$), the result is a vector $C_v^{t+1} = (c_1, ..., c_k', ..., c_H)$, with $k \in 1, ..., H$, and

$$c_k' = \begin{cases} c_k + \triangle(t, c_{kr} - c_k) & \text{if } a = 0, \\ c_k - \triangle(t, c_k - c_{kl}) & \text{if } a = 1, \end{cases}$$

where a is a random number that can have a value of zero or one, and, the function $\triangle(t,y)$ returns a value in the range $[0, y]$ such that the probability of $\triangle(t,y)$ being close to 0 increases as t increases:

$$\triangle(t, y) = y(1 - r^{(1-\frac{t}{T})^b})$$

where r is a random number in the interval $[0, 1]$, T is the maximum number of generations and b is a parameter chosen by the user, which determines the degree of dependency with the number of iterations.

This property causes this operator to make an uniform search into the initial space when t is small, and very locally at later stages.

In relation with the crossover operator, we employ the max-min-arithmetical crossover presented by us in [9]. If $C_v^t = (c_1, ..., c_k, ..., c_H)$ and $C_w^t = (c_1', ..., c_k', ..., c_H')$ are to be crossed, we generate the four following offsprings,

$$C_1^{t+1} = aC_w^t + (1-a)C_v^t$$
$$C_2^{t+1} = aC_v^t + (1-a)C_w^t$$
$$C_3^{t+1} \text{ with } c_{3k}^{t+1} = \min\{c_k, c'_k\}$$
$$C_4^{t+1} \text{ with } c_{4k}^{t+1} = \max\{c_k, c'_k\}$$

and select the two best offspring chromosomes.

4.1.5. Parameters

We propose to carry out experiments with the following parameters:

- Population size: 60
- Elitist selection and linear ranking [2]
- Probability of crossover: $P_c = 0.6$
- Max-Min-Arithmetical crossover, $a = 0.35$
- Probability of mutation: Probability of chromosome update $P = 0.6$
- Non-uniform mutation, $b = 5$
- Stop conditions: N iterations.

4.2. Covering method

The covering method is developed as an iterative process. It permits to obtain a set of rules covering the set of examples.

1. Initialization:
   - To introduce τ and ϵ.
   - To assign $CV[k] \leftarrow 0$, $k = 1, ..., p$.
2. Over the set of examples E_p we apply the RCGA.
3. We select the best chromosome C_r with R_r the associated fuzzy rule.
4. We introduce R_r in the set of rules R, which initially is empty.
5. We obtain $E^+(R_r)$ from E_p.
   For every $e_k \in E^+(R_r)$ do

   $$CV[k] \leftarrow CV[k] + R_r(e_k),$$

   If $CV[k] \geq \epsilon$ then to remove it from E_p.
6. If $E_p = \emptyset$ then Stop else return to Step 2.

5. Experimental Results

To test this approach we have carried out experiments with a simulation of the inverted pendulum problem. On the assumption of $|\theta| \ll 1$ (radian) the nonlinear differential equation that leads the behavior of pendulum can be simplified to,

$$m\frac{L}{3}\frac{d^2\theta}{dt^2} = \frac{L}{2}(-F + mg\sin\theta - k\frac{d\theta}{dt})$$

where $k\frac{d\theta}{dt}$ is an approximation of the friction force.

A pendulum with 5 kg in weight, and 5 m in length has been considered. Under these parameters the universes of discourse of the variables are the following:

$\theta \in [-0.5240, 0.5240]$ rad $\quad \omega \in [-0.8580, 0.8580]$ rad/sc $\quad f \in [-2980.0, 2980.0]$ N

Experimentally, we have obtained a training data set with 68 input-output data in the intervals $[-0.275, 0.275]$, $[-0.454, 0.454]$ and $[1576.681, 1576.681]$ for θ, ω and f, respectively.

We have applied the proposed generating process to this set of examples. We compare the results with the fuzzy rules set proposed in [20] according to the discourse universes partition presented in [11]. The comparison is done using the measure of medium square error over the set of 68 examples. The generating process is applied with the parameters $\tau = 0.1$ and $\epsilon = 1.0$.

As any optimum values always depend on specific models, we use the Max-Min inference system, the conjunction operator $T = MIN$, and the defuzzification strategy center of area.

The results are presented in the following table:

	Triangular Mem. Func.		Trapezoidal Mem. Func.	
	Medium SE	Number of Rules	Medium SE	Number of Rules
Expert Base	9.758,42	7	9.844,22	7
Fitness Function 1	4.836,07	10	7.350,16	10
Fitness Function 2	3.098,59	10	32.199,51	8

Table 1. Medium Square Error and Number of Generated Rules

The differences between the two fitness functions are in the rate wide functions used. The functions are

$$g_1(x) = e^{1-x} \text{ and } g_2(x) = e^{1-4x},$$

for the first, and the following for the second:

$$g_1(x) = e^{-|1-3x|} \text{ and } g_2(x) = e^{-|1-4x|}.$$

The results show a good behavior of the set of rules obtained by our generation process. Taking into consideration that the generation process is a first component of a complete learning process, we can think that the use of the learning process can improve the final results in a considerable way.

5. Conclusions

We have focused this paper on the description of a method for generating desirable fuzzy rules from examples by using a RCGA and a covering process of a set of examples. This is the first component of the learning process described in section 1.

As future work we will develop the complete learning process and compare this with other learning methods by testing them on more complex problems.

REFERENCES

[1] Araki, S., Nomura H., Hayashi, I. and Wakami, N. (1991) A self-generating Method of Fuzzy Inference Rules, Proceedings of IFES'91, 1047-1058.

[2] Back, T., and Hoffmeister, F. (1991) Extended Selection Mechanisms in Genetic Algorithms. Proc. of the Fourth Int. Conf. on Genetic Algorithms, Morgan Kaufmmann, San Mateo, 92-99.

[3] Binaghi, E. (1991) Learning of uncertainty classification rules in medical diagnosis in Symbolic and Quantitative Approaches to Uncertainty. Lectures Notes in Computer Science 548, (R. Kruse, P. Siegel, Eds.), Springer-Verlag, 115-119.

[4] Delgado, M. and González, A. (1993) An Inductive Learning Procedure to Identify Fuzzy Systems. Fuzzy Sets and Systems, 55, 121-132.

[5] Delgado, M. and González, A. (1994) A frequency model in a fuzzy environment. evidences. Int. Journal of Approximate Reasoning, 11, 159-174.

[6] Goldberg, D.E. (1989) Genetic Algorithms in Search, Optimization, and Machine Learning. Addison-Wesley, New York.

[7] Holland, J.H, Adaptation in Natural and Artificial Systems. Ann Arbor, 1975. (MIT Press, 1992).

[8] Herrera, F., Lozano, M., and Verdegay, J.L. (1993) Crossover Operators and Offspring Selection for Real Coded Genetic Algorithms, Dept. of Computer Science and Artificial Intelligence, Technical Report #DECSAI-93113.

[9] Herrera, F., Lozano, M., and Verdegay, J.L. (1995) Tuning Fuzzy Logic Controllers by Genetic Algorithms. To appear in: International Journal of Approximate Reasoning.

[10] Lee, C.C. (1991) A self-learning rule-based controller employing approximate reasoning and neural net concepts, International Journal of Intelligent Systems, 6, 71-93.

[11] Liaw, C-M., Wang, J-B. (1991) Design and Implementation of a Fuzzy Controller for a High Performance Induction Motor Drive. IEEE Transactions on Systems, Man, and Cybernetics, 21, 921-929.

[12] Michalewicz, Z. (1992) Genetic Algorithms + Data Structures = Evolution Programs. Springer-Verlag, New York.

[13] Nomura, H., Hayashi I., and Wakami, N. (1992) A Learning Method of Simplified Fuzzy Reasoning by Genetic Algorithm. Proc. of the Int. Fuzzy Systems and Intelligent Control Conference, 236-245, Lousville, KY.

[14] Pedrycz, W. (1984) An identification algorithm in fuzzy relational systems. Fuzzy Sets and Systems, 13, 153-167.

[15] Peng, X., and Wang, P. (1988) On generating linguistic rules for fuzzy models. In: Uncertainty and Intelligent Systems, Lectures Notes in Computer Science 313, (B. Bouchon, L. Saitta, R.R. Yager, Eds.), Springer-Verlag, 185-192.

[16] Sugeno, M., and Kang, G.T. (1988) Structure identification of fuzzy model. Fuzzy Sets and Systems, 28, 15-33.

[17] Sugeno, M., and Tanaka, K. (1991) Successive identification of a fuzzy model and its applications to prediction of a complex system. Fuzzy Sets and Systems, 42, 315-334.

[18] Takagi, T., and Sugeno, M. (1985) Fuzzy Identification of Systems and Its Applications to Modeling and Control. IEEE Trans. Sys. Man and Cyber., Vol. SMC-15 No. 1, 116-132.

[19] Yager, R.R., and Dimitar, P.F. (1993) Generation of Fuzzy Rules by Mountain Clustering. Iona College, TEch. Report #MII-1318.

[20] Yamakawa, T. (1991) Stabilization of an Inverted Pendulum by a High-Speed Fuzzy Logic Controller Hardware System, Fuzzy Sets and Systems, 32, 161-180.

[21] Yamaoka, M. and Mukaidono, M. (1991) A learning method of fuzzy inference rules with a neural network. Fourth IFSA Congress, Vol. Engineering, 222-225.

[22] Yoshinari, Y., Pedrycz, W., and Hirota, K. (1993) Construction of fuzzy models through clustering techniques. Fuzzy Sets and Systems, 54, 157-165.

AUTOMATIC DESIGN OF FUZZY SYSTEMS BY GENETIC ALGORITHMS

Henning HEIDER*, Viktor TRYBA*, Eike MÜHLENFELD**

* SIBET GmbH,
Garbsener Landstraße 10,
D-30419 Hannover, Germany

** Technische Universität Clausthal,
Institut für elektrische Informationstechnik,
Leibnizstr. 28,
D-38678 Clausthal-Zellerfeld

ABSTRACT

A new method for the automatic design of fuzzy control and decision/diagnosis systems is described. The fuzzy system is extended by a learning ability without changing the traditional fuzzy rule framework. A genetic algorithm is used to find optimal fuzzy rules and membership functions. By choosing the right parameters, the final fuzzy system will use a minimal number of rules and membership functions (fuzzy sets).

1. Introduction

The use of fuzzy systems is mainly required when conventional systems prove insufficient, or if acceptable solutions require considerable vast expenditure. This applies mainly to:

- non-linear systems
- systems with a large number of input ports
- and systems with unknown mathematical models

Even if conventional solutions do exist, the application of fuzzy logic is quite often justified, as long as it helps reducing the 'Time to Market' -time of a product in a considerable way.

Due to the available convenient tools of development and the definition of the system's reaction by linguistic terms, which by their make up complies with the nature of human thinking, even engineers without an expert-knowledge on fuzzy logic can reach acceptable results, doing a development of a fuzzy-controller, for example.

However, the optimization of the controller is usually rather extensive and requires an expert-knowledge. Time of development will be prolonged due to the complexity of the problem. Difficulties with the development will emerge especially in cases suitable for fuzzy logic systems.

You can well imagine the output-behaviour of a system with two input- and one output-ports. You will have difficulties though, to express the rules and optimize the fuzzy membership functions, working with systems with three or more input-ports. If there are a lot more entries, it will be practically impossible to develop an optimal fuzzy system within an acceptable period of time.

It will be as difficult to develop a complex controller for some processes where the mathematical model is not known. There, you will have to fall back on the knowledge of the expert who normally controls the process. This expert however will often have problems to express his knowledge of processing in terms of fuzzy-rules. In order to shorten times of development you must have tools that carry out the design of the fuzzy system automatically.

Currently, you can get good results developing automatically fuzzy systems with entire rule-bases, where the specific behaviour results out of the reassessment of different rules. These fuzzy systems however are no longer interpretable and for the application of a designed fuzzy system big rule-bases are hard to handle.

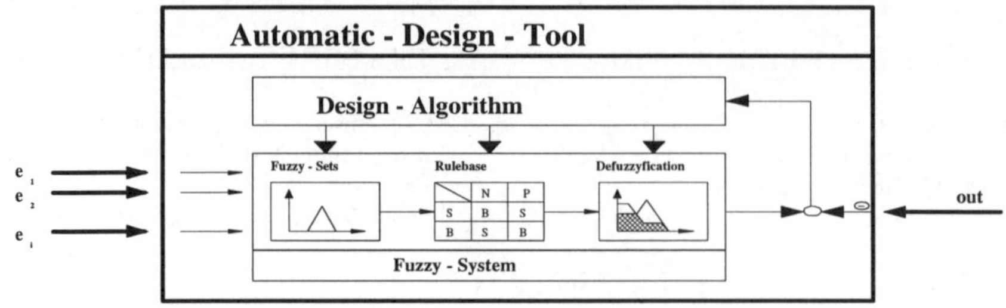

Figure 1: The Automatic - Design - Tool

Various algorithms have been proposed that generate the fuzzy rules automatically [3,4,5], which is a good solution if the optimal fuzzy sets are known.

A fast method for the automatic development of fuzzy systems is to use neural networks [6,7]. However, most of these tools use special calculation procedures and therefore it is not possible to work on the developed fuzzy systems with the available tools only, so as to generate a special fuzzy processor code, for example.

The author proposes a one step design method which extends a traditional fuzzy system so that available tools can be used to work on the design and the extended system is still interpretable. The complexity of the designed system (number of the fuzzy-rules and membership functions) will be minimized automatically by the design tool.

The tool creates the fuzzy sets and the rule-base, so that you hardly need knowledge about the technical system that is to be automated.

In the next chapters the author describes how a genetic algorithm works and how in this case the algorithm is used to design a fuzzy system with a reduced complexity. In section five he shows the first results of the automatic design tool by developing a fuzzy system. Some final remarks on the future development will conclude this paper.

2. Automatic Design of a Fuzzy System

In order to develop a technical system, you have to define the interfaces and the marginal conditions at first. Figure 1 shows a tool's principle, optimizing or automatically generating a fuzzy system.

For the training period with automatic generating, you will need learning data sets, gained from the process. These data sets consist of the input and the corresponding output data of the fuzzy system that is to develop. The learning sets must contain the entire information about the systems behaviour which is necessary to configure the fuzzy system. Moreover, you have to introduce to the algorithm the structure of the input data and the utmost extent of the designated system. In that respect there are often restrictions due to the hardware the system is to be applied on. From that, the input of the data following result:

- number of input ports
- internal scale of data (where an exactness of 8 bit is common)
- maximum number of membership-functions per port

The automatic acquisition of knowledge, for which neural networks and genetic algorithms are particularly suitable, takes place in the cell marked 'Design Algorithm'. As the neural networks frequently run into local minima and moreover the acquired knowledge tends to be hardly extractable from the neural network, there has been developed an algorithm that takes advantage of genetic optimizing strategies. The process of optimization will be explained in the case of a fuzzy controller.

3. Genetic Algorithm

The generic algorithm chosen for an automatic acquisition of knowledge is designed to copy the strategy that is applied by nature in order to allow organisms to adapt to their environment. This strategy of nature being subject to optimization itself [9], a corresponding algorithm for optimization of technical systems represents a very efficient tool. The essential mechanisms used by nature are mutation, crossing-over and selection.

3.1. Selection

The number of individuals belonging to a population in nature is obviously limited. Subsequently, the production of a new representative must, on the average, go along with a sorting out of individual [1]. This effects generally an individual, which has least ability for adaptation and is, therefore, least able to survive.

In terms of a fuzzy controller, it means that a fixed number of controllers, a population, has to be kept in memory.

When a new controller is generated, it will have to be compared with the representatives of the population respectively to its assessment (fitness).

If the new controller assessed is not as good as the others, it will be forgotten and another cycle is introduced, generating a new controller. But if it is better then just one representative of the population, it will be taken into the population. As the number of individuals in the population remains constant, one of its representatives has to be sorted out.

The probability for an individual to be sorted out depends on its fitness, which is described later. The higher the fitness of an individual is, the higher is its ability to survive.

The new controllers are generated by a combination of mutation and crossing-over, described in the following.

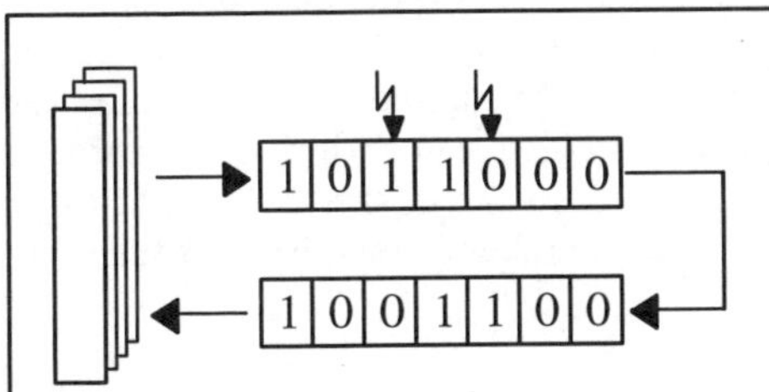

Figure 2 : Mutation

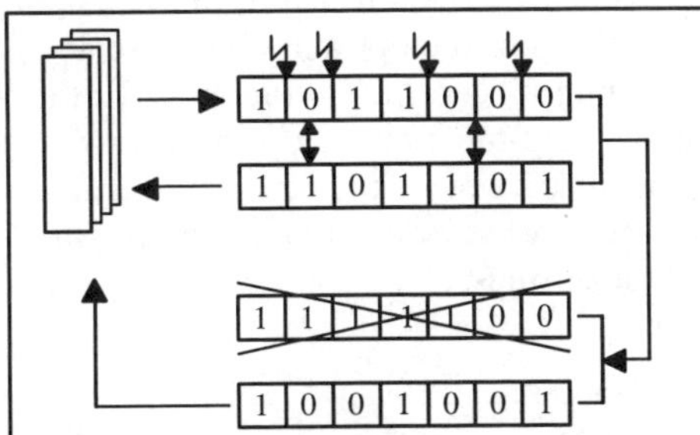

Figure 3 : Crossing - Over

3.2. The generation of a representative by means of mutation

To start mutation, the bit-string of an incidentally chosen controller out of the population is copied at first. Then a part of the genes (bits) belonging to the copied string is inverted. This will say that a '1' is replaced by a '0' and a '0' is replaced by a '1' (see figure 2).

The choice of the bits to be inverted, is taken over by a random process. The mutation rate that you have to set before starting the genetic algorithm, indicates the amount of bits to be changed.

3.3. Producing a representative by means of crossing-over

There are two distinguished methods, the two-point and the m-point crossover. In both cases during recombination, two representatives out of a population are chosen and copied in an incidental process. In this process representatives with a good fitness are chosen more often than representatives with a lower fitness.

In case of the m-point crossover method, the bit-strings belonging to the copied controllers are cut at 'm' different points and every second cut-sequence will be exchanged between the two bit-strings. That way, two new controllers exist, one of them being assessed and handed over to the mechanism for selection. This process is shown in figure 3. The number of interjections 'm' is fixed by a system parameter called crossover-rate.

In case of the two-point crossover method, the bit-strings are cut only at two different incidentally chosen points and the cut-sequence between these points will be exchanged to create a new controller for the population.

In the presented tool a two-point crossover mechanism is used. The crossover-rate fixes the number of representatives from the old population that are processed by the crossover method to create a new population.

Using these three mechanisms, the genetic algorithm generates a bit-string that contains the fuzzy controllers structure in form of a binary code. The bit-string is decoded by the tool and the extracted controller is simulated. The edited process-measurement-data (ei) from the learning-data sets are lined up to the controller's input ports. The out-values, thus calculated, are to be compared with the given out-values from the learning data sets. The controller is to be assessed respectively to the calculated deviations by the fitness-function.

The controllers velocity of convergence towards its optimum is a great deal influenced by the fitness-function.

Extensive research will have to be done to reach the ideal fitness-function. Furthermore is the position of the optimum in the scale of parameters, and therefore the resulting controller, depending on the fitness-function. In the version submitted, the number of fuzzy-rules and membership-functions per port, belonging to the resulting system, can be influenced through the input of weight-factors. Best results have been reached so far in the assessment of an average quadratic deviation, including the assessment of the number of membership-functions and fuzzy-rules.

$$err = \sqrt{\frac{\sum_{i=1}^{n}(st_i - out_i)^2}{n} + (r \cdot g_1) + (s \cdot g_2)}$$

fitness-function

err = assessment of a representative
n = number of learning data sets
st = calculated output
out = pretended output
r = number of rules
s = number of sets
g1 = weight of a rule
g2 = weight of a membership-function

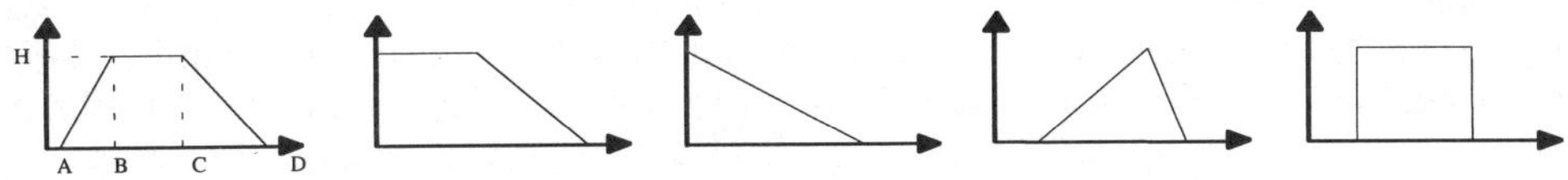

Figure 4 : Different types of 'Trapezoidal - Functions'

A quadratic fitness-function ensures that severe deviations from the rated value have a particularly negative effect on the assessment of a representative. Both additive terms produce a better assessment of controllers, which bring about the wanted behaviour with a diminished number of fuzzy-rules and membership-functions.

4. Bit-string representation

The main part of the tool is the representation of the fuzzy system by a bit-string. To achieve a solution of the problem, all sorts of demands on behalf of the genetic algorithm as well as the technical process have to be taken into account. For example a genetic algorithm converges faster towards its optimum if a close technical context of the data to be optimized is transformed into a close local context of it's representing bits. The reason is, that two very distant bits are separated through recombination more often then two bits close to each other.

The fuzzy-system's transformation brings about limits to the fuzzy-system. If not, the string to be optimized would become to long and the time needed for the process of optimization, would enlarge too much. However, if possible such limitations should have no effects on the fuzzy-systems efficiency, that is, the output-behaviour shouldn't be affected.

Therefore a limitation of the fuzzy-sets to trapezoidal-functions is a useful convention. Trapezoidal-functions are many-sided in their shape and they are suited to approach asymmetrical membership functions used for many technical processes. Figure 4 shows some of the multiple shapes a trapezoidal-function can take. Moreover, trapezoidal-functions are easy to deal with, and they cover very little memory. Most fuzzy tools allow the input and processing of trapezoidal-functions. As the exact shape of a membership-function, in comparison with the position and the location of the intersection of different functions has very little influence on the output-behaviour, the use of trapezoidal-functions does not bring about much restriction to the fuzzy-system.

Concerning the rules, the following conventions are appointed, having no functional restrictions on the fuzzy system either. At linking terms, only 'AND'-operators are admitted within a rule. Different rules that have an impact on one and the same output-membership-function, have to be 'OR'-linked.

A rule in the form

IF e1 IS warm 'OR' e2 IS small
THEN out IS high

has to be replaced by the following rules:

IF e1 IS warm THEN out IS high.
IF e2 IS small THEN out IS high.

By keeping these limitations you can code the fuzzy system in the following way. The trapezoidal fuzzy sets are represented throughout the four values A B C D (see figure 4), the height H and a flag which contains the information whether the set is used or not. The fuzzy rules are coded by fixing the number of the output fuzzy sets they are forcing upon. That means, that if the parameter mem_func allows seven membership functions per port, you need three bits (0 - 8) for every rule to fix the corresponding output set (conclusion). A rule that does not 'fire' is represented throughout the code (000). The condition of a rule is kept by the position the three 'conclusion-bits' take in the genetic bit-string. Because the author starts out from entire rule bases, every rule (condition) has a fixed position in the bit-string. A rule base contains

$$r = (s + 1)^e - 1$$

rules, with s = number of sets per port
and e = number of input ports.

5. Example

The present version of the automatic design tool supports the development of fuzzy-systems that deal with the 'Min-Max' inference method and systems that take average of the 'product' inference method. In addition you can choose between the 'Center of Gravity' defuzzification method (COG) and the 'Height' -method. In the center of gravity method, the overlapping sections between two fuzzy sets are calculated twice. That means, that different rules causing equal output-values are able to amplify each other.

You can show that the automatic design tool is able to learn nearly any randomly created data-pattern and that the tool minimizes the rulebase automatically if you choose the right design-parameter.

The development of a fuzzy-system with three input ports and one output port will be given as an example. The example is to allow seven membership-functions per port. Out of these marginal conditions, a fuzzy system represented by a string 2684 bit long will result .

The technical systems behaviour is presented by only 8 coincidentally generated learning data sets, to show, that the tool is able to minimizes the rulebase automatically (you normally need less rules to learn only a few data-pattern than learning a complex systems behaviour). For one optimization step, the algorithm needs 188 ms on a 'HP 730' workstation. This period of time depends on the fuzzy systems complexity and the number of learning-data-sets. If you have the same problem calculated with a number of 4 fuzzy-sets allowed per port, you will have as a result a 1031-bit-string to be optimized and the time of 33 ms for each step of optimization. Figure 5 shows the course of the process of optimization, typical for a genetic algorithm.

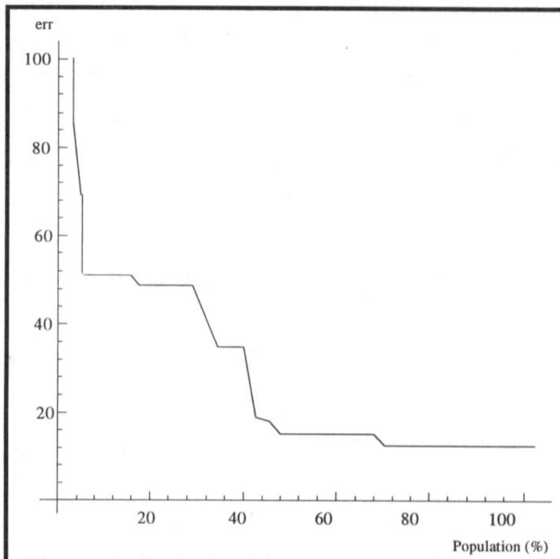

Figure 5: Optimization

The 22614th population contains a representative providing the following results. Input port '1' does not provide any information about the calculation of the output value and therefore is not taken into account in the rule base. The largest possible number of 7 membership-functions per port for this example has been reduced for port '2' to 1 and for port '3' to 2 functions.

The system needs 2 output-functions. The number of the rules being used has been reduced to 3 out of 512 possible. The greatest deviation from the given data is no more than 1.77%. The average deviation is 0.61%.

In example two, a developed fuzzy controller for a thermal process with 3 input ports and one output port is used to generate the learning data sets for the 'Automatic Design Tool'. For this, the existing fuzzy controller is used to present the behaviour of the thermal process. Thereby it is possible to use a fuzzy system simulator initialised with the rules and fuzzy sets of the existing fuzzy controller to calculate the expected output-values which the automatically designed controller shall produce for any input pattern which represent a special process situation.

The input pattern are randomly generated and stored together with the accompanying output values, calculated by the fuzzy controller simulator. This process is like the data collection at the real process, controlled by an expert. The fuzzy controller which is automatically developed by the described Design Tool trained with this learning data sets, shows an average output deviation from 0.79%. In the table below you can see the used number of rules and fuzzy-sets. The automatically developed system uses only 15 of 28 possible fuzzy sets. The number of rules is reduced to 17 out of a possible 512.

	number of sets	number of rules
example two		
real fuzzy controller	16	21
GA - fuzzy controller	15	17
possible number	28	512
example one		
GA - fuzzy controller	7	3
possible number	28	512

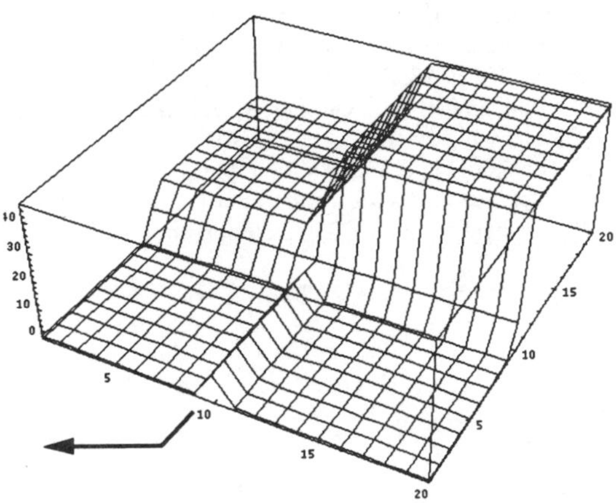

3D-sector out of the 4-D field of the manually developed fuzzy controller

6. Conclusion

The presented algorithm is able to develop or optimize the membership functions and the rule base of fuzzy systems. Doing this, a clustering of the acquired process-knowledge is carried out automatically, bringing about very little rule bases that can be transformed into correspondingly quick systems.

In the presented example the designed controller needs no more than 3 out of 512 possible rules. The developed fuzzy systems can be processed with most of the available fuzzy-tools, because it does not use any special calculation methods. The described tool is also able to develop fuzzy systems with asymmetric membership-functions. Since the tool written in C++ has not been optimized with regard to the program's running time, it will still be able to accelerate the algorithm. The velocity of convergence for the process of optimization can also be highly improved through research concerning the parameter of the genetic algorithm and through slight modifications at interpreting the bitstring that represents the controller in the genetic algorithm.

REFERENCES

[1] Goldberg, D.E. (1989) Genetic Algorithms in Search, Optimization and Machine Learning, Reading Massachusetts, Addison-Wesley.

[2] Buckley, J.J. (1992) Theory of the Fuzzy Controller: An introduction, Fuzzy Sets and Systems 51/1992 S.249-258.

[3] Trift, P. (1991) Fuzzy Logic Synthesis with Genetic Algorithms, Proc. of the Fourth International Conference on Genetic Algorithms, San Diego, California, July 13-16, 450-457.

[4] Valenzuela-Rendon, M. (1991) The Fuzzy Classifier System: A Classifier System for Continuously Varying Variables, Proc. of the Fourth International Conference on Genetic Algorithms, San Diego, California, July 13-16, 346-353.

[5] Lee, C.C., Berenji, H.R. (1989) An intelligent controller based on approximate reasoning and reinforcement learning, Proc. IEEE Int. symp. on intelligent control, Albany, NY, 200-205.

[6] Kahn, E., Venkatapuram, P. (1993) Neufuz: Neural Network Based Fuzzy Logic Design Algorithms, FUZZ-IEEE `93, San Francisco, California.

[7] Halgamuge, S.K., Glesner, M. A Fuzzy-Neural Architecture with a new Method for Rule Extraction, ITG Fachbericht 122 VDE.

[8] Langari, G., Tomizuka, M. (1990) Stability of Fuzzy Linguistic Control Systems, Proceedings of the 29th Conference of Decision and Control, Honolulu, Hawaii, 2185-2190.

[9] Rechenberg, I. (1973) Evolutionsstrategie, Frommann-Holzboog, Stuttgart.

LEARNING

MACHINE LEARNING FROM EXAMPLES UNDER ERRORS IN DATA

Janusz Kacprzyk and Grażyna Szkatuła

Systems Research Institute, Polish Academy of Sciences
ul. Newelska 6
01-447 Warsaw, Poland
E-mail: kacprzyk@ibspan.waw.pl

ABSTRACT

An algorithm for inductive learning from erroneous examples is presented. It is assumed that the errors may occur in the attributes' values. However, their location (in which example, and in which attribute) is unknown. Moreover, the errors are assumed incorrigible as it is often the case in practice. A modification of the start-type algorithm is proposed. Importance of the attributes - reflecting, e.g., the attributes' relevance, their proneness to errors, reliability of methods for determining their values, etc. - is elicited from the experts, and weights are determined by Saaty's analytical hierarchy process (AHP). An oncological example illustrating the method proposed is mentioned.

1. Introduction

Knowledge elicitation and acquisition are key issues in, and a prerequisite for any knowledge based system. Machine learning is here one of widely used paradigms, with learning from examples being probably the most natural and straightforward approach. Learning from examples is basically meant here as a process of inferring a classification rule (concept description, rule, hypothesis, ...) of a class (concept) from descriptions of some individual elements of the class called *examples, or* - more appropriately - *positive examples*. Moreover, some elements from outside of the class, called *negative examples* are also used, mainly for narrowing the solution space, hence for enhancing the efficiency of algorithms.

The examples are commonly assumed, also here, by a set of "attribute - value" pairs which, in terms of a convenient Michalski's (1983) variable-valued logic (VL) formalism, may be conveniently written as

$$e = \bigwedge_{i=1}^{K} \left[a_i \# r_i\right] \tag{1}$$

where a_i denotes attribute i with value $r_i \in d_i$, and # is a relation exemplified by =, <, >, ≈, ≥, etc.

For instance, in a banking context, the examples (customers) may be described by the attributes: "assets", "total_debt", and "annual_growth_rate", and the concept "good_customer" may be described by the following classification rule (to be derived by machine learning):

$$[\text{assets} > \$1{,}000{,}000] \wedge [\text{total\_ debt} < \$250{,}000] \wedge$$
$$\wedge[\text{annual\_growth\_ rate} > 10\%] \rightarrow [\text{class: good\_customer}] \tag{2}$$

to be read as: if "his or her assets exceed $1,000,000" and "his or her debt is less than $250,000" and "his or her annual growth rate is higher than 10%", then "he or she is a *good customer*". Notice that the connective "∧" (and) is explicitly put into (1) and (2) but it is omitted for simplicity in many works.

Procedures for deriving the classification rules should satisfy some requirements among which the following ones are the most relevant for our considerations:

- *completeness*, i.e. that the classification rule must correctly describe *all* the positive examples,
- *consistency*, i.e. that the classification rule must describe *none* of the negative examples,
- *convergence*, i.e. that the classification rule must be derived in a *finite* number of steps,

- a classification rule of *minimal length* (e.g., involving the minimum number of attributes) is to be found.

In this paper we additionally assume that the (positive and negative) examples contain errors in data, whose location (in which example, and concerning which attribute) is unknown, and which cannot be corrected for practical reasons. Evidently, by errors we mean here those which "remain" after some routine tests of, e.g., whether a value belongs to some interval of admissible values. Needless to say that such a case is present in virtually all nontrivial practical cases.

In case of errors traditional methods of learning from examples as, e.g., due to Michalski (1973, 1983), cannot be directly applied. First, it is obvious that the requirements for a machine learning procedure should be redefined. This concerns mainly the *completeness* and *consistency* requirements. In case of errors, if we assumed the traditional requirement of completeness and consistency (i.e., the covering of *all* the positive and *none* of the negative examples), the classification rules would most probably be very long and "illegible" to domain experts, in the extremal case without any generalization, i.e. involving descriptions of all examples; this would evidently be unacceptable in any practical case.

Thus, we need to weaken these completeness and consistency conditions, and replace them by:

- *partial completeness,* i.e. that the classification rule must describe a (considerable) part of positive examples,
- *partial consistency*, i.e. that the classification rule must not describe a (considerable) part of negative examples.

Notice that this bears a close resemblance to a fuzzy-linguistic-quantifier based approach by Kacprzyk and Iwański (1990, 1991a, b, 1992) but the meaning is different.

The other requirements, i.e. that of convergence and minimum length rule remains the same. The requirement of the minimum length is crucial here. Namely, roughly speaking, a short rule is beneficial, first, because it is legible to the expert, and, second, because its shortness lower the chance that it may contain errors, because, with a lack of evidence to the contrary, we may assume that the errors are evenly distributed, and hence the chance of their occurring increases as the number of attributes involved increases. We would therefore prefer shorter rules, and those which contain relevant attributes. These will be clarified while formulating the problem.

The method proposed here is a modification of Michalski's (1973, 1983) star-type method. In particular, some additional information from the domain expert will be used. Saaty's (1980) AHP (analytical hierarchy process) will be employed.

2. Formulation of the problem

Suppose that the set of positive examples is S_p and the set of negative examples is S_n, $S_p \cap S_n = \varnothing$, by assumption. Each example is described by K attributes, $a_1,\ldots,a_K$, and is represented in the VL formalism shown in (1), i.e.

$$e = \bigwedge_{i=1}^{K} [a_i \# r_i] \tag{3}$$

where a_i denotes attribute i with value $r_i \in d_i$, where d_i is the set of (admissible) values of a_i, and # is a relation exemplified by =, <, >, ≈,≥, etc. In this work "=" will be used for simplicity, and other relations will be considered in further works.

The element of example *e* given by (3), $s_i = [a_i \# r_i]$, is called a *selector*, and a conjunction of selectors, i.e.

$$\bigwedge_{i \in I \subseteq \{1,\ldots,K\}} s_i = C \tag{4}$$

is called a *complex*.

A complex is said to *cover* an example if all the conditions on attributes given as selectors are the same as the values of the respective attributes in the example. For instance, the complex $[a_1 =$ "woman"$]\wedge[a_3 =$ "35 years"$]$ cover the example $[a_1 =$ "woman"$]\wedge[a_2 =$ "married"$]\wedge[a_3 =$ "35 years"$]$ but does not cover the example $[a_1 =$ "man"$]\wedge[a_2 =$ "married"$]\wedge[a_3 =$ "35 years"$]$.

We consider the classification rules being the disjunction of "elementary" rules consisting of complexes of type (4), i.e.

$$(s_{i1} \wedge .. \wedge s_{im}) \cup ... \cup (s_{l1} \wedge .. \wedge s_{lm}) \rightarrow \text{[example: } \textit{positive}\text{]} \quad (5)$$

where $\{i_{i1},...,i_{im}\},\{il1,...,i_{lm}\} \subseteq I = \{1,...,K\}$, and "$\cup$" corresponds to the connective "or"; the meaning of (5) is clear.

We need now some technical concepts to facilitate our next discussion. First, suppose that for a positive example, denoted by e^P, (for simplicity the upper index P will be omitted in our next discussion if it will not lead to confusion)

$$e = [a_1 = v_1] \wedge .. \wedge [a_K = v_K] \quad (6)$$

we consider the complex

$$C = [a_{i_1} = v_{i_1}] \wedge .. \wedge [a_{i_l} = v_{i_l}] \quad (7)$$

whose set of indices is $I = \{i_{i_1},...,i_{i_l}\}$; this complex (or its corresponding set of indices) is clearly equivalent to a vector $x = [x_j]^T$, such that $x_j = 1$ if a selector $[a_j = v_j]$ occurs in the complex (7), and 0 otherwise.

Now, the *length* of complex (7) is defined to be

$$d(C) = \frac{1}{K} \sum_{j=1}^{K} x_j \quad (8)$$

i.e. is equal to the (normalized) number of selectors.

On the other hand, if we assign to each selector occurring in complex (7), i.e. to each $x_j = 1$ a weight $w_j \in [0,1]$, and denote such a weighted complex by C_W, then we can define its *weighted length* as

$$d'_W(C_W) = \frac{1}{K} \sum_{j=1}^{K} w_j x_j \quad (9)$$

Now, for our purposes it is expedient to introduce the *augmented weighted length* of complex C_W as

$$d_W(C_W) = \frac{1}{K-1} \left(\sum_{j=1}^{K} x_j - \sum_{j=1}^{K} w_j x_j \right) = \frac{1}{K-1} \sum_{j=1}^{K} (1 - w_j) x_j \quad (10)$$

Notice that such a definition of an augmented weighted length is motivated by our desire to find short rules (with a small number of complexes and selectors) including elements (complexes and selectors) of a higher weight (relevancy).

Example 1. If the five attributes are as given below as arguments in the particular $d(.)$'s and the respective weights are $w_1 = 1/5, w_2 = 1/10, w_3 = 3/5, w_4 = 1/20, w_5 = 1/20$, then

$$d_W(s_1) = 1/5$$
$$d_W(s_3) = 1/10$$
$$d_W(s_2 \wedge s_3 \wedge s_5) = 9/16$$
$$d_W(s_1 \wedge s_2 \wedge s_3 \wedge s_4 \wedge s_5) = 1$$

Then, the *length of the classification rule* composed of L complexes, $R = (C^1 \cup \cdots \cup C^L)$, is defined as

$$d'_W(C^1 \cup \cdots \cup C^L) = \max_{i \in \{1,...,L\}} d(C^i) \quad (11)$$

while the *length of the augmented weighted classification rule*, $R_W = (C_W^1 \cup \cdots \cup C_W^L)$ is defined as

$$d_{R_W}(C_W^1 \cup \cdots \cup C_W^L) = \max_{i=1,\ldots L} d_W(C_W^i) \tag{12}$$

For Example 1, for the rule composed of 4 complexes, we evidently have $d_{R_W} = 1$.

Notice that the definitions (11) and (12) are clearly of a pessimistic, safety-first type that seems to be adequate in the case considered, in particular since our knowledge as to how many and where the errors occur is very rough..

As we have already mentioned, examples used in our learning process contain errors in values of attributes. Our aim is now to find a classification rule that would be less susceptible to errors. The line of reasoning is basically as follows. First, the errors may occur in all the attribute's values. However, it is reasonable to assume that some attributes are more prone to errors as, e.g., their values are determined by less reliable procedures. Thus, a shorter rule involving those less prone to errors attributes would be better from our point of view. Evidently, one should not forget that some attributes, though prone to errors, may be crucial from the point of view of the particular area of application. These aspects may be reflected by assigning some weights associated with the particular attributes (i.e. selectors as in (9)).

We can therefore formulate the problem of learning from examples under errors as to find an optimal (weighted) classification rule $R_W^* = (C_W^1 \cup \cdots \cup C_W^L)^*$ such that

$$R_W^* \to \min_{R_W} d_{R_W}(C_W^1 \cup \cdots \cup C_W^L) = \min_{R_W} \max_{i \in \{1,\ldots,L\}} d_W(C_W^i) \tag{13}$$

that is, which minimizes the augmented weighted length of the classification rule. Moreover, some technical conditions related to the partial completeness and consistency conditions need to be included.

To be able to solve this problem, we need:

- some means to define the weights W so that they reflect the idea and intent of our approach,
- an algorithm for solving problem (13) or, in fact, since the source problem (13) is difficult to solve, a suboptimal, heuristic algorithm for solving some simpler problem.

and these two issues will be discussed below.

3. Weights associated with the attributes

The weights associated with the attributes (selectors) are used in the approach presented to reflect some additional knowledge concerning our expectancy as to errors. First, the position, exact magnitude, etc. of the errors is unknown (and, needless to say, incorrigible). We assume that only some general knowledge may be available as to:

- which attributes are considered crucial so that they should be taken into account,
- which attributes are "robust" to errors due to, e.g.,:
  - a lower subjectivity of value assessment,
  - a higher reliability of {"objective") measuring methods,
  - a higher reliability of registration procedures (e.g., "by hand", by less qualified personnel), etc.

Each of the above elements corresponds to some criterion, that is:

$K1$ - for how crucial the attributes are,
$K2$ - for the error robustness due to a lower subjectivity of assessments,
$K3$ - for the error robustness due to a higher reliability of measuring methods,
$K4$ - for the error robustness to a higher reliability of the measuring and registration procedure, etc.

and notice that they are consistent in the sense that higher/lower values of each of them are advantageous/disadvantageous from the point of view of our approach.

We use Saaty's (1980) AHP (analytical hierarchy process) approach to derive weights W needed for the problem formulation from the experts' testimonies. This approach is well known and has proved to be useful in many applications. Basically, the experts perform pairwise comparisons of the attributes involved with respect to, consecutively, $K1$, $K2$, $K3$, $K4$, ... These judgments are then aggregated, and - finally - we obtain the weights needed, W, belonging to the unit interval. Details on the method can be found in Saaty's (1980) book or in other works presenting an application of this method; and will not be presented here for lack of space.

Now, when the weights W are available, let us briefly present a suboptimal, heuristic algorithm to solve the problem considered (13).

4. Algorithm

For obvious reasons the only practical option is to devise a heuristic algorithm for solving problem (13). The basic steps of the algorithm are as follows:

Step 1. Set initial values, mainly $S = S_p$, i.e. the whole set of examples is initially assumed to contain the positive ones.; $R_W = \varnothing$, i.e. the initial set of complexes is assumed empty.

Step 2. Use the AHP procedure to determine the weights W.

Step 3. Determine a *centroid* which is defined as an artificial (nonexistent) example that basically contains selectors with attribute values that most often occur in the real, positive and negative, examples (cf. Kacprzyk and Iwański, 1988 - 1992 for a related concept of a *typoid* which includes the positive examples only; in the case considered in this paper the centroid is however more efficient). The concept of a centroid is crucial for the efficiency of the algorithm that is greatly influenced by the starting point in each iteration.

Step 4. Determine the starting point, i.e. by a (real) example that is the closest (in the sense of some measure of similarity) to the centroid.

Step 5. Find a complex, C_W^*, with the minimal augmented weighted length starting from the example determined in Step 3, and using a greedy algorithm based integer programming procedure by Iwański and Szkatuła (1991).

Step 6. Include complex C_W^* found in Step 5 into the classification rule sought R_W^* (i.e. that with the minimal augmented weighted length), $R_W^* := R_W^* \cup C_W^*$, and discard from the set of positive examples S all examples covered by complex C_W^*

Step 7. If the set of positive examples S remaining is empty, STOP and the rule R_W^* is the one sought; otherwise, proceed to Step 3.

The approach presented was tested on an oncological example concerning patients with soft tissue sarcoma. Each patient is described by the following 22 attributes: sex, age, radiotherapy performed before first surgery, chemotherapy, duration of disease, tumor depth, infiltration of bones, vessels, etc., involvement of lymph nodes, tumor type, tumor location, tumor size, multicentric type of tumor growth, infiltration of skin, ulceration, etc., tumor stage, surgery type, surgical margin type, interoperative tumor damage, postoperative complications, temperature after surgery, sarcoma type determined after first surgery, histological tumor grading, and staging due to AJC. The set of positive examples contained patients who survived two years, while that of the negative examples contained the other ones.

The results obtained, i.e. rules for the survival up to two years, and for the opposite case were tested in two ways. First, they were presented to an oncologist whose assessment was quite positive (in the sense of confirming the state of domain knowledge). Second, a simulation experiment was performed to check how random changes in data (i.e. errors as meant in this paper) influence the form of the rule determined by the algorithm. Here, again, the results were good.

Just to give some examples, in case of data disturbed in 10% (in the sense of the percentage of attributes and examples) the classification error (defined as a weighted sum of the errors of omitting a positive example and of including a negative example) for the rule obtained was less than 4%, while for data disturbed in 30%, that error was less than 9%.

5. Concluding remarks

In the paper we proposed an approach to inductive learning under errors in data. The number and location of these errors are unknown. A modification of Michalski's start method was employed, with an integer programming based heuristic procedure. Additional information was elicited from the domain experts, and aggregated using the AHP method. the results are promising.

REFERENCES

[1] Iwański, C. and Szkatuła, G (1991) Inductive learning supported by integer programming. Computers and Artificial Intelligence 10, 57–66.

[2] Kacprzyk, J. and Iwański, C. (1990) Machine learning from incomplete and imprecise examples. In: Proceedings of 8th Italian-Polish Symposium on Systems Analysis and Decision Support in Economics and Technology (M. Fedrizzi and J. Kacprzyk, eds.):, Omnitech Press, Warsaw, p. 209–223.

[3] Kacprzyk, J. and Iwański, C. (1991a) Inductive learning from incomplete and imprecise examples. In: Uncertainty in Knowledge Bases (B. Bouchon-Meunier, R.R. Yager and L.A. Zadeh, eds.), Springer - Verlag, Berlin, p. 424–430.

[4] Kacprzyk, J. and Iwański, C. (1991b) Inductive learning under imprecision and errors, Proceedings of the Fourth IFSA Congress (Brussels, Belgium), Ser. Artificial Intelligence, pp. 78–81.

[5] Kacprzyk, J. and Iwański, C. (1992) Fuzzy logic with linguistic quantifiers in inductive learning. In: Fuzzy Logic for the Management of Uncertainty (L.A. Zadeh and J. Kacprzyk, eds.), Wiley, New York, p. 465–478.

[6] Michalski, R. (1973) Discovering classification rules using variable-valued logic system VL1. Proceedings of the Third International Joint Conference on Artificial Intelligence (IJCAI), p. 162–172.

[7] Michalski, R. (1983) A theory and methodology of inductive learning. In: Machine Learning (R. Michalski, J. Carbonell and T.M. Mitchell, eds.) Tioga Press, Palo Alto, CA.

[8] Saaty, T.L. (1980) The Analytic Hierarchy Process. McGraw-Hill, New York.

A CONNECTIONIST APPROACH FOR TEMPORAL REASONING

Hiroshi NARAZAKI
Process Tech. Res. Lab.
Kobe Steel, Ltd.
5-5 Takatsukadai 1-chome
Nishi-ku, Kobe, 651-22 Japan

Anca L. RALESCU*
Lab. for International Fuzzy Engineering Research (LIFE)
Siber Hegner Bldg., 89-1 Yamashita-cho
Naka-ku, Yokohama, 231 Japan

ABSTRACT

We propose a connectionist approach for temporal reasoning using knowledge such as "If p then q in the future." The proposed method consists of a network and a logical mechanisms. The knowledge is translated into constraints, and the network mechanism repeatedly searches for solutions consistent with those constraints. Among them, the logical mechanism selects a solution based on preference criterion. Further, we propose an "observe-and-edit" strategy to shorten the time to reach the preferred solution. This division of labor between the network and logical mechanisms contributes to simplifying control in the reasoning process.

1. Introduction

We often reason about events over different time periods. For example, we often guess possible causes in the past of the present trouble, probable consequences that present actions may yield in the future, or necessary actions that we should take to achieve a goal in the future, etc. Temporal reasoning (see [1] for survey) provides a formal basis for this kind of problems based on logical techniques. In this paper, we explore an alternative direction based on a connectionist approach. This work extends our method proposed in [2].

For illustration purpose, consider the following well-known Yale Shooting Problem(YSP)[3]. Informally, the *surface* knowledge for YSP is stated as follows:

R1: If we load the gun, then the gun becomes loaded.
R2: If a man is shot, then the man will die.

Two issues arise in connection with this knowledge base: (1) a prediction issue and (2) a planning issue.

Suppose that we know that John is alive during (0, 1), Mike loads the gun during (1, 2) and shoots at John during (3, 4) after waiting during (2, 3) (here (t_1, t_2) denotes a time interval). The prediction issue concerns the question "Is John alive or dead during (4, 5)?" This question is about the events that may take place in the future.

*On leave from Computer Science Department, University of Cincinnati, U.S.A. Research partially supported by NSF Grant INT91-08632

Suppose that we know that John is alive and that the gun is unloaded in (0, 1) and we want John to be shot to death in (3, 4). The planning issue concerns the question "How can we achieve the goal?" This question is about the actions necessary in order to obtain a desirable consequence in the future.

We may view the predication and planning problems as forward and backward (or abductive) temporal reasoning problems, respectively. Conventional approaches treat these problems separately under different paradigms, generating different reasoning "engines." In our connectionist approach, there is virtually no need for such distinction.

The surface knowledge stated explicitly is, at times, insufficient, and in practice, our *commonsense* provides implicit assumptions and preference of the reasoning process to obtain an intuitively acceptable belief. Conventional logic approaches embed such assumptions and preference into axioms and reasoning strategies (e.g. circumscription[4]). However, alternative views argue that logic is not sufficient for describing such commonsense knowledge[3]. In our method, as shown later, the preference is represented as a function that maps a truth value distribution to a real number to order solutions obtained.

We first describe the outline of our method. Then, we explain how the preference is represented and how the reasoning is controlled in our method. We show that our approach simplifies the control in reasoning.

2. Outline of the connectionist inference engine

Generally, the connectionist approach is based on *simultaneous constraint satisfaction.* As in the Hopfield neural network for an optimization problem[5], the knowledge is translated into a set of constraints. The degree of constraint violation is measured by an *energy function*, and the relaxation method updates the current truth value distribution so as to decrease the energy. Thus, a reasoning problem is formulated as an energy minimization problem, or an optimization problem.

Along these lines, we present a method having the following features (Refer [2] for details):

1. Our method consists of two modules, (1) the *network* mechanism and (2) *"logical"* mechanism.

2. The network has three kinds of nodes; (1) *constraint nodes* representing constraint equations, (2) *proposition nodes* representing facts (e.g. "Tweety is a bird"), and one *global node* which plays an important role in avoiding a local solution as explained later.

3. Knowledge given in the knowledge base is translated into constraint equations. The energy function is defined as the total square sum of violations of those constraints.

4. A *(feasible) solution* is a truth value distribution that makes the energy 0. If the energy attains a minimum with a nonzero value, we say that we have attained a *local solution.* A feasible solution corresponds to a consistent belief.

5. The *relaxation mechanism* searches for a feasible solution and converges to either a feasible solution or a local solution. When a local solution is attained, the value of the global node is designed to go to infinity. When it exceeds a pre-specified upper limit, a perturbation is triggered to jump out of the local solution.

6. A *snapshot* of the network state is taken when attaining a feasible or local solution. A snapshot is sent to the logical mechanism (LM) whose main task is the *control* of the search process based on the *observation* of the network state.

In our method the truth values are allowed to be *fuzzy*, i.e. somewhere in [0,1]. If we want only crisp truth values, we can use an integer programming approach[6], or the Hopfield neural network[5]. The continuous truth values are useful in this context to deal with conflicting knowledge as previously discussed in [2].

As seen from above, the connectionist approach is basically a mathematical programming approach as in [6] and [7]. The advantage of this approach is "holistic nature" and we can avoid several "inference control issues" as stated in [6]. The "neural network" for an optimization problem proposed by Hopfield provides a basis for an effective implementation that exploits parallelism. This may lead to a "thinking machine" as efficient as human brain as discussed in [8] and [9].

Viewing side by side the conventional logical approach and the connectionist approach, we notice the following:

1. The logical approach *derives* a belief while the connectionist approach *resolves conflict* by reducing constraint violations. The connectionist approach is more suitable for belief revisions and nonmonotonic reasoning problems.

2. The deductive reasoning approach searches for what is true in all models.In contrast, the connectionist approach converges to *a model* among many. (A *model* needs to be understood in a more general manner in this work so that it allows intermediate truth values. For example, with the constraint $p \rightarrow q$, $p = q = 0.5$ can be a solution which reflects that there is no reason to favor either p or q in particular).

3. The time required to reach a solution does not depend on the *size* of the knowledge base (if the relaxation process is carried out in parallel). If initial values of the relaxation process are chosen randomly, this time depends mainly on how long it takes to allocate the initial values in the *attractor* of the solution.

To any given problem, *multiple solutions* are likely, and for humans *good thinking* is often concerned with selecting the *preferred solution* among them. To achieve this, we use an *iterative* approach: We iterate the relaxation processes by changing initial truth values when a local or feasible solution is obtained. As explained later, the logical mechanism selects the best solution based on the preference criterion reflecting our commonsense.

During the iterative process, it is useful to maintain upper and lower bounds of the truth values. Let's denote the interval for p as $I(p)$. We can also maintain the average or variance information of the truth values, if necessary. If $I(p)$ has collapsed to a single point v, p has *necessarily* the truth value v. For example, if $I(p)$ has converged to [1,1] in the end, it means that p is true in all models and hence *deductively true.* If $[l(p), u(p)]$ has converged to $[0, 1]$, p can be anything from false to true.

In the iteration process, it is impossible to know whether we have already exhausted all (feasible and local) solutions. For example, when the relaxation process keeps finding only local solutions, it is impossible to know whether no feasible solution really exists *or* we are just having trouble to reach a feasible solution and more search time is required. Therefore, this judgment concerning the extent of the search is necessarily of a *heuristic* nature; it may make use of a motivation function $m(t)$ which may correspond to our *patience* for continuing the search as explained later.

In order to choose the preferred solution, we need a criterion, say Π, for ordering solutions. The circumscription uses *minimality* as such a criterion. In our method, Π can be any function that maps a truth value distribution to a scalar index value to order the solutions obtained so far. Our method is based on the *division of labor* between network and logical mechanisms: The network concentrates on providing snapshots and it is the task of the logical mechanism to order and select the best solution among them.

One way of defining $m(t)$ is to use $m(t) = e^{-\alpha\tau}$ where τ shows how many times the relaxation processes have been iterated since the current best solution is obtained. $\alpha > 0$ is a parameter that represents our patience, i.e. we give up search when $m(t)$ becomes lower than a prespecified threshold. Notice that the best solution may be either a feasible or a local solution. If the best solution is a local solution, $m(t) \to 0$ concludes that *the problem is not likely to be feasible.* If the best solution is a feasible solution, $m(t) \to 0$ concludes that *the current best solution is likely to be really the best solution.* This motivation function is related to the utility function and performance profile as in the *anytime algorithm*[10], and $m(t)$ should reflect the *complexity* of the problem, not the *size* of the knowledge base. Roughly speaking, the complexity of the problem depends on the number of (local) solutions, not the number of IF-THEN rules. A detailed investigation of these problems is part of our future work.

To shorten the time required for reaching the best solution, we use an *observe-and-edit* strategy which generates a new initial value distribution by *editing* the obtained solution. The edit is done such that we can get closer to the most preferred solution, but we do not need to generate a consistent truth values distribution because this inconsistency is remedied by the relaxation method.

3. Outline of our method

3.1. Truth values over temporal intervals

Given a truth value v of the proposition p, we define the structure $(v, (l,\ u), Flag)$ to represent "p takes a truth value $v \in [0,\ 1]$ during the time interval $(l,\ u)$." We assume that l and u are integers such that $l < u$. For example, *present*, *future*, and *past* could be represented as (0, 1), (1, 2), and (-1, 0), respectively. If $Flag = T$, p has a constant truth value v over $(l,\ u)$ during the course of inference. Otherwise, p has a variable truth value. Each proposition p has a database $V(p)$ to store truth values over several time intervals.

$$V(p) = \{(v_i,\ (l_i,\ u_i),\ Flag_i),\ i = 1, 2, .., t_p\}$$

We define the following two kinds of constraints.

Atemporal constraint $(AC(p, q))$: An atemporal constraint should be satisfied across *all* time intervals; it doesn't refer to the future, the present or the past; For example, an atemporal if-then rule is expressed as "If p then q" where p and q are propositional logic sentences.

Evolutionary constraint ($EC(p,q;t)$): An evolutionary constraint is described as "If p then q in the time lag t (an integer) from p" where p and q are propositional logic sentences. In this paper, we assume that the antecedent always refers to the present. More complicated constraints are considered in our future work. $EC(p,q;t)$ is evaluated by referring to the premise over (l,u) and the consequent over $(l+t, u+t)$. ($t > 0$ refers to the future and $t < 0$ refers to the past. Atemporal and evolutionary constrains are related by $EC(p,q;0) = AC(p,q)$).

3.2. Relaxation on admissible temporal intervals

The relaxation method in [2] is extended to deal with evolutionary constraints.

Let $[T_p,\ T_f]$, $T_p < T_f$, (p and f indicate *past* and *future*, respectively) denote the time interval in which we are interested to make inferences. We call *primitive time intervals*, the collection $PS = \{(l_i,\ u_i), i = 1,2,..,N\}$, if $\forall\ i\ T_p \le l_i < u_i \le T_f$, which satisfy the following two properties: (1) the sets in PS are mutually disjoint, and (2) any proposition takes a constant value on $(l,u) \in PS$. Atemporal constraints can be evaluated for the intervals in PS while evolutionary constraints can be evaluated only for the subset of PS, PS', such that for $(l,\ u) \in PS'$, $(l \pm t, u \pm t) \in PS$.

After the primitive temporal sets are generated, the truth values of p over $(l,\ u) \in PS$ are updated using the following gradient $\phi(p)$:

$$\phi(p) = \sum_{i=1}^{n1} \rho_{1,i} EC(q_i,p;t_i)\frac{\partial EC(q_i,p;t_i)}{\partial p} +$$
$$\sum_{j=1}^{n2} \rho_{2,j} AC(q_j,p)\frac{\partial AC(q_j,p)}{\partial p} +$$
$$\sum_{k=1}^{n3} \rho_{3,k} EC(p,q_k;t_k)\frac{\partial EC(p,q_k;t_k)}{\partial p}$$

where $\rho_{i,j}$s are weights for the constraint aggregation. The meaning of $\phi(p)$ is as follows: if $t > 0$, the first and third terms of $\phi(p)$ represent the forward and backward propagations along the time axis by the evolutionary constraints; the second term is the gradient of the atemporal constraints. Thus, we reason about the past, present and future *simultaneously*. For a proposition whose truth value is given as fixed in (l,u), we decide that $\phi(p) = 0$ in $(l,\ u)$. It should be also noted that the value of the global node (see [2] for details) needs to be extended so that the value should be calculated on each primitive time interval. If an explosion of the global node value occurs on one of the intervals, we trigger the perturbation for the truth values on all primitive temporal intervals. This is because the truth values in the past or future may be the cause of getting stuck in the current local solution.

4. Knowledge representation and search strategy

We need appropriate rules which *translate* the verbal description of our knowledge.

Here we discuss an action rule which can be verbally described as *"If Precondition(PC) holds (and the Intended Effects(IE) are absent), then the application of Action(A) yields Intended*

Effects(IE) and Side Effects(SE) in the future." We assume that PC, A, IE, and SE are all propositional logic sentences and we further assume that they are disjoint and do not share the same proposition. For example, in the above knowledge base, "John is dead($\neg Alive$)" is an intended effect of an action "Shoot", and hence $Alive$ is not included in a precondition PC. Hereafter, we call propositions corresponding to state descriptions and actions as *state* and *action variables*, respectively.

We interpret the above rule about the action A as follows: (1) We are motivated to apply A only when PC is present and IE is absent *now*, (2) the application of A should lead to IE and SE *in the future* if PC holds *now*, and (3) we are allowed not to apply A in whatever situation. This can be translated into the following two material implications.

$$PC \wedge A \rightarrow IE^{fut} \wedge SE^{fut} \tag{C1}$$

$$A \rightarrow PC \wedge \neg IE \tag{C2}$$

$C1$ is an evolutionary constraint read as "If PC and A then IE and SE in the future." $C2$ is an atemporal constraint. In some cases, indicating lack of the intended effects is not necessary for the application of action. In such a case, we can change $C2$ to $A \rightarrow PC$, omitting $\neg IE$.

The above representation allows *voluntary* state changes which is sometimes counter intuitive (e.g. "a dead man becomes alive again"). To avoid this, we can add the following implication.

$$S \wedge \neg A \rightarrow S^{fut} \tag{C3}$$

This is also an evolutionary constraint read as "If S and $\neg A$ then S in the future." In other words, only A can cause the state change of S. "A" can be a disjunction or conjunction of actions if several actions are responsible for the change of S. $C3$ is a frame axiom and some may want to avoid the explicit representation of this rule as usually done using circumscription technique[4]. We will show later how we can incorporate frame rules into the reasoning mechanism as an intervention strategy by the logical mechanism.

In summary, an action rule is translated into two rules of the type $C1$ and $C2$, and each state variable needs to be associated with two rules of type $C3$ (for changes from $True$ to $False$ and from $False$ to $True$, if necessary). Thus if we have n action rules and m state variables, then the knowledge is translated into (at most) $2(m+n)$ rules.

Furthermore, we define another constraint stating that we apply at most one action over a primitive temporal interval: Let $\{A_i, i = 1, 2, .., m\}$ be a set of propositions for actions. Then we add the following action selection rule:

$$\sum_{i=1}^{m} a_i \leq 1 \tag{C4}$$

where a_i is a truth value of A_i. (The equality constraint can be used to express "exactly one action." However, the inequality constraint is still sufficient because we can easily fill "No action" over those intervals such that $\sum_{j=1}^{m} a_i = 0$.)

Based on the above discussion, it is now easy to translate the knowledge in YSP: $PC = SE = True$ here, and $R1$ is translated as: $Load \rightarrow Loaded^{fut}$, and $Load \rightarrow \neg Loaded$. Similarly, $SE = True$ here, and $R2$ is translated as: $Loaded \wedge Shoot \rightarrow \neg Alive^{fut}$, and

$Shoot \rightarrow Loaded \wedge Alive$. We use a time lag equal to 1 for state propagation, namely, we have $EC(Load, Loaded; 1)$, etc.

Next let's define rules of type C3 for the state variables *Loaded* and *Alive*. We assume that a gun becomes loaded and unloaded only by loading and shooting, respectively. Thus, we have $(\neg Loaded) \wedge (\neg Load) \rightarrow \neg Loaded^{fut}$, and $Loaded \wedge (\neg Shoot) \rightarrow Loaded^{fut}$. Similarly, for *Alive*, we consider only the death by shooting, i.e. only *Shoot* can change *Alive* from *True* to *False*(e.g. *Death by heart attack before being shot* is eliminated). Further no action can make a dead man alive again, i.e. *False* is assigned to A in C3. Thus, we have $Alive \wedge (\neg Shoot) \rightarrow Alive^{fut}$, and $\neg Alive \rightarrow \neg Alive^{fut}$

Finally, we use a dummy *Wait* action for representing "do nothing," and the unique action constraint in $C4$ is given by $\alpha(Load + Shoot + Wait - 1) \leq 0$ where α is a normalizing factor to ensure (C4). Here we use $\alpha = 0.5$.

The above knowledge can be compiled into the network structure and we can start the relaxation process with $T_p = 0$, $T_f = 4$. That is, the inference is valid over (0, 4). We apply a simple *defuzzification* which approximates the truth value v as 1 when $v \geq 0.5$ and 0 otherwise after a solution is found. In general, care is needed about the defuzzification. For example, for quantified statements such as "Q of .... are true", we need set 1 to the propositions having Q largest truth values. The relaxation results in the solutions described below when a simple *defuzzification* is applied after the convergence.

Prediction: Given the fact that $V(Alive) = \{(0, (0,1), T)\}$, $V(Load) = \{(1, (0,1), T)\}$, $V(wait) = \{(1, (1,2), T)\}$, $V(Shoot) = \{(1, (2,3), T)\}$ (John was alive, Mike loaded a gun and shot at John after waiting during (1,2)). We select initial truth values randomly. When the relaxation method converged to a feasible or local solution, the perturbation mechanism again chooses truth values randomly and performs the relaxation process again. We iterate this relaxation/perturbation sequence, and the logical mechanism samples the snapshots just before the perturbation is triggered. In a simulation, after visiting 4 solutions, the truth value intervals show that John should be dead during (3,4). However, for the state variable *Loaded*, the truth value lies in [0, 1] during the time interval (3, 4) because nothing in the knowledge base decides whether a gun is loaded or not after shooting.

Planning: Given $V(Alive) = \{(1, (0,1), T), (0, (3,4), T)\}$, and $V(Loaded) = \{(0, (0,1), T)\}$, we start the relaxation from randomly chosen initial values. This is a planning process to shoot John to death by the time interval (3,4) the latest. In a simulation, we find a feasible solution where we have to *load* the gun during (0, 1), and *shoot* at John during (2, 3). Another feasible solution is to shoot at John during (1, 2). This solution could be eliminated if the knowledge base contains additional information specifying that John should do something during the interval (1, 2).

As seen from above, our method makes no distinction between the prediction and planning mechanisms, and the only difference is what kind of query we make to the inference mechanism.

5. Logical mechanism

As stated before, our method finds the best solution by iterating the relaxation processes, starting with various initial values. We propose the following *observe-and-edit* strategy to determine an initial truth value distribution: *"When a feasible solution is found, edit the solution to improve the preference criterion. Start the relaxation process again from the edited result."* The editing policy should be domain-specific. Here we propose a domain-independent strategy suitable for the prediction problem based on *a persistence assumption.*

First, we describe the preference criterion used by the *observe-and-edit* method.

The preference criterion decides which of the two solutions is more preferable. Here we explain a domain-independent preference criterion $\Pi(V(p_1), ..., V(p_m))$ over the databases $M = \{V(p_i), i = 1, 2, .., m\}$.

Let $V(p_i) = \{(v_{i,1}, (l_{i,1}, u_{i,1})), ..., (v_{i,k_i}, (l_{i,k_i}, u_{i,k_i}))\}$ be a history of the truth values of p_i. (The third component, *Flag*, of each element in $V(p_i)$ is omitted). Here $\{(l_{i,j}, u_{i,j}), j = 1, 2, .., k_i\}$ are assumed to be chronologically ordered primitive temporal intervals, i.e. $u_{i,j} \leq l_{i,j+1}, j = 1, 2, .., k_i - 1$.

We first define $P(p_i)$, a preference criterion for p_i, as follows:

Action variables: If p_i is an action variable, $P(p_i)$ is defined as a weighted sum of $v_{i,j}$, i.e. $P(p_i) = \sum_{j=1}^{k_i} w_j v_{i,j}$ where w_i is a weight.

State variables: If p_i is a state variable, $P(p_i)$ is defined as a weighted sum of the truth value changes, i.e. $P(p_i) = \sum_{j=1}^{k_i-1} w_j d(v_{i,j}, v_{i,j+1})$ where w_i is a weight and $d(v_{i,j}, v_{i,j+1})$ is a distance between $v_{i,j}$ and $v_{i,j+1}$. The distance can be defined in various ways. Here we use an absolute difference $d(v_{i,j}, v_{i,j+1}) = |v_{i,j} - v_{i,j+1}|$.

Then, $P(p_i)$ is aggregated as $\Pi(M) = \sum_{i=1}^{m} P(p_i)$. We prefer the solution having the smaller value of $\Pi(M)$, namely, the solution with less actions and less state changes is preferred. This reflects the popular persistence assumption.

Further, the chronological preference as addressed in [3],[11] can be expressed in our method as follows: We weigh less the state change in the future than the current change by using an exponentially decaying weight such as $w_i = w_0 e^{-\lambda t}$ where w_0 is a constant, t is a time, and λ is a decay ratio.

With the above preference criterion, the observe-and-edit strategy is stated as follows:

Step 1: Evaluate how necessary the current result is by looking at the difference $\delta(p, (l, u))$ between the initial and final truth value of a proposition p for a relaxation process over the interval (l, u).

Step 2: By chronologically scanning the truth value history, detect the points where a significant state change occurs. Let v_1 and v_2 be the significantly different truth values over the neighboring intervals (l_1, u_1) and (l_2, u_2), $u_1 \leq l_2$. (We need a threshold test for the significance judgment).

Step 3: If v_1 has no preceding interval and not a fixed value, and, further, if $\delta(p, (l_1, u_1))$ is not significant, then copy the future value v_2 to v_1. If v_1 has a preceding value and not a fixed value, and if $\delta(p, (l_2, u_2))$ is not significant, copy v_1 to v_2.

There is no guarantee that the above *observe-and-edit* strategy yields a feasible solution. However, such new constraint violations can be taken care of by the relaxation process. The observe-and-edit strategy is much easier to implement than directly writing down all search strategies because we can escape from some of the consistency requirements owing to the division of labor between the network and logical mechanisms.

It should be recollected that the prediction problem in the previous section has solutions which are different according to whether the gun is loaded or unloaded after the shooting. With the above observe-and-edit strategy, the solution where the gun remains loaded is preferred. In some sense this could be justified as follows: the preferred solution is that which allows the action to be carried out again. For the evaluation of the observe-and-edit strategy, we omit the explicit frame rules in C3. Thus a gun is allowed to mysteriously get loaded or unloaded and a man is allowed to die and get alive freely. We retain a list to store 3 best solutions, and every time a new solution is added to this list (i.e. the new solution is exceeding one of the existing solutions), we edit the new solution. Otherwise if the new one does not exceed the old solutions or when the relaxation attains a local solution, we perturb the truth values by assigning random value in [0,1]. For a preference criterion, we use a decaying ratio $\lambda = 0.5$, $w_0 = 1$, and t is given by $(l+u)/2$ for the interval (l, u). For comparison, we repeat 150 steps of the relaxation 10 times for the cases with and without the edit operation. With the edit operation, 8 out of 10 cases yield $\Pi(M) = 4.73$ where the victim is shot to death and the gun remains loaded after the shooting. The remaining 2 cases have $\Pi(M) = 4.81$ where the gun is mysteriously unloaded and the victim is alive after shooting. They fail to find the best solution within 150 steps. In contrast, without the edit operation, the average value of $\Pi(M)$ is 5.12, ranging from 4.72 to 5.57. This shows that the observe-and-edit strategy is effective in improving the search efficiency and contributes to the reduction of the variation in the quality of the solution.

6. Conclusion

We presented an outline for a connectionist reasoning mechanism for temporal knowledge. The division of labor of our method gives a flexibility and simplicity of the reasoning process.

REFERENCES

[1] Shoham, Y., Goyal, N. (1988) Temporal reasoning in artificial intelligence. In: Exploring Artificial Intelligence, (H.E.Shrobe and AAAI eds.), CA, Morgan Kaufmann, p.419-438.
[2] Narazaki, H., Ralescu, A.L. (1992) A connectionist approach for rule-based inference using an improved relaxation method, IEEE Trans. NN, 3(5), 741-751.
[3] Hanks, S., McDermot, D. (1986) Default reasoning, nonmonotonic logic, and frame problem, Proc. 5th Annual National Conf. on Artificial Intelligence, 328-333.
[4] McCarthy, J. (1986) Application of circumscription to formalizing common-sense knowledge, Artificial Intelligence, 28, 89-116.
[5] Hopfield, J.J., Tank,D.W. (1985) "Neural" computation of decisions in optimization problems, Biological Cybernetics, 52, 141-152.
[6] Yager, R.R. (1988) A mathematical programming approach to inference with the capability of implementing default rules, Int. J. Man-Machine Studies, 29, pp.685-714.
[7] Genesereth, M.R., N.J.Nilsson, N.J., (1987) Probabilistic entailment. In: Logical Foundations of Artificial Intelligence, CA, Morgan Kaufmann, p. 193-197
[8] Hillis, W.D. (1989) The connection machine, MIT Press, MA.

[9] Feldman, J.A. (1985) connectionist models and their applications: Introduction, Cognitive Science, 9, 1-2.
[10] Russel, S.J., Zilberstein S. (1991) composing real-time Systems, Proc. IJCAI 91, 1, 212-217.
[11] Shoham Y. (1986) chronological ignorance: An experiment in nonmonotonic temporal reasoning, Proc. of 5th National Conf. on AI, 389-393.

A LEARNING PROCEDURE TO IDENTIFY WEIGHT OF RULES USING NETWORKS*

Armando BLANCO, Miguel DELGADO, Ignacio REQUENA
Department of Computer Sciences and Artificial Intelligence
University of Granada
Adv Andalucia n.38
10872 Granada Spain

ABSTRACT

In many cases the identification of systems by means of fuzzy rules is given by taking these rules from a predetermined set of those possible. In this case, the correct description of the system is to given by a finite set of rules, each with an associated weight which assesses its correctness or accurancy. Here, we present a method for learning this consistence level or weight using neural networks, for identifying the system, with which to associate, each possible rule with its weight .

Keywords: Neural network, Weight of rule.

1 Introduction

Many systems have been represented by fuzzy rules, for instance, in the fuzzy control setting, however it is not always possible to obtain this identification. Often all the rules in fuzzy logic controllers are obtained by analyzing the expert's experience or by a trial-and error approach, although an interesting alternative is being developed, the automatic technicals, which go from the statistical methods to the news genetic algorithms, which solve the problem of identifying fuzzy systems, when doing so by experts is not possible.

In several cases the rules are to taken from a given set of possible ones(for instance when linguistic variables are assessed in a fixed term set).

In this case, in order to conceive the right description of the system as given by a finite set of rules, each with a weight which assesses it correctness or accuracy, seems reasonable.

To obtain such an identification, several approaches have been developed. We present a technique, based on the ability learning of neural networks to learn the weight of all the rules which can identify a system.

In this paper by "continuous system" we refer to any system described by $y = f(x)$ where $x \in R^n$ $y \in R^m$ $f : R^n \longrightarrow R^m$ being continuous. The systems based on fuzzy rules that we consider, are characterized by a finite set of fuzzy rules, these rules have the following form:

R_i : If X_1 is A_{1i} and $\cdots$ X_n is A_{ri} then Y_1 is B_{1i} and $\cdots$and Y_n is B_{ti}; $i = 1, \cdots, p$ A_{ji} y B_{ki} are fuzzy sets, being defined on referentials X and Y respectively, X_j are input variables and Y_j are output variables of X and Y respectively.

In short, we can express these rules by

R_i : If A_{1i} and $\cdots$ A_{ri} then B_{1i} and $\cdots$ B_{ti}.

By "fuzzy approximator of system" we refer to a process represented by a measurement system between the input and the output with crisp values, obtaining this output in order to apply it to the input, a fuzzyfication, a fuzzy system and a defuzzyfication. We consider a continuous relation between the input and the output.

$\xrightarrow{X}$ [fuzzyfication] $\rightarrow$ [S.D.] $\rightarrow$ [defuzzyfication] $\xrightarrow{Y}$

*This work has been developed under project PB 89-0463 of DGICIT. MADRID

We can approximate a network to a fuzzy approximator of system, because this is a subset of the continuous systems.

Castro and Delgado in [1] have proven that the fuzzy approximator of systems are universal approximators, so from a continuous system it is possible to find a fuzzy approximator of system closely arbitrarily approximated.

Several research papers [2], [4] have shown that neural networks with this topology are universal aproximators and that these neural networks can be trained by supervised learning. These neural networks are a continuous overlap from $[0,1]^n$ to $[0,1]^m$

Buckley and cols, in [3] show, that from a feedforward network (continuous function) it is possible to obtain a system based on rules closely arbitrarily approximated.

Therefore, from a theoretical point of view, the systems based on rules are universal approximators.

Using a multilayer network with a big number of neurons in the hidden layer, we can obtain a continuous overlap from the input variables to the output variables.

This learning is obtained from knowledge of about the empirical database, such as training sets.

In this paper we propose a practical method to identify the continuous systems using systems based on fuzzy rules.

In the fuzzy system, the inputs and outputs are fuzzy sets, so the systems based on rules are fuzzy systems.

From the point of view of practical identification below we will denote any fuzzy set and its membership with the same capital letter, and we discretize the fuzzy sets (input-output of fuzzy system) in the following way:

If A is a fuzzy set, it is the membership function of a fuzzy set, whose domain is the interval $[u_0, u_1]$, we choose a fixed natural k and we make a partition of the domain, obtaining the elements: $s_i = u_0 + (i-1)\frac{u_1-u_0}{k-1}$, $i = \{1, 2, \cdots k\}$; we associate the vector $\vec{a} = (a_1, a_2, \cdots, a_k) = (A(s_1), A(s_2), \cdots, A(s_k))$, $a_i = A(s_i) \in [0,1]$ to vector $s = (s_1, s_2, \cdots, s_k)$

So, we shall model the inputs of fuzzy system by means of an n-dimensional $(x_1, x_2, \cdots, x_n)$ vectors whereas the output will be an m-dimensional one $(y_1, y_2, \cdots, y_m)$, with x_i, $y_j \in [0,1]$, where n and m denote the number of elements chosen in the domain sets of fuzzy input-output sets.

Therefore, we can express a fuzzy system as an overlap from $[0,1]^n$ to $[0,1]^m$

By this discretation, each rule can be considered as an overlap from $[0,1]^{n_1+\cdots+n_r}$ to $[0,1]^{n_1+\cdots+n_t}$

The system based on rules is considered as the intersection of the overlaps of each rule.

More specifically, with this discretation we can talk about the continuous system based on rules, or with a finite number of discontinuous.

2 Discussion of the problem

Let us show our method on a system with one input variable and a single output variable, although similar procedure can be used for more complex system.

Let us consider a continuous system directed by an input-output equation in the following way $v = f(u)$ $u \in R^n$ $v \in R^m$

We suppose that f is unknown, but we have a finite number of observations (u_i, v_i) $i = 1 \cdots N$ obtained through an automatic system.

In the literature there are models where u_i and v_i can be fuzzy sets, but in most real cases they will be $u_i \in R^n$, $v_i \in R^m$, because the sensors available give no fuzzy data.

We are interested in the problem of identifying the system using fuzzy rules.

From the continuous system defined in $U \longrightarrow V$, we only know a series crisp input-output (u_i, v_i) $u_i \in U, v_i \in V$ $i = 1 \cdots N$, which generate a set of rules in this way:

R_1: If u is u_1 then v is v_1
R_2: If u is u_2 then v is v_2
$\vdots$
R_N: If u is u_N then v is v_N

Defining the variables H_U and H_V, like the linguistic variables associated to the u and v input and output variables, respectively, we obtain the system:

R_1: If H_U is u_1 then H_V is v_1
R_2: If H_U is u_2 then H_V is v_2
$\vdots$
R_N: If H_U is u_N then H_V is v_N

This system shall be called Initial.

The set of all rules, which can express the continuous system defined in $U \longrightarrow V$ will be R_i: If H_U is h_{ui} then H_V is h_{vi} $i = 1 \cdots k'$ where h_{ui} (values of H_U in the rules) and h_{vi} (values of H_V in the rules) are to taken from fixed sets of labels $\{h_{u1}, h_{u2}, \cdots h_{us}\}$, $\{h_{v1}, h_{v2}, \cdots h_{vt}\}$

Each label will have its semantics given by a fuzzy subset of the corresponding framework of discernment (U for H_U, V for H_V) $k' = s * t$, that is, k' is the number of all possible rules associated with the Cartesian product $L = \{\{h_{u1}, h_{u2}, \cdots h_{us}\} \times \{h_{v1}, h_{v2}, \cdots h_{vt}\}\}$.

Having chosen a fixed system, we can associate a weight or consistence level witho each rule of R_i, a real number $\lambda \in [0,1]$, this weight may be interpreted as the strength of implication of the fuzzy rule to the determined system.

So, our first objective is to find the weights of each rule of R_i associated with the continuous system defined in $U \longrightarrow V$

Once we have all the consistence levels of each rule, our second objective, in the end, is to find a subset of L selecting all rules which weights which are one or near to one, obtaining a system, which will be expressed through a set of fuzzy rules like the following:

R_i: If H_U is h_{ui} then H_V is h_{vji} where $j1 \cdots jk$ is a list of the "convenient" elements of $\{1, 2, \cdots, t\}$ and h_{ui}, h_{vji} $i = 1 \cdots k$ denote sets of labels, whose semantics will be given by fuzzy sets of R^n and R^m respectively.

This system, that we shall call the Last system, obviously describes the Initial system.

3 The Identification procedure

We solve this problem as follows:

- We discretize the referential sets U and V, thereby obtaining $U = \{u_1, u_2, \cdots, u_n\}$ and $V = \{v_1, v_2, \cdots, v_m\}$.
- Let's suppose that the variables H_U and H_V can take values in the sets $\{h_{u1}, h_{u2}, \cdots h_{us}\}$, $\{h_{v1}, h_{v2}, \cdots h_{vt}\}$ of linguistic labels.
- We associate with each crisp rule, a real number that we shall call "weight of rule" $w_i \in [0,1] \subset R$ which reveals their consistence level.
- We build the referential $W = U \times V$ defined by ordered pairs (u_l, v_r) $l = \{1, 2, \cdots, n\}$, $r = \{1, 2, \cdots, m\}$.
- We establish the set $\mathcal{R} = \{h_{u1} \times h_{v1}, h_{u1} \times h_{v2}, \cdots, h_{us} \times h_{vt}\}$ Cartesian product from the sets where the linguistic labels take values, that is, H_U and H_V.
- On the new referential W we define the variable $H_{U \times V}$ which takes values on the sets $\mathcal{R}$ or W.
- We define the variable H_I (weight of a rule) which takes values in the interval $[0,1]$.

So, we have defined the set of all possible rules from the establishment of the variable $H_{U\times V}$ associated with the Cartesian product set $\mathcal{R}$.

Once we have defined these new concepts, we build a new system that we shall indicate as the Intermediate system that will be defined in $W \longrightarrow [0,1]$, by the set of rules

R_1: If $H_{U\times V}$ is (u_1, v_1) then H_I is 1
R_2: If $H_{U\times V}$ is (u_2, v_2) then H_I is 1
⋮
R_N: If $H_{U\times V}$ is (u_N, v_N) then H_I is 1

The objective that now we pursue, is to identify the Intermediate system using a system based on rules such as

R_1: If $H_{U\times V}$ is (h_{u1}, h_{v1}) then H_I is λ_1
R_2: If $H_{U\times V}$ is (h_{u2}, h_{v2}) then H_I is λ_2
⋮
$R_{k'}$: If $H_{U\times V}$ is $(h_{uk'}, h_{vk'})$ then H_I is $\lambda_{k'}$
where $\lambda_1, \lambda_2, \cdots, \lambda_{k'} \in [0,1]$, $k' = s * t$

Once we have achieved this new system, we have all the feasible rules that can identify the Initial system and the level of consistence associated with each rule, hence the first objective.

3.1 Identification of the Intermediate system.

With the same procedure used, we can talk about the continuous system based on rules
$R_{k'}$: If H_U is $h_{uk'}$ then H_V is $h_{vk'}$ by overlapping from $[0,1]^n$ to $[0,1]^m$, we can talk about the continuous system as
$R_{k'}$: If $H_{U\times V}$ is $h_{uk'} \times h_{vk'}$ then H_I is $\lambda_{k'}$ by overlapping from $[0,1]^{n+m}$ to $[0,1]$

To this effect, the Initial system is continuous, since small deviations on the input-output will produce small deviations on the aforesaid level of consistence of rule by the variables H_I, so the Intermediate system is a continuous system.

Therefore we can use a feedforward network with a hidden layer and sigmoidal activation functions to identify the Intermediate system.

To identify the system by neural networks, we need to define the net topology and the training models.

3.2 Net topology

We propose a feedforward neural network with 3 hidden layers and sigmoidal activation functions in the hidden layer and the output layer as $f(x) = \frac{1}{1+e^{-x}}$

The system that we have identified is an overlapping from one $[0,1]^{n+m}$ to $[0,1]$, so we shall use a network with $n+m$ neurons in the first layer and one neuron in the output layer.

So the input in the first layer will be an element of $[0,1]^{n+m}$ and the output from the last layer will be an element of $[0,1]$ according to the next figure

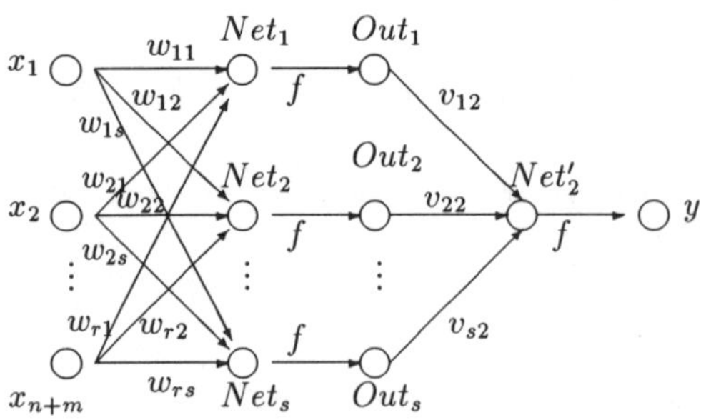

We can carry out the training of the network by choosing an appropriate method such as the back-propagation algorithm in Rumelhart et al. [5] with previous models, and we can find a function that comes as close as we want to the system.

3.3 Training models

The construction and training of the network require the selection of models; for each input the models are $(n + m)$-dimensional vectors.

We propose to choose the models for training the network from the next ordered pairs:

I- A finite number of ordered pairs $((u_i, v_i), \lambda_i)$, which are the crisp rules representing the system.

II- Because the Initial system is continuous, the Intermediate system also is, therefore we have ordered pairs such as $((u_i, v_i \pm \epsilon), \lambda_i)$, where ϵ is chosen as a small number.

III- By fuzzification, we obtain the step to fuzzy values with pairs $((u_i, [v_i - \epsilon, v_i + \epsilon]), \lambda_i)$

Of course, the λ_i from types I, II and III are 1, because the consistence of rules is the highest one.

IV- From type I, we can get pairs of training negatives $((u_i, v_i \pm \beta), \lambda_i)$, which have been chosen by an expert or have been inferred from the crisp rules; β is chosen so that $v_i \pm \beta$ will go away from v_i, in this case the consistence of rules will be nil, so $\lambda_i = 0$.

V- From fuzzyfication about the pair negatives, we get pairs like the type III pairs $(u_i, [v_i - \beta - \epsilon, v_i - \beta + \epsilon], \lambda_i)$ and $(u_i, [v_i + \beta - \epsilon, v_i + \beta + \epsilon], \lambda_i)$, in this case $\lambda_i = 0$ as in case IV.

VI- Finally, because of continuity, we choose the input pairs $(u_i, v_i \pm \rho)$ where ρ is a real number that is not very small, where $\lambda_i = 1 - k\rho$, and $k\rho$ is a real number proportional to ρ.

The training models ($n + m$-dimensional vectors) are chosen for each ordered pair, so that their components will all be null except the ones that occupy the position given by the ordered pair, which are assigned to 1.

Once the neural network has been trained, we can obtain the weight of each rule, showing in the first layer the appropriate vectors to the antecedent and consequent rule, linked together, obtaining the weight of this rule.

So we achieve a system where the right output, weight or the accurate level of consistence for each input or fuzzy rule is assigned.

Finally, using this system to achieve our second objective shown at the beginning, we take only those rules, the level of consistence of which is one or near to one, so we shall have the Initial system identified by rules, that is, the Last system.

We can generalize the previous process in the case that universes U y V, were a product of a finite number of universes, where there is more than one variable.

4 Example

Let us consider an example that is associated to the relation $X = Y$ that is previously unknown, defined on the universes $U = \{1, 2, \cdots, 11\}$ $V = \{1, 2, \cdots, 11\}$, and where the sets of linguistic variables $H_U = H_V$ are limited to the following values: Large, Medium and Small

From this system, we know only the crisp input-output, and we are interested in identifying of this system, using a fuzzy system based on rules.

4.1 Training the network.

By the procedure of discretization, each label has an associated membership function like these

Small $\longrightarrow$ (1.00, 0.81, 0.60, 0.41, 0.19, 0.01, 0.00, 0,00, 0.00, 0.00, 0.00)

Medium $\longrightarrow$ (0.01, 0.20, 0.39, 0.60, 0.80, 1.00, 0.80, 0,60, 0.41, 0.20, 0.00)

Large $\longrightarrow$ (0.00, 0.00, 0.00, 0.00, 0.00, 0.00, 0.21, 0,40, 0.61, 0.80, 1.00)

Let us suppose that the system is characterized by input-output like (1,1), (2,2)... (11,11). Then we express the ordered pairs of previous types

I.- (1, 1), (2, 2) ... (11, 11).
II.- (1, 2), (2, 3) ... (10, 11), (2, 1), (3, 2) ... (11, 10).
III.- (2, (1, 2, 3)), (3, (2, 3, 4) ... (10, (9, 10, 11).
IV.- (1, 5), (2, 6) ... (11, 1), (1, 9), (2, 10) ... (11, 3).
V.- (1, (4, 5, 6)), (2, (5, 6, 7)) ...(1, (9, 10, 11))...
VI.- (1, 3), (2, 4) ... (8, 10), ... (3, 1), ... (10, 8).

It is not necessary to get all models to trained, except when we increase the number of elements to be trained we shall achieve better effects.

For each pair of elements the training model will be chosen as we indicated before.

Then we show the codification about some models

(3,4) ⟶ (0, 0, 1, 0, 0, 0, 0, 0, 0, 0, 0, 0, 0, 0, 1, 0, 0, 0, 0, 0, 0, 0)
(3, (2, 3, 4)) ⟶ (0, 0, 1, 0, 0, 0, 0, 0, 0, 0, 0, 0, 0, 1, 1, 1, 0, 0, 0, 0, 0, 0, 0)
(3, (6, 7, 8)) ⟶ (0, 0, 1, 0, 0, 0, 0, 0, 0, 0, 0, 0, 0, 0, 0, 0, 0, 1, 1, 1, 0, 0, 0)
(2, 4) ⟶ (0, 1, 0, 0, 0, 0, 0, 0, 0, 0, 0, 0, 0, 0, 1, 0, 0, 0, 0, 0, 0, 0)

The output of the last layer from previous models is 1, 1, 0, 0.3, respectively

The remaining of inputs of the network will be codified in the same manner.

4.2 Obtaining the system based on rules.

After training the network, the outputs and the feasible rules are

Small ⟶ Small
Input: (1.00, 0.81, 0.60, 0.41, 0.19, 0.01, 0.00, 0,00, 0.00, 0.00, 0.00, 1.00, 0.81, 0.60, 0.41, 0.19, 0.01, 0.00, 0,00, 0.00, 0.00, 0.00)
Output (Level of consistence): 0.92

Small ⟶ Medium
Input: (1.00, 0.81, 0.60, 0.41, 0.19, 0.01, 0.00, 0,00, 0.00, 0.00, 0.00, 0.01, 0.20, 0.39, 0.60, 0.80, 1.00, 0.80, 0,60, 0.41, 0.20, 0.00)
Output (Level of consistence): 0.00

Small ⟶ Large
Input: (1.00, 0.81, 0.60, 0.41, 0.19, 0.01, 0.00, 0,00, 0.00, 0.00, 0.00, 0.00, 0.00, 0.00, 0.00, 0.00, 0.00, 0.21, 0,40, 0.61, 0.80, 1.00)
Output (Level of consistence): 0.01

Medium ⟶ Small
Input: (0.01, 0.20, 0.39, 0.60, 0.80, 1.00, 0.80, 0,60, 0.41, 0.20, 0.00, 1.00, 0.81, 0.60, 0.41, 0.19, 0.01, 0.00, 0,00, 0.00, 0.00, 0.00)
Output (Level of consistence): 0.07

Medium ⟶ Medium
Input: (0.01, 0.20, 0.39, 0.60, 0.80, 1.00, 0.80, 0,60, 0.41, 0.20, 0.00, 0.01, 0.20, 0.39, 0.60, 0.80, 1.00, 0.80, 0,60, 0.41, 0.20, 0.00)
Output (Level of consistence): 0.98

Medium ⟶ Large
Input: (0.01, 0.20, 0.39, 0.60, 0.80, 1.00, 0.80, 0,60, 0.41, 0.20, 0.00, 0.00, 0.00, 0.00, 0.00, 0.00, 0.00, 0.21, 0,40, 0.61, 0.80, 1.00)
Output (Level of consistence): 0.08

Large ⟶ Small
Input: (0.00, 0.00, 0.00, 0.00, 0.00, 0.00, 0.21, 0,40, 0.61, 0.80, 1.00, 1.00, 0.81, 0.60, 0.41, 0.19, 0.01, 0.00, 0,00, 0.00, 0.00, 0.00)
Output (Level of consistence): 0.07

Large ⟶ Medium
Input: (0.00, 0.00, 0.00, 0.00, 0.00, 0.00, 0.21, 0,40, 0.61, 0.80, 1.00, 0.01, 0.20, 0.39, 0.60, 0.80, 1.00, 0.80, 0,60, 0.41, 0.20, 0.00)
Output (Level of consistence): 0.04

Large ⟶ Large
Input: (0.00, 0.00, 0.00, 0.00, 0.00, 0.00, 0.21, 0,40, 0.61, 0.80, 1.00, 0.00, 0.00, 0.00, 0.00, 0.00, 0.00, 0.21, 0,40, 0.61, 0.80, 1.00)
Output (Level of consistence): 0.99

Then the rules are such that ...
Rule-1: Small ⟶ Small
Rule-2: Medium ⟶ Medium
Rule-3: Large ⟶ Large

5 Concluding Remarks

We have developed a methodology for learning rules and their consistence level, in a fuzzy environment which uses only empirical information, using a feedforward network.

The extension to more complex problems is performed in a direct way. When we consider a system where there are several input-output variables, then we have to build the Cartesian product of the sets where they are defined and then apply the same method.

REFERENCES

[1] Castro, J. Delgado, M. (1993). Fuzzy systems with defuzzification are universal approximators. submitted to *IEEE Transaction on Systems Man, and Cybernetics.*

[2] K. Hornik, M. Stinchcombe, and H. White, "Multilayer feedforward neural networks are universal aproximators" *Neural Networks*, vol. 2, 359-366 (1989).

[3] Buckley, J., Hayashi, I., Czogala, E. (1993). On the equivalence of neural nets and fuzzy expert systems. *Fuzzy Sets and Systems*, 53, 129-134.

[4] Kreinovich, V. (1991). Arbitrary nonlinearity is sufficient to represent all functions by neural networks: A theorem. *Neural Networks*, 4, 181-200.

[5] Rumelhart, D., Hinton, G. and Williams, R. (1986). Learning internal representations by error propagation *Parallel distributed processing* Vol. 1, pp. 318-362. Cambridge, MA: MIT Press.

A PROPOSAL OF IMPLICIT KNOWLEDGE ACQUISITION BY F-CBR

Atsushi INOUE, Shun'ichi TANO, Wataru OKAMOTO, Toshiharu IWATANI

Laboratory for International Fuzzy Engineering Research (LIFE)
Siber-Hegner Bldg., 4F., 89-1 Yamashita-cho,
Naka-ku, Yokohama, KANAGAWA 231 JAPAN

ABSTRACT

A method to acquire knowledge commonly expressed among several cases to make human-computer communication more intelligent is proposed. New design concepts of a natural language communication system in which this method is to be implemented, a new reasoning method, Fuzzy Case-Based Reasoning (F-CBR), used in the system to generate responses, and a problem of human-computer communication by F-CBR causing the necessity of this method are described respectively as well as the proposal.

1. Introduction

The Laboratory for International Fuzzy Engineering Research (LIFE) is currently developing a new natural language communication system, **F**uzzy **Lin**gual **S**ystem (FLINS). It can communicate and learn by a natural language with fuzziness and try to be a faithful and intelligent servant of the user. It has some different features from conventional natural language systems such as an unified representation and a consistent treatment of knowledge, rules and data, positive use of fuzziness in natural languages, and dynamic construction of deep semantic structures. All of them are most likely opposed to the features of the conventional systems. However, it is believed that such features are going to be a breakthrough in unachieved issues of natural language communication systems.

The procedure to generate responses to the user in FLINS is just like other conventional systems as follows:

1. Get natural language inputs
2. Transform them into a semantic representation
3. Process it to generate responses in the semantic representation by reasoning
4. Transform them back into natural languages

It is obvious that how FLINS produces natural, intelligent responses depends on the features of the reasoning engine used in step 3. FLINS thus hires a new reasoning method, Fuzzy Case Based Reasoning (F-CBR), described in Chapter 3. This method realizes a heuristic non-logical reasoning and offers higher degrees of communication flexibility than other conventional systems.

Unfortunately, offering the communication flexibility implies the cause of the large number of misinterpretations. In the conversations among of humans, it is well handled even if such misinterpretation occurs, and they intend not to repeat the same type of misinterpretation, i.e. a learning process cooperated with the reasoning is involved. In the conversations between a human and a system, however, it is handled by using some algorithmic or rule based methods in the conventional systems. Therefore, the same type of misinterpretations are often occurred and make the user feel that the systems are not so intelligent, flexible, and natural as expected. This problem is described in Chapter 4.

As a consequent, a method to acquire knowledge commonly expressed among several cases which cooperates with the reasoning is proposed and under development to make FLINS more intelligent.

2. Design Concepts of FLINS

There are two significant design concepts in order to develop FLINS[1]. These are mutually involved to realize features such as an unified representation and a consistent treatment of knowledge, rules and data, a positive use of fuzziness in natural languages, and dynamic construction of deep semantic structures.

2.1 Text-Based Architecture

Text-based architecture is a concept that all knowledge, meta-knowledge, rules, and data are represented in the same manner, texts. The system uses a basic pattern matcher controlled by a schema described by texts to represent all knowledge. This feature enables the user to use texts to teach even meta knowledge because the process of problem solving is also controlled by texts.

This concept includes the following subconcepts, which play very important roles, when Text Based architecture is implemented.

2.1.1 Case-based Structure and Text-Base[2]

Most of conventional natural language systems such as machine translation systems hire case-based structures[3] as their knowledge representations.

The problem is that the case-based structure is too sensitive to vocabulary and sentence structures because semantics of words in inputs appear in the case-based structure directly and the sentence structure affects the case-based representation, i.e. it is a static deep semantic structure. Therefore, even if the meaning of two texts is the same, their case-based representations sometimes differ, and the systems are enforced to be domain dependent.

What if lexical knowledge, knowledge of word meanings, are also given in the same representation? It is considered that the meaning of inputs are derived by reasoning, exactly the same way as the domain dependent knowledge is used for problem solving, with the lexical knowledge in such cases.

The key idea here is to treat all knowledge as surface semantic structures at the same fixed level of meaning, and the deep semantic structures are derived dynamically by reasoning depending on other domain specific knowledge. The image of such process is shown in Figure 1.

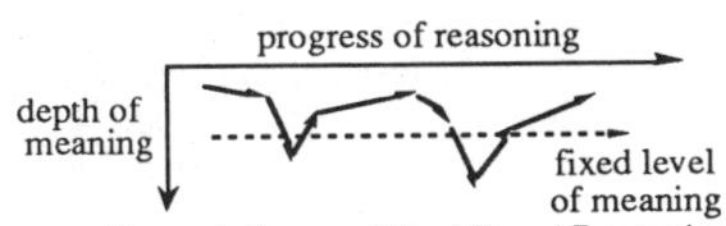

Figure 1 Image of Text Based Reasoning

2.1.2 Semantic and Procedural Primitives

The initial entries in Text Base are semantic primitives and procedural primitives. Note that the semantic primitives are completely different from those of the conceptual dependency(CD) theory[4].

Semantic primitives are words that have a special role in reasoning. For example, 'imply', 'if' and 'when' are semantic primitives. They are evaluated through an reasoning when paraphrasing a text or making a new goal event.

Procedural primitives connect subroutines and execute primitive functions such as text pattern matchings and database calls. Words like 'system-call' and 'event-match' are examples of procedural primitives. If these procedural primitives are evaluated under certain conditions, the procedures associated with the primitives are called automatically. Reasoning can be controlled by activating a procedural primitive. This means that how to infer or how to answer can be specified by using the primitive. This feature enables the systems to be taught meta knowledge by using texts because the process of problem solving is also controlled by the texts.

The inherent knowledge of the systems represented in a special form are made up of only these two types of primitives. All the other knowledge acquired through interactions with users is represented as texts.

2.2 Fuzzy Centered Architecture

One of the inherent characteristics of natural languages is fuzziness. This fuzziness varies from the ambiguity that exists in multi-parse trees for a single sentence and multi-meanings for a single word to the fuzziness of the meaning itself. Disambiguation algorithms are developed to cope with the former types of fuzziness, but the later types are not considered in studies of natural language understanding systems.

Fuzzy theory[5] provides a promising way to represent and process such fuzziness in meanings of natural languages such as fuzzy predicates (e.g., 'tall', 'old'), fuzzy modifiers (e.g., 'very' , 'more or less'), fuzzy modality (e.g., 'most of', 'usually'), and fuzzy inferences (e.g., generalized modus ponens, gradual rule).

A rough categorization of the fuzziness is shown in Figure 2. Basically, it can be divided into ambiguity and fuzziness of meaning. The ambiguity arises from the multiple meanings of a word. For example, the word 'execute' means 'to start a program' or 'to kill a person'. A sentence becomes ambiguous when there is a multi-parse tree for one sentence. For example, 'A girl saw the boy with telescope' can be interpreted in two ways, depending on who has the telescope.

	Word	Sentence
Ambiguity	Multi-meaning	Multi-structure
Fuzziness of Meaning Simple	"tall" well-structured	"the more, the more" universe of discourse
Complex	"handsome" unknown uni	"proverb" verse of discourse

Figure 2 Classification Of Fuzziness

The fuzziness in the meaning of a word can be classified as 'simple' or 'complex' fuzziness. The simple fuzziness is a fuzziness which can be represented with a well-structured universe of discourse. For example, 'tall' can be defined on a height axis. The universe of discourse for 'height' is a continuous real number.

On the other hand, it is difficult to define such a universe of discourse for a word like 'handsome'. This is an example of the complex fuzziness. Similarly, fuzziness in the meaning of a sentence can also be classified as simple fuzziness and complex fuzziness. An example of a simple fuzzy sentence is 'the more xxx, the more yyy'. In this sentence, the relationship between xxx and yyy is fuzzy. In other words, 'the more ... the more' is a fuzzy sentence structure. A more complex sentence such as a proverb exhibits the complex fuzziness.

Figure 3 shows the corresponding fuzzy theories that can be applied to the types of fuzziness in Figure 2. The simple fuzziness of a word can be processed by the fuzzy set or the fuzzy symbol theory. The generalized modus ponens and the gradual inference are applicable for the simple fuzziness of a sentence[6]. The fuzzy theory thus provides us a lot of methodologies for coping with the various types of fuzziness found in natural languages.

	Word	Sentence
Ambiguity	disambiguation b	y world knowledge and user interaction
Fuzziness of Meaning Simple	Fuzzy Set Fuzzy Symbol	Generalized Modus Ponens Gradual Inference
Complex	Fuzzy Rule Fuzzy Inference	(Fuzzy CBR ?)

Figure 3 Application Of Fuzzy Theory

3. Overview of Fuzzy Case-Based Reasoning

Most of conventional natural language processing systems hire logical reasoning methods to have interactions with users. However, they do not generate appropriate answers if there does not exist knowledge which logically apply to the inputs. It makes users feel that the systems are not flexible, unnatural, and unintelligent, i.e. some more flexible non-logical reasoning methods are necessary.

Fuzzy Case-Based Reasoning (F-CBR) is a new non-logical reasoning method dealing with the fuzziness of meanings that meets such criteria[7].

3.1 Fuzzy Case-Based Reasoning

The conceptual mechanism of the reasoning method in FLINS is an integration of logical and non-logical reasoning methods. Figure 4 shows its hierarchy, and Figure 5 illustrates the formalizations of these reasoning methods: (a)the non-fuzzy logical reasoning, (b)the fuzzy logical reasoning, and (c)fuzzy non-logical reasoning. The basic reasoning procedure is as follows:

1 try reasoning by non-fuzzy logical reasoning method
2 If it fails, try fuzzy logical reasoning method

3 If it fails, try fuzzy non-logical reasoning method

Fuzzy Case-Based Reasoning (F-CBR) plays roles of the fuzzy non-logical reasoning method in FLINS; It is a combined technology of fuzzy theory, case-based reasoning[8], and analogy-based reasoning[9].

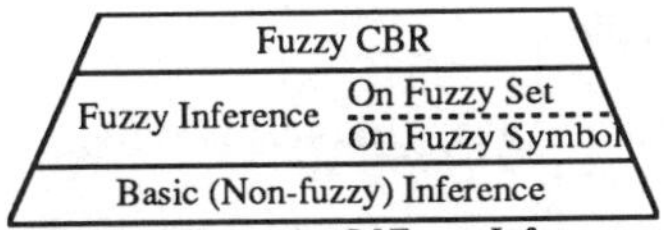

Figure 4 Hierarchy Of Fuzzy Inference

(a) non-fuzzy logical	(b) fuzzy logical	(c) fuzzy non-logical
A → B A ——— B	A → B A' ——— B'	A → B C ——— D

Figure 5 Formalization Of Reasoning Methods

The basic procedure of F-CBR is conceptually the same as CBR. There are two significant extensions: Fuzzy Analogy Based Reasoning (F-ABR) to search common features among cases and attributes of cases are not fixed like conventional CBR. Figure 6 illustrates the basic flow of F-CBR.

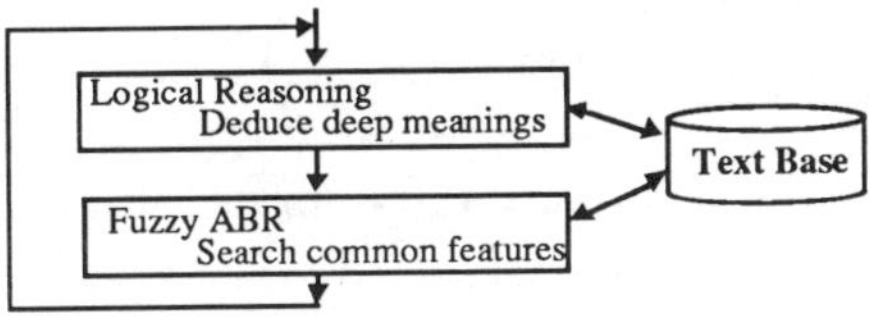

Figure 6 Basic Flow Of F-CBR

3.2 <u>Fuzzy Analogy-Based Reasoning</u>

Fuzzy Analogy-Based Reasoning(F-ABR) is the analogy-based reasoning method extended by fuzzy theory.

3.2.1 Defining Analogy-Based Reasoning(ABR) by an Example

Here is a typical example of analogy-based reasoning.

Question: When a transformation U1 is applied to S1, the result is shown in Figure 7. If U1 is applied to T1 what is the result? S1, T1 and U1 are given as follows:
S1={square(a), triangle(b), above(a, b)},
T1={arrow(x), square(y), above(x, y)},
U1={shrink(a)}.

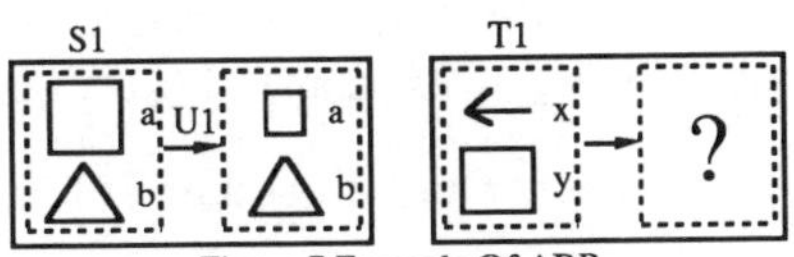

Figure 7 Example Of ABR

Examples of Answer: If the analogy is formed by an associative pair such as (<Common-features>, <Substitution>) where Common-features = {square(R)} and Substitution = ({R←a}, {R←y}), U1 can be adjusted to {shrink(y)}. Hence, the answer is shown in Figure 8.

Figure 8 Answer By ABR

3.2.2 Treatment of Fuzziness in Predicates

The non-fuzzy ABR can treat non-fuzzy predicates only, i.e. it cannot treat predicates with fuzziness shown in Figure 9 as follows:

S2={big_square(a), triangle(b), above(a,b)}
T2={arrow(x), very_big_square(y), above(x,y)}
U2={shrink(a)}

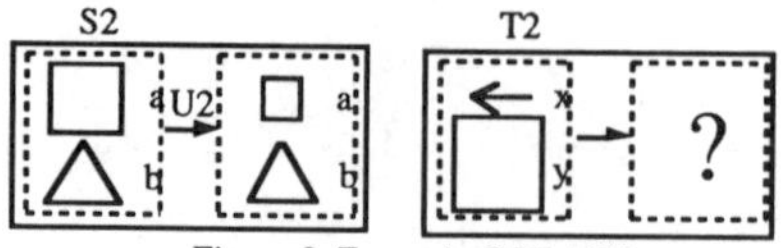

Figure 9 Example Of F-ABR

In this example, the predicates, big_square() and very_big_square(), can not be matched because the name of the predicates is different. We resolve the problem by introducing the following three types of fuzzy matching between fuzzy predicates.

[a] Matching by Grade of Membership

When one predicate has a membership function μ_A and the other is a value x on the membership function μ_A's universe of discourse, the matching degree can be calculated as $\mu_A(x)$. For example, the matching degree between two predicates, tall() and 180-cm-height(), is given by $\mu_{tall}(180)$.

[b] Matching by Shape of Membership

When two predicates have membership functions μ_A and μ_B, the matching degree can be calculated by the converse truth qualification[5]. For example, the matching degree between huge() and large() is the center of gravity of the linguistic truth value τ which satisfies the (x is μ_{huge}) is τ $\Leftrightarrow$ (x is μ_{large}).

[c] Matching by Fuzzy Modifier

When two predicates consist of the same predicate name plus the different fuzzy modifiers, such as 'very_tall' and 'rather_tall', the matching degree can be estimated only by the definition of fuzzy modifiers. For example, the matching degree between 'tall' and 'very_tall' is 0.6. This can be given by the converse truth qualification in the similar way of above [b]. Note that the definition of 'tall' is not needed. All information needed here is the definition of fuzzy modifiers, such as $\mu_{very\_tall}(x)=(\mu_{tall}(x))^2$ (this equation represents the meaning of 'very').

These extensions enables to get the answer for the question of Figure 10. Figure (a) and (b) are the possible answers.

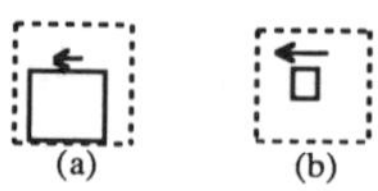

Figure 10 Answers By F-ABR

4. The Problem of Human-Computer Communication

One of easily expected problems of human-computer communication is misinterpretations of the natural language inputs caused by generating many reasoning results and its poor, inappropriate filtering methods. Such misinterpretations happen especially by using flexible non-logical reasoning methods such as F-CBR. This chapter describes the study of the problem by comparing the misinterpretation handling between humans and between a system and a user.

Consider the situation where many answers are generated through interpreting a question, and the best answer is to be filtered then responded, but a wrong one is filtered for some reason.

4.1 Handling Misinterpretations Between Humans

When communication is taken place between humans, and one misinterprets another's question, some of possible methods of handling misinterpretations are as follows:

•by being taught

Another tells that one misinterprets the question and generates an inappropriate response. Another may teach the interpretation rule as well. One usually accepts this fact and remembers the rule if it is given or asks the question how it should be interpreted. This is the most simple and direct method.

•by recognizing unnatural flow of discourse

The misinterpretation is easily detected by recognizing unnatural flow of discourse, i.e. the flow of discourse becomes unsmooth or terminated since a misinterpretation implies a sudden change of the topic in communication between humans. One asks another the question how it should be interpreted or learns knowledge common to the known similar cases explained next.

•by learning knowledge common to all known similar cases

Human beings become more intelligent by learning through their experience. They try not to make the same types of mistakes again. Applying this principle, one tries to learn knowledge common to all similar cases experienced before to filter the most appropriate response when one interprets another's question and generates responses. The responses by instanciating common knowledge are the right responses. The more similar cases exist, the better common knowledge one learns.

4.2 Handling Misinterpretations Between a System and a User

Suppose that the user asks a question, The system generates many responses, and picks a wrong one by using some algorithmic or rule based filtering method. The following methods of handling misinterpretations are expected:

•by the user's guidance

The user tells the system that a inappropriate response is generated. The system asks the user farther directions.

•by recognizing an unnatural flow of discourse

It is possible for the system to detect that itself misinterprets the user's question by recognizing some unexpected changed in a flow of discourse if a discourse management utility is implemented.

•by having the user filter the most appropriate response

Asking the user to select the most appropriate response instead of using some filtering method is the most simple and certain.

Using "by the user's guidance" method, the system may generate the same misinterpreted response, which cause the user irritation. The method of "by recognizing an unnatural flow of discourse" seems to work properly, but discourse management is still a big research issue of natural language processing, and no applicable discourse management utilities exist now. Using "by having the user filter the most appropriate response" method, misinterpreted responses are never generated but forces the user messy, frustrated filtering process. It also make the flow of communication unsmooth and unnatural.

5. Implicit Knowledge Acquisition by F-CBR

5.1 Needs

Due to the study of handling misinterpretations, FLINS generating responses only by reasoning causes misinterpretations and thus makes the user irritated and frustrated. The flow of discourse becomes unsmooth and unnatural. Therefore, It is not intelligent enough to be a faithful servant.

Considering the difference of handling misinterpretations between humans and systems, a function of learning common knowledge like humans is necessary to make FLINS more intelligent. Consequently, a method to acquire knowledge common and implicitly expressed among several experienced cases which cooperates with F-CBR is proposed to meet the necessity.

5.2 Basic Concept

Figure 11 shows the basic concept of the implicit knowledge acquisition cooperated with F-CBR. The inputs are several cases indicating some common implicit knowledge. Suppose that F-CBR is taken place on each case. Then, analogical knowledge between all pairs of cases are acquired by using F-ABR. The common implicit knowledge is acquired by integrating these analogies. The integration of analogies is a key to acquire the most appropriate knowledge and thus should be heavily researched and proposed. The set intersection operation is hired as its default.

Even if a similar case is input to FLINS after the common implicit knowledge is generated, any misinterpretation is no longer generated because of logical deduction of the case with the acquired implicit knowledge.

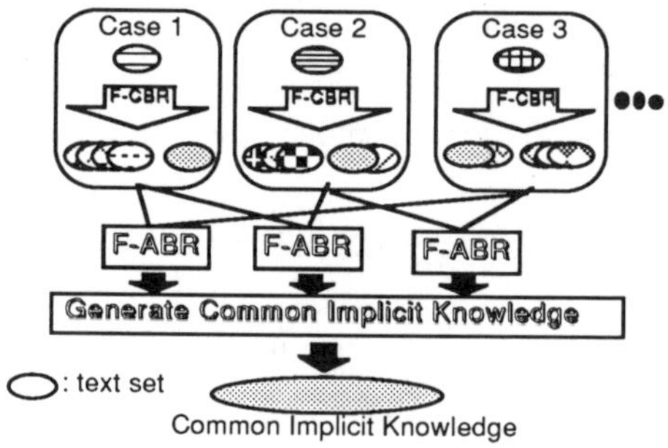

Figure 11 Concept of Implicit Knowledge Acuqisition

This method is considered as a sort of explanation-based learning[10] method where the inputs are considered as experience and F-CBR is considered as an explaining process.

5.3 Studies

The first ongoing research issue to be conducted is how to integrate the analogies generated by F-ABR. It generates multiple analogies for each pair of cases so that the integration of analogies between all pairs of cases is a complex process.

A mock experimental non-fuzzy reasoning, non-fuzzy analogical reasoning, and learning environment is implemented in Common LISP[11] to study the integration method. A simple experiment with about 50 rules stored in Text Base and a set of inputs, idiomatic cases as follows, is ongoing now:

Case 1 : A monkey falls from a tree.
Case 2 : Budda fails to read a sutra.
Case 3 : Kappa drowns in a river.
(Kappa is a Japanese monster living near water and swims well.)
Case 4 : Genie fails to fly.

The common implicit knowledge among these above cases, which should be acquired, is "one sometimes fails one's expertise." Figure 12 shows the result of this experiment. A simple set integration operation is used as the analogy integration operation at this time. Note that the implicit knowledge includes more than expected such as the predicate '(strong ?avar30)'. It is an incident that a monkey in case 1, Budda in case 2, Kappa in case 3, and Genie in case 4 has this predicate as its common feature. Therefore, adding other cases implying the implicit knowledge except this predicate takes care of this problem. It implies that the more cases are input, the better knowledge is acquired.

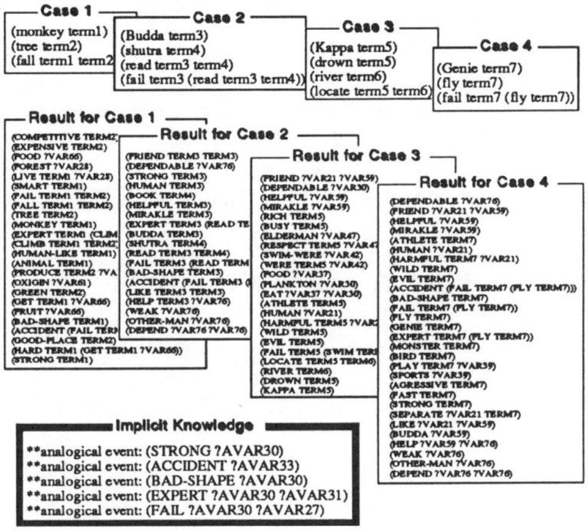

Figure 12 A Result Of The Simple Experiment

It is known by the experiment that the non-fuzzy set intersection operation works most likely fine but needs more adjustment and extension. The followings are the experiments for the analogy integration method coming on next:

- Experiments with other cases.
- Apply F-ABR instead of ABR.
- Use F-CBR instead of the simple experimental non-fuzzy reasoning.
- Study to apply Fuzzy theory to the default integration operation, i.e. intersection operation.
- Study to hire reasoning involved operations as the analogy integration.

6. Conclusion

First, the two significant new design concepts of FLINS are summarized, and the overview of F-CBR and F-ABR is described. Then, the problem of handling misinterpretations between a human and a system is studied. Finally, The implicit knowledge acquisition by F-CBR is proposed.

The key to implement this implicit knowledge acquisition is the analogy integration operation, and some experiments in a simple non-fuzzy operational environment to study this operation is held. The following experiments are listed and will be held with fuzzy operational environment which is under development now.

This knowledge acquisition method is considered as a new explanation based learning and expected to be a breakthrough of the knowledge acquisition bottleneck issue of knowledge based systems in general as well. The followings are remaining research issues besides the analogy integration operation:

- Selection of similar cases stored in Text Base.

- Study of the method to validate acquired common implicit knowledge.
- Study of the method to control the reasoning process of F-CBR by using this knowledge acquisition method.
- Applications.

REFERENCES

[1] S. Tano, W. Okamoto, and T. Iwatani. "New Design Concepts for the FLINS-Fuzzy Lingual System: Text-based and Fuzzy-centered Architectures." Proc. of ISMIS'93, pp.285-294, 1993.

[2] S. Tano, W. Okamoto, T. Iwatani, and A. Inoue. "Basic Structure of Three-layered Fuzzy Inference in FLINS - Fuzzy Lingual System." FUZZ/IEEE-'94, 1994.

[3] C.J Fillmore. "The Case for Case." Universals in Linguistics. Ed. Bach, E. and Harms, R.T. New York : Holt Rinehart and Winston, pp. 1-90, 1968.

[4] R.C. Shank. "Conceptual Dependency: A Theory of Natural Language Analysis." Cognitive Psychology, Vol. 3, 1972.

[5] L.A. Zadeh. FUZZY SETS AND APPLICATIONS: Selected Papers by L.A. Zadeh. Ed. R.R. Yager. JOHN WILEY & SONS, 1988.

[6] W. Okamoto, S. Tano, and T. Iwatani. "Estimation of Fuzzy Quantifiers in FLINS." 1st Asian Fuzzy System Symposium. 1993.

[7] T. Iwatani, S. Tano and W. Okamoto. "F-CBR (Fuzzy Case-based Reasoning) in natural language communication system." ANZIIS'93, pp.619-623, 1993.

[8] K.J. Hammond. Case-Based planning. Academic Press, 1989.

[9] M. Haraguti. "Toward a Mathematical Theory of Analogy." Bulletin of information and Cybernetics, vol.21, no.3-4, pp.29-56, 1985.

[10] P.H. Winston. ARTIFICIAL INTELLIGENCE 3rd ed. Addison-Wesley, 1992.

[11] P.H. Winston and B. K. P. Horn. LISP 3rd ed. Addison-Wesley, 1989.

SEMANTICS OF FUZZINESS REDUCTION COMBINATION FUNCTION AND LEARNING OF THE PARAMETERS

Shun'ichi TANO, Takuya OYAMA, Thierry ARNOULD and Atsushi INOUE

LIFE: Laboratory for International Fuzzy Engineering Research,
Siber Hegner Building 3 Fl., 89-1 Yamashita-cho, Nakaku, Yokohama, 231 JAPAN

Abstract

It is a well known problem with conventional fuzzy reasoning that the fuzziness of inferred results gradually increases according to the progress of the inference. Essentially, the problems are that (1) the combined result obtained with a conventional combination function becomes close to one of the two non-fuzzy values, that is, grade 0 or 1, and never approaches the other non-fuzzy value, and (2) the lack of a reinforcement property. We propose a new combination function which resolves those problems by introducing equilibrium E and dependence factors α and β.

1. Introduction

At the Laboratory for International Fuzzy Engineering Research in Japan, we have developed an expert system called FOREX (FOReign exchange trade support EXpert system) as an application of fuzzy theory to a real problem[1]. The experience proved that fuzzy methodology is feasible and quite effective. However, at the same time, we found several serious problems with fuzzy inference. We classified the problems and developed six research themes to study at LIFE and are now developing fuzzy inference environment software called FINEST[11]. The six research themes include (1) generalization of and/or/aggregation function (2) categorization of implication functions and parameterization, (3) fuzziness reduction using a combination function, (4) formalization of fuzzy backward reasoning, (5) automatic tuning of an inference method including (1) to (4) above and, (6) software for a fuzzy reasoning environment equipped with (1) to (5) above.

This paper describes number (3) fuzziness reduction method for a combination function, as a solution to the well known problem called explosion of fuzziness. This problem arises because the inferred results constantly increase according to the progress of the inference. First, the basic idea and the calculation flow are briefly explained. Second, semantics of the parameters and the coverage of the function, two important features are described. Finally, the process of learning the parameters is demonstrated by using the fuzzy inference environment software, FINEST.

2. Overview of the Fuzziness Reduction Combination Function [10]

2.1. Fuzziness of Fuzzy Set

First of all, it is necessary to clarify the fuzziness of a fuzzy set before discussing the problems of a combination function. It is said that a fuzzy set has two types of fuzziness which come directly from the shape of the membership function, i.e. height and width of the set.

The fuzziness tends to be measured by the following two criteria.

(a) if the height of a fuzzy set is tall, the fuzziness is small.

(b) if the breadth of a fuzzy set is narrow, the fuzziness is small.

However, a fuzzy set should be judged to have small fuzziness even when the membership value is near zero. Since the criterion (a) is not precise enough to represent this fuzziness, it is broken into two criteria (a1) and (a2), where (a1) is equivalent to (a) and (a2) is newly added.

(a1) if the membership value is close to 1, the fuzziness is small.

(a2) if the membership value is close to 0, the fuzziness is small.

As a measure of such fuzziness, fuzzy entropy[2,3] and specificity[4] were proposed, corresponding to criteria (a1, a2) and (b), respectively.

However, the breadth that criterion (b) takes account of is not suitable for measuring fuzziness. In the extreme case of a fuzzy set in which the membership grade is everywhere equal to 1, although the fuzzy set is often denoted as total ignorance, generally speaking, it is impossible to affirm that it is vague. Therefore, (b) should be removed from the criteria, leaving criteria (a1) and (a2) to measure the fuzziness of a fuzzy set.

2.2. Problems with a Combination Function

A combination operation is defined as a method of getting one result A''' when two fuzzy sets A' and A'' are deduced by two inference processes. We call A''' the combined result, A' and A'' the premise results. Since it is obvious that $\mu_{A'''}(x)$ depends solely on $\mu_{A'}(x)$ and $\mu_{A''}(x)$, a combination function can be represented as follows:

$$\mu_{A'''}(x)=\text{combine}(\mu_{A'}(x),\ \mu_{A''}(x))$$

As the combination process can be considered to be a sort of logical disjunction, T-conorms are preferably used and, in many cases, the Max operator is chosen as the combination function. Using Max, the following formulas hold:

$$\mu_{A'''}(x) \geqq \mu_{A'}(x),\quad \mu_{A'''}(x) \geqq \mu_{A''}(x)$$

Judging from criterion (a1), the fuzziness of the result is reduced. It seems to be no problem at all. However, the combined result obtained with Max always becomes close to total ignorance. In other words, the combined result obtained with a conventional combination function becomes close to one of the two non-fuzzy values, that is, grade 1, and never approaches the other non-fuzzy value, that is, grade 0. Basically, conventional functions do not possess property (a2). Because Max is a lower bound of T-conorms, this analysis is true for any T-conorm function. T-norm functions, such as MIN, are sometimes used as a combination function. In this case, the analysis shown above is still true, that is, the combined result becomes close to one of the two non-fuzzy values, grade 0, and never approaches the other, grade 1.

The next problem is caused by the idempotent property of Max. The fuzziness of the combined result is not reduced, even if the same results are deduced by two inference processes which are independent of each other. This happens due to the idempotent nature, i.e., max(x,x)=x, for any x,, contrary to human intuition. When a person solves a problem using fuzzy data and knowledge, he/she considers the problem from many points of view and synthesizes all the results in order to reduce the fuzziness of the result. The Max function is used in all conventional fuzzy applications, whereas, in possibility theory, Min is used for reliable knowledge and Max for unreliable knowledge[5,6]. The use of Min and Max means nothing more than the simple selection of the grade of one result. We never use the result synergistically. In other words, the results never reinforce each other, meaning the fuzziness is not reduced as expected for a human. All idempotent combination functions have this problem.

The primary problems discussed here can be summarized as follows:

(P1) the combined result becomes close to one of the two non-fuzzy values and never approaches the other one.

(P2) a reinforcement property is lacking

2.3. Introduction of Two New Parameters

To resolve the problems (P1) and (P2) discussed above, we introduce two new parameters, equilibrium E, and dependence factors α and β.

Equilibrium E : Degree of Positive/Negative Belief

In this paper, "$\mu_S(x)=a$" means that the degree of belief of "an element x is S" is a,, This is an interpretation based on the degree of belief where grade 1 expresses complete belief and grade 0 expresses complete disbelief.

Now assume that "$\mu_S(x)=a$" and "$\mu_S(x)=b$" are derived from two independent groups of evidence. If both degrees, a and b, are close to 1, each result should be reinforced, giving combine(a,b) > max(a,b).

On the contrary, if the degrees, a and b, are close to 0, we would expect each result to be reinforced, but the direction of reinforcement depends on the following interpretation.

(1) "an element x is S" is a little bit believed.
By this interpretation, combine(a,b) > max(a,b).
(2) "an element x is S" is strongly disbelieved.
By this interpretation, combine(a,b) < min(a,b).

The interpretation of (1) is the degree of positive belief, while the interpretation of (2) is the degree of negative belief. Although a grade value close to 1 can be interpreted as a degree of positive belief, a value close to 0 can not be uniformly interpreted. In some cases, it may represent a degree of positive belief. In other cases, it may represent a degree of negative belief.

An equilibrium parameter E is introduced to provide a tradeoff point which is interpreted as neither a positive belief nor a negative belief. The meaning of the grade is divided at E. Values greater then E correspond to positive belief, whereas values lower than E correspond to negative belief. When E is equal to 0, only positive meaning is represented. When E is equal to 1, only negative meaning is represented.

Dependence Factors α and β : Dependence Represented by Stochastic Rules

As mentioned above, the property of reinforcement is indispensable for combination operations. It is also important to take into account of the dependence of evidence. For example, premise results derived from independent groups of evidence can be reinforced, but if the evidence is not independent, reinforcement should not be done. This dependence is an important factor in reinforcement.

Since it is possible to think of dependence as the underlying derivation relation, a pair of stochastic rules can represent it. The following stochastic rules represent the dependence between the evidence in groups A and B.

$$A \to B : \alpha$$

$$B \to A : \beta$$

The first rule says that the evidence in A induces the evidence in B with the probability α. The values α and β are interpreted as the ratio of the shared portion to A and B, i.e. $|A \cap B| / |A|$ and $|A \cap B| / |B|$, respectively. For example, "A -> B : 0.7" implies that 70% of the evidence in A is contained in B.

2.4. Calculation Flow

Assumption

"$\mu_S(x)=a$" was deduced from evidence in A, and the equilibrium is E1. "$\mu_S(x)=b$" was deduced from the evidence in B, and the equilibrium is E2. The dependence between A and B is represented by A -> B : α, B -> A : β. The equilibrium of the combined result is E3.

Step 1

The degrees of belief of the premises, a and b, are transformed into normalized degrees of positive or negative belief. A grade bigger than E, [E,1], is transformed into a PB value in [0,1] by the function PB. The function NB transforms a grade in [0, E] into an NB value in [0,1]. These functions are defined as follows:

$$PB(x: E) = (x - E)/(1 - E)$$
$$NB(x: E) = (E - x)/E$$

In this step, a and b are transformed into u and v, respectively.

Step 2

In this step two cases have to be considered.

Case I: Both u and v are PB values or NB values.

Assume that both u and v are PB values. If the evidence in A and B is independent, the result is u + v - uv, according to DS theory[7] or probability theory. For example, $\mu_S(x)$=a can be interpreted as $m_A(\{x\}) = u$, $m_A(\Omega) = 1 - u$, where Ω is the universe of discourse. By applying Dempster's combination rule, we get m({x}) = u + v - uv.

Next, four possible approaches are shown to cope with the dependence in case the evidence is not independent:

(1) Use only the result derived from the evidence in A. The combined PB value becomes u.
(2) Use only the result derived from the evidence in B. The combined PB value becomes v.
(3) Remove the shared evidence from group B and combine the modified B and A.

The evidence in $A \cup B$ is divided into A and B'. B' is what remains of group B after throwing away the shared evidence. This guarantees that A and B' are independent. Because the ratio of the shared portion to B is b, the result derived from B' is modified as (1-β)v. The other result is still u, so the combined grade is u + (1-β)v - (1-β)uv.

(4) Remove the shared evidence from group A and combine the modified A and B.

Similarly, the combined PB value is (1-α)u + v - (1-α)uv.

We choose the least fuzzy result as the combined PB value. Then, the combined PB value is given by the following REINFORCE function. This also covers the case when both u and v are NB values.

$$\begin{aligned} &REINFORCE(u{:}\,\alpha,\, v{:}\beta) \\ &= \max((1-\alpha)u + v - (1-\alpha)uv,\; u + (1-\beta)v - (1-\beta)uv) \end{aligned}$$

Case II: Either u or v is a PB value and the other one is an NB value.

Assume that u is a PB value and v is an NB value. This case is a sort of contradiction. We use the intuitive combination method, similar to MYCIN's method[8], because there is no standard rational theory.

IF the evidence in A and B is independent, it is natural to conclude that the combined result is a PB value with the value u-v, if u is bigger than v.

However, even if the evidence in A and B is identical, it again seems reasonable to conclude that the combined result is a PB value with the value u-v, if u is bigger than v.

Step 3

PB^{-1} and NB^{-1}, defined below, transform the PB and NB values into grade values.

$$PB^{-1}(x{:}\,E) = E + (1-E)x$$

$$NB^{-1}(x{:}\,E) = E - Ex$$

The combination operation explained here is shown below as the function COMBINE (a:E1:α, b:E2:β, E3).

$$COMBINE(a{:}\,E_1{:}\,\alpha,\, b{:}\,E_2{:}\beta,\, E_3)$$

$$= \begin{cases} a \ge E_1,\, b \ge E_2 \\ \quad PB^{-1}(REINFORCE(PB(a{:}\,E_1){:}\,\alpha,\, PB(b{:}\,E_2){:}\beta){:}\,E_3) \\ a \ge E_1,\, b \le E_2 \\ \quad PB^{-1}(PB(a{:}\,E_1) - NB(b{:}\,E_2){:}\,E_3) \quad if\ PB(a{:}\,E_1) \ge NB(b{:}\,E_2) \\ \quad NB^{-1}(NB(b{:}\,E_2) - PB(a{:}\,E_1){:}\,E_3) \quad if\ NB(b{:}\,E_2) \ge PB(a{:}\,E_1) \\ a \le E_1,\, b \ge E_2 \\ \quad NB^{-1}(NB(a{:}\,E_1) - PB(b{:}\,E_2){:}\,E_3) \quad if\ NB(a{:}\,E_1) \ge PB(b{:}\,E_2) \\ \quad PB^{-1}(PB(b{:}\,E_2) - NB(a{:}\,E_1){:}\,E_3) \quad if\ PB(b{:}\,E_2) \ge NB(a{:}\,E_1) \\ a \le E_1,\, b \le E_2 \\ \quad NB^{-1}(REINFORCE(NB(a{:}\,E_1){:}\,\alpha,\, NB(b{:}\,E_2){:}\beta){:}\,E_3) \end{cases}$$

3. Semantics of Parameters a, b, and E

Equilibrium E

Figure 1 shows the influence of equilibrium E by illustrating COMBINE (μ_A(x):E:0, μ_A(x):E:0, E). E is the turning point at which the strengthening effect turns into the weakening effect.

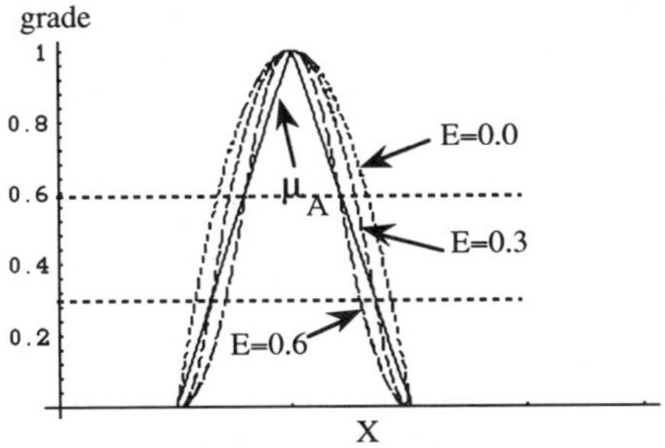

Fig.1 Influence of equilibrium E

Dependence Factors α and β

The REINFORCE function evolves from the logical sum(max) to the algebraic sum(u+v-uv), depending on the dependence parameters α and β. In cases where α=β=1, the REINFORCE function is identical to max(u,v), in cases where α=β=0, it is u+v-uv.

Figure 2 illustrates COMBINE (x:0.5:y, x:0.5:y, 0.5) - x. The weakening and strengthening effects are inverted at E. The degree of weakening and strengthening increases with the degree of independence. For α=β=1, the effect disappears.

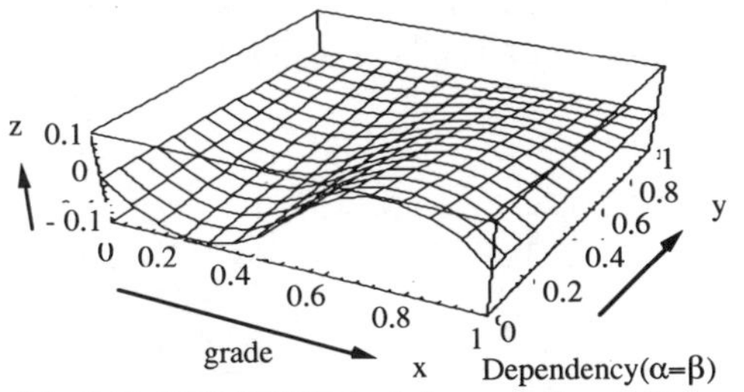

Fig.2 Z=COMBINE (x:0.5:y, x:0.5:y, 0.5) - x

4. Coverage of Conventional Methods

To analyze the coverage of the fuzziness reducing combination function proposed above, we show how our function covers four major combination methods, the max-based combination function, min-based combination function, voting-model-based combination function, and probability-based combination.

Max-based combination function

In cases where E=0 and α=β=1, the combination function is identical to the popular max-based combination function used in most applications. In possibility theory, Max is used in cases of unreliable knowledge [5,6].

Min-based combination function

In cases where E=1 and $\alpha=\beta=1$, the combination function is identical to the conjunctive combination used in possibility theory when the knowledge is reliable (min). This analysis suggests that a possibility distribution can be interpreted as a fuzzy set whose equilibrium E is equal to 1.

Voting-model-based combination function[9]

Baldwin proposed interpreting the fuzzy set based on the voting model[9]. As an example, the fuzzy set shown Fig. 3 (a) is transformed into a mass assignment (Fig. 3 (b)).

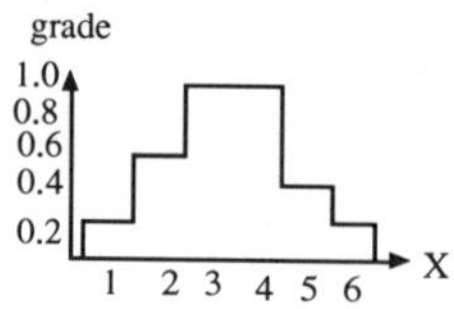

(a)

m({ 3, 4 }) = 0.4
m({ 2, 3, 4 }) = 0.2
m({ 2, 3, 4, 5 }) = 0.2
m({1, 2, 3, 4, 5, 6}) = 0.2

(b)

Fig. 3 Example of the voting model

Once the fuzzy set is transformed, the combination rule of DS theory can be applied to get the combined fuzzy set. Now we have two fuzzy sets, F1 = a / 1 + b / 0.8 + c / 0.3 and F2 = 1 / 0.8 + b / 1 + c / 1. These two sets are transformed into the mass assignments, M1 = {a}:0.2, {a,b}:0.5, {a,b,c}:0.3 and M2 = {b,c}:0.2, {a,b,c}:0.8. Finally, using the DS theory combination rule, we get the combination of M1 and M2 as M = {a}:0.16, {b}:0.1, {a,b}:0.4, {b,c}:0.06, {a,b,c}:0.24.

Figure 4 shows two examples of voting-model-based combination. In Fig. 4 (a), μ_A is combined repeatedly, as the graph shows: μ_A, combine(μ_A, μ_A) and combine(μ_A, combine(μ_A, μ_A)). Due to the repetition, the combined result constantly becomes smaller. Figure 4 (b) shows the combination of two fuzzy sets μ_A and μ_B.
Note that the combined result can be calculated as a simple product of two grades.

In cases where E=1 and $\alpha=\beta=0$, our function is identical to the voting-model-based combination function.

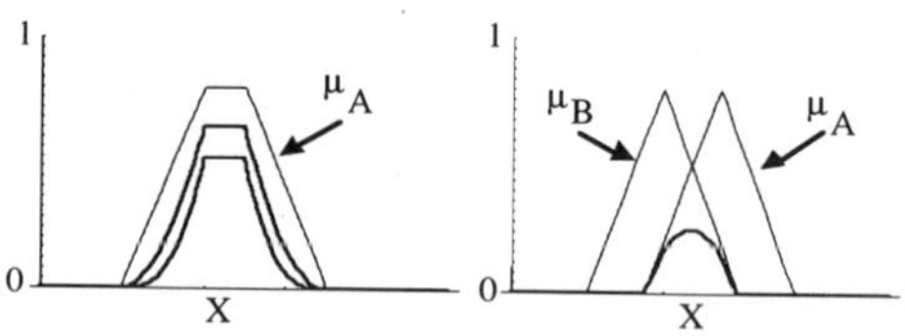

(a) recursive combination (b) combination of μ_A and μ_B

Fig. 4 Examples of voting-model-based combination

Probability-based combination

From the probabilistic point of view, when the evidence used to deduce the grade, x and y, is independent, the combined probability is $1 - (1 - x)(1 - y) = x + y - xy$. Figure 5 shows the result of probability-based combinations for the same examples given in Fig. 4 to clarify the difference between the methods. Although the grades are weakened in Fig. 4, they are strengthened in Fig. 5.

In cases where E=0 and $\alpha=\beta=0$, our function is identical to the probability-based combination function.

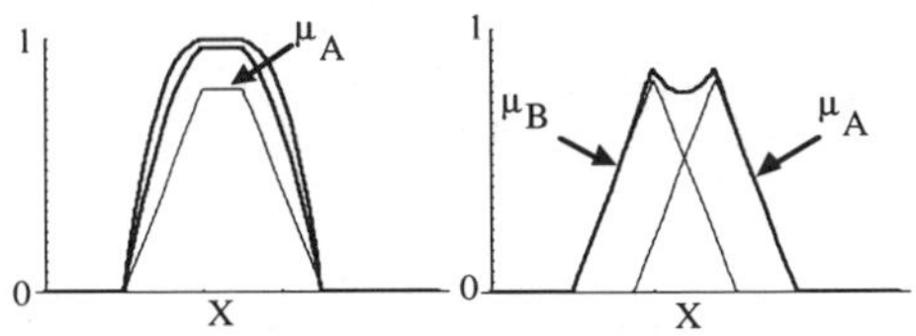

(a) recursive combination (b) combination of μ_A and μ_B

Fig. 5 Examples of probability-based combination

The analysis mentioned above is summarized in Table 1. These four cases are regarded as extreme cases from the viewpoint of the equilibrium and dependence.

Table 1 Coverage of fundamental functions

		E	
		0	1
$\alpha = \beta$	0	x + y - x y probablistic combination	x y voting-model-based combination
	1	max (x, y) max-based combination	min (x, y) min-based combination

5. Tuning the Parameters with FINEST

The fuzziness reducing combination function has parameters E, α, and β. It is very important to know the parameters because, by knowing the parameters, you know the semantics of the fuzzy sets and the dependence of the underlying evidence. In this section, the process of learning the parameters is demonstrated using Fuzzy Inference Environment Software with Tuning, called FINEST.

5.1. Overview of FINEST[11]

FINEST can be seen as a tool for building fuzzy knowledge-based systems since all the knowledge is represented by units and the final system is a collection of units.

A unit is defined as an object, which simply gets inputs through its input interface, processes this input in a specified way, and sends the results through its output interface. Therefore, each unit consists of three parts, i.e. the input interface, the data processing body, and the output interface. Notice that the configuration of units is independent from the way data is processed in the unit. At this point, FINEST has the following five types of units.

Rule unit

A rule unit is composed of one or more rules. The rules inside a rule unit use input data for inference and the result is output through the rule unit's output interface. Rule-type knowledge is expressed in the form " If x is A and/or y is B, then z is C", where x, y, and z are attribute names and A, B, and C are values. These attribute names and attribute values (linguistic labels, etc.) are also defined in the rule unit. The inference method for each rule of the rule unit is also specified in the unit. Normally, the parameters of the aggregation operators, the implication function, and the combination function can also be defined within the unit.

Function unit

A function unit is different from a rule unit insofar as calculations are done using LISP functions instead of rules. A function unit evaluates the data from the input interface as arguments, and outputs the evaluated value through the output interface. A function unit can contain some parameters which can be tuned if the derivative functions are given.

External unit

An external unit corresponds to an executable UNIX file, such that the calculation process of the external unit is executed as a UNIX process. An external unit can be combined with an existing system, but cannot be tuned.

Memory Unit

A memory unit stores information relative to the status of the system and the intermediate results of inference. For example, rules in one rule unit refer to and update data in memory units. Similarly, functions in a function unit sometimes read from and write to a memory unit. Data in a memory unit is expressed in the form "x is A", where "x" represents an attribute name, and "A" is its attribute value. The attribute value is a numeric value, character, fuzzy set, or linguistic label.

Composite unit

Units can be combined with each other. A combination of several units is called a composite unit. Using a composite unit, the user can build a system step by step and hierarchically.

FINEST provides us with the necessary tuning and inference mechanisms. The parameters of the fuzzy predicates, the aggregation operators, the implication functions, and the combination function in the rule units can be tuned. Moreover, inner parameters of functions which are not expressed in the form of a rule, as well as the parameters of the rules can be learned if the derivative functions with respect to these parameters are given.

5.2. Examples of the Tuning Process

Although it is possible to use the tuning mechanism of the combination function in the rule unit, we utilize the tuning mechanism of the function unit to focus our attention on tuning the combination function. The fuzziness-reducing combination function is implemented as a function unit and the derivation functions with respect to E, α, and β are given in the function unit as additional information.

Given a single pair of fuzzy sets input and one desired combination result, FINEST can learn the parameters E, α, and β which minimize the difference between the desired combination result and the actual calculated result.

For example, Case 1 in Fig. 6 (a) shows a pair of fuzzy sets (μ_A and μ_B) and a desired combination result μ_{Result}. Once the data set is given, FINEST continues tuning the parameters E, α, and β, as shown Fig. 7, following the gradient descent method.

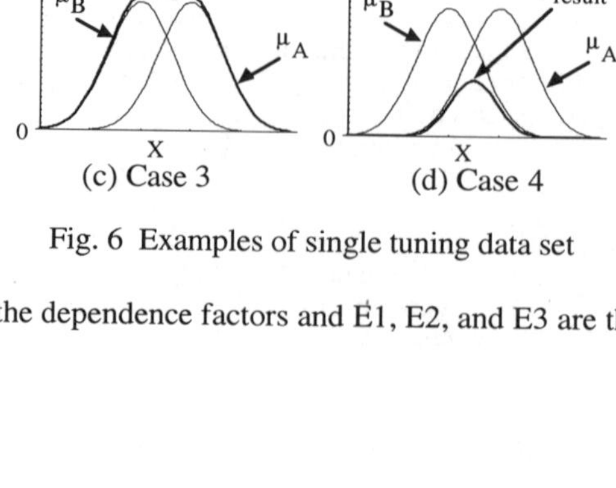

Fig. 6 Examples of single tuning data set

In Fig. 7, α and β are the dependence factors and E1, E2, and E3 are the equilibrium of μ_A, μ_B, and μ_{Result}, respectively.

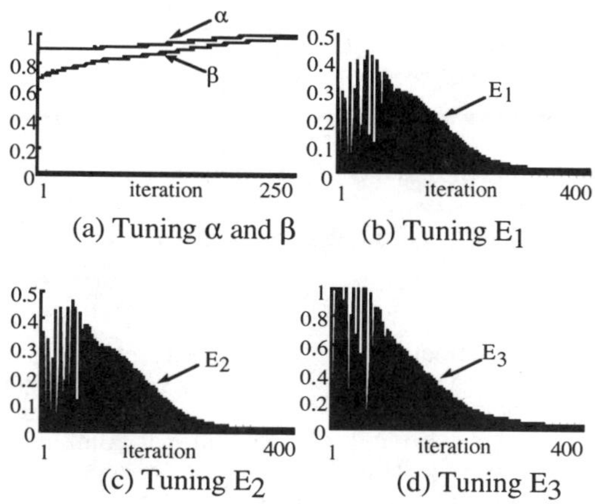

Fig. 7 Tuning process for Case 1

As shown in Table 2, most parameters are adequately tuned. However, we have to point out the difficulty in selecting the initial values of E, α, and β, and the parameters of the tuning algorithm itself, such as a learning factor of the gradient descent method. Note that each case in Fig. 6 corresponds to the cases in Table 1.

Table 2 Tuning results for Cases 1 to 4

		E			
		0		1	
$\alpha = \beta$	0	Case 3	α=0.065559 β=0.074501 E1=0.000012 E2=0.000007 E3=0.001345	Case 4	α=0.0 β=0.037333 E1=1.00000 E2=1.00000 E3=1.00000
	1	Case 1	α=1.00000 β=1.00000 E1=0.000008 E2=0.000008 E3=0.000008	Case 2	α=0.996979 β=0.99691 E1=0.961878 E2=0.961905 E3=0.962697

In contrast to the above examples, Figs. 8 and 9 demonstrate a more complicated situation in which three pairs of input fuzzy sets and one desired combination result are given. Here FINEST is used to learn the parameters E, α, and β which minimize the total difference between the three desired combination results and the actual calculated results. In this case, β, E1, E2, and E3 were adequately tuned, (that is, β=0.0, E1=0.4, E2=0.4, E3=0), however, α does not converge over 1000 iterations.

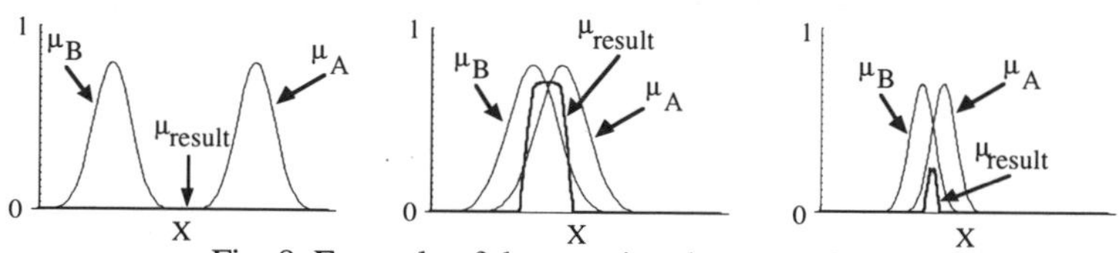

Fig. 8 Example of three tuning data sets: Case 5

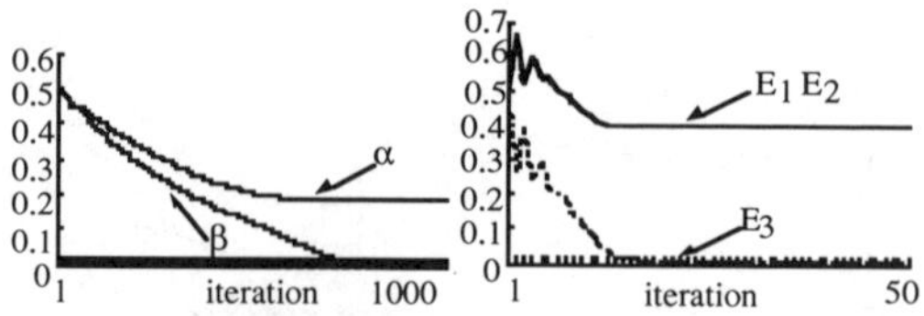

(a) Tuning of α and β (b) Tuning E_1, E_2, and E_3

Fig. 9 Tuning process in Case 5

6. Summary

We have proposed a new combination function which resolves the problem arising when the combined result becomes close to one of the two non-fuzzy values and never approaches the other one (P1). It also addresses the lack of a reinforcement property. Its behavior is consistent with that of a human. Moreover, it covers the range of conventional combination functions. From the viewpoint of human intuition, it is quite natural to adopt the concept of equilibrium E, which is also quite significant for the other components of inference.

The process of learning the parameters of this function was demonstrated successfully using fuzzy inference environment software, FINEST. FINEST can be used as a tool for quantifying the fuzzy meaning of sentences expressed in the form of fuzzy rules. It should be stressed here that FINEST is not only a tool for developing fuzzy controllers, but also now for qualifying the fuzziness of natural language. For example, aggregation operators, implication methods, combination methods, as well as fuzzy predicates can be tuned with FINEST, and as a result, the nature of the sentences is clarified. The interpretation is, for example, "the 'and' in this rule has a strong synergistic nature", "the 'or' has a weak cancellation property", "this rule expresses knowledge of the form 'the more ..., the more...' ", etc. That is to say, this work can be regarded as a quantification of the fuzzy meaning of natural language.

References

[1] H. Yuize, T. Yagyu, M. Yoneda, Y. Katoh, S. Tano, M. Grabisch, and S. Fukami: Decision support system for foreign exchange trading - practical implementation -, Proc. of IFES'91 , pp. 971-982 (1991).
[2] A. De Luca and S. Termini: A definition of a nonprobabilistic entropy in the setting of fuzzy set theory, Information Control, Vol. 20, pp. 301-312 (1972).
[3] B. Kosko: Fuzzy entropy and conditioning, Information Science, Vol. 40, pp. 165-174(1986).
[4] R. R. Yager: Measuring tranquility and anxiety in decision making : an application of fuzzy set, Int. J. General System, Vol. 8 (1982).
[5] D. Dubois and H. Prade: On the combination of evidence in various mathematical frameworks, Proc. NATO Advanced Research Workshop on Knowledge Based Control (1988).
[6] D. Dubois and H. Prade: On the combination of uncertain or imprecise pieces of information in rule-based systems - a discussion in the framework of possibility theory, Int. J of Approximate Reasoning, Vol. 2, pp. 65-87 (1988).
[7] G. Shafer: A mathematical theory of evidence, Princeton University Press (1976).
[8] B. Buchanan and E. H. Shortliffe: Uncertainty and evidential support, in Rule-Based Expert System (B.Buchanan and E.H.Shortliffe, Eds,) Addison-Wesley, Reading, Mass., pp. 209-232 (1984).
[9] J. F. Baldwin: A Mass Assignment Theory for Uncertainty Reasoning, FUZZ-IEEE'92, pp. 1367-1374 (1992).
[10] S. Tano, T. Arnould, Y. Kato, and T. Miyoshi: Fuzziness reduction method for a combination function, IFSA-93, pp. 62-65 (1993)
[11] S. Tano, T. Oyama, Y. Kato, T. Miyoshi, T. Arnould, and A. Bastian: Overview and Special Features of FINEST: Fuzzy Inference Environment Software with Tuning, 1st Asian Fuzzy Systems Symposium, pp. 294-302 (1993).

FUZZY AND
HYBRID SYSTEMS

AUTONOMOUS FUZZY INTELLIGENT SYSTEMS FOR IMAGE PROCESSING

Abraham Kandel*, Mordechay Schneider† and Gideon Langholz‡

★Department of Computer Science & Engineering,
University of South Florida,
Tampa, FL. 33620, USA.

†Department of Computer Science,
Florida Institute of Technology,
Melbourne, FL 32901, USA

‡ Department of Electrical Engineering
Tel-Aviv University
Tel-Aviv, Israel

ABSTRACT

In this paper we propose a novel framework for the design of autonomous fuzzy intelligent systems. The system integrates the following modules into a single autonomous entity:

(1) A fuzzy expert system
(2) Artificial neural network
(3) Genetic algorithm
(4) Case-base reasoning

We describe the integration of these units into one intelligent structure and discuss potential applications to image processing.

1 Introduction

This paper addresses some of the issues involved in developing a technology that supports the implementation of an autonomous fuzzy intelligent systems. The proposed technology is based on the premise that integrated solution architectures will be much more effective and highly flexible in their ability to successfully handle a broad base of applications with a wider scope of problem variations.

Hybrid systems [13] in artificial intelligence represent a new field of research that deals with the synergism of expert systems and neural networks technologies. The integration of the computational paradigms of these two highly complementary knowledge representation techniques is imperative to the process of developing effective robust intelligent systems for a large number of important applications.

Machine learning in an uncertain or unknown environment is of vital interest to those working with intelligent systems. The ability to garner new information, process it, and increase the understanding/capability of the machine is crucial to the performance of autonomous systems. The field of artificial intelligence provides two major approaches to the problem of knowledge learning - expert systems and neural networks. Harnessing the power of these two techniques in a hybrid, cooperating system holds great promise.

Expert systems and fuzzy expert systems are strongly tied to knowledge-based techniques for gathering and processing information. Knowledge representation in such systems is most often in the form of rules garnered through consultation with human experts. Coupling the methods of approximate reasoning with knowledge-based techniques yields systems which model human decision making. There are many examples of rule-based systems which function as experts in a given domain, e.g., trouble-shooting for complex mechanical processes, medical diagnosis systems, and financial risk assessment. Expert systems provide a ready mechanism for explanation why certain decisions are made, even when the human expert is unable to articulate the chain of reasoning leading to a decision. This trace of the reasoning process is often crucial to those maintaining the system.

A major disadvantage of knowledge-based system is their reliance upon consultation with human experts for new information. Furthermore, *autonomous* learning in an expert system does not usually include the capability to synthesize new knowledge but is limited instead to dependence upon structures the designer builds in to assess the similarity between situations or to generalize upon sets of similar rules.

Neural networks are data-driven systems based on an architecture of many simple processing units which are interconnected. The knowledge of a neural net resides in the connections between these processing units and in the strengths of these connections. Neural networks are especially applicable to problems which involve large number of weak constraints. They have been successfully applied to perceptual tasks such as pattern recognition, vision processing, and speech synthesis.

The ability to gracefully handle minor inconsistencies or conflicts in the data is an advantage that neural network systems hold over most expert systems. A robust intelligent system must be able to handle conflicting information from different experts, or some degree of contamination in incoming data, without too much degradation in performance.

In this paper we carry this idea of hybridization one step further by incorporating also genetic algorithms and case-base reasoning into the autonomous fuzzy intelligent system.

In the following sections we briefly describe the nature of the modules and the principles involved in their integration. We conclude by indicating a high SIQ (System Intelligent Quotient) in the performance of the system as applied to image analysis.

2 Autonomous fuzzy intelligent system architecture

The autonomous fuzzy intelligent system (AFIS) architecture integrates the computational paradigms of fuzzy expert system, neural network(s), genetic algorithm(s), and case-base reasoning. Expert systems and neural networks represent complementary approaches to knowledge representation; the logical, cognitive, and mechanical nature of the expert system versus the numeric, associative, and self-organizing nature of the neural network.

The expert system component of AFIS is a fuzzy expert system [10], [26]. It reflects the fact that human expertise embedded in the knowledge base of the expert system is imprecise, incomplete, or not totally reliable. The fuzzy expert system offers knowledge-based techniques for gathering and processing information coupled with methods of approximate reasoning to better emulate human decision-making processes.

We use fuzzy logic, the logic underlying approximate reasoning [11], [29], to approach the task of managing uncertainty in the fuzzy expert systems component of the AFIS. Fuzzy logic is of particular importance to the management of uncertainty in expert systems because it provides a systematic framework for dealing with fuzzy quantifiers. It subsumes both predicate logic and probability theory and makes it possible to deal with different types of uncertainty within a single conceptual framework [11].

Neural networks consist of densely interconnected simple nonlinear devices, called processing elements (or neurons). The knowledge of a neural network resides in the interconnections between these elements and in the strengths (weights) of the interconnections. Collectively, neurons with simple properties, interacting according to simple rules, can accomplish complex functions such as generalization, error correction, information reconstruction, pattern analysis, and learning [25]. Neural networks provide a greater degree of robustness, or fault tolerance, than conventional von-Neumann sequential machines. Indeed, damage to a few neurons or interconnections, and/or minor variabilities in the characteristics of neurons, do not impair overall network performance significantly. In addition, neural networks also possess the ability to gracefully handle inconsistencies or conflicts in the data.

To quote Lee and Takagi [21], who have experimented with integrating design stages of fuzzy systems using genetic algorithms, "A genetic algorithm is a probabilistically guided optimization technique modeled after the mechanics of genetic evolution. Unlike many classical optimization techniques, genetic algorithms do not rely on computing local derivatives to guide the search process. Genetic algorithms also include random elements, which helps avoid getting trapped in local minima.

Genetic algorithms explore a population of solutions in parallel. The size of the population is a free parameter, which trades off coverage of the search space against the time required to compute the next generation. Each solution in the population is coded as a binary string or gene, and a collection of genes forms a generation. A new generation evolves by performing genetic operations, such as reproduction, crossover, and mutation, on genes in the current population and then placing the products into the new generation."

It is, therefore, of paramount importance to develop methodologies for integrating the computational paradigms of fuzzy expert systems (FES), neural networks (NN), genetic algorithms (GA), and case-base reasoning (CBR) tools and for exploring the similarities between the underlying structure of these tools of knowledge representation. Such architecture, which allows for imprecise information and/or uncertain environments, is more powerful than either of its components standing alone.

3 General organization

Several models for the hybrid system have been proposed in the literature [13]. As a general premise, we believe that the most effective model is one in which the expert system begins with a base of knowledge which is necessarily incomplete, a neural network layer takes the knowledge from the expert system and modifies it through learning, and all information can be passed easily and transparently from one part of the system to another as needed.

A variety of investigations in present literature [13] show several configurations of this basic model which have different uses:

1. Everything learned in the neural network is passed back to the expert system. In effect, the neural network is training the expert system. In this verion, the user of the system is always able to trace the decision-making process via the expert system.

2. When a problem is presented to the hybrid system, it is partitioned into segments which are evaluated to be appropriate for solution by either the expert system layer or by the neural network layer. The solution to the problem is a hybrid of the segment solutions. In this

version, the two layers act as cooperating partners, each doing what it does best, keeping functional overlap to a minimum.

3. An entire network of smaller systems, expert systems and neural networks, cooperate and communicate to learn in different modes, or in different domains. Each part is designed with a different part of the problem solution process in mind.

From the practical point of view, we use the neural network to acquire new knowledge from the environment and replace a computationally intensive form of processing with a more efficient one. On the other hand, the expert system makes use of knowledge it has already acquired (e.g., from experts) and thus relieves the neural network from learning things it already knows and and by that makes the entire system more effective and efficient.

The fluid transfer of information from one type of system to the other obviously is crucial to the system. Having thoroughly checked the interface with off-the-shelf packages, we utilize an expert system which may incorporate fuzzy linguistic quantifiers, hedges, and weights, such as FEST [27].

The AFIS allows for a bidirectional (transparent) transfer of knowledge between its components. Initially, the fuzzy expert system is invoked with a base of knowledge which is necessarily incomplete. The neural network as well as the genetic algorithm and the CBR tool, take the knowledge from the fuzzy expert system and modifies it through learning. (Since the fuzzy expert system supplies the metaknowledge to begin the learning process, learning in the neural network is implemented more efficiently). Based on the information acquired from the learning agents, the fuzzy expert system can modify certainty factors, modify existing rules, or infer new rules.

Thus, the ability of the AFIS to learn in an imprecise environment enables the fuzzy expert system to modify and enrich its knowledge structures autonomously. Consequently, among other things, the AFIS can overcome some of the major drawbacks of conventional intelligent systems:

1. Their reliance upon consultation with human experts for knowledge acquisition (the knowledge acquisition bottleneck);

2. Their inability to synthesize new knowledge; and

3. Their inability to allow for dynamic environments by changing knowledge whenever this becomes necessary.

The AFIS is functionally decomposed into four separate units which communicate by means of common data structures. The rule base is the common data structure between the fuzzy expert system and the neural network, whereas the collection of state arrays is the common data structure between the neural network and the learning unit.

Several papers [1] - [9], [20] - [21], [23] have focused on tuning fuzzy membership functions. For example, Takagi and Hayashi [28] use neural networks as a membership values generator and Nomura et.al. [22] treat fuzzy systems as networks and use back-propagation techniques to adjust membership functions. Nodes in these networks perform parameterized functions. These parameters are tuned by computing derivatives of the network, with respect to these parameters, and then back-propagating the error as in traditional neural networks.

Other methods use genetic algorithms to determine the number of fuzzy rules [15] - [19]. Karr has developed a method for determining membership functions and number of fuzzy rules using genetic algorithms [16]. In his paper, Karr uses a genetic algorithm to determine the number of rules according to a predefined rule base. Following this stage, he applies a genetic algorithm for tuning the fuzzy membership functions.

Several paradigms for the transfer of knowledge from expert systems to neural networks have been proposed (e.g., [4], [5], [6], [12], [14]). Most of these models initially break the expert system rule-base clauses into three classes: query clauses requiring input from the user, consequent clauses which yield the final output, and intermediate clauses which do not fall into either category. The certainty factors are identified with weights in the neural network, and antecedents and consequents are associated with nodes in the neural network. These classes give rise to a mapping onto a neural network consisting of an input layer, an output layer, and a special type of hidden layer whose entities characteristically have interconnections resulting from chaining in the expert system.

The nodes of the neural network are classified in terms of rule-base operations. Nodes resulting from conjunctions are different from nodes arising from disjunctions, whereas negation yields yet another type of node. However, no cycles occur among these nodes since circular logic is typically disallowed in expert system rule bases. Each of these types of node has distinct type of combining and activation functions. In order for the neural network to yield results similar to those of the expert system, these combining and activation functions must reflect the workings of the inference engine of the expert system. To guarantee translation of the knowledge back to the rule base, the neural network must rigidly conform to this topology and, therefore, learning requires non-traditional algorithms.

The neural network functions in three phases: (1) A forward pass following from inputs to final outputs. Each node waits for all its input lines to be activated before applying its combining function; (2) A backpropogation of error using the reverse flow through the network. Error is assigned to nodes using the reverse of the connections; and (3) A second forward pass to adjust the weights of the connections. To invoke learning, the weights of the connections are adjusted using the error assigned to the node, the inputs to the node, and the derivative of the activation function for the node.

The structural similarity suggests the functional similarity between the two types of syatems.

It should be emphasized that, since the transfer of knowledge between system components is bidirectional, it is the learning capabilities of the neural network that enable the hybrid intelligent system to infer new rules or modify existing rules based on neural network performance. The division of labor, by providing the system with whatever knowledge is available a priori through the expert system and the knowledge-base, and by developing optional learning strategies for the neural network, is precisely the technology that provides us with fast, autonomous effective learning on top of previously acquired knowledge.

4 Multi-domain intelligent image processing environment

Real world image processing and recognition is an important problem for many advanced military and industrial applications. Automatic recognition of images has been an area of active research for over two decades. The problem is difficult because a complete and robust system must consider many different image variations including: position, scale, orientation, contrast, perspective, background occlusion, and noise. The ultimate goal is to develop a flexible system which can extract critical information from complex and uncertain data and which can classify objects in varying positions, orientations, and dimensions in the image plane [24]. Furthermore, such systems should be tolerant to a certain degree of random variations in the shape.

To date, the various approaches to the problem, stemming from classical image recognition and computer vision technologies, have been only partially successful at addressing a few of these issues. Conventional statistical image recognition algorithms require a set of data to train the classifier. The overall system design usually includes off-line feature selection and/or extraction, error analysis, and decision rule construction. The resulting decision rule is then applied online to classify unlabeled pattern vectors. A major deficiency of these algorithms is their lack of flexibility to adapt to changing environments. Statistical decision rules do not support incremental learning. To update decision rules, the classifiers must be re-trained using all the old and new samples. Further, there is lack of guidelines for selecting or implementing classifiers for real-world problems.

Hence, an effective, adaptive, intelligent image processing environment should support the integration of multiple process domains, such as numeric, model-based, knowledge-based, and neural-based, through a process of close interaction. Such an environment provides an intelligent strategy to efficiently exploit modern sensor image data between multiple process domains, and to take maximum advantage of the strengths of each process as they relate to knowledge extraction from sensor image data.

4.1 Architecture

The architecture of the integrated multi-domain intelligent image processing environment is shown in Figure 1. The environment integrates knowledge-based, neural-based, model-based, and algorithmic-based processes. Each of these domain technologies offers specific advantages to the generalized image exploitation process.

The role of the domain technologies within the integrated environment is discussed in the following paragraphs.

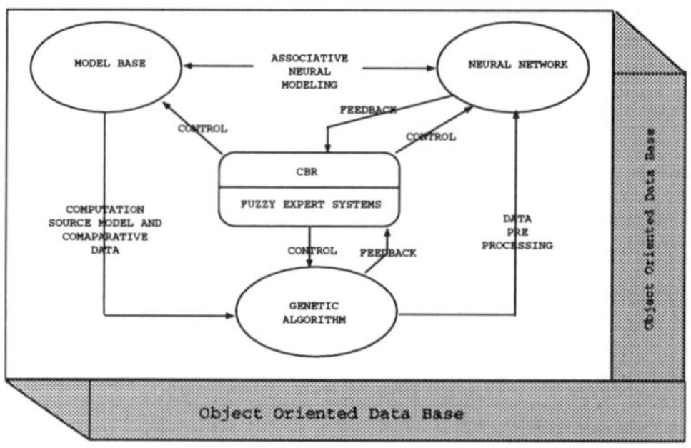

Figure 1: Integrated Image Processing Environment

The neural network component of the AFIS consists of relatively small networks, each limited to performing very specific cognitive and recognition tasks. In this way, highly optimized network functions are achieved within manageable training and computational constraints. A variety of neural network paradigms are used according to the capability of each paradigm to optimally contribute to some facet of the exploitation problem. In particular, associative neural networks are used to provide high performance model storage mechanisms for model-driven recognition and classification.

The AFIS provides a bidirectional transfer of knowledge between its components. Initially, the fuzzy expert system is invoked with a base of knowledge which is necessarily incomplete. The neural network takes the knowledge from the fuzzy expert system and modifies it through learning. Based on the information acquired from the neural network, the genetic algorithm and the CBR tool the fuzzy expert system can modify certainty factors, modify existing rules, or infer new rules.

Thus, the learning capability of the system, provided by the neural network and the genetic algorithm, as well as with examples derived from the CBR, is used to acquire new knowledge from the environment, whereas the fuzzy expert system makes use of knowledge it had already acquired. Therefore, the fuzzy expert system relieves the other component from learning things already known.

4.1.1. Model-based Processes

Model-based techniques are key for highly automated real-time image processing. Numerically explicit models or models trained for associative neural recall provide the experiential expertise necessary for real-time image exploitation. Such models also improve significantly the capability to derive meaningful results in highly noise-corrupted data collection environments.

4.1.2. Algorithmic-based Processes

Algorithmic-based Processes are used to carry out low level segmentation and analysis functions, as well as to precondition collected data prior to presenting that data to neural classifiers. Data preconditioning is used to accentuate facets of data sets so that certain characteristics can be more finely discriminated. The choice between algorithmic or neural processes is guided by the higher level knowledge-based decision system.

4.1.3. Object Oriented Data Base

The object oriented data base is the primary integrating platform for implementing the multi-domain intelligent image processing environment. Object oriented data base technology provides flexibility in setting up data and process structures to support intelligent image exploitation. Access to common data by multiple processes is accomplished more easily and greatly reduces the complexity that would otherwise be attendant in a highly process-interactive system. Since, in an object oriented system, data and processes to interpret that data are stored together in a common object, interfacing the data from multiple processes is simplified. Hard-coded rules structure establishes the inter-relationships between data, conditions, and processes with relational data bases providing an adequate measure of distributed influence over a span of parameters and conditionally variable data.

5 Integrated system structure

The traditional approach for transitioning from low level processing of images to high level synthesis of knowledge relies on a number of decision processes and iterative application of image processing functions. As the system attempts to maneuver through this complicated and complex structure, the chances for errors increase. The control system must be capable of synthesizing a knowledge pyramid which is rational from the initial processing outputs to the final system output.

The integrated system shown in Figure 1 merges knowledge-based, neural-based, algorithmic-based, and model-based paradigms, all used to exploit the image data. These paradigms are intimately bound and distributed throughout the general process of image exploitation. The elements of the system are integrated within an object-oriented process and data base environment.

Processes within the integrated image processing environment are freely selected and concatenated based on the output data produced by the processes themselves. The image exploitation process is guided by the knowledge base which uses the feedback generated by individual process execution to drive the choice and strategy of the next process initiated by the exploitation process.

Each of the processes acts in concert with the others. For example, algorithmic preprocessing of a given type may be initiated prior to the specific selection of an appropriate neural network architecture and trained weight set. A neural network process may feedback information to the knowledge-based system that would trigger an exploitation initiative using another more specialized neural network architecture or training set. Specific models and the manner in which they interplay with both neural network and algorithmic processes are guided by the knowledge-based system.

6 Conclusions

This paper addressed some of the issues involved in developing a technology that supports the implementation of an adaptively intelligent image exploitation environment. The proposed technology is based on the premise that integrated solution architectures are much more effective and highly flexible in their ability to successfully handle a broad base of applications with a wider scope of problem variations.

The hybrid system component of the multi-domain image processing environment integrates the computational paradigms of a fuzzy expert system and neural network(s). It provides uncertainty management capabilities through the fuzzy expert system which is capable of properly handling fuzzy knowledge and data. The learning capability of the system is provided by the integrated neural network(s). The transfer of knowledge between the fuzzy expert system and the neural network is bidirectional so that the neural network is used to acquire new knowledge from the environment whereas the fuzzy expert system makes use of knowledge it had already acquired. Therefore, the fuzzy expert system relieves the neural network from learning things already known.

The key to the integrated image processing environment is that the knowledge-based system most predominantly drives the strategy of exploitation within the facilitating object-oriented environment. The object-oriented approach effectively implements a flexible and adaptable intelligent system that allows the flow and sharing of information and coordination of processes to enhance significantly the overall objective of the image.

References

[1] Araki, S., Nomura, H., Hayashi, I., and Wakami, N., Self-generating method of fuzzy inference rules, *Int'l Fuzzy Engineering Symposium (IFES'92)*, 1992, pp. 1047-1058.

[2] Davis, L. (Ed.), *Handbook of Genetic Algorithms*, Van Nostrand, Reinhold, 1991.

[3] Dayhoff, J., *Neural Network Architectures*, Van Nostrand, Reinhold, New York, 1990.

[4] Fu, L. (1989), Integration of neural heuristics into Knowledge-based inference, *Connection Science*, Vol. 1, No. 3.

[5] Gallant, S. I. (1988), Connectionist expert system, *Comm. ACM*, Vol. 31, No. 2, pp. 152-169.

[6] Hall, L. and Romaniuk, S. (1990), FUZZNET toward a fuzzy connectionist expert system development tool, *Proc. Int. Joint Conference on Neural Networks* (Washington, D.C., Jan. 15-19), Vol. II, 483-486.

[7] Ichikawa, R., Nishimura, K., Kunugi, M., Shimada, K., Auto-Tuning Method of Fuzzy Membership Functions Using Neural Network Learning Algorithm, *Proc. of the 2nd Int. Conf. on Fuzzy Logic and Neural Networks (IIZUKA'92)*, 1992, pp. 345-348.

[8] Jang, R., Fuzzy Controller Design without Domain Experts, *Proc. IEEE Int. Conf. on Fuzzy Systems (FUZZ-IEEE '92)*, 1992, pp. 289-296.

[9] Jang, R., Self-Learning Fuzzy Controllers Based on Temporal Back Propagation, *IEEE Trans. on Neural Networks*, Vol. 3, No. 5, 1992, pp. 714-723.

[10] Kandel, A., M. Schneider, and G. Langholz, The use of fuzzy logic for the management of uncertainty in intelligent hybrid systems, in *Fuzzy Logic for the Management of Uncertainty* (L. A. Zadeh and J. Kacprzyk, Editors), J. Wiley & Sons, New York, 1991.

[11] Kandel, A., *Fuzzy Mathematical Techniques with Applications*, Addison-Wesley, Reading, MA., 1986.

[12] Kandel, A., *Fuzzy Expert Systems*, CRC Press, 1992.

[13] Kandel, A. and G. Langholz (Eds.), *Architectures for Hybrid Intelligent Systems*, (Eds.), CRC Press, 1992.

[14] Kandel, A. and G. Langholz (Eds.), *Fuzzy Control Systems*, CRC Press, 1994.

[15] Karr, C., Freeman, L., Meredith, D., Improved Fuzzy Process Control of Spacecraft Autonomous Rendezvous Using a Genetic Algorithm, *Proc. of the SPIE Conf. on Intelligent Control and Adaptive Systems*, Orlando, FL., 1989, pp. 274-283.

[16] Karr, C., Applying Genetics to Fuzzy Logic, *AI Expert*, Vol. 6, No. 2, 1991, pp. 26-33.

[17] Karr, C., Design of an Adaptive Fuzzy Logic Controller using a Genetic Algorithm, *Proc. of the Int. Conf. of Genetic Algorithms (ICGA '92)*, 1992, pp. 450-457.

[18] Karr, C., Gentry, E., A Genetics-Based Adaptive pH Fuzzy Logic Controller, *Proc. of the Int. Fuzzy Systems and Intelligent Control Conf. (IFSICC '92)*, Louisville, KY., 1992, pp. 255-264.

[19] Karr, C., Sharma, S., Hatcher, W., Harper, T., Control of an Exothemic Chemical Reaction using Fuzzy Logic and Genetic Algorithms, *Proc. of the Int. Fuzzy Systems and Intelligent Control Conf. (IFSICC '92)*, Louisville, KY., 1992, pp. 246-254.

[20] Katayama, R., Kajitani, Y., Nishida, Y., A Self Generating and Tuning Method for Fuzzy Modeling using Interior Penalty Method, *Proc. of the 2nd Int. Conf. on Fuzzy Logic and Neural Networks (IIZUKA '92)*, 1992, pp. 349-352.

[21] Lee, M. A., and H. Takagi, Integrating design stages of fuzzy systems using genetic algorithms, *Proceeding of 2nd IEEE-Fuzzy Conference*, pp. 612-617, San Francisco, CA., 1993.

[22] Nomura, H., Hayashi, I., Wakami, N., A Self-Tuning Method of Fuzzy Reasoning By Genetic Algorithm, *Proc. of the Int. Fuzzy Systems and Intelligent Control Conf. (IFSICC '92)*, Louisville, KY., 1992, pp. 236-245.

[23] Qian, Y., Tessier, P., Dumont, G., Fuzzy Logic Based Modeling and Optimization, *Proc. of the 2nd Int. Conf. on Fuzzy Logic and Neural Networks (IIZUKA '92)*, 1992, pp. 349-352.

[24] Roth, M.W., Survey of neural network technology for automatic target recognition, *IEEE Trans. on Neural Networks*, Vol. 1, pp. 28-43, 1990.

[25] Simpson, P.K., Artificial Neural Systems, Pergamon Press, New York, 1990.

[26] Schneider, M., and A. Kandel, COFESS - cooperative fuzzy expert systems for intelligent recognition on small computers, *Proc. ACM Symp. on Personal and Small Computers*, Cannes, France, May 4-6, 1988.

[27] Schneider, M., A. Kandel, G. Langholz, and G. Chew, FEST: Fuzzy Expert System Tools, John Wiley and Sons, London, 1995.

[28] Takagi, H. and Hayashi, I., NN-driven Fuzzy Reasoning, *Int'l J. Approximate Reasoning (Special Issue of IIZUKA '88)*, Vol. 5, No. 3, 1991, pp. 191-212.

[29] Zadeh, L. A., The role of fuzzy logic in the management of uncertainty in expert systems, *Fuzzy Sets and Systems*, Vol. 11, pp. 199-227, 1983.

OBTAINING FUZZY CLASSIFICATION RULES IN SEGMENTATION

Lawrence O. Hall*, Tina L. Majchrzak*, and Martin S. Silbiger**

* Department of Computer Science and Engineering
University of South Florida
4202 Fowler Avenue
Tampa, Florida 33620

** Department of Radiology
University of South Florida
4202 Fowler Avenue
Tampa, Florida 33620

ABSTRACT

A fuzzy instance-based connectionist learning system is described. Its ability to segment magnetic resonance images of the brain into tissue types is analyzed. Three competing methods are considered for partitioning each real-valued attribute of the given data set into fuzzy subsets of each attribute's range. The system is capable of generating fuzzy rules that classify the regions of an image. We discuss the derivation and interpretation of such rules. The results obtained by applying this learning strategy to a ground truthed image segmented by radiologists are very promising.

1. Introduction

The process of learning fuzzy rules usually involves a domain in which some or all of the attributes are real-valued. Learning fuzzy rules requires that the real-valued attributes be partitioned into a covering set of fuzzy sets. These fuzzy sets are then used in the antecedents of the learned rules which are associations between the attribute values and the outputs. Usually, the attribute values are described in the rule antecedent by the fuzzy set in which they have a membership of one.

To learn fuzzy classification rules from labeled examples involves two steps. The steps are the process of partitioning real-valued attributes into fuzzy sets and the process of associating the fuzzy subsets with output classes through a set of disjunctions and/or conjunctions. The two steps cannot be clearly separated. The fuzzy sets may need to be adjusted in order to develop accurate rules. The rules developed are constrained by the choice of fuzzy set partitions of the real-valued attributes.

A number of approaches to learning fuzzy rules have been recently proposed [1, 6, 11, 10]. They range from modifying an initial guess by an expert to discovering the fuzzy rules solely from training data. Here, we are interested in approaches that discover fuzzy rules from the training data.

In this paper, we explore several approaches to partitioning real-valued attributes for classification problems. After the attributes are partitioned, fuzzy rules are generated by an instance-based learning approach embedded in a connectionist network. The partitions are dynamically adjusted during the incremental learning process. Results are reported from the domain of magnetic resonance image (MRI) segmentation. There are three real-valued attributes available: proton density, T1 relaxation time, and T2 relaxation time. The tissues of visual interest for volunteers or normals consist of three classes: gray matter, white matter and cerebro-spinal fluid (csf). In the images, fat, bone and muscle also exist and are used in accuracy counts.

Excellent segmentations are obtained with less storage required than in a k-nearest neighbor technique. We discuss the differences between the segmentations obtained with several methods of partitioning the real-valued attributes into sets of fuzzy subsets.

2. Fuzzy instance-based learning

SC-net acquires its knowledge through the recruitment of cells algorithm (RCA). Unlike other neural networks which begin with a fixed structure and learn by something like the back-propagation algorithm [4], SC-net begins with an empty structure and augments it incrementally to incorporate unique instances. Specifically, SC-net uses an error threshold value, ϵ, supplied by the user to choose one of three courses of action. This action depends on the difference, α, between the expected output and the actual one. The expected output is supplied in the training data, while the actual output is the output of the network given the attribute values of the current instance. Depending upon the value of α, SC-net will do one of three things.

1. If $\alpha \geq 5\epsilon$, the instance will be considered unknown and SC-net will add a new information collector cell (ICC) to its structure along with appropriate links. This cell acts as an exemplar for all instances sufficiently similar to the current one.
2. If $\epsilon \leq \alpha < 5\epsilon$, the instance is close to one previously seen, and SC-net merely alters the bias of the appropriate ICC. Adjusting the bias allows this ICC to better cover the instances it represents.
3. If $\alpha < \epsilon$, the instance is considered already known, and the structure of the network is not altered.

3. Representing real-valued attributes

SC-Net requires that each real-valued attribute be partitioned into a collection of fuzzy sets which cover the range of the attribute. Each of these fuzzy sets is represented by a trapezoidal membership function. The membership function associated with a real-valued attribute is simply the one with the highest membership value for the given attribute's actual value. For example, consider the real-valued attribute Age which might run from 0 to 125 years. Suppose this range is partitioned into the three fuzzy sets young, middle aged, and old as shown in Figure 1. Now, given an instance with a value of 22 for the attribute Age, the fuzzy membership function associated with Age for this instance is young, since the membership value of 0.9 is the highest one across all three functions. Note that the *normal* range for young in this example is 0 to 20. Specifically, the normal range of a membership function includes those values for which the function is one.

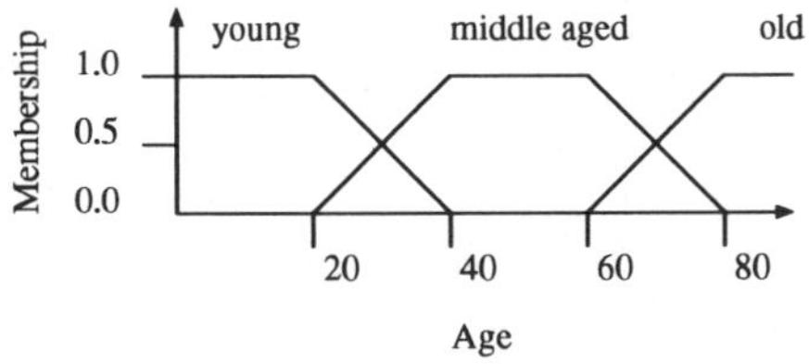

Figure 1: A fuzzy partition for the attribute Age.

3.1. Creating a fine grained fuzzy partition

The fuzzy set partition used for a real-valued attribute has a significant impact on classification accuracy and on the number and quality of the rules generated in the learning scheme described here. Currently, SC-Net can either incorporate a partition supplied by the user (possibly an expert) into its structure, or it can generate its own based on the set of training instances. When building its own partition for an attribute, SC-Net first orders the values a real-valued attribute is assigned in the training set from lowest to highest. Along with each value, it records the class(es) of the example(s) in which the value appears. Each partition is then created as follows. The first fuzzy membership function is a single element partition containing only the first value on the ordered list. It has a single element normal range. The normal range of the next fuzzy membership function always begins at the end of the previous function's normal range. The actual endpoint of the normal range currently being considered is the midpoint between the last consecutive value on the list having the same class(es) and the first one having a different class(es). The normal ranges of these fuzzy membership functions abut. If the last element in the list is encountered before one associated with a different class(es) is reached, then the current partition's normal range ends at the last element on the list. Finally, a single element partition is created for the last element.

The arms of each fuzzy membership function initially extend from a membership of one at the endpoints of the normal range to a membership of zero just beyond the lowest and highest values on the ordered list such that a trapezoidal membership function results. In other words, the left arm of each function goes to zero just left of the smallest attribute value, while the right arm goes to zero just right of the largest attribute value. At this point, SC-Net has created a partition of fuzzy sets for a given real-valued attribute. The values (from the training instances) covered by each fuzzy set have membership values of one for the corresponding function and membership values less than one for all other functions.

Unfortunately, when an attribute value is quite far from the normal range of a membership function, it still possesses non-zero membership in the function due to the extensive spanning of the function's arms. Hence, a method of modifying the arms is often necessary, as discussed in Section 3.4.

3.2. Uniform fuzzy partitions

In another partitioning scheme, the range of each attribute value is divided uniformly into as many non-trivial fuzzy subsets as there are classes in the training data. In this domain there are three attributes, T1, T2, and PD, and six classes, white matter, gray matter, csf, fat, muscle, and background. If the values of T1 range from say 1 to 120, then the following partitions are created:

p0: [1,1]; p1: [1,20]; p2: [20,40]; p3: [40,60];

p4: [60,80]; p5: [80,100]; p6: [100,120]; p7: [120,120].

This method of partitioning is along the lines of the genetically determined set of partitions in [6]. However, it is tied to the range of the data in each dimension.

3.3. Partitioning based on decision tree splits

The C4.5 learning system [9] is an inductive learner which represents learned knowledge in decision trees and can generate rules from the decision trees. At each branch of the decision tree, an information theoretic measure is used to choose an attribute whose values provide the direction (branch) to take. Binary splits along two branches are made for real-valued attributes.

If we assume each binary split indicates the end of the normal range of one fuzzy subset [2] and the beginning of another, we can use the split points to define our fuzzy partitions of the real-valued attributes. Again, the fuzzy subsets go to zero just beyond the leftmost and rightmost endpoints of the corresponding attributes range. A partitioning is done for each real-valued attribute based upon the split points created for that attribute during the decision tree learning process of C4.5. This partitioning was used in the learning process, and results from it are discussed in section 6.

3.4. Tuning the fuzzy partitions

SC-Net currently tunes attribute partitions by adjusting the arms of the membership functions during training. Tuning is done independent of the method of partition creation. The process of storing instances is incremental using the recruitment of cells algorithm. After training on all the instances, each of the training instances is matched against all of the stored prototypes (many of which may exactly match training instances). If the output of an instance associated with a class is active (i.e. above a threshold), and it is not one of the classes of the current example, the membership function of the attribute with the highest value for that class has its appropriate arm (on the left or right side of the trapezoidal membership function) brought in. The arm is adjusted to reduce the membership of the attribute to below the activation threshold for the output(s). If the attribute membership functions are combined with a min operator, deciding the appropriate new membership value for the offending function is quite easy. It is simply something just below the fuzzy threshold value required for an output to be active. Other t-norms will require different methods of membership function adjustment.

As an example of tuning, consider a real-valued attribute in a two class domain. The first class, C1, has a membership value of 0.9 for a given example, and the other class, C2, has a membership value of 0.8. Let the firing threshold be 0.75. The classes C1 and C2 will be turned on, since they both have membership values above the threshold. If we assume C2 should not be turned on, then the appropriate membership function of the fuzzy (real-valued) attribute must be modified. We further assume that C1 and C2 have different fuzzy membership functions associated with the real-valued attribute. Recall that SC-Net's automatic partitioning will guarantee this assumption is true. Figure 2 shows a possible initial fuzzy membership function and how it is changed. It is assumed that the actual attribute value for the instance is four and that the membership function in the figure is the fuzzy set with the highest membership for this value associated with class C2. The actual value corresponds to a fuzzy membership value of 0.8. This membership value is changed to 0.74 by the arm adjustment shown in the figure.

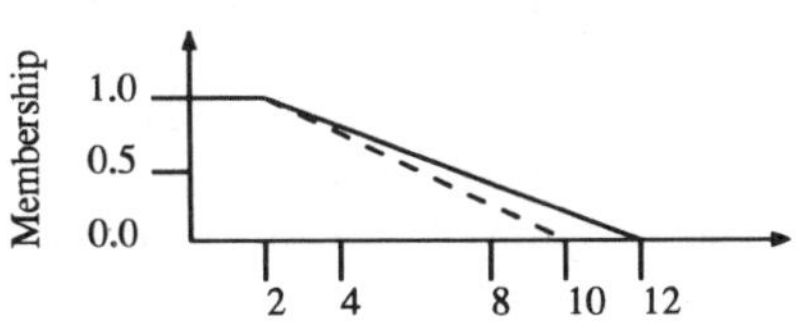

Figure 2: An Example of adjusting the right arm of a trapezoidal membership function.

4. Pruning and rule generation

Another feature of SC-net, the Global Attribute Covering Algorithm (GAC), prunes the network. This pruning not only manages SC-net's growth, which in the worst case is linear in the number of instances, but also facilitates the generation of a concise set of rules which represent what it has learned. These rules are read directly from the pruned structure of the network and are beneficial for at least two reasons. First, they may be incorporated into an expert system. Second, they shed light on what has actually been learned. This may lead to the conclusion that the learning focussed on irrelevant attribute combinations or it may lead to the discovery of a nuance in the domain not previously considered. In comparison with rules obtained from other approaches to learning in a non-image domain, experts believe SC-Net's rules compare favorably [8].

5. The nature of magnetic resonance images of the brain

SC-Net is currently being used to classify brain tissue from MR images. These images are obtained by placing a human subject in a magnet, sending in radio waves, terminating the radio waves, and then measuring the signal emitted by the subject. Three images, T1-weighted, T2-weighted, and proton density weighted, are created for each subject. Each image contains 256×256 pixels. In the training set, four values are associated with each pixel or example. The three images each contribute one value, while the fourth value is the class of the pixel. The first three values can range between 0 and 415. The latter value is one of ten classes: background, muscle, csf, fat, gray matter, white matter, tumor, edema, necrosis, or unknown. The unknown category corresponds to those pixels not classified in the ground truth, an image hand labeled by a radiologist. Each pixel represents an instance. The three values from the MR images are the real-valued attributes of the instance, while the class value is the class assigned to the instance by an expert.

6. Results

Results are reported from each partitioning method for a single normal brain slice (no tumor), for which we have ground truth. The training and testing sets for the slice are based on the ground truth for the image. The majority of the pixels in the ground truth were hand labeled by a very experienced practicing radiologist. Those few that are not labeled are not considered in the calculation of SC-Net's accuracy rate. However, the segmented images, judged quite good by a domain expert, do include SC-Net's classification of the unlabeled pixels. The training set consists of a subset of the ground truth, 304 pixels hand selected by an expert. The testing set consists of all pixels labeled by the expert.

Although SC-Net is capable of basing its classification on the K neighbors nearest of a training instance, results indicate that SC-Net performs best for K = 1. This is due to the way SC-Net learns and the nature of the MRI data. As discussed above, the network grows dynamically during training. When a new instance is seen that is significantly different from all previously seen instances, a new information collector cell (ICC) is added to its structure. This ICC represents all instances in the training set which are very near one another with respect to their attribute values and class. Because many of the instances (pixels) in the MR images share very similar attribute values, few ICCs are created. The K nearest neighbors are the K ICCs nearest in output value to a given testing instance. If each ICC represents an average of say ten instances, and there are only twenty instances for a particular class, then there will be only two ICCs for that class. If K is set to five, then the classification is likely to be wrong. Even if the first twenty nearest training instances share the same class as the testing instance, only two ICCs will be among the top five neighbors. If the other three nearest ICCs come from a common different class, SC-Net will give the wrong classification.

6.1. Fine grained partitioning

Before pruning, SC-Net has an accuracy rate of 96.2% with respect to ground truth. After pruning, the accuracy rate is 94.6%. The advantage of pruning is that the rule set is reduced from a maximum of 304 rules, the worst case of a rule for each instance. In fact, after training and before pruning there are 73 rules. The 54 rules that remain after pruning may be further reduced by trivially combining some of the rules. However, accuracy is lower and the effect is noted by the domain expert as a significant introduction of noise into the image. After showing an expert radiologist the segmented images produced before and after pruning in Figures 3A and 3B respectively, he found that the separation between white and gray matter is good for both images. He further stated that the image generated prior to pruning had little noise and was centrally sharp, having clean ventricles. The raw images are shown above the segmented images in Figure 3.

6.2. Uniform partitioning

The segmented image produced using the uniform partitioning method for the real-valued attributes is considered the best. It is shown in Figure 3E before pruning and 3F after pruning. Although it is only a little better than the fine grained partitioning, it has slightly more detail in it. These details are clearest in the ventricular area (center x-shaped region). Furthermore, the segmented image prior to pruning has very little noise in it.

Again, after pruning, the results are noisy. Apparently, the pruning process is causing tissues close in the feature space to be confused with one another. Since a single pixel can actually cover several tissues (due to the quality of image resolution), and since the classes are not well-separated in the feature space, the result is plausible. However, because the current pruning method is sensitive to the order of the training examples, this result does not necessarily imply that pruning is not useful in this domain.

This method of partitioning generates less rules (28 before pruning and 22 after pruning), since more training instances are associated with each information collector cell. A smaller rule set is normally easier to work with and therefore preferable. In Figure 4, we show the rules generated for csf and white matter before pruning. The fuzzy membership functions referenced by these rules appear in Figure 5. It is interesting to note that a single rule per tissue type allows for effective classification of these two tissue types.

6.3. Decision tree based partitioning

Decision tree based partitioning of the attributes produces the fewest fuzzy subsets. This means it may be less fine in discriminating among the training and testing examples. The segmentation results using this method are noisy as shown in Figure 3C before pruning and 3D after pruning. That is, tissues are mixed in regions where they are not mixed in reality. Also, the clarity of small details, such as blood vessels in the the ventricular area, are lacking when this model of partitioning is used. Less rules (just 13) are generated, but they are also less effective. The rules for csf and white matter differ from the rules in Figure 4 because of the partitions, which are now broader. After pruning, the segmentation is considered very noisy. The gross detail is still available, but fine detail is completely obscured.

7. Conclusions

SC-Net's performance on the normal brain slice is promising. The segmentations obtained with the fine grained partitioning and the uniform partitioning are both very good and come

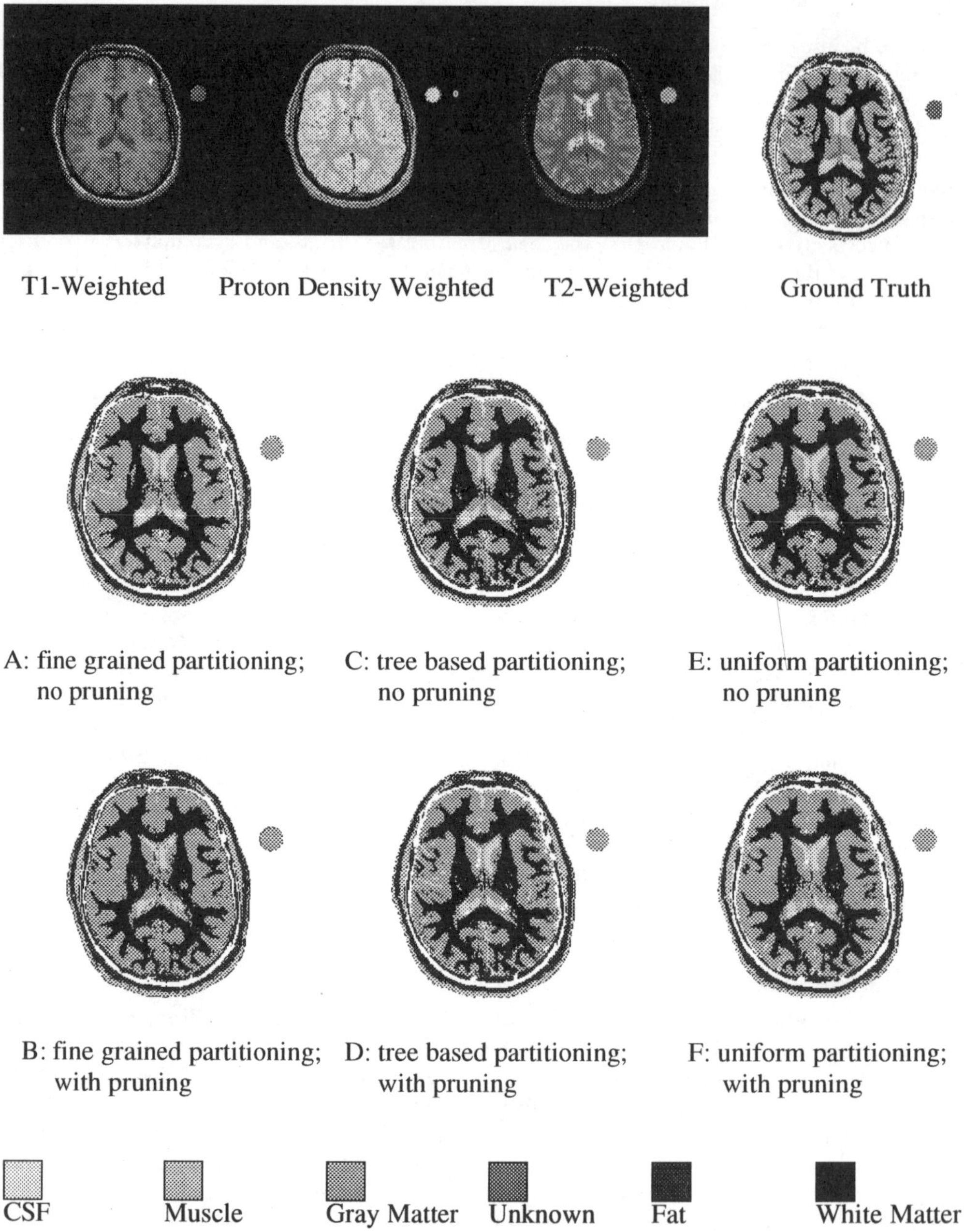

Figure 3: Segmented brain images with different partitionings.

Rule 1: if T2 is P6,
PD is P5, and
T1 is P2,
then csf (1.00).

Rule 25: if T2 is P3,
PD is P5, and
T1 is P4,
then white matter (1.00).

Figure 4: Rules generated for csf and white matter using uniform partitioning.

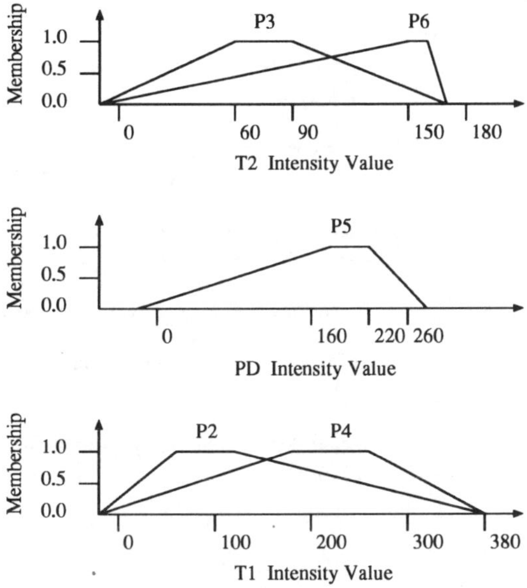

Figure 5: Fuzzy functions referenced in rules.

very close to matching the segmentation produced by an expert radiologist. The uniform partitioning segmentation is classified as excellent. Although the segmentation results after pruning (with GAC) in this domain are not as good as those produced by the unpruned network, the potential advantages of storage reduction as well as past success in other domains have led to the development of a more robust pruning algorithm. This algorithm performs quite well on the promoter and Iris domains [3].

In the uniform partitioning, the single rules for white matter and csf are quite effective. It is clear, for example, that low values in T1 indicate csf. Although not shown, very low values of T1 (just greater than zero) indicate air (background). The rules, thus, corroborate what has been found in other work [7].

Acknowledgements: This research was partially supported by National Cancer Institute grant CA59 425-01 and a grant from the Whitaker Foundation.

REFERENCES

[1] Berenji, H.R. and Khedkar, P. (1992) Learning and tuning fuzzy logic controllers through reinforcements, IEEE Transactions on Neural Networks, vol. 3, no. 5, p. 724-740.

[2] Berenji, H.R. (1990) Machine learning in fuzzy control, Proceedings of International Conference on Fuzzy Logic and Neural Networks (IIZUKA'90) Iizuka, Japan, p. 231-234.

[3] Hall, L.O. and Majchrzak, T.L. (1995) Instance-based learning with fuzzy set attribute values and pruning, to appear.

[4] Hecht-Nielson, R. (1990) Neurocomputing, Addison-Wesley, Reading, MA.

[5] Ishibuchi, H., Nozaki, K. and Tanaka, H., et.al (1993) Empirical study on learning in fuzzy systems, FUZZ-IEEE'93, p. 606-611.

[6] Ishibuchi, H., Nozaki, K., and Yamamoto, N. (1993) Selecting fuzzy rules by genetic algorithm for classification problems, FUZZ-IEEE'93, p. 1119-1124.

[7] Li, C., Goldgof, D., and Hall, L.O. (1993) Knowledge-based classification and tissue labeling of MR images of human brain, IEEE Transactions on Medical Imaging, vol. 12, no. 4, 740-750.

[8] Perez, R.A., Hall, L.O., Romaniuk, S., Lilkendey, J.T. (1992) Evaluation of machine learning tools using real manufacturing data, International Journal of Expert Systems, vol. 5, no. 4, p. 299-318.

[9] Quinlan, J.R. (1992) C4.5 programs for machine learning, Morgan Kaufmann, San Mateo, CA.

[10] Romaniuk, S.G. and Hall, L.O. (1995) SC-net: a hybrid connectionist, symbolic system, to appear in Information Sciences.

[11] Sugeno, S. and Yasukawa, T. (1993) A fuzzy-logic-based approach to qualitative modeling, IEEE Transactions on Fuzzy Systems, vol. 1, no. 1, p. 7-31.

TYPE II FUZZY SET INTERPRETATION OF LINGUISTIC EXPRESSIONS

I. Burhan TÜRKŞEN

Department of Industrial Engineering
University of Toronto
Toronto, Ontario M5S 1A4 Canada

ABSTRACT

There are two models for the representation of linguistic expressions. These are Type I and II fuzzy models. It is suggested that Type II representation of linguistic expressions with fuzzy normal forms do provide a more flexible and richer interpretation of complexities and uncertainties embedded in combination of concepts. However, it adds a computational expense and makes a user responsible for the selection of an appropriate membership function in the "non-specificity" interval defined by fuzzy normal forms. In this context, "implication" and "excluded middle" expressions are reinterpreted with their Type II models. A new "non-specificity" measure is defined for the assessments of semantic uncertainty of Type II fuzzy sets.

1. Introduction

Linguistic expressions of a natural language may be interpreted with two related but different models in fuzzy logic. They are known as Type I and Type II representations.

The essence of this paper is to suggest that Type II Fuzzy set models are a more suitable representation of complex and uncertain system behaviours.

A model is a human construction. It is to help our understanding. No arrogance should allow us to expect that anything we build should last forever or that any "understanding" should be final. Models arise in, and in response to, living circumstances, not divorced from them in any way either that allows those circumstances usefully to be ignored or, consequently, that poses a genuine problem in a "return" to such circumstances. If we want to be able to get on with our lives, what use is it to hide in theories devoid of complexity and uncertainty?

It is the role of linguistic expressions of a natural language in our lives, we want to look at, not some brittle alien ornament. It is "the swarming underworld" [3] of meaning that we want to investigate, not some disembodied, and so dead, distant echo. Granted, we have to start somewhere with some model or "representation" however crude or "approximate" it may be. But it is a mistake to start somewhere that is dead or tells part of the story, when what one wants to investigate is the life of linguistic expressions of a language. Perhaps it is impossible for a model by itself to be anything but dead. Thus we have to infuse it with life, first, by making sure there is room in it for us, and second by staying awake in it. The first is a matter of design. The second is a matter of attitude.

Humans are able to parse and produce novel sentences in finite time on a finite language base. The basic problem for us is to be able to model this process as closely as possible for a better representation of our expert knowledge. The question to be asked is: how do linguistic expressions of a language, its words and sentences, etc., hook up together so that they can be reused in different and novel configurations?

The Type I fuzzy theorist go one way. They start with linguistic terms of a language, i.e., fuzzy sets, then they assume that their compositions are formed directly by any one of the crisp connectives that comes from the subclasses of min-max, t-norms-conorms, and pseudo t-norms-conorms, together with a suitable complementation operator.

I propose to go another way. It has been found out experimentally that the connectives in combination of concepts do not correspond to crisp connectives which has lead to "compensatory 'and'" connectives to be proposed [20]. Later, it was shown that compensatory operator $\gamma \in [0,1]$ is included within the normal form bounds of fuzzy combinations [14]. These results suggest that the structure of Type I fuzzy set models for the representation of the combination of concepts does not allow sufficient room for the richer expression of complexities and uncertainties embedded in living language use.

A way to introduce needed flexibility in the structure of the combined concepts in fuzzy logic is to investigate a generalization of the twin normal forms known as disjunctive and conjunctive normal forms, DNF and CNF, in the classical logic. But this ought to be done with their suitable fuzzification. With suitable fuzzification, it is meant the fuzzy normal forms, i.e., fuzzy disjunctive and conjunctive normal forms, FDNF and FCNF, be derived form fuzzy truth tables [13] with suitable interpretations. In this manner, it is found that FDNF and FCNF constructs form suitable bounds to contain richer interpretive expressiveness that arise out of the complexity and uncertainty of combined fuzzy concepts. These bounds allow us to form interval valued fuzzy sets which are particular constructs in Type II fuzzy set theories. There are naturally other ways where Type II fuzzy sets arise. For example, Type II fuzzy sets are generated in measurement experiments of vague linguistic terms [7]. With FDNF and FCNF bounds, the responsibility for the determination of a specific meaning, i.e., a selection of a specific fuzzy set membership function, is shifted to the users of the language rather than left with the model of a language "itself" or its derived constructs. That is the responsibility of choosing a particular value within the interval of uncertainty identified by FDNF and FCNF bounds of the complex combination of vague concepts rests with users of the language.

This is in harmony with our general acceptance that both of the users meanings, i.e., the speaker's meaning and receiver's meanings, need be incorporated both in representation and selection. Within the context of "fuzzy expert systems", we stress the fact that both of the so-called experts meaning and the so-called system designers meaning needs be incorporated into the representation of the complexity of meaning transformation. The image that needs to be suggested is the network of associations in the residual memories of at least two or more persons. This incorporation process is to be arrived at by a suitable selection of fuzzy normal forms with an appropriate resolution of the network of associations triggered in the residual memories of both an expert and a system analyst; and naturally the final assessment ought to be determined with experimental validation exercises.

2. Fuzzy Normal Forms

In a recent study [13], it is shown that there are at least four subclasses of Fuzzy Normal Forms, $FNF^{(i)}$, i = 1, 2, 3, 4, in four subclasses of fuzzy logic where the fourth subclass turns out to be the Boolean normal forms in Boolean logic. As an illustration, three subclasses of these normal forms, i.e., $FNF^{(i)}$, i = 1, 2, 3, are identified next together with the axiomatic restrictions put on their propositional connectives for the particular linguistic expression known as "AND", "OR" and "→", i.e., implication". In a similar manner, $FNF^{(i)}$, i = 1, 2, 3, can be defined for the remaining eleven linguistic combination of concepts.

2.1 Class I: FNF's of Connectives Corresponding to Pseudo t-norms and conorms

These FNF's are constructed with the weak propositional De Morgan connectives that are constrained by boundary monotonicity and the involutive complementation axioms, as stated above. At the computational level, the corresponding connectives are the pseudo t-norms and s-norms. These connectives are non-commutative, non-associative and non-idempotent. For this subclass it can be shown [13] that

$$FDNF^{(1)}(A_1 \text{ AND } A_2) = (A_2 \cap A_1) \cup (A_1 \cap A_2)$$

$$FDNF^{(1)}(A_1 \text{ AND } A_2) = (A_2 \cup A_1) \cap (c(A_2) \cup A_1) \cap (A_2 \cup c(A_1))$$
$$\cap (A_1 \cup A_2) \cap (c(A_1) \cup A_2) \cap (A_1 \cup c(A_2))$$

$$FDNF^{(1)}(A_1 \text{ OR } A_2) = (A_1 \cap A_2) \cup (A_1 \cap c(A_2)) \cup (c(A_1) \cap A_2)$$
$$\cup (A_2 \cap A_1) \cup (A_2 \cap c(A_1)) \cup (c(A_2) \cap A_1)$$

$$FCNF^{(1)}(A_1 \text{ OR } A_2) = (A_1 \cup A_2) \cap (A_2 \cup A_1)$$

$$FDNF^{(1)}(A_1 \to A_2) = (c(A_2) \cap c(A_1)) \cup (A_2 \cap c(A_1)) \cup (A_2 \cap A_1)$$
$$\cup (c(A_1) \cap c(A_2)) \cup (c(A_1) \cap A_2) \cup (A_1 \cap A_2),$$

$$FCNF^{(1)}(A_1 \to A_2) = (A_2 \cup c(A_1)) \cap (c(A_1) \cup A_2).$$

Let these two normal forms define an interval-valued fuzzy sets:

$$I^{(1)} = \left[FDNF^{(1)}(A_1 \bullet A_2), FCNF^{(1)}(A_1 \bullet A_2)\right]$$

where $\bullet$ stands for "AND", "OR" and "$\to$", etc.

The interval-valued fuzzy set so constructed identify a Type II fuzziness, i.e., a second order semantic uncertainty. Thus, for example, combinations of two Type I fuzzy sets with "vague (fuzzy) implications" corresponding to pseudo t-norms and conorms at the computational level specify a subclass of Type II fuzzy set representations of fuzzified implications.

A particular example of this subclass of fuzzy implications is "Fuzzified Gödel" implication where $\cap$ and $\cup$ correspond to

$$\hat{T}_{min} = \begin{cases} 0 & a_1 + a_2 \leq 1 \\ a_2 & \text{otherwise} \end{cases} \text{ and } \hat{S}_{max} = \begin{cases} 1 & a_1 + a_2 \geq 1 \\ a_2 & \text{otherwise} \end{cases}, \text{ respectively.}$$

2.2 Class II: FNF's of Connectives Corresponding to t-norms and Co-norms

These FNF's are constructed by propositional De Morgan connectives corresponding to the well-known t-norms, s-norms, and standard complement at the computational level and are constrained by boundary, monotonicity, commutativity, associativity, and the involutive complementation axioms, but they are non-idempotent .

For this class, it can be shown [13] that:

$$FDNF^{(2)}(A_1 \text{ AND } A_2) = (A_1 \cap A_2) \cup (A_1 \cup A_2)$$

$$\mathrm{FCNF}^{(2)}(A_1 \text{ AND } A_2) = (A_1 \cup A_2) \cap (A_1 \cup c(A_2)) \cap (c(A_1) \cup A_2)$$
$$\cap (A_1 \cup A_2) \cap (A_1 \cup c(A_2)) \cap (c(A_1) \cup A_2).$$

$$\mathrm{FDNF}^{(2)}(A_1 \text{ OR } A_2) = (A_1 \cap A_2) \cup (A_1 \cap A_2) \cup (A_1 \cap c(A_2)) \cup (A_1 \cap c(A_2))$$
$$\cup (c(A_1) \cap A_2) \cup (c(A_1) \cap A_2)$$

$$\mathrm{FCNF}^{(2)}(A_1 \text{ OR } A_2) = (A_1 \cup A_2) \cap (A_1 \cup A_2)$$

$$\mathrm{FDNF}^{(2)}(A_1 \to A_2) = (c(A_1) \cap c(A_2)) \cup (c(A_1) \cap c(A_2))$$
$$\cup (c(A_1) \cap A_2) \cup (c(A_1) \cap A_2)$$
$$\cup (A_1 \cap A_2) \cup (A_1 \cap A_2),$$

$$\mathrm{FCNF}^{(2)}(A_1 \to A_2) = (c(A_1) \cup A_2) \cap (c(A_1) \cup A_2).$$

Again an interval-valued fuzzy set is defined by these fuzzy normal forms which identify a Type II fuzziness, i.e., a second order semantic uncertainty. This interval is defined as:

$$I^{(2)} = \left[\mathrm{FDNF}^{(2)}(A_1 \bullet A_2), \mathrm{FCNF}^{(2)}(A_1 \bullet A_2)\right]$$

where again • stands for "AND", "OR" and "→", etc.

Thus, for example, the combinations of two Type I fuzzy sets with vague (fuzzy) implications corresponding to t-norms and conorms at the computational level specify an interval-valued Type II fuzzy set representation of fuzzified implications. "Fuzzified Lukasiewicz" implication falls into this category. It should be pointed out that in current literature of Type I fuzzy set theory and logic, Lukasiewicz implication, $I_Ł$, "$A_1 \to A_2$" is modeled as "$c(A_1) \cup A_2$" where $\cup$ corresponds to Bold union operator. It can be shown that $I_Ł \in I^{(2)}$. Thus "Fuzzified Lukasiewicz" implication contains the point valued Lukasiewicz implication.

2.3 Class III: FNF's of Zadeh's Connectives

This subclass of FNF's are constrained by Zadeh's propositional De Morgan connectives. They correspond to Max-Min and standard complement at the computational level over the interval [0,1] which preserve the boundary, monotonicity, commutativity, associativity and idempotency conditions. For this class it can be shown [13] that

$$\mathrm{FDNF}^{(3)}(A_1 \text{ AND } A_2) = A_1 \cap A_2 = \mathrm{DNF}(A_1 \text{ AND } A_2)$$

$$\mathrm{FCNF}^{(3)}(A_1 \text{ AND } A_2) = (A_1 \cup A_2) \cap (A_1 \cup c(A_2)) \cap (c(A_1) \cup A_2) = \mathrm{CNF}(A_1 \text{ AND } A_2)$$

$$\mathrm{FDNF}^{(3)}(A_1 \text{ OR } A_2) = (A_1 \cap A_2) \cup (A_1 \cap c(A_2)) \cup (c(A_1) \cap A_2)$$

$$\mathrm{FCNF}^{(3)}(A_1 \text{ OR } A_2) = (A_1 \cup A_2)$$

$$FDNF^{(3)}(A_1 \to A_2) = (c(A_1) \cap c(A_2)) \cup (c(A_1) \cap A_2) \cup (A_1 \cap A_2)$$
$$= DNF(A_1 \to A_2),$$

$$FCNF^{(3)}(A_1 \to A_2) = c(A_1) \cup A_2 = CNF(A_1 \to A_2).$$

With expressions $FDNF^{(3)}(A_1 \to A_2)$ and $FCNF^{(3)}(A_1 \to A_2)$, the interval-valued Type II "Fuzzified Kleene-Dienes" implication is defined.

It should be observed that this third class of fuzzy normal forms turn out to be equivalent *in the propositional form only* to the two-valued logic expressions of DNF and CNF. Furthermore, for all of the classical sixteen combinations of concepts, this equivalence is preserved, however, in the *propositional form only*, i.e.,

$$FDNF_i^{(3)}(\bullet) \equiv DNF_i(\bullet) \text{ and } FCNF_i^{(3)}(\bullet) \equiv CNF_i(\bullet),\ i = 1, \ldots, 16. \qquad (7)$$

It is shown [15] that $FDNF^{(3)}(\bullet) \subseteq FCNF^{(3)}(\bullet)$ in fuzzy logic where as it is well known that $DNF(\bullet) \equiv CNF(\bullet)$ in the two valued logic. (It should be noted that fuzzified versions of $DNF(\bullet)$ and $CNF(\bullet)$ were investigated in 1986 writing and at that time the containment relationship was expressed as $DNF(\bullet) \subseteq CNF(\bullet)$.) It is clear that the Type I-Kleene-Dienes implication is the upper bound of the "Fuzzified Kleene-Dienes" implication.

3. Excluded Middle Reinterpreted

The recognition that every linguistic expression leads to two propositional logic expressions in fuzzy logic suggests rather strongly that we ought to reassess some of the special linguistic expressions.

As an example, in this section we will reassess the meaning that ought to be associated with the well-known linguistic expression: "Excluded Middle," i.e., "A OR NOT A," which is a special case of the linguistic expression "A_1 OR A_2," when we substitute "A" in place of A_1 and "NOT A" in place of A_2.

3.1 Excluded Middle for Zadeh Class

In order to be able to carry out the required analysis, we need to resolve the issue of "NOT A" when we substitute it in place of A_2. The linguistic expression "NOT A" is represented by its DNF and CNF as in the third class of connectives.

$$DNF(NOT\ A) = (c(A_1) \cap A_2) \cup (c(A_1) \cap c(A_2)) \subseteq (c(A_1) \cup A_2) \cap (c(A_1) \cup c(A_2)) = CNF(NOT\ A)$$

at the propositional level. However, this is a relative complement with respect to any fuzzy set A_2. What we need to find is the complement with respect to the universal set I. Thus, in terms of DNF and CNF expressions by substituting I and $\mathbf{c}(I)$ in place of A_2 and $\mathbf{c}(A_2)$, and A and c(A) in place of A_1 and $c(A_1)$, respectively, we obtain:

$$(\mathbf{c}(A) \cap I) \cup (\mathbf{c}(A) \cap \mathbf{c}\,(I)) \subseteq (\mathbf{c}(A) \cup I) \cap (\mathbf{c}(A) \cup \mathbf{c}\,(I)).$$

Since $\mathbf{c}(I) = \varnothing$, $\mathbf{c}(A) \cap I = \mathbf{c}(A)$, $\mathbf{c}(A) \cap \varnothing = \varnothing$, $\mathbf{c}(A) \cup I = I$, and $\mathbf{c}(A) \cup \varnothing = \mathbf{c}(A)$, we get $\mathbf{c}(A) = \mathbf{c}(A)$. Thus, "NOT A" in linguistic expressions is to correspond to $\mathbf{c}(A)$ in the fuzzy propositional expressions whenever we are interested in its relative complement with respect to I, the universal set, in the third class of connectives. Therefore, we need to substitute **c**(A) in place of A_2 in the fuzzy propositional expression of "Excluded Middle" and other such expressions whenever we are analyzing such expressions with respect to the relative complement of the universe of discourse. Similarly it can be shown that DNF(A) = CNF(A) = A.

With this result, when we substitute "NOT A"'s equivalent, **c**(A), in place of A_2 and A in place of A_1 in the " A_1 OR A_2" expression, we get $(A \cap c(A)) \cup (A \cap c(c(A)) \cup (c(A) \cap c(A)) \subseteq (A) \cup c(A)$ corresponding to the DNF and CNF expressions of "Excluded Middle." The corresponding computational level expression in the membership domain is obtained as:

$$(a \wedge N(a)) \vee (a \wedge a) \vee (N(a) \wedge N(a)) \leq a \vee N(a)$$

with $\vee = \max(a,b)$, $\wedge = \min(a,b)$, $N(a) = 1\text{-}a$. This results in $0.5 \leq a \vee N(a) = a \vee N(a) \leq 1.0$ for $a \in [0,1]$. Thus, the linguistic expression known as "Excluded Middle" in classical logic needs to be reinterpreted as the degree of "A OR NOT A" which is realized as $0.5 \leq \mu_{DNF(A\ OR\ NOT\ A)} = \mu_{CNF(A\ OR\ NOT\ A)} \leq 1$ with Zadeh's De Morgan Logic. That is, for specific values of $a \in [0,1]$ the "Excluded Middle" expression is satisfied to a degree specified by "$a \vee N(a)$." For example, if a=1, then we get "$a \vee N(a)=1$." Thus the expression holds in the classical sense. If, however, a=.4, or a=.6, then "$a \vee N(a)=.6$," and hence it is satisfied to the degree (.6). For a=.5, "$a \vee N(a) = .5$." Thus it is satisfied to the degree (.5), etc.

It is clear that, in Zadeh's fuzzy logic, we ought not to state that the "Excluded Middle" expression holds or does not. We should instead state that the "Excluded Middle" expression is satisfied to the degree it is specified by "$a \vee N(a)$" for $a \in [0,1]$.

3.2. <u>Excluded Middle for t-norm-conorm class.</u>

Again in order to be able to carry out the required analysis, we need to resolve the issue of "NOT A" when we substitute it in place of A_2 and A in place of A_1 in the "OR" expression of fuzzy normal forms for this class of operators.

It can be shown that [13]

$$c(A) \cap c(A) \subseteq c(A) \subseteq c(A) \cup c(A) \text{ and } A \cap A \subseteq A \subseteq A \cup A$$

for the t-norm-conorm class of operators.

Therefore, in substituting it in the "OR" expression, we get

$$\begin{aligned} FDNF^{(2)}(A\ OR\ NOT\ A) = &(A \cap A \cap c(A) \cap c(A)) \cup (A \cap A \cap c(A) \cap c(A)) \\ &\cup (A \cap A \cap A \cap A) \cup (A \cap A \cap A \cap A) \\ &\cup (c(A) \cap c(A) \cap c(A) \cap c(A)) \cup (c(A) \cap c(A) \cap c(A) \cap c(A)) \end{aligned}$$

$$FCNF^{(2)}(A\ OR\ NOT\ A) = [(A \cup A) \cup c(A) \cup c(A)] \cap [(A \cup A) \cup c(A) \cup c(A)]$$

For example, let us next evaluate this "Excluded Middle" expression of the second class of fuzzy normal forms for the specific De Morgan Triple known as "bold intersection-union-standard complement, i.e.,

$(\wedge_L, \vee_L, N(a) = 1 - a)$ where

$$\wedge_L(a,b) = \max(0, a+b-1), \quad \vee_L(a,b) = \min(1, a+b).$$

First it can be shown that

$$0 \le \mu^{(L)}_{FDNF(A \text{ OR NOT } A)} \le \mu^{(L)}_{FCNF(A \text{ OR NOT } A)} = 1$$

for the bold intersection-union and standard complement De Morgan Triple for $a \in [0,1]$.

That is, for example,

when $a = 1$, we get: $\mu^{(L)}_{FDNF(A \text{ OR NOT } A)} = \mu^{(L)}_{FCNF(A \text{ OR NOT } A)} = 1$;

when $a = .8$, $\mu^{(L)}_{FDNF(A \text{ OR NOT } A)} = .4 \le \mu^{(L)}_{FCNF(A \text{ OR NOT } A)} = 1$;

when $a = .6$, $\mu^{(L)}_{FDNF(A \text{ OR NOT } A)} = 0 \le \mu^{(L)}_{FCNF(A \text{ OR NOT } A)} = 1$, etc.

It is clear that, based on Type II interpretation we cannot state that the "Excluded Middle" expression holds or not. We can only state that it holds to the degree contained within the interval specified by $\mu^{(L)}_{FDNF(A \text{ OR NOT } A)}$ and $\mu^{(L)}_{FCNF(A \text{ OR NOT } A)}$ bounds.

These results explain why there has been incomplete and inaccurate statements in the literature. Statement that the "Law of Excluded Middle" holds for "bold union-intersection and standard complement" is based only on the $\mu^{(L)}_{FCNF(A \text{ OR NOT } A)} = 1$ which is the upper bound of the Type II representation which is, as it turns out, a limited perspective on the result obtained here.

4. Semantic Uncertainty of Type II Fuzzy Sets

With Type I fuzzy set representation, one can compute, for example, a generalization of the uncertainty measure known as "non-specificity" measure which is dependent on α-cuts, as

$$U(A) = \int_0^1 \log_2 |A_\alpha| d_\alpha$$

[6]. The α-cut is needed because non-specificity is based on the cardinality of a set as it is defined originally by the Hartley measure. With the interval-valued Type II fuzzy sets, one obtains an interval. The cardinality of the interval is measured by its length [9]. Given an interval-valued Type II fuzzy set, we get an interval

$$I = [\mu'_j(\bullet), \mu''_j(\bullet)], \text{ where } \mu'_j(\bullet) = FDNF^{(j)}(\bullet) \text{ and } \mu''_j(\bullet) = FCNF^{(j)}(\bullet), \quad j = 1, 2, 3,$$

for each of the three subclasses of fuzzy logics at every $x \in X$ of a base variable as described in Section 2 above. The uncertainty information content of this interval is interpreted as follows: one is assured that the membership function of, say, $A_1 \to A_2$ is in I, but one is not sure where it is in that interval. This means that the derived interval-valued Type II fuzzy set representation provides a "non-specific" information. It is proposed here that this interval-valued Type II fuzzy set representation should lead to a new formulation of "non-specificity" measure. Let this new "non-specificity" measure be defined as

$N(I)=[-\log_2|I|]^{-1}$, for $|I|\in[0,1)$, and $N(I)=+\infty$ for $|I|=1$ where $|I|$ is the length of the interval and is computed as:

$$|I|=|\mu'_j(\bullet)-\mu''_j(\bullet)|.$$

In this definition, Hartley measure is reinterpreted for use in the interval [0,1] such that $0\leq N(I)\leq+\infty$.

4.1. Specific Cases of Implication

Let us now illustrate the computational domain representation of a specific implication expression for a particular realization. That is let us investigate $A_1\rightarrow A_2$ for the particular case of $A\rightarrow A$. For this particular case, let us look at the very special case of "Fuzzified Lukasiewicz" based fuzzy implication. This implication is generated by a particular case of the second class of De Morgan connectives known as "Bold Intersection-Union and Standard Complement". Thus the $FDNF^{(2)}$ and $FCNF^{(2)}$ are to be used for its propositional expression. In these expressions , A is substituted in place of A_1 and A_2. That is, we get:

$$FDNF^{(2)}(A\rightarrow A) = (c(A)\cap c(A)) \cup (c(A)\cap c(A)) \cup (c(A)\cap A) \cup (c(A)\cap A) \cup (A\cap A) \cup (A\cap A), \text{ and}$$

$$FCNF^{(2)}(A\rightarrow A) = (c(A)\cup A) \cap (c(A)\cup A).$$

In the computational domain, when we substitute the bold intersection-union operators, $T_B(a,b) = \max(0,a+b-1)$ and $S_B(a,b) = \min(0,a+b-1)$ together with the standard complement $N(a) = 1-a$, it can be shown that

$$0.0\leq\mu_{FDNF^{(2)}(A\rightarrow A)}\leq\mu_{FCNF^{(2)}(A\rightarrow A)}=1$$

Furthermore, computed results show that $FDNF^{(2)}(A\rightarrow A) = FCNF^{(2)}(A\rightarrow A) = 1$ indicating that there is no second order semantic uncertainty when $\mu_A(x) = a \in [0.0, 0.25]$ or when $a\in [0.75, 1]$. But when $a \in (0.25, 0.75)$, an interval-valued Type II fuzzy set is generated as:

$$[\mu_{FDNF^{(2)}(A\rightarrow A)}, \mu_{FCNF^{(2)}(A\rightarrow A)}=1]$$

which specifies the content of the second order semantic uncertainty.

This interval has its largest value at a = 0.5,because $\mu FDNF^{(2)}(A\rightarrow A) = 0$ whereas $\mu FCNF^{(2)}(A\rightarrow A) = 1$. Thus at a = 0.5, the second order semantic uncertainty for $A\rightarrow A$ is the largest signifying a "non-specificity" covering the whole of the [0,1] interval. This is in agreement with our expectation that a =0.5 is the most uncertain value in fuzzy set representation. At other points, the "non-specificity" is smaller. For example, at a = 0.3 we compute the interval to be:

$$[\mu_{FDNF^{(2)}(A\rightarrow A)}=0.8, \mu_{FCNF^{(2)}(A\rightarrow A)}=1].$$

Therefore the second order semantic uncertainty interval is I = [0.8,1.0] at a = 0.3. In other words, the confidence interval is I = [0.8, 1.0] meaning that we are confident with $Bel(A\rightarrow A) = 1.0$ that the membership values $\mu(A \rightarrow A)$ are in [0.8, 1.0] but our information is "non-specific", i.e., we do not know where $\mu(A \rightarrow A)$ is located exactly. Thus the non-specificity measure of A→A at a = 0.3 is

$$N(I) = [-\log_2(0.2)]^{-1} \cong 1.43 \text{ bits.}$$

Similar analysis can be made to appreciate the essence of Type II semantic uncertainty for other "implications" as well as for other linguistic expressions .

5. Conclusions

The meaning of words is its network of associations, associations a word has for us, in virtue of how we treat it. As users, we make the associations. There are two kinds: (i) associations with personal experiential memories, and (ii) associations with other words, i.e., affects of residual words of our own experimental results and their analysis and synthesis with other residual words from prior experiences. There are also associations with objects, as names are associated with their 'bearers', for example, in object-oriented designs. There need be no strict boundary for a word, past which all associations cease; only actual use need be recognizably finite. Here, one can think of some of the feature of Wittgenstein's [16] 'family resemblance' picture. Use need not be dependent on 'complete knowledge' of all (actual or possible) associations. All that is needed, for example, is finite habits of use, finite sets of habitual associations. One can think of associations as being different strengths, i.e., membership functions, starting with full strength and fading off into effective null strength. Besides habit, strength might depend on degree of trauma or surprise content of the associated experience, and so on strength of memory, or frequency of use of associated words, etc., etc. Not all words a speaker is able to use need have personal 'direct experience' memory associations; but every word must have been 'grounded' in some speaker's experiential memory.

Associations of the second kind are thus parasitic, indirect, on associations of the first kind, in some sense. However, no special unusurpable authority need be given to coiners of words. One is not proposing a causal theory of meaning and meaning representation here. An association can be extremely strong, having deep social and historical roots, or roots in ancient survival tactics, for example; but as an association, its strength is always variable, whether by will or happenstance. We can forget; we can change the meaning of words, smoothly or brusquely; "deviations" can become new habits, etc. When words are combined, their already effectively finite association sets are restricted through the proximity of other words in the combination and through the connectives that structure the combination; some regions of the set become weaker while others become stronger. The proposed fuzzy normal form structures specify the weaker and stronger associations depending on strength of the combination operators and involutive complement selected for the construction of the fuzzy normal forms as well as the Type I definitions of the words that are combined.

Specific knowledge transfer context, not just the combined concept by itself, will also contribute to the strengthening and weakening by a suitable selection of min-max or a particular t-norm-conorm or a particular pseudo t-norm-conorm pairs together with an involutive complement. Thus, the underlying suggestion is that meaning is, and hence meaning representation ought to be, almost always variable, rather than constant, from context to context, and has a different set of uncertainty boundaries depending on context in contrast to Type I view and naturally in contrast to classical view.

It should be clear that there is no particular claim to originality with the basic idea of associations. The idea of association in memory has a precedent in the Stoic notion of 'suggestive' or commemorative sign [11,17]. What is proposed and suggested here is the role of memory in the determination of meaning, and the open-endedness and variability at the level of word meaning. This in particular distinguishes the interval valued Type II fuzzy set models proposed above from the Type I fuzzy set models. It should be noted that the open-endedness of the variability and uncertainty content only occurs at the level of the combination of concepts, i.e., in the second order lattice structure, and hence the need for Type II fuzzy sets models.

While Type I fuzzy set theory has avoided the temptation and tendency of gross oversimplification, it needs to be extended with a healthy conscious respect for the abundance, variety and variability of human life, replete with uncertainties, and enriched with natural language to express it requisite varieties. I do not believe that such an extension in knowledge representation would lead to an "abject surrender to chaos", so much feared by those who harbour secret or not so secret desires for "secure knowledge". Such desires only add needless anxiety to the task of getting on with every day life. However, it does add a burden of an additional computational complexity which can be handled with currently available computational power of computers and highlights the responsibility of the users, decision-makers, scientists, engineers, for the appropriate selection among the alternative choices that are exposed with a measure of uncertainty over the interval of "non-specificity".

Thus I believe it is a task of fuzzy theory researchers to foster Type II fuzzy set theory as the practical faith in the face of acknowledged complexity. This is not an illusory hope and desire but easily reachable with the suggested extension of Type I fuzzy set theory with fuzzy normal forms. This would bring us closer to the complexity and uncertainty of real life situations much beyond the ridiculous simplicity of classical theory.

REFERENCES

[1] Carnap R.(1942), Introduction to Semantics. Harvard University Press, Cambridge, Mass.
[2] Davidson D.(1990), The Structure and Content of Truth. J. Phil, 87 , p. 279-328.
[3] Hacking S. (1979), What is Logic? J. Phil., LXXXVI, 6 , p. 285-319.
[4] Horwitch P.(1993), Theories of Truth. In: A Philosophical Companion of First Order Logic. (R.I.G. Hughes ed.), Hackett Publishing Company, Inc., Annapolis.
[5] Klir, G.J. and Folger, T.A. (1988), Uncertainty and Information. Prentice Hall.
[6] Klir G. J. (1993). Developments in uncertainty-based information. Advances in Computers, 36 p. 255-332.
[7] Norwich A.M., and Türkşen I.B. (1984), A Model for the Measurement of Membership and the Consequences of its Empirical Implementation. Fuzzy Sets and Systems, 12, p. 1-25.
[8] Peirce C.S. (1995), Philosophical Writings of Peirce. (J. Buchler ed.), Dower Public Inc. New York.
[9] Ramer A. (1990), Concepts of fuzzy information measures on continuous domains. I.J. of General Systems, 17, p. 241-248.
[10] Schweizer, B. and Sklar, A. (1983), Probabilistic Metric Spaces. North Holland, Amsterdam.
[11] Sextus Empiricus (1933), Outlines of Pyrrhonism. Harvard University Press, Cambridge Mass.
[12] Squires A.C. (1994), Memories of Truth, Habits of Following. Ph.D. Dissertation, University of Toronto, Toronto, Ontario.
[13] Türkşen I. B. (1994), Fuzzy normal forms. Fuzzy Sets and Systems (Invited Special Issue) (in press).
[14] Türkşen I.B. (1992), Interval-Valued Fuzzy Sets and 'Compensatory AND'. Fuzzy Sets and Systems, 51, p. 295-307.
[15] Türkşen I.B. (1986), Interval-Valued Fuzzy Sets Based on Normal Forms. Fuzzy Sets and Systems, 20, p. 191-210.
[16] Wittgenstein L. (1972), On Certainty. Harper and Row, New York.
[17] Wittgenstein L. (1976), Philosophical Investigations. Basil Blackwell, Oxford , London.
[18] Zadeh, L.A., (1965), Fuzzy Sets, Information and Control, 8, p. 338-353.
[19] Zadeh, L.A., (1975), Concept of a Linguistic Variable and its Application to Approximate Reasoning, I, II, III. Information Sciences, 8, p. 199-249, 301-357; 9, p. 43-80.
[20] Zimmermann H.J., and Zysno P. (1980), Latent Connectives in human decision making. Fuzzy Sets and Systems, 4, p. 37-51.

FUZZY NEUROCOMPUTATIONS IN THE DESIGN OF EXPERT SYSTEMS

Witold PEDRYCZ*, Eugene ROVENTA**

*Dept. of Electrical and Computer Eng.
University of Manitoba
Winnipeg, Manitoba
Canada R3T 2N2

** Dept. of Computer Science
Glendon College, York University
Toronto, Ontario
Canada M4N 3M6

ABSTRACT

The paper discusses the use of techniques of fuzzy neurocomputations in the construction of expert systems. It is revealed how fuzzy sets utilized in this context become advantageous in bridging some fundamental concepts of learning stemming from neural networks and the ideas of explicit knowledge representation residing within symbolic computations. We accentuate an issue of numerical quantification of qualitative relationships being usually available as a part of any domain knowledge encountered in knowledge-based systems. A new class of logically-inclined processing units constructed exclusively with the aid of fuzzy set operators is advantageous in supporting both the explicit form of knowledge representation and learning capabilities. We study generic architectures including those capturing dynamics of the individual features. The issue of handling uncertain and incomplete information is addressed as well.

1. Introduction

The array of techniques utilized in the development of expert systems depart from a diversity of schemes of explicit knowledge representation [5] [8] [9] about the application domain (frames, scripts, schemes, rules, etc.). This definitely enhances and simplifies all the processes of knowledge manipulation (such as knowledge modification/update, inference procedures, etc.). The technology of neural networks and algorithms encountered there [1] [6] offer significant learning capabilities that could be definitely well received by all those developers who has already suffered a lot from quite lengthy and error prone activities of knowledge acquisition. Despite some achievements in this area, cf. [3], there are still some obstacles in taking full advantage of the learning capabilities of the networks and proceeding with their effortless embedding into the hybrid structures embracing both neural networks and expert systems. The main stumbling block resides within the nature of neurocomputing itself. It is quite evident that most of the problems supported by expert systems convey less or more visible logical structure. Neural networks are neither good at revealing these pieces of knowledge nor learning these relationships. It is worth mentioning that even learning of simple and dimensionally modest Boolean functions like a two-dimensional exclusive OR (XOR) usually requires hundreds of learning epochs. Obviously, the configuration of the produced connections do not facilitate any further interpretation that could lead towards a better understanding of the logical relationships existing in the problem. Neural networks operate on a purely numerical level and as such are not well equipped with any architectural features that could enhance "readability" of the underlying logical structure manifesting in the data. Additionally, the distributed model of the network promotes an implicit form of the acquired knowledge.

We proceed with an introduction of generic processing units (logical neurons) [4] illuminating how these constructs directly pertain to knowledge representation of disjunctive and conjunctive pieces of data. In sequel, the discussion will be focussed on representing and combining "if-then" rules by constructing a relevant neural network (logic processor). Finally we study the problem of representation of uncertainty residing within available data and its propagation along the network.

2. Logic-based neurons

Let us consider a collection of inputs x_i, i=1,2,...,n arranged in a vector form as $\mathbf{x} \in [0,1]^n$ The first class of the neurons (known as aggregative logic neurons) realizes a well-defined logical aggregation of the input signals. The second category of the neuron carries out some referential processing. All the computational models are presented along with their interpretation placed in the context of expert systems.

2.1. Aggregative logic neurons

The OR neuron is described as

$$y = \mathrm{OR}\,(\mathbf{x};\, \mathbf{w}) \tag{1}$$

whose coordinatewise version reads as

$$y = \mathrm{OR}[x_1 \text{ AND } w_1, x_2 \text{ AND } w_2, \ldots, x_n \text{ AND } w_n]$$

where $\mathbf{w}= [w_1, w_2, \ldots, w_n] \in [0,1]^n$ constitutes a vector of the connections of the neuron.
The standard implementation of the fuzzy set connectives involves a formal apparatus of triangular norms [2] in which the OR and AND logical operators are realized by some s - and t-norms, respectively. This produces the formula,

$$y = \mathop{S}_{i=1}^{n} [x_i \, t \, w_i] \tag{2}$$

The AND neuron is expressed as

$$y = \mathrm{AND}(\mathbf{x};\, \mathbf{w}) \tag{3}$$

which reads as

$$y = \mathop{T}_{i=1}^{n} [x_i \, s \, w_i] \tag{4}$$

The AND neuron realizes a conjunctive (AND-type) aggregation of the subconditions of a given rule. Its inputs are interpreted as some confidence levels associated with the corresponding terms (subconditions). The connections are introduced to discriminate between different levels of influence that the individual subconditions might have on the confidence attached to the overall condition. For the OR neuron the following monotonicity condition is fulfilled: the higher the connection w_i, the stronger the contribution of this input to the outcome generated by the neuron. The converse behavior holds true for the AND neuron where higher values of the connections imply less essential role of their associated inputs.

2.2. Referential logic-based neurons

In referential computations, the input signals are not directly aggregated as this has been realized in the aggregative neuron but rather than that they are analyzed first (e.g., compared) with respect to the given reference point.The results of this computation (including operations like matching, inclusion, difference, dominance) are afterwards summarized in the aggregative part of the neuron along the way that has been described in Section 2.1. In general, one can describe the reference neuron as

$$y = \mathrm{OR}(\,\mathrm{REF}(\mathbf{x};\ \mathbf{reference\_point}),\mathbf{w})$$

(that develops a disjunctive form of the aggregation)
or in the form

$$y = \mathrm{AND}(\mathrm{REF}(\mathbf{x};\ \mathbf{reference\_point}),\mathbf{w})$$

building a conjunctive form of aggregation; REF(.) stands for the reference operation carried out with respect to the given point of reference, **reference_point**.

Depending upon the reference operation, the functional behavior of the neuron is described accordingly,
(i) MATCH neuron:

$$y = \mathrm{MATCH}(\mathbf{x};\mathbf{r},\mathbf{w}) \tag{5}$$

or equivalently

$$y = \underset{i=1}{\overset{n}{S}} \left[w_i\ t\ (x_i \equiv r_i)\right]$$

where $\mathbf{r} \in [0,1]^n$ stands for a reference point defined in the unit hypercube. The matching operator is defined as

$$a \equiv b = \frac{1}{2}\left[(a\ \varphi\ b) \wedge (b\ \varphi\ a) + (\bar{a}\ \varphi\ \bar{b}) \wedge (\bar{b}\ \varphi\ \bar{a})\right]$$

where $\wedge$ denotes minimum and $a\ \varphi\ b = \sup\ \{c \in [0,1] \mid a t c \le b\}$. To emphasize the referential character of the processing realized by the neuron, one can rewrite (5) as

$$y = \mathrm{OR}(\mathbf{x} \equiv \mathbf{r};\ \mathbf{w})$$

The use of the OR neuron indicates an "optimistic" (disjunctive) character of the final aggregation.The pessimistic form of the aggregation can be realized using AND operation.
(ii) difference neuron. The neuron combines degrees to which $\mathbf{x}$ is different from the given reference point $\mathbf{g} = [g_1, g_2, ..., g_n]$. The output is interpreted as a global level of difference observed between the inputs and the reference point,

$$y = \mathrm{DIFFER}\ (\mathbf{x};\ \mathbf{w},\ \mathbf{g}) \tag{6}$$

i.e.,

$$y = \underset{i=1}{\overset{n}{S}}\ [w_i\ t\ (x_i \equiv\!|\ g_i)]$$

where the difference operator $\equiv\!|$ is defined as a complement of the equality index,

$$a \equiv\!| b = 1 - a \equiv b.$$

As before, the referential character of processing is emphasized by noting that

$$\text{DIFFER}\ (\mathbf{x};\ \mathbf{w},\ \mathbf{g}) = \text{OR}\ (\mathbf{x} \equiv \mathbf{g};\ \mathbf{w})$$

(iii) the inclusion neuron summarizes the degrees of inclusion to which the individual coordinates of **x** are included in the corresponding coordinates of the reference point (**f**),

$$y = \text{INCL}\ (\mathbf{x};\ \mathbf{w},\ \mathbf{f})$$

$$y = \underset{i=1}{\overset{n}{S}}\ [w_i\ t\ (x_i \rightarrow f_i)]$$

The relationship of inclusion is expressed in the sense of the pseudocomplement operation (implication) defined above.
(iv) the dominance neuron expresses a relationship dual to that carried out by the inclusion neuron

$$y = \text{DOM}\ (\mathbf{x};\ \mathbf{w},\ \mathbf{h})$$

where **h** is a reference point. In other words, the dominance relationship generates the degrees to which **x** dominates **h**. The coordinatewise notation of the neuron reads accordingly,

$$y = \underset{i=1}{\overset{n}{S}}\ [w_i\ t\ (h_i \rightarrow x_i)]$$

3. Mapping the rules onto the neural network -development of logic processors

3.1. Rule-induced topologies of the networks

We first proceed with some examples illuminating how different rules could be easily accommodated in the corresponding architectures of the networks. Let us first consider two simple rules of the format

-if A(k) and B(k) then C,

k=1, 2,..., N, where A(k) and B(k) are antecedents while C stands for a conclusion. Each A(k) and B(k) takes on values from a finite collection of symbolic values, say $\{A_1, A_2, \ldots, A_{n_1}\}$ and $\{B_1, B_2, \ldots, B_{n_2}\}$, respectively. An example of two rules leading to the same conclusion is provided in the form

-if A_1 and B_2 then C
-if A_2 and B_1 then C

The network realization of these rules is completed in two essential steps

(i) construction of the condition part of the rule being built out based on a family of subconditions,
(ii) aggregation of the rules.

The first step is completed by AND-ing the successive subconditions of the rules. This step is accomplished with the aid of the AND neurons (AND-wise combination of the subconditions). By default, the aggregation of the rules which follows this phase, is completed OR-wise.

Each neuron in the network is equipped with its own collection of the connections that are adjusted during learning. This reflects a variety of importance levels of the individual subconditions in the rules as well as diverse relevance of the rules themselves. The above example suggests a general architecture (logic processor) encapsulating a collection of the "if-then" statements to be composed of the following layers:

- the input layer (subcondition layer) distributes inputs (**x**) to all the elements in the hidden layer (condition layer),
- the hidden layer composed of the AND neurons builds up the condition parts out of the subconditions situated in the input layer,
- the conclusion parts (conclusion layer) are accommodated by the OR neurons distributed as the output layer.

The input layer can be also augmented by the complements of x_i's, $\bar{x}_i = 1 - x_i$. The role of these complements is to enhance the basic representation capabilities of the condition layer (this extension allows for representing a disjunctive type of subconditions aggregation) and come up with a more condensed description of the rules. We will be referring to this structure as a logic processor (LP).

It should be noted that the network may consist of several layers whereas the connections between them could involve not only the elements (subconditions) distributed within any two successive layers.

The variety of problem - driven architectures may be substantial. For instance, the condition part can be formed in a more complex way requiring the use of some referential neurons. The rule of the form,

-if A_1 is included in G_1 and A_2 is similar to P_1 then C_1

calls for the use of some referential neurons (more precisely, the inclusion and matching neuron).

The "if-then" rules of the format studied so far are essentially not capable of expressing dynamical (time-induced) relationships between the subconditions and actions. This aspect could be, however, essential to a proper problem description. Take, for instance, a following diagnostic problem in which a decision about system's failure should be issued while one of the system's sensors provides information about abnormal (elevated) temperature. Observe that the duration of this phenomenon (that is a raise of the temperature) has a primordial impact on how one can articulate a level of confidence about the failure. If the elevation of the temperature prolongs, our confidence about the failure rises up. It might well be that some short temporary temperature elevations (spikes) reported by the sensor could be almost ignored and need not to have any impact on the decision about failure. To properly capture this effect in the model, one has to equip the basic logic neuron with some feedback link. The neuron is described accordingly

$$x(k + 1) = [b \text{ OR } u(k)] \text{ AND } [a \text{ OR } x(k)] \tag{7}$$

and its dynamics is uniquely defined by the feedback connection (a). The initial condition x(0) expresses a priori confidence associated with failure x.The level of accumulation of evidence as well as a speed at which this accumulation occurs is defined by the values of the feedback connection. For sufficiently long period of time x(k+1) could take on higher values in comparison to the level of the original evidence being available at the input.

3.2. Learning

The learning of the logic processor is worked out in a supervised mode[1]. For a given collection of input-output pairs of data $(\mathbf{x}_1, \mathbf{t}_1),\ldots, (\mathbf{x}_N, \mathbf{t}_N)$ that summarizes the confidence levels of the subconditions and the associated conclusions, the procedure of the parametric learning modifies the parameters of the network (including the connections as well as the respective reference points) so that a certain performance index Q becomes minimized. The general scheme of learning can symbolically be expressed as

$$\Delta\_\text{parameter} = - \alpha \frac{\partial Q}{\partial \text{parameters}}$$

where α denotes a learning rate. The parameters of the network are adjusted following these increments, namely

$$\text{new\_parameters} = \text{parameters} + \Delta\_\text{parameters}$$

The relevant details of the learning scheme can be fully worked out once the topology of the network as well as some other details regarding the form of triangular norms have been made available.

Within the same mode of supervised learning one can start the training within quite diverse initial

[1]One can look at this learning as constructing (approximating) a fuzzy function through a series of generalized minterms or maxterms, cf. [7]

configurations of the network. These configurations depend primarily upon availability of the available domain knowledge. Two among possible versions are worth underlining:
(i) We are provided with a family of associations (conditional statements) between subconditions and conclusions; the rules have to be induced. Since the prior domain knowledge about the structure is primarily nonexistent, the network should be fully connected.
(ii) On the other side, the rules might be given and the learning session reduces then to parametric modifications of the connections. These changes are driven by the cases provided in the training data set.

Obviously, we could be confronted with many situations situated in between these two extremes in which some rules have been supplied however new ones could be yet induced based on the training data set.

In the context of the learning we make a clear distinction between the conditional statements and rules. The conditional statement can contain any combination of the available subconditions however the confidence attached to it might be quite low. It could also be quite restricted in its generality. In contrast, the rules need to be more meaningful and general - we come up with a formal characterization in the next section.

3.3. Rules induction

Interestingly enough, the network developed as discussed in the previous sections, can be utilized to generate (induce) "if-then" statements for some conditions specified in advance.They could be utilized as a part of a consistent domain knowledge and implemented with the use of some standard expert system shells. In this sense one can look upon the fuzzy neural network as an example of a tool facilitating a systematic way of knowledge acquisition. For each input **x** of the logic processor we compute

$$\mathbf{y}=\mathrm{LP}(\mathbf{x}; \mathbf{connections})$$

where **y** contains an information about the activation levels of the resulting conclusions. The choice of **x** might be carried out in various application-dependent ways. In particular, one can consider a Boolean form of **x**'s having only 0-1 entries, e.g.,

$$\mathbf{x}=[0\ 0...0\ 1\ 0...0]$$

or

$$\mathbf{x}=[0\ 0...0\ 1\ 0..0\ 1\ 0.0]$$

etc., where a single or several positions of **x** are set to 1. In this fashion the condition part of the statement is equivalent to the statement

" if x_i then ... "

or

“if x_i and x_j then...”.

This condition invokes several actions that are activated to degrees (levels of confidence) equal to $y_1, y_2, \ldots, y_m$, respectively. Of course not every "if-then" statement becomes a rule viz. a proposition conveying a meaningful, general enough and yet substantially supported piece of evidence. There are at least two reasons why this distinction makes sense:
(i) firstly, **x** can invoke outputs with too low levels of confidence thus the conclusion cannot be fully justified (as it becomes weakly supported),
(ii) secondly, **x** may be associated with too many outcomes (conclusions) at a same time so that the statement is not specific enough and therefore cannot be viewed as a rule (for instance, a statement " if **x** then *anything* " is not informative at all).
To express these two requirements in a formal way, we introduce two constraints:
- confidence requirement stating that a height of **y** (viewed as a fuzzy set) should be at least equal to γ_1 , $\mathrm{hgt}(y)=\max_{i=1,2,..,m} y_i=\gamma_1$,
-specificity requirement expressing that a total averaged activation of all y_i's should not exceed γ_2, with γ_2 resulting from the following expression

$$\frac{1}{m}\sum_{j=1}^{m} y_j \leq \gamma_2,$$

where γ_1 and γ_2 are two threshold levels, $\gamma_1, \gamma_2 \in [0,1]$.

4. Representing and processing uncertainty in fuzzy neural networks

Situations could emerge in which some of the pieces of evidence considered as the inputs of an expert system might not be fully specified or could be provided with a certain quite limited precision. To express this fact one has to come up with a relevant knowledge representation scheme. In particular, this requires a development of some matching mechanisms that are capable of coping with this phenomenon. We will tackle this problem of matching fuzzy quantities using possibility and necessity measures.

Let A denotes a fuzzy set viewed as a reference. Any input datum X (despite of its character) is then "translated" into an internal logical format being utilized within the network as follows, cf [10]

$$\text{Poss}(X|A) = \sup_{x \in \mathbf{X}} [\min(X(x), A(x))] \tag{8}$$

$$\text{Nec}(X|A) = \inf_{x \in \mathbf{X}} [\max(1-X(x), A(x))] \tag{9}$$

The basic properties of these measures have been thoroughly studied in the literature; for more details the reader can refer e.g., to [2]. Briefly speaking, the possibility measure evaluates a degree of overlap (coincidence) of X and A while the necessity measure is involved in expressing a degree of inclusion of X in A

The possibility and necessity measures processed together can be useful in handling uncertainty, in particular the aspects of ignorance and conflict manifested in the available input information X. Again, these two notions are context-dependent and as such should be analyzed with respect to the given fuzzy set A. The context-dependency implies also that the numerical qualifications of these phenomena depend upon the environment (the frame of cognition) within which they are embedded. Let us define two indices

$$\lambda = \text{Poss}(X|A)$$

$$\xi = 1 - \text{Nec}(X|A)$$

as expressing relationships occurring between X and A. For a pointwise character of X, the quantities λ and ξ are linked together via the straightforward relationship

$$\lambda + \xi = 1$$

(for any numerical information X both the measures coincide).

In general, when the datum is of a general (viz. nonpointwise) character then we end up having one of these inequalities

$$\lambda + \xi < 1, \quad \lambda + \xi > 1$$

These cases are worth studying since they tackle the situations including information ignorance and conflict:

Let $\lambda + \xi > 1$ that can be expressed as $\lambda + \xi = 1 + \gamma$ where $\gamma \in [0,1]$. The higher the value of γ, the higher the level of conflict emerging out of X placed in the context of A.

The case in which $\lambda + \xi < 1$, with $\lambda + \xi = 1 - \gamma$, $\gamma \in [0,1]$, articulates a situation of ignorance arising when expressing X via A. More precisely, γ is utilized to express this level of ignorance.

Thus the representation scheme applied to the network calls for an extension of the input layer by linking each objective with the lower (necessity) and upper (possibility) values of its bounds. The same knowledge representation extension applies to the output layer. The level of ignorance or conflict recorded at the level of the objectives is then transmitted via the network producing the corresponding qualification of these input deficiencies in the produced decisions. Of course, the learning scenarios should also include some cases exhibiting ignorance or conflict so that these aspects are properly accommodated in the network in the course of its learning.

5. Conclusions

We have discussed logic-oriented neural networks and their role in designing expert systems. Several important aspects are worth underlining:

- ☐ the networks constitute a scheme of explicit knowledge representation: the outcomes can be easily interpreted as well as any pieces of knowledge given a priori can be easily deployed onto the network the topology of whose is determined by the architecture of the problem itself.
- ☐ the network retains learning capabilities. The results of learning lead towards a calibration of the available surface knowledge as well as revealing new rules.
- ☐ the aggregative and referential neurons can be used directly in mapping rules and facts into a coherent structure.

Acknowledgment

Support from the Natural Sciences and Engineering Research Council of Canada and MICRONET is gratefully acknowledged.

6. References

[1] IEEE Trans. on Neural Networks(1992)- Special issue on fuzzy logic and neural networks, 3.

[2] D. Dubois, H. Prade(1988), Possibility Theory - An Approach to Computerized Processing of Uncertainty, Plenum Press, New York.

[3]C.G. Looney(1993), Neural networks as expert systems, Expert Systems with Applications, 6, 126-136.

[4] W.Pedrycz(1993), Fuzzy neural networks and neurocomputations, Fuzzy Sets and Systems, 56, 1-28.

[5] E. Shortliffe, B.Buchanan(1975), A model of inexact reasoning in medicine, Math Biosciences, 23, 351-375.

[6] P.K. Simpson(1990), Artificial Neural Systems: Foundations, Paradigms, Applications and Implementations, Pergamon Press, New York.

[7]L.G.Valliant (1985), Learning disjunctions of conjunctions, Proc. IJCAI-85, Los Angeles, 560-566.

[8] S.Weiss, C.Kulikowski (1994), A Practical Guide to designing Expert Systems, Rowman and Allanheld, Totowa, NJ.

[9]S.Weiss, C.Kulikowski(1991), Computer Systems That Learn, Morgan Kaufmann, San Mateo,CA.

[10] L.A. Zadeh (1978), Fuzzy sets as a basis for a theory of possibility, Fuzzy Sets and Systems, 1, 3-28.

ON A NEURO-FUZZY APPROACH IN A DATA FUSION CONTEXT BASED ON THE THEORY OF EVIDENCE

Marie-Claude Perron-Gitton

ONERA
29, avenue de la Division Leclerc
92320 Châtillon, France

ABSTRACT

Uncertainty management techniques offer a formalism well appropriate to deal with data variability. Nonetheless, the multisensor data fusion processing methods developed using uncertainty theories have emphasized the practical difficulty in taking context into account. The question is to exploit the available contextual knowledge so that its effect can be modeled through parameters usable by the decision criterion. The neuro-fuzzy approach proposed here is based on the interpretation of an analytical reference criterion issued from the theory of evidence, so that it can be implemented by a neural network.

1.Introduction

Developing a centralized multi-sensor decision system raises the problem of establishing a data fusion process that will take advantage of the complementary of the available sensors, taking into account the uncertainties that might affect the knowledge handled. The complexity of the information managed in a fusion process may come from the heterogeneity of the sensors and/or contextual data, the lack of knowledge or incompleteness of the underlying models...

In particular, it is advisable that a multi-sensor fusion system takes into account the advantages and disadvantages of the various sources, according to the context (weather conditions, temporary failure of one of the sensors, and so forth).

One stage prior to the fusion of disparate measurements is the local evaluation of quantities reflecting the relative importance of the occurrence of each pre-stated decision hypothesis, H_i, given the measurement m_j. The fusion efficiency is related to the quality of the modeling used to evaluate, on line, the probability of observing m_j under H_i. The uncertainties resulting from statistical learning show some probabilistic approaches like the Bayesian to fail. Specific fusion techniques based on uncertainty theories, allow one to deal with a certain number of problems difficult to solve by classical techniques. Nevertheless, modeling uncertainty is still not obvious in practice.

Some techniques that might offer hypothesis likelihood criteria [APPR91] are based on modeling elementary basic probability assignments from pre-established distributions and associated confidence coefficients. However, the theory provides no direct solution to estimate parameters able to reflect the variability of the context. The development of dedicated learning methods is thus required.

The hybrid approaches combining fuzzy logic with neural methods have learning capacities. A "neuro-fuzzy" approach is proposed here in order to learn decision rules involving coefficients expressing contextual knowledge. The development of an appropriate neural structure is based on the interpretation of an analytical fusion criterion, issued from the theory of evidence, in a form that can be easily implemented by a multilayer network.

2.Reference analytical approach

Given an exhaustive set of N hypotheses listed as H_i (i = 1, ..., N), the question is to recognize the most likely of them in a given situation described by M measurements m_j (j = 1, ..., M) generated by the multisensor system. We assume prior learning of the probability distributions $P(m_j/H_i)$ of each of the measurements under different hypotheses.

The analytical approach proposed by A. Appriou [APPR91] is based on an *a priori* modeling of the probabilities $P(m_j/H_i)$, and associated confidence factors denoted d_{ij}, in the form of a set of basic probability assignments according to the Dempster-Shafer theory, satisfying consistency with the Bayesian decision.

The resulting models lead to the formulation of two asymptotically equivalent decision criteria. The decision procedure consists in choosing the most likely hypothesis as the one of maximum plausibility. The following expression is one possible formulation of the plausibility to maximize:

$$Pl(H_i) = \pi_j \left[1 - d_{ij} \times (1 - R_j \times P(m_j / H_i)\right]$$

where d_{ij} belongs to [0,1]
and R_j is a normalization factor:

$$R_j \in [0, (\max_i^N(P(m_j / H_i)))^{-1}]$$

When all the d_{ij} are 1, this approach is equivalent to a maximum likelihood approach.

The following scheme illustrates how the decision procedure is implemented, assuming that the confidence factors related to the available contextual information have been previously determined.

Decision implementation
Reference approach

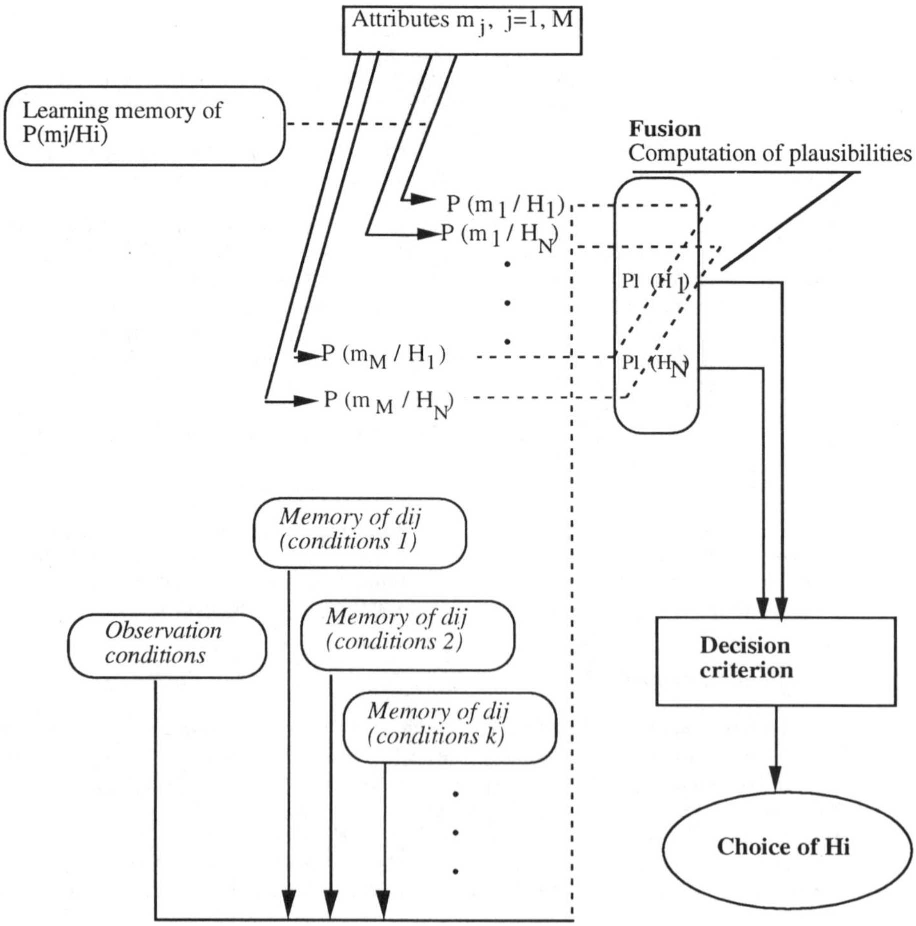

The development of a neuro-fuzzy solution for learning coefficients comparable to the d_{ij} memories is based on the neural implementation of a formalism related to the reference analytical approach.

The reference criterion can be formulated in terms compatible with fuzzy logic assuming the membership functions are known:

$\mu X_{ji} : xj \rightarrow [0,1]$

characterizing the fuzzy subsets X_{ji} corresponding here to the measurement sets m_j observable under the hypothesis H_i.

Since the membership functions μX_{ji} can be likened to the $P(m_j/H_i)$ functions, to within a normalization factor, the criterion C_i to be optimized can be expressed:

$$C_i = [\prod_j C_i^j(xj)]$$

in which $C_i^j(xj) = 1 - dij + dij\ \mu_{Xji}(xj)$

3.Fuzzy formalism stemming from analytical formalism

The fuzzy formalism used for implementing the analytical approach described above is defined from fuzzy relations, assuming appropriate operators are chosen.

3.1 Expression of a rule in fuzzy relations. Definition

Let $x = (x_1, x_2, ..., x_M)$ be the input vector, and $y = (y_1, y_2, ..., y_M)$ the output vector. Then X = X1xX2x...xXM (resp. Y=Y1xY2x...xYN) denotes a fuzzy set defined on the reference space R^X (resp. R^Y) included in R^M, and *R* is the fuzzy relation defined on $R^X \times R^Y$. When the composition rule o (expressed in disjunction and conjunction operators, V and Λ) is applied to fuzzy sets X and *R*, we can infer the conclusion Y:

$$\mu Y(y) = \mu X(x) \circ \mu R(x,y)$$

$$\mu Y_i(yi) = \bigvee_{xj} [(\mu X_j(xj) \wedge \mu R(xj, yi))]$$

3.2 Criterion formulation in fuzzy logic

X_{ji} being a fuzzy subset defined in section 2, we denote by Y_i the set on which the output variable y_i, corresponding to hypothesis H_i is defined. We introduce the fuzzy subset Y_{ij} defined on Y_i as a linguistic variable by which the degree of belonging of measurement x_j to reference set X_{ji} can be refined, under the conclusion y_i relative to hypothesis H_i. Let R_{ij} be a fuzzy relation to be defined, expressing the dependency between fuzzy subsets X_{ji} and Y_{ij} according to a formalism appropriate to approximate $C_i^j(x_j)$ (see section 2).

Let us consider the following fuzzy rule:

Rule R_{ij} If x_j is X_{ji} and x_{oj} is X_{oji}
then y_i is Y_{ij},

in which X_{oji} is a linguistic variable corresponding to "conditional distribution $P(m_j/H_i)$ is correct". The associated membership function corresponds to a conventional set.

By using μY_{ij} to denote the membership function associated with the conclusion of a local R_{ij} rule, the problem is to determine the appropriate fuzzy operators for obtaining the expression $C_i^j(x_j)$. For an input vector $x = [x_j, x_{oj}]$ belonging to the cartesian product $X = X_{ji} \times X_{Oji}$, μY_{ij} is found by:

$$\mu Y_{ij}(yi) = \mu X(x) \circ \mu R_{ij}(x,yi)$$

in which μR_{ij}, defined on $X \times Y_i$, is the membership function characterizing the fuzzy relation R_{ij}, and

$\mu X(x) = [\mu Xji(xj), 1]$

Using the V-Λ composition rule:

$\mu Yij\,(yi) = [\mu Rij(x_j,y_i) \wedge \mu Xji(xj)] \vee [\mu Rij(x_{0j},y_i) \wedge 1]$

and by letting $\mu Rij(x0j,yi)=1 - \mu Rij(xj,yi)$

$\mu Yij\,(yi) = [\mu Rij(x_j,y_i) \wedge \mu Xji(xj)] \vee [(1 - \mu Rij(x_j,y_i)) \wedge 1]$

$\mu Yij(yi)$ can be likened to $C_i^j(xj)$ if the Λ operator is chosen to be equal to the product and V to the "OR" of Lukaviewicz, with:

$\mu_A(x) \underset{\text{Lukaviewicz}}{\vee} \mu_B(x) = \min(\mu_A(x) + \mu_B(x), 1)$

If the following constraints are complied with:

$\mu Xji(xj) \leq 1$

$0 \leq \mu Rij(xj,yi) \leq 1$

by setting the condition $\mu Rij(xj,yi)=d_{ij}$,

we can write $\mu Yij(yi)$ as:

$$\mu Yij(yi) = 1 - \mu Rij(x_j,y_i) + \mu Rij(x_j,y_i) \cdot \mu Xji(x_j)$$
$$= 1 - dij + dij\, \mu Xji(x_j)$$

Given the coefficient d_{ij}, the expression $\mu Y_{ij}(yi)$ depends only on the variable x_j, whence a more adequate notation used hereafter is $\mu X'_{ji}(xj)$

3.3 Fuzzy reasoning applied

The application of the analytical criterion C_i is equivalent to applying rules of the type:

Rule R_i If x_1 is X'_{1i} and x_2 is X'_{2i} ... and x_M is X'_{Mi}

then y is Y^{Ri}

where the proposition "x_j is X'_{ji}" results from the application of the rule R_{ij} in the presence of the premise "x_j is X_{ji} and x_0 is X_{0ji}", the local conclusion inferred by the rule R_{ij} being characterized by a membership function such that :

$\mu X'ji(xj) = C_i^j\,(xj)$

As the variable y in the conclusion part is such that $y = [y_1, y_2, ..., y_N]$, the fuzzy set Y^{Ri} is a cartesian product of fuzzy subsets:

$Y^{Ri} = Y1^{Ri} x Y2^{Ri} x....x YN^{Ri}$

in which the local conclusions Y_k^{Ri} (relative to the decision hypotheses H_k (k = 1 ..., N)) of the rule R_i are characterized by the membership functions μY_k^{Ri} (k = 1, ... N).

If the fuzzy relation associated with rule R_i is characterized using the product operator (Larsen implication) we get:

$\mu_{Y'}{}^{Ri}(y) = \alpha_i\, \mu_Y{}^{Ri}(y)$

or

$\mu_{Y'k}{}^{Ri}(yk) = \alpha_i\, \mu_{Yk}{}^{Ri}(yk)$, k=1,N

where α_i is the degree of adequacy of the example presented, $x = [x_1, x_2, ..., x_M]$ with the conditional part of the rule R_i .

The local conclusion $Y'k^{Ri}$ of the rule R_i can be deduced from the veracity α_i of the condition part and from the "real" conclusion Y_k^{Ri}.

α_i is determined by applying a conjunction operator to the membership functions characterizing the individual premises. The appropriate fuzzy operators are chosen to simplify the implementation of a neural structure based on classical backpropagation. In practice, it is sufficient to choose operators that can be expressed by sigmoid functions operating on a weighted sum of the inputs. The α_i approximating the reference criterion C_i is obtained here by applying the AND logic operator, chosen as the Lukaviewicz operator:

$$\alpha_i = \bigwedge_{\substack{\text{Luka-}\\\text{viewicz}\\ j}} (\mu_{X'ji}(xj))$$

with:

$$\mu_A(x) \underset{\text{Lukaviewicz}}{\wedge} \mu_B(x) = \max (\mu_A(x) + \mu_B(x) -1 , 0)$$

whence, according to section 3.2,

$$\alpha_i = \bigwedge_{\substack{\text{Luka-}\\\text{viewicz}\\ j}} [(1 - \mu_{Rij}(x_j,y_i)) \underset{\substack{\text{Luka-}\\\text{viewicz}}}{\vee} \mu_{Rij}(x_j,y_i) \underset{\text{produit}}{\wedge} \mu_{Xji}(xj))]$$

The neural advantage is due to the self-adapting capacity of the network weights in new contextual situations, usually expressed in symbolic form.

4.An appropriate neuro-fuzzy structure

4.1 Possible architecture for fusion

A neural structure is used for learning classification rules in a supervised framework. Given M measurements delivered by observation sensors and N decision hypotheses, the proposed structure is equivalent to considering M x N elementary modules. Each module includes a layer consisting of M input cells and one neuron for introducing a bias, and one output layer reduced to a single neuron. Each of these networks, corresponding to the modeling of the conjunctive premise of a "local" fuzzy rule, is learned independently at first. These N outputs are totally connected to a final layer including as many cells as conclusion hypotheses H_i. The weights assimilated to confidence coefficients are refined by a global learning phase. Adding this additional layer furthermore endows the proposed structure with a character of generality, in that it allows the generation of more complex rules.

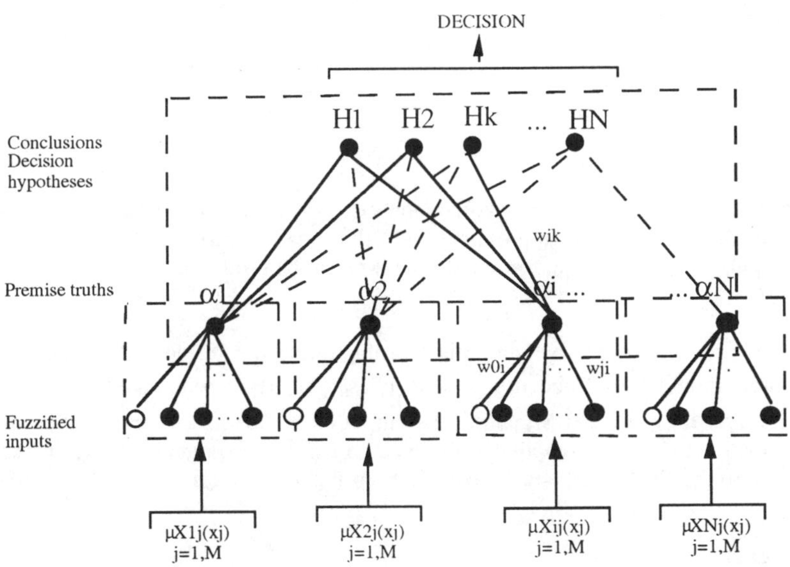

4.2 Practical introduction of fuzzy logic

In a problem of hypothesis evaluation, the fuzzy subsets on which the measurements are evaluated can be related directly to the hypotheses H_i of the frame of discernment. Let X_j be the domain of variation to which the input variable x_j (corresponding to measurement m_j) is assumed to belong. For N decision hypotheses, N fuzzy subsets X_{ji} (i = 1, ... N) constitute the fuzzy partition associated with the reference space X_j. The sets X_{ji} are none other than the definition intervals of the conditional distributions $P(m_j/H_i)$.

When lacking appropriate heuristics, we can obtain the membership functions μX_{ji}:xj->[0,1], j=1,M, i=1,N, by normalizing the frequency histograms of the measurements observed under each of the hypotheses (fuzzification stage).

4.3 Learning of rules with fuzzy premises

Once the fuzzification stage is completed, the following step is to grant a degree of truth to the R_i rules whose premise part is a conjunction of fuzzy propositions of the "x_j is X'_{ji} " type, represented by $\mu X'_{ji}(x_j)$, if x_j is a system input. This confidence is characterized by applying a conjunction operator based on Lukaviewicz logic.

The function retained (see section 3.3) can be likened to a nonlinear sigmoid type function applied to a weighted sum of inputs. Other operators may be chosen, as long as substantial modifications are made in the error gradient backpropagation algorithm.

Learning confidence coefficients

In the framework of the reference approach of section 2, the product operator applied to inputs of the form $(1 - d_{ij} + d_{ij}\, \mu X_{ji}(x_j))$, j=1,M where d_{ij} is a real belonging to [0, 1], characterizes the likelihood of the hypothesis H_i. A set of d_{ij} coefficients may be estimated from the synaptic weights relating inputs of the form ($\mu X_{ji}(x_j)$, j=1,M) to the cell α_i.
That is, using Lukaviewicz AND operator (section 3.3) yields:

$$\alpha_i = \max[\, 0,\ \sum_{j=1}^{M} \mu X'_{ji}(x_j) - (M-1)\,]$$

with

$$\mu X'_{ji}(x_j) = 1 - d_{ij} + d_{ij} \cdot \mu X_{ji}(x_j)$$

whence

$$\alpha_i = \max(0,\ \sum_j d_{ij}\, \mu X_{ji}(x_j) + d_{0i}\,)$$

with

$$d_{0i} = 1 - d_{1i} - d_{2i} - \dots - d_{Mi}$$

Let the w_{ji} be the synaptic weights relating the fuzzified inputs $\mu X_{ji}(x_j)$, j=1,M to neuron α_i, and w_{0i} be the bias. If the w_{ji} belong to [0, 1], the approximation of the function max (0, x) acting on the weighted sum of the inputs by an appropriate sigmoid function, can be used to approximate the d_{ij} coefficients.

The premise truths α_i are, at the end of an initial learning phase, quantified with respect to a conclusion $Y_i^{R_i}$, which should be refined in a second phase. To begin with, the subsystems consisting of the input layer $\mu X_{ji}(x_j)$, j=1,M, and of the output α_i are learned by minimizing the errors on the outputs relative to a given hypothesis, in this case H_i. The estimated weights w_{ji} are then refined with regard to the final conclusions deduced from the rule set generated in the initial phase.

4.4 Learning the conclusions

The number of output cells is assumed to be equal to the number of decision hypothesis considered. The intermediate layer comprising the α_i cells is totally connected to the output layer by connections whose weights w_{ik} (k = 1, ..., N) representing the "real" conclusions of the R_i rules.

The global conclusion Y'_k relative to hypothesis H_k resulting from the activation of the N rules R_i corresponds to the union of the partial conclusions:

$\mu_{Y'k}(yk) = \vee_{i} (\alpha_i \ wik)$

in which α_i qualifies the premise part of the R_i rule in terms of veracity.

A global learning phase using a sigmoid type activation function leads to a refining of the α_i, and consequently the w_{ji}, comparable to the d_{ij}.

4.5 Simulations

Rudimentary simulations are presented in the purpose to give an idea of the feasibility of a tool for learning coefficients that might reflect a degradation of the context. Although they lack realism concerning the nature of the contextual knowledge being managed, the experiments allow one to anticipate the potential of the method in the presence of *ad hoc* information.

The application framework considered is the discrimination between two hypotheses, H_1 and H_2, which might for example be representative of the identity of a target in a classification problem. The variability of the context is reflected in an artificial degradation of one or more sensors towards a reference situation. The examples retained for illustrative purposes come from [APPR91]. The configuration is limited to two sensors, S_1 and S_2, generating measurements m_j, j = 1, 2. The pre-learned distributions $P(m_j/H_i)$ preset for a reference situation are assumed to be normal laws:

$P(m_j/H_1) = N(0,1)$

$P(m_j/H_2) = N(M_{aj},1)$

The m_j measurements simulating the behavior of the sensors in a different contextual situation are generated in equal numbers according to laws supposed to reflect the context variation:

$P_r(m_j//H_1) = N(0,1)$

$P_r(m_j/H_2) = N(M_{rj},1)$

The neural structure considered consists of two elementary modules whose outputs are connected to a final layer consisting of two cells associated with hypotheses H_1 and H_2. A certain number of operating constraints can be applied during the learning phase. The estimated coefficients may integrate an idea of intrinsic sensor quality, in terms of performance. The sensor confusion matrices are said to be well-conditioned if they are assumed to be correctly fixed, i.e. typically when the diagonal terms representative of the aptitude of sensor S_j to recognize hypothesis H_i are maximum. When these individual confusion matrices are well-conditioned, the coefficients may, under certain constraints established during learning, be interpreted in terms of detection probabilities and false alarms.

For the different scenarios tested, the use of the d_{ij} coefficients, estimated by neuro-fuzzy learning, in a decision process using the reference analytical criterion, verifies the validity of the proposed approach.

The first configuration associates a sensor of good but uncertain quality ($M_{a1} = 6$; $M_{r1} = M$) with a sensor of lesser but more reliable quality ($M_{a2} = M_{r2} = 2$). Simultaneous learning of the confidence factors leads to satisfactory recognition performance. This is especially verified in the least favorable case where M = 0, which is a scenario where the maximum likelihood criterion yields a mean well-classified rate of 50%. The performance then exceeds 80%. Such a set of d_{ij} performs at least as well as when only the uncertainty affecting $P(m_1/H_2)$ is considered using the analytical approach, in the form of a coefficient $d_{21} = D$, such as D = 0.9. It should be noted that the choice of D is not critical because of the stability of the analytical performances observed when D is varied, with $d_{11} = d_{12} = d_{22} = 1$.

Another configuration simulated is a sensor S_1 combined with another sensor S_2 of the same but uncertain quality ($M_{a2} = 6$; $M_{r2} = 2$). Letting $d_{11} = d_{12} = 1$ and $d_{21} = d_{22} = 0.9$, the analytical simulations reported in [APPR91] show that the performance curve corresponding to a parametric variation of M results in a substantial gain over the usual probabilistic fusion in the least favorable cases (major error simultaneously on two sensors). For this configuration, a simultaneous optimization of the d_{ij} by the neural approach again achieves performances at least equal to those reported in [APPR91] in the least favorable configuration (M = 0), and even better in some other cases.

These few very simple simulations tend to bring out the advantage of a fine learning of the interest coefficients in terms of recognition performances. The results, within the bounds of the scenarios tested, show the aptitude of the proposed neuro-fuzzy approach to capture the effect of the environment in the form of parameters that can satisfactorily be used by the reference analytical criterion. The knowledge of the variation of the synaptic coefficients according to new contextual data allow one to adjust the useful fusion criteria.

Fine optimization of the quality factors may be justified as the number of sensors and uncertain data increases, or when the application requires special error conditions.

5.Conclusion

One of the purposes of the proposed approach is to manage contextual data, *a priori* knowledge expressed in symbolic form, in a numerical fusion context.

On a methodological level, the neuro-fuzzy approach proposed here in a supervised context is based on the neural implementation of a formalism stemming from uncertainty theories, thereby immediately providing it with the required adaptive learning capacity. The proposed solution is defined in a framework federating different ideas handled by uncertainty theories such as theory of evidence and fuzzy subset theory.

The implementation of this approach is based on the generation of learning memories corresponding to the various listed contextual situations. These memories are used in operational phase to define appropriate processing strategies in light of the available contextual knowledge.

REFERENCES

[APPR89] Appriou A., "Intérêt des théories de l'incertain en fusion de données", Conférence invitée, Colloque International sur le Radar, Paris, Avril 1989.

[APPR90] Appriou A., "Perspectives liées à la fusion de données", Science et défense, Paris, Mai 1990.

[APPR91] Appriou A., " Probabilités et incertitude en fusion de données multi-senseurs", Revue Scientifique et Technique de la Défense, no 11, 1er trimestre 1991.

[BOUC93] Bouchon-Meunier B., "La logique floue", Presses universitaires de France, 1993.

[DUBO85] Dubois D., Prade H., "Théorie des possibilités", Masson 1985.

[GLOR92] "Application of neuro-fuzzy networks to identification and control of nonlinear dynamical systems", Glorennec P.Y., Barret C., Brunet M.,IPMU92.

[LEE90] Lee C. C., " Fuzzy logic in control systems: fuzzy logic controller", IEEE Transactions on Systems, Man and Cybernetics, Vol. 20, no 2, 1990.

[PEDR89] Pedrycz W., "A fuzzy cognitive structure for pattern recognition", Pattern Recognition Letters 9, 1989.

[PEDR91] Pedrycz W., "Neurocomputations in Relational Systems", IEEE Transactions on Pattern analysis and machine intelligence, Vol 13, No 13, 1991.

[SHAF76] Shafer G., A mathematical theory of evidence, Princeton University Press, New-Jersey, 1976

[TERA91] Terano T., Asai K., Sugeno M., "Fuzzy systems theory", Academic Press, 1991.

LINEAR LOGIC FOR IMPRECISE FIRINGS IN OBJECT PETRI NETS

Janette CARDOSO*, Robert VALETTE**, Brigitte PRADIN-CHEZALVIEL**

* LCMI-EEL-UFSC,
88040-900 Florianópolis (SC), Brasil
e-mail janete@lcmi.ufsc.br

** LAAS-CNRS,
F-31077 Toulouse Cedex, France
e-mail {robert,chezalvi}@laas.fr

ABSTRACT

After a short introduction of a Linear logic based representation of Petri nets and of Object Petri nets we will propose a characterization of disjunctive sets of sequences using Linear additive connectives and study their relationships with Petri net imprecise markings. An example is considered: it will show pseudo-firings and how imprecision will be decreased on receiving precise information.

1. Introduction

Current work combining Petri nets and fuzzy sets is being developed along three lines:

- computing fuzzy sequences given an initial marking and a likelihood degree for each transition [5, 9],
- defining fuzzy markings based on a notion of fuzzy token location [3, 13],
- attaching fuzzy variables to tokens and conditions on these variables to the transitions which displace these tokens [2].

It seems clear that a comprehensive theory of fuzzy Petri nets should address these three aspects in a consistent way. Given a process being executed, what are the interactions between an ill-known knowledge of its behavior (fuzzy sequences), an ill-known knowledge of its current state (fuzzy marking) and an ill-known knowledge of its internal data (fuzzy variables)?

The objective of this paper is to try to address the consistency of the first interaction, i.e. the one which should exist between fuzzy markings and fuzzy firing sequences.

An important point is the underlying formal approach. In [5, 9], the computation of fuzzy sequences is done using first order logic and only acyclic unbounded Petri nets can be interpreted in this logic without trouble. On the opposite, in this kind of nets it is not possible to define a fuzzy marking as in [3, 13]. In fact, the definition of a fuzzy location for a token is only consistent when tokens denote object class instances and are, in consequence, individuals.

In [5, 9] a token is a mere pointer in a reasoning process. When a token is removed from a place, this does not mean that the corresponding proposition turns false, this means that the proposition is not currently being considered. A place which has been marked in the past can always be marked again without removing any other token. It is why a formal modeling of such cases implies unbounded Petri nets as in [10].

In the sequel of the paper we shall implicitly consider that we work in the context of object Petri nets [12] and therefore the first problem is to correctly characterize the firing sequences. Places are states and tokens are object instances or resources passing from states to states. We are consequently handling bounded cyclic Petri nets and using Linear logic formalism, because it is very well-suited for resources and sequences or sets of sequences [11].

We will use "*multiplicative connectives* " for sequences and "*additive connectives* " to express disjunctive sets of sequences or markings: such sets permit to characterize imprecise sequences or markings.

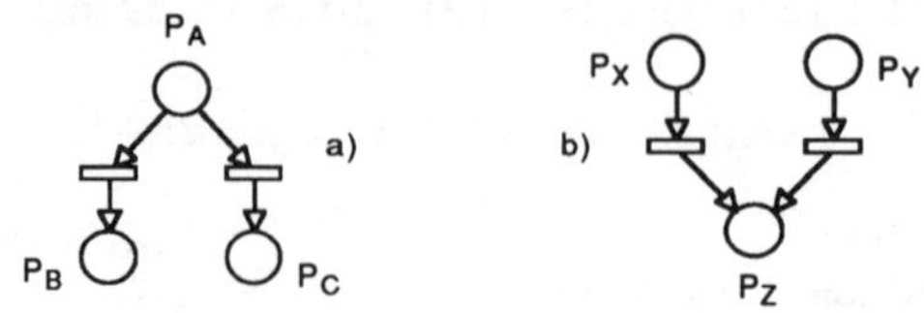

Figure 1: External choices

2. Linear Logic and Petri Nets

2.1. Linear logic: overview

Linear logic, proposed by Girard [7], is a "*resource conscious*" logic: it differs from classical logic because some usual rules are no more valid (contraction and weakening rules). As a consequence, Linear logic has 3 sets of connectives:

- the multiplicative ones : $\otimes$, $⅋$ and $\multimap$ that expresses causal dependency between resources,
- the additive ones: & and $\oplus$ which permit external and internal choices.
- the exponential ones: ! and ? .

Linear logic also manipulates negation $^{\perp}$. It does not express truth/falsehood properties, but rather consume/produce duality[11].

2.2. Petri nets description

Any Petri net can be described by a collection of Linear logic well-formed formulas: markings are described by consumable formulas and transitions by remanent ones [1, 8]. Formula $\vdash A \otimes D$ describes a current distribution of tokens in a Petri net, which may be a partial marking. A transition t with $pre(t) = \{A\}$ and $post(t) = \{B, C\}$ is not consumed by firing, so it is represented by the formula $\vdash !(A \multimap B \otimes C)$

2.3. Sequence characterization

A very important point to underline is that the Linear logic characterization of firing sequences is more accurate than the usual Petri net one [11]. A sequence characterization is derived from transitions ones by using the "*cut* " rule.

3. The additive connectives in Linear logic

In our Petri net interpretation of Linear logic, additive connectives & (with) and $\oplus$ (plus) are used to build disjunctive sets of firing sequences or of markings. Our purpose is to use them to characterize imprecise firing sequences or imprecise markings. One of the sequences (or markings) belonging to the set is (or will be) actual but we do not exactly know which one.

Linear logic differentiates "*internal*" and "*external*" choices. In the first case, the choice of the sequence to be fired is internal to the system, independently from its environment. In the second case, the choice of the sequence to be fired is free and can be made by the system environment.

3.1. External choice

There are two types of external choices: they are represented in figure 1 and they are described by two different Linear logic formulas.

a) With the Petri net in figure 1.a by consuming resource A, it is possible to produce either resource B or resource C, but not both simultaneously. A choice is required and remains free. This formula is obtained by applying the $\&R$ rule (see annex) to the two transitions formulas:

$$\frac{\vdash\ !(A \multimap B) \quad \vdash\ !(A \multimap C)}{\vdash\ !(A \multimap B \& C)}\ \&\mathrm{R} \tag{1}$$

It is a Linear logic representation of the whole net template. Actually, this formula characterizes a set of sequences which exactly require the same minimal initial marking and produce alternative final markings. As a consequence, it can be considered as an imprecise sequence which transforms a precise marking $\vdash A$ into an imprecise one $\vdash (B\&C)$. Either we have a token in place P_B or in place P_C. It is important to point out that from one token in P_A we can freely choose the firing sequence and therefore the eventual marking. We can say it is an "*externally chosen imprecise marking*" or a "*freely produced imprecise marking*".

b) Let us now consider the Petri net in figure 1.b. It describes that it is possible to produce resource Z by consuming either resource X or resource Y. Whatever the initial marking, a token in P_X or (exclusively) in P_Y, it is possible to find a sequence leading to a token in P_Z. The Linear logic description is obtained by the $\oplus L$ rule:

$$\frac{\vdash\ !(X \multimap Z) \quad \vdash\ !(Y \multimap Z)}{\vdash\ !(X \oplus Y \multimap Z)}\ \oplus\mathrm{L} \tag{2}$$

This formula describes a set of firing sequences as a whole but unlike in the previous example, there is no conflict; no choice is made by the token player. Once an initial marking has been chosen, the token player proceeds forward without making any decision. It is another example of "*external choice*" in Linear logic. Indeed, we are free to execute any sequence of the set (but only one because the set is disjunctive) by choosing the correct initial marking. Whatever the initial one, the final marking is the same. Formula $\vdash (X \oplus Y)$ denotes a "*freely consumed imprecise marking*".

3.2. Internal choice

As for external choice, there are two types of internal choices represented on figure 2.

a) The first one can be seen as the result of applying rule $\oplus$R to a transition formula:

$$\frac{\vdash\ !(A \multimap B)}{\vdash\ !(A \multimap B \oplus C)}\ \oplus\mathrm{R} \tag{3}$$

It is described by figure 2 a. If there is a firing sequence leading from place P_A to P_B, then we can say that there exists at least one of the following firing sequences: one from P_A to P_B or one from P_A to P_C. We have introduced imprecision (or uncertainty) in our knowledge because we no longer know which of the two sequences actually exists. We interpret such a formula in the following manner: the Petri net model describes the normal behavior of the system whereas the formula takes into account some bad functioning that is not described (B would be the normal state and C the abnormal one). As for external choice, an imprecise marking $\vdash (B \oplus C)$ is obtained by firing such a sequence, from a marking $\vdash A$.

It clearly differs from the case in figure 1.a. Only one firing sequence is consistent with the Petri net model which was built before imprecision was introduced. From a token in place

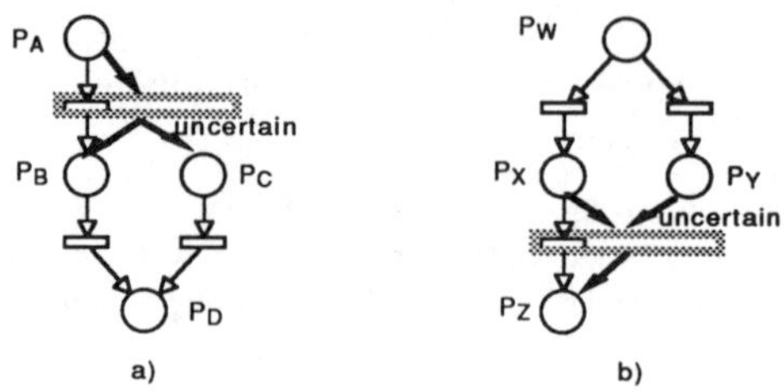

Figure 2: Internal choices

P_A we usually obtain a token in P_B but obtaining a token in P_C is also possible. We can say it is an "*internally chosen imprecise marking*" or an "*imposed produced imprecise marking*".

b) Similarly, by means of rule &L, we derive another type of internal choice:

$$\frac{\vdash\ !(X \multimap Z)}{\vdash\ !(X\&Y \multimap Z)}\ \&\mathrm{L} \tag{4}$$

This formula means that there exists at least one firing sequence which leads from P_X to P_Z or from P_Y to P_Z (see figure 2 b). As we do not know which one is correct, we work with imprecision or uncertainty. In this context, $\vdash (X\&Y)$ is an "*imposed consumed imprecise marking*". We will interpret it in next sections.

Such situations are called "*internal choices*" because in case of uncertainty, you cannot know exactly what the behavior of the system would be. In a way, you can consider that the system autonomously chooses its behavior. So it is a choice within the system (a failure for instance) and not a decision freely made by the human operator or by the system environment.

3.3. Decreasing uncertainty

Depending on the type of uncertainty considered, there will be some rules for decreasing it.

a) Consider first the net on figure 2.a. We previously saw that the Linear logic formula represents the normal behavior (B) and the abnormal one, when there is some failure (C). Such a formula will be useful if we have a way of decreasing this imprecision (recovery after failure), for example a place P_D such that:

$$\vdash\ !(B \oplus C \multimap D)$$

This formula is an example of the second type of external choice. Not surprisingly we note that the condition part of this formula is equal to the consequent part of formula 3 and that the uncertainty will disappear by applying a cut rule. From P_A we are certain that a firing sequence leads to P_D. We can prove that, whatever the behavior (normal or abnormal), place P_D is eventually reached. It will be very useful for verifying if sufficient failure recovery procedures are defined in the system [4].

b) Consider now the net on figure 2.b. Formula 4 means that there exists at least one sequence leading to Z. This makes sense if we are able to keep this uncertainty under control. For example, if P_X and P_Y have a common ancestor (P_W in figure 2.b). The view of the model is dual of the former one. We explicit both normal and abnormal behavior (P_X and P_Y) but we just know that the recovery mechanism exists, without detail.

3.4. Conclusion

We have just seen in this section that Linear logic offers a good formalism to define imprecise firing sequences (we have a set of possible sequences without a preference ordering i.e. all are equally possible) which is the first step before the introduction of fuzzy sequences (among the set of possible sequences some of them are more likely than the others).

Another major point is that it is possible to decrease uncertainty when certain information is received, for example. Now we are going to compare this approach to that of imprecise markings as defined in [3, 13].

4. Petri Nets with imprecise markings

4.1. Imprecise markings

First of all, let us remind that in order to define imprecise markings in a Petri net it is required to work with Object Petri nets as defined in [12]. The chief feature of Petri nets with objects is that a token is a tuple of unique instances of some classes (or sub-classes). In this way it is impossible to have two identical tokens in a Petri net (in case of a precisely known marking, of course).

Traditionally, the marking of a Petri net is defined as the mapping M of the set of places P to the set $\mathcal{N}$ of natural numbers, $M : P \longrightarrow \mathcal{N}$. If O is the set of objects and O^* the set of tuples of objects, the marking M of a PNO can be represented by:

$$M : O^* \text{ x } P_\phi \longrightarrow \{0, 1\}.$$

where P_ϕ is the set of places P augmented by a place ϕ which is supposed to contain all the tuples of tokens not yet built or already destroyed.

Given a place p, if $M(o^*, p) = 1$ and if for all the places q such that $q \neq p$ $M(o^*, q) = 0$, then the tuple of object instances o^* is certainly in the place p.

Given two places p_1 and p_2, if

$$M(o^*, p_1) = M(o^*, p_2) = 1$$

then we have an imprecise information. We are certain of the existence of the tuple of object instances o^* but its location is imprecise (either place p_1 or p_2).

Let us assume that the token o^* is the only one for which the location is imprecise. This means that we are considering two possible markings M_1 and M_2. They are identical for all the tokens excepted o^*. For M_1 o^* is in place p_1 and for M_2 o^* is in place p_2.

Considering now the Linear logic syntax, we have either

$$\vdash M_1 \& M_2$$

that expresses that we have an ill-known state because of imprecise information about the system state, or

$$\vdash M_1 \oplus M_2$$

that expresses that we have an ill-known state because of a bad functioning (internal choice). In the first case, both behaviors are described by the Petri net model but we don't know the actual one. In the second case, only one behavior is modeled by the Petri net and we consider consequences of some unmodeled sequence.

4.2. Pseudo firing

In [3, 13], the concept of pseudo-firing transitions has been introduced in order to increase imprecision (i.e. for example to pass from a precise marking to an imprecise one). When we have an imprecise information about an event occurrence (we do not know if it has actually occurred or not), we want to take into account both markings:

- the marking M existing before the firing of t, and

- the marking M', with $M \xrightarrow{t} M'$, reached after the firing of transition t corresponding to the event.

It has been proposed to put the tokens into the output places of the transition without deleting them from its input places. This is the pseudo-firing and can be considered as the beginning of a firing.

The pseudo-firing of t from a marking M leads to an imprecise marking $\mathcal{M} = \{M, M'\}$, which is a disjunctive set of markings. To be consistent with the notation, we write $M \xrightarrow{\lambda} M$, where λ is the empty sequence. This means that no firing was done. We use the notation $M \xrightarrow{\{\lambda,t\}} \mathcal{M}$ to describe the fact that we consider the disjunctive set of firing sequences $\{\lambda, t\}$.

This kind of firing does not correspond to an evolution of the system modeled by the Petri net, it rather means that we are making some *inference* about its state (for instance for some monitoring purpose). It must be pointed out that the imprecision concerns only some objects and that the location of the others is known with certainty. This means that the part of the manufacturing shop which is not altered by the abnormal operation operates normally.

With a Linear logic notation, doubts about firing a transition are expressed by an application of the $\&R$ rule. Let us consider a transition t_1 with input place P_A and output place P_B, and a transition t_2 with input place P_B and output place P_C. We want to characterize the imprecise sequence defined by the two following alternatives: t_i has been fired or t_i has not (*firing* of empty sequence λ).

$$\frac{\vdash !(A \overset{\lambda}{\multimap} A) \quad \vdash !(A \overset{t_1}{\multimap} B)}{\vdash !(A \overset{\{\lambda,t_1\}}{\multimap} A\&B)}\ \&\text{R} \tag{5}$$

Label $\{\lambda, t_1\}$ on $\multimap$ indicates that we have a disjunctive set of firing sequences. Either the firing of λ or (exclusive) t_1 occurred, but not both. After this first step, P_A and P_B are marked (imprecise marking), and t_2 also can be pseudo-fired.

To characterize the imprecise sequence $\{\lambda, t_1, t_1t_2\}$ we apply $\&R$:

$$\frac{\vdash A \overset{\{\lambda,t_1\}}{\multimap} A\&B \quad \vdash A \overset{t_1t_2}{\multimap} C}{\vdash A \overset{\{\lambda,t_1,t_1t_2\}}{\multimap} A\&B\&C}\ \&\text{R} \tag{6}$$

The imprecise marking reached is obtained by:

$$\frac{\vdash A \quad \vdash A \overset{\{\lambda,t_1,t_1t_2\}}{\multimap} A\&B\&C}{\vdash A\&B\&C}\ \text{cut} \tag{7}$$

4.3. Recovery

Imprecision cannot increase indefinitely. At some time point we have to receive a message containing certain information as for example: transition t' has been fired. In the pure Petri net based approach [3], we have to elaborate a firing sequence leading from the last well-known marking to a possible marking enabling transition t'. The algorithm used to do this, eliminates from the imprecise marking (i.e. the set of possible markings) all the markings which are not such that transition t' is enabled before firing it. All "Pseudo firings" are considered and are either canceled (take the tokens out of the output places) or terminated (take the tokens out of the input places).

What about the Linear logic approach? Let us consider figure 2.b. By "*pseudo-firing*" the two conflicting transitions we derive:

$$\vdash !(W \overset{\{\lambda,t_1,t_2\}}{\multimap} W\&X\&Y) \tag{8}$$

This formula characterizes an imprecise firing sequence consisting of no transition fired ("*firing*" of λ) or transition t_1 fired or transition t_2 fired. The current state is an imprecise one: $\vdash W\&X\&Y$.

If we receive the information that transition t_1 has been fired, (it is consistent with our knowledge), we have to restore our knowledge about the system state. In order to do this, we have to apply $\&L$ rule to t_1: we have to "*merge*" this certain information into an imprecise context.

$$\frac{\vdash !(W \overset{t_1}{\multimap} X)}{\vdash !(W\&X\&Y \overset{t_1*}{\multimap} X)}\ \&\mathrm{L} \tag{9}$$

By applying rule $\&L$, we can derive that λ and t_2 were not fired because it was necessary to "merge" X with W (the output set of λ) and Y (the output set of t_2). Doing this merging in the imprecise context, we derive now a precise information (there is no more & connective):

$$\frac{\vdash W\&X\&Y \quad \vdash !(W\&X\&Y \overset{t_1*}{\multimap} X)}{\vdash X}\ \mathrm{cut} \tag{10}$$

So, we see that rule $\&R$ expresses disjunctive sets and permits to introduce imprecision while rule $\&L$ consists in decreasing this uncertainty and deduce what was the executed sequence.

5. Example

5.1. Description

Let us now consider the example of a transport system based on automated guided vehicles (AGV) as in [3]. We may proceed with Linear logic exactly in the same way as with Petri nets with imprecise markings. The Petri net in figure 3 depicts all the possible behaviors of the system when it is automatically controlled but the failure recovery procedures are not described.

These vehicles follow automatically some circuits, their locations are only known at some points that are called *contacts*. They can only receive a command (stop, go, itinerary change) when they are on such a contact. In order to avoid collision, the circuits are decomposed into sections and the vehicles have to be controlled in such a manner that at a given time each section can only contain one vehicle.

The Petri net in figure 3 describes the coordination of such a transport system: transitions a, b and the places MS (*vehicle moving along a section*), C (*vehicle stopped on a contact*) and S (*section free*) represents the normal system operation. The variable v represents a vehicle, s is some section and ns is the next section. Place MS contains the tuples $< v, s >$ such that $< v >$ moves along $< s >$. Place C contains the vehicles $< v >$ on the contact of section $< s >$. Place S contains the free sections $< s >$ and $< ns >$ (the next section). Place MOS corresponds to the movement out of section and place BL corresponds to the battery loading station. From this station it is possible to re-introduce a vehicle into the section where it was running (this failure recovery procedure is not described here). Linear logic description of this transport system is:

$$a :\vdash !(MS_{v,s} \multimap C_{v,s}) \quad b :\vdash !(C_{v,s} \otimes S_{ns} \multimap MS_{v,ns} \otimes S_s)$$
$$f :\vdash !(MOS_v \multimap BL_v) \quad d :\vdash !(MS_{v,s} \multimap MOS_v \otimes S_s)$$

for the transitions and,

$$\vdash (MS_{v_1,s_1} \otimes S_{s_2})$$

for initial marking (with two sections s_1 and s_2 and a vehicle v_1).

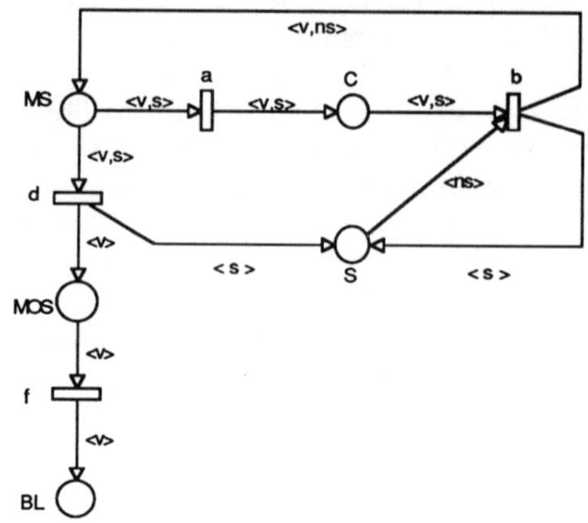

Figure 3: Transport system

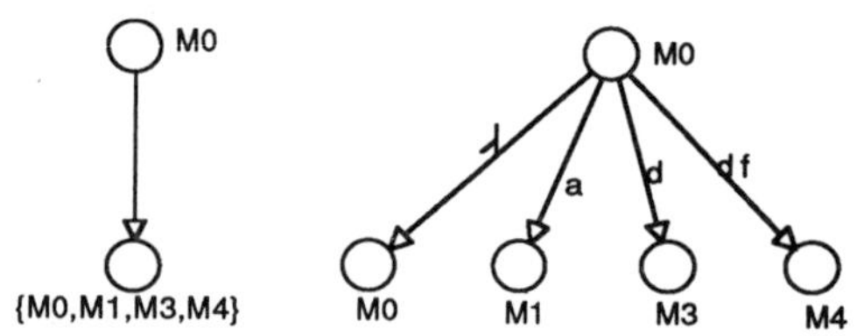

Figure 4: Disjunctive set of sequences

5.2. The disjunctive set of firing sequences

Let us suppose that vehicle v_1 is in section s_1 (token (v_1, s_1) in place MS). The next section is free (token s_2 in place S). The sequent $\vdash MS_{v_1 s_1} \otimes S_{s_2}$ represents this marking (M_0 in figure 4). A fragment of the reachable marking graph is depicted in figure 4 :

$$M_0 = MS_{v_1,s_1} \otimes S_{s_2} \quad M_3 = MOS_{v_1} \otimes S_{s_1} \otimes S_{s_2}$$
$$M_1 = C_{v_1,s_1} \otimes S_{s_2} \quad M_4 = BL_{v_1} \otimes S_{s_1} \otimes S_{s_2}$$

Sometimes the vehicle stops because it has detected something in front of it, for example a human operator crossing the shop floor. It can be manually reset and will eventually reach the next contact (firing of a). Sometimes the battery is discharged and the vehicle has to be driven manually to the battery loading station (firing of d and f). The transport system coordination has to tolerate the two situations and release the concerned section when it is certain that the vehicle has been driven away.

So, after a time-out, if the vehicle has not reached the contact, transitions a, d and f are pseudo-fired (see [3]). The pseudo-firing of a, d and f from M_0 leads to the imprecise marking $\{M_0, M_1, M_3, M_4\}$. The imprecise marking $\mathcal{M}$ is a connected disjunctive set of markings, where each marking is added after a pseudo-firing. Only one of these markings corresponds to the actual state, but we do not know the actual state of the system (the marking). So we have a disjunctive set of sequences from the initial marking (known) to the imprecise marking $\mathcal{M}$.

5.3. Using Linear logic

The disjunctive set of sequences is $\{\lambda, a, d, df\}$. This imprecise sequence is represented in Linear Logic by a formula obtained from the rule $\&R$.

$$\vdash !(M_0 \overset{\{\lambda,a,d,df\}}{\multimap} M_0\&M_1\&M_3\&M_4) \tag{11}$$

Uncertainty will decrease when some certain information will be received. At this moment, the only certain information is M_0.We are considering two cases:

a) Let us suppose we receive information about the firing of transition d. The intersection of the imprecise sequence and d is $\{d, df\}$. With Linear logic, merging d with the imprecise knowledge context results in:

$$\vdash !(M_0\&M_1\&M_3\&M_4 \overset{d*}{\multimap} M_3) \tag{12}$$

Merging df in the imprecise context results in:

$$\vdash !(M_0\&M_1\&M_3\&M_4 \overset{df*}{\multimap} M_4) \tag{13}$$

From these two formulas 12 and 13, we derive:

$$\frac{\vdash M_0\&...\&M_4 \overset{d*}{\multimap} M_3 \quad \vdash M_0\&...\&M_4 \overset{df*}{\multimap} M_4}{\vdash M_0\&M_1\&M_3\&M_4 \overset{\{d*,df*\}}{\multimap} M_3\&M_4}\ \&\mathrm{R} \tag{14}$$

and then, by a "*cut*" with 14 and the imprecise marking $\vdash M_0\&M_1\&M_3\&M_4$, we derive $\vdash M_3\&M_4$. Uncertainty was decreased but we still characterize an imprecise sequence $\{d, df\}$ because the firing of d does not permit to derive which precise marking has been reached.

b) Let us suppose we receive information about the firing of transition f. After merging, we derive:

$$\frac{\vdash M_0\&M_1\&M_3\&M_4 \quad \vdash M_0\&M_1\&M_3\&M_4 \overset{f*}{\multimap} M_4}{\vdash M_4}\ \mathrm{cut}$$

In this case, derivation is simpler: we immediately know the executed sequence df (because only df contains f) and derive the precise marking.

6. Conclusion

We have seen that by using the additive connectives we can characterize imprecise firing sequences in Linear logic. This concept is consistent with that of imprecise markings as defined in [3]. Indeed, given a (precise) marking, firing an imprecise sequence of transitions results in an imprecise marking. It is also an imprecise firing sequence which can produce the reverse transformation: i.e. which allows to pass from an imprecise marking to a precise one.

What would be the advantage of working with the two formalisms? In Linear logic we explicitly keep the set of possible markings characterizing an imprecise marking. It might be cumbersome sometimes. In addition, the token view is absent. The object oriented view of a token denoting a dynamic relation among a tuple of object instances is out of the scope of the Linear logic formalism, and it is not possible to respond to a query such as: what is the location (i.e. state) of the object instance o. As a consequence for any reasoning purely based on the markings the Petri net model is richer.

On the other hand, Linear logic offers a clear characterization of imprecise sequences of transitions. In addition, the notion of internal and external events is rich and important. This notion exists in many approaches such as LOTOS but is not present in Petri nets (although labeling functions might be used).

What is common to the two approaches (imprecise markings and imprecise sequences) is that they are two attempts to cope with flexibility and exceptions. Manufacturing as well as workflow systems are socio-technical systems and Petri nets may be too rigid in some cases [6]. The purpose of our work is to introduce some possibility of relaxing the sequential constraints defined by the Petri net in some off normal situations.

REFERENCES

[1] C. Brown: Relating Petri Nets to Formulas of Linear Logic, *Edimburg Tech. Report ECS-LFCS-89-87*, June,1989

[2] T. Cao, A. C. Sanderson: Sensor-based Error recovery for robotic task sequences using Fuzzy Petri nets, *IEEE International Conference on Robotics and Automation*, Nice, France, May 1992.

[3] J. Cardoso, R. Valette, D. Dubois: Petri nets with uncertain markings, in *Lecture Notes in Computer Science "Advances in Petri nets 1990"*, Vol.483, Ed. G. Rozenberg, Springer Verlag, 1991, pp.64-78.

[4] J. Cardoso, R. Valette, B. Pradin-Chézalviel: Fuzzy Petri nets and Linear logic, *IEEE/SMC Int. Conf. on Systems, Man and Cybernetics: Systems Engineering in the Service of Humans*, Le Touquet, France, Oct 17-20 1993.

[5] Shyi-Ming Chen, Jyh-Sheng Ke, Jin-Fu Chang: Knowledge representation using fuzzy Petri nets, *IEEE Trans. on Knowledge and Data Engineering*, Vol 2, N° 3, p. 311-319, 1990.

[6] Clarence A. Ellis, Gary J. Nutt: Modeling and enactment of workflow systems, *Application and Theory of Petri Nets 1993*, Lecture Notes in Computer Science 691, Springer Verlag, 1993, pp.1-16.

[7] J.Y. Girard: Linear Logic; *Theoretical Computer Science*, 50, 1987.

[8] C. Gunter, V. Gehlot: Nets as tensor theories, *10th International Conference on Application and Theory of Petri nets*, Bonn, Germany, 1989.

[9] C. G. Looney: Fuzzy Petri nets for rule-based Decision making *IEEE Transaction on Systems , Man and Cybernetics* Jan-Feb 1988, Vol 18, N 1, pp 178-183.

[10] T. Murata, D. Zhang: A Predicate-Transition net model for parallel interpretation of logic programs, *IEEE Transaction on SoftwareEngineering*, Vol. 14, No. 1, (April 1988).

[11] B. Pradin-Chézalviel, R. Valette: Petri nets and Linear logic for process oriented diagnosis, *IEEE/SMC Int. Conf. on Systems, Man and Cybernetics: Systems Engineering in the Service of Humans*, Le Touquet, France, Oct 17-20 1993.

[12] C. Sibertin-Blanc: High-level Petri nets with data structures, 6th *European Workshop on Application and Theory of Petri nets*, Helsinki, Finland, (June 1985).

[13] R. Valette, J. Cardoso, D. Dubois: Monitoring Manufacturing Systems by means of Petri Nets with Imprecise Markings *IEEE International Symposium on Intelligent Control* 1989, 25-26 sept. Albany N.Y., USA.

ANNEX: STRUCTURAL RULES for ADDITIVE CONNECTIVES

$$\frac{\Gamma \vdash A, \Delta \quad \Gamma \vdash B, \Delta}{\Gamma \vdash A \& B, \Delta} \&R \qquad \frac{\Gamma, A \vdash \Delta}{\Gamma, A \& B \vdash \Delta} \&L$$

$$\frac{\Gamma \vdash A, \Delta}{\Gamma \vdash A \oplus B, \Delta} \oplus R \qquad \frac{\Gamma, A \vdash \Delta \quad \Gamma, B \vdash \Delta}{\Gamma, A \oplus B \vdash \Delta} \oplus L$$

Acknowlegement: This work has been done in the framework of PRC-GDR IA, the French Research Programme for Artificial Intelligence, sponsored by the Ministery for Research and by CNRS.

HIGH LEVEL FUZZY PETRI NETS AND BACKWARD REASONING

Heloisa SCARPELLI*, Fernando GOMIDE**

* UFSCar/CCT/DC
Cx. Postal 676 - CEP 13565-905
São Carlos - SP - Brazil

** Unicamp/FEE/DCA
Cx. Postal 6101 - CEP 13081-970
Campinas - SP - Brazil

ABSTRACT

In this paper we introduce an automated procedure for extracting information from knowledge bases that contain fuzzy production rules. A backward reasoning algorithm based on the High Level Fuzzy Petri Net model is developed. The algorithm consists of the extraction of a subnet and an evaluation process. We informally verify that the proposed algorithm is similar to another procedure suggested earlier by Yager, with advantages concerning the efficiency in knowledge base information search.

1. Introduction

The ordinary and high level Petri nets [11] have been proposed as a knowledge representation scheme through several approaches. Some researches focus on the relation between Petri nets elements and the implication statements in ordinary logic. In [7], a very interesting approach can be found, where Giordana and Saitta propose the use of Predicate/Transition (Pr-T) nets to represent rules in a production system, considering binary predicates. Murata and co-workers present [10] [12] a model, based on the same type of high level net, to represent logic programs consisting of Horn clauses.

A representation in the form of Petri Nets was proposed in [9], to allow for fuzzy rule based reasoning as worked out in the framework of propositional logic. In a later work [4], the representation of fuzzy production rules with certainty factors was considered. A reasoning algorithm based on the certainty factors approach was proposed and the truth value of each proposition is represented by a single value between zero and one. A refinement of the FPN was developed by Chung and Bien [5] in a very interesting work where the algebraic form of a state equation of the FPN can be systematically derived, using a matrix representation that has not been established in any one of the previous works. Both forward and backward reasoning are performed by using the state equations. Another different representational model of fuzzy production systems based on Petri Nets, which attains a high level of expressive power was presented in [1]. Other approaches can be found in [2], [3], [14].

The High Level Fuzzy Petri Net (HLFPN) model, introduced by the authors in [15] and [19], as opposite to other approaches found in literature, derives from High Level Petri Nets such as Pr-T Nets [6] and colored Petri Nets [8]. The HLFPN is related to a detailed model proposed by the authors [17] [18], in the same way as High Level Nets are related to ordinary Petri Nets.

In the construction of fuzzy systems and in general knowledge based systems, a problem of considerable importance is the systems ability to answer questions. In order to provide a means for manipulating information when a query is posed to a fuzzy knowledge base, Yager has suggested [21] an automated procedure for questioning large knowledge bases.

Here we present an structured approach for the solicitation of information concerning a property V of a knowledge base represented in the HLFPN model. The proposed algorithm is performed in two phases. It uses a backward chaining strategy in the first phase and a forward chaining strategy in the second. The main advantage of our method is that all the knowledge base(KB) is organized in a net structure. As a consequence, the partial sources of information concerning a particular variable are readily found, requiring only direct accesses to the incidence matrix of the associated Petri net of the HLFPN.

In the next section we briefly review some essentials of the theory of approximate reasoning and fuzzy inference methods. Section 3 contains a short description of the HLFPN and the techniques to model the basic inference patterns. The reasoning algorithm is presented in section 4, together with an example which illustrates both phases of the procedure. In the section 5, conclusions and future work are addressed.

2. Fuzzy Inference Methods

Assume V is a variable with values in the set X. The set X is called the base set of V. The value of a variable V with base set X can be expressed as a fuzzy subset A of the base set X in a data proposition of the form V *is* A. Data propositions are also called *facts.*

The implication statement

$$\text{IF } V \text{ is } A \text{ THEN } U \text{ is } B$$

where A and B are fuzzy subsets of X and Y respectively, V is a variable with base set X and U is a variable with base set Y, induces a fuzzy relation over $X \times Y$, conventionally defined in Zadeh's theory [23] as $F_{B/A}(x,y) = min[1, 1 - A(x) + B(y)]$

The implication statements are also called *fuzzy production rules.* The equation used to evaluate the fuzzy relation induced by an implication statement is called *implication relation.* In general, the implication relation can be given in terms of T-norms and T-conorms, by extending any of the conventional implication methods [13]. We then denote the fuzzy relation F induced by the implication IF V is A THEN U is B as $F_{A\to B}(x,y) = f_{\to}(A(x), B(y))$.

The most used inference rule of approximate reasoning is the *fuzzy compositional inference.* An important instance of this rule occurs when we have the following type of information:

$$\begin{array}{l} V \text{ is } A' \\ \underline{\text{IF } V \text{ is } A \text{ THEN } U \text{ is } B} \\ U \text{ is } B' \end{array} \tag{1}$$

In this case the inference is called *generalized modus ponens* and allows to infer the proposition U is B', where $B' = A' \bullet F_{A\to B}$, such that

$$B'(y) = sup_x \mathbf{T}\ (A'(x), F_{A\to B}(x,y)) \tag{2}$$

where the symbol $\mathbf{T}$ denotes a T-norm.

3. Fuzzy Production Rules and the HLFPN Model

The basic concepts concerning HLFPN are provided bellow. For details, see [15] and [19]. Formally, a HLFPN consists of:

1. A bipartite directed graph, defined by a triple $(P, T, \mathcal{F})$ such that P and T are two disjoint sets of vertices ($P \cap T = \emptyset$) called places and transitions respectively, and $\mathcal{F}$ is a set of directed arcs, each one connecting a place $p \in P$ to a transition $t \in T$ or vice-versa ($\mathcal{F} \subset P \times T \cup T \times P$).

2. A labeling of the arcs with formal sums of t-uples of variables;

3. A structure Σ, defining a collection of typed objects together with some operations and relations applicable to them. Formulae built up in Σ can be used as *inscriptions* inside some transitions.

Each place may or may not contain a *token*. In HLFPN, the value carried by the tokens is a structured object representing a fuzzy subset. A token $\langle a'_1, a'_2, \cdots, a'_n \rangle$ in a place $p \in P$ denotes the fact that the variable V on base set $X = \{x_1, \cdots, x_n\}$, corresponding to that place has as its value the fuzzy subset A', such that $A'(x_i) = a'_i, x_i \in X$.

A distribution of tokens over the places of a HLFPN is called a *fuzzy marking M of the network.*

For each transition $t \in T$ and each place $p \in P$, the sets $I(t) = \{p \in P \mid (p,t) \in \mathcal{F}\}$ and $O(t) = \{p \in P \mid (t,p) \in \mathcal{F}\}$ are called the sets of the *input* and *output* places of the transition t, respectively.

A transition $t \in T$ is enabled whenever each place $p \in I(t)$ contains at least one token. An enabled transition t can be fired by removing from each place $p \in I(t)$ a set of tokens as specified by the label on the arc (p,t) and by adding to each place $p' \in O(t)$ a token specified by the label on the arc (t, p'). The token value is evaluated by the inscribed formulae in t. The lack of any formula inscribed in a transition means that the token value in p' remains the same as in p. The inscriptions inside transitions are expressions that perform compositional modus ponens, conjunction of multiple antecedents, aggregation of parallel rules [15] and partial sources of information [16].

The rule and fact in (1) where A and A' are fuzzy subsets of base set $X = \{x_1, \cdots, x_n\}$, B is a fuzzy subset of base set $Y = \{y_1, \cdots, y_m\}$ and the induced fuzzy relation $F_{A \to B}$ is denoted by $F_{A \to B}(x_i, y_j) = f_{ij}$, can be modeled as shown in figure 1(a). The fact V is A' is represented by putting a token $\langle a'_1, a'_2, \cdots, a'_n \rangle$ in the place V. The conclusion U is B' is reached by firing t. The formulae inscribed in t in figure 1 are the expressions used to calculate $B' = A' \bullet F$ as defined in (2). The marking yielded by firing transition t is shown in figure 1(b), where each b'_i means the membership value of y_i in B'.

The representation of other forms of fuzzy production rules such as rules with multiple antecedents, rules with quantifiers in the antecedent, rules with certainty factors, parallel rules and conflicting rules have been addressed in [15] and [18]. The modeling techniques to represent knowledge base portions that provide partial sources of information of the type considered by Yager [21] have been discussed in [16].

When we have more than one data proposition related to the variable V, we represent them in HLFPN placing one token for each proposition in place V. The place is denoted by a double circle and, during the net evaluation, is replaced by a subnet that performs conjunction of token values. A detailed example is shown is section 4.

The HLFPN model has also been shown to be useful in the field of knowledge based verification [20].

4. Backward Reasoning Algorithm

Two types of fuzzy reasoning algorithms can be defined based on the HLFPN model, to answer different kinds of questions: a forward reasoning algorithm and a backward reasoning

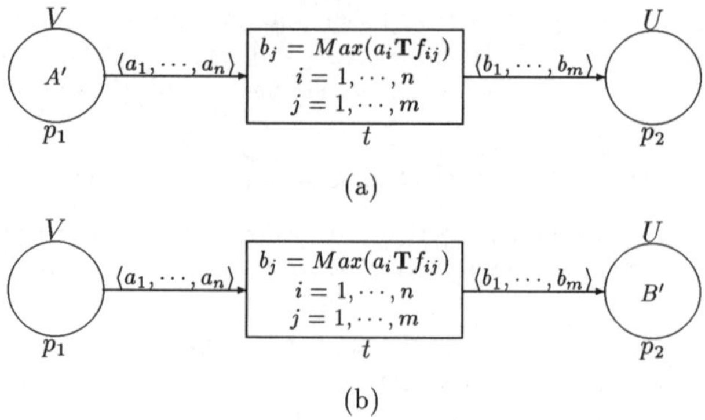

Figure 1: High Level Fuzzy Petri Net for a conditional rule.(a) Before transition firing, where $A' = \langle a'_1, a'_2, \cdots, a'_n \rangle$. (b) After transition firing, where $B' = \langle b'_1, b'_2, \cdots, b'_m \rangle$.

algorithm. The forward reasoning algorithm has been defined in [18]. We describe now an algorithm which uses a backward reasoning strategy. It should be used to answer questions asking for the value of a particular variable, which we denote by *V is ?*. Yager [21] proposed an organized approach for the solicitation of information concerning a property V from a KB, which involves the construction of a tree structure relating all the partial information about the desired variable and the evaluation of the tree structure. The fuzzy reasoning algorithm presented here is also executed in two phases. However, in the first phase, instead of building a knowledge tree, we extract a subnet from the HLFPN, corresponding to all the partial information about the desired variable. In the second phase, the subnet is evaluated. We note that the algorithm introduced here is similar to the procedure presented by Yager, with the advantage that the KB is already structured.

The concept of incidence matrix C^* of the associated Petri net of a HLFPN is used, to find out which places and transitions are connected to the goal place.

Given a HLFPN $\mathcal{N}$, we call the *associated Petri net* (APN) of $\mathcal{N}$, the ordinary Petri net $\mathcal{N}^*$ obtained from $\mathcal{N}$ by means of the following substitutions:

1. The labels of arcs in $\mathcal{N}^*$ are set to 1.
2. The tokens in the places lose their identity and become ordinary anonymous tokens.
3. The inscribed formulae are ignored.

The incidence matrix of $\mathcal{N}^*$ is denoted by C^*.

The algorithm uses two queue structures. The expression $x{:=}$ *delete(Q)* represents the operation of deleting the first element of queue Q and setting it to x. The expression $insert(x, Q)$ represents the operation of inserting x as the last element of the queue Q.

Let us suppose that a KB is modeled in a HLFPN $\mathcal{N}$ and that the question V is ? has been asked. The place p with label V in $\mathcal{N}$ such that $O(p) = \emptyset$ is called a *goal place*. The algorithm can be described as follows:

- Phase 1:

  1. Insert place p in the set of places $P'$and in queue Q_P.
  2. While $Q_P \neq empty$ do

     $p_1 =: delete(Q_P)$

     Let t be the element in $I(p_1)$

     If t is not in T' then insert t in T'

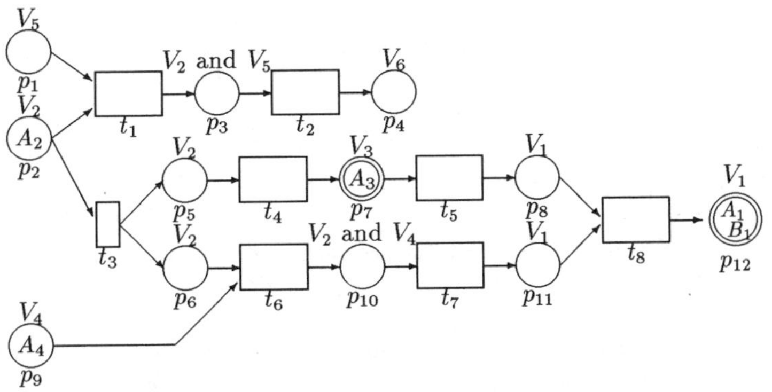

Figure 2: Example of a knowldege base modeled in HLFPN.

and $insert(t, Q_T)$

3. While $Q_T \neq empty$ do
   $t_1 =: delete(Q_T)$
   Let C_l = elements in $I(t)$
   For each $p \in C_l$ do
   If p is not in P' then insert p in P'
   and $insert(p, Q_P)$
4. If $Q_P \neq empty$ then go to step 2, else continue.
5. Consider the subnet of C containing the places in P' and transitions in T'.

- Phase 2:
  1. Apply the reasoning algorithm described in [18] to the subnet derived in phase 1 (see next section).

For example, assume our KB K consists of the following information:

d_1:V_1 is A_1 d_2:V_1 is B_1 d_3:V_2 is A_2
d_4:V_3 is A_3 d_5:V_4 is A_4 d_6:V_5 is A_5

I_1 : IF V_2 is C_2 THEN V_3 is C_3
I_2 : IF V_2 is D_2 AND V_4 is D_4 THEN V_1 is D_1,
I_3 : IF V_3 is E_3 THEN V_1 is E_1
I_4 : IF V_2 is F_2 AND V_5 is C_5 THEN V_6 is C_6

The HLFPN model representing the KB K is shown in figure 2, where the inscribed formula have been omitted for simplification.

Suppose the question V_1 is ? has been asked. It then initiates the extraction of subnet $\mathcal{N}'$, which contains all partial sources of information related to V_1. Step 1 inserts place p_{12} in P' and makes $Q_P = \{p_{12}\}$. Applying successively steps 2 and 3, we end up with $P' = \{p_2, p_5, p_9, p_6, p_7, p_8, p_{10}, p_{11}, p_{12}\}$ and $T' = \{t_3, t_4, t_6, t_5, t_7, t_8\}$. Looking up in matrix C^* we find the arc relations between places and transitions in P' and T' and finally have the subnet $\mathcal{N}'$, which can be easily observed in in figure 2.

The next step is to apply an evaluation algorithm to the subnet extracted from the overall net as described in the preceding section, to give us the information about variable V. We do

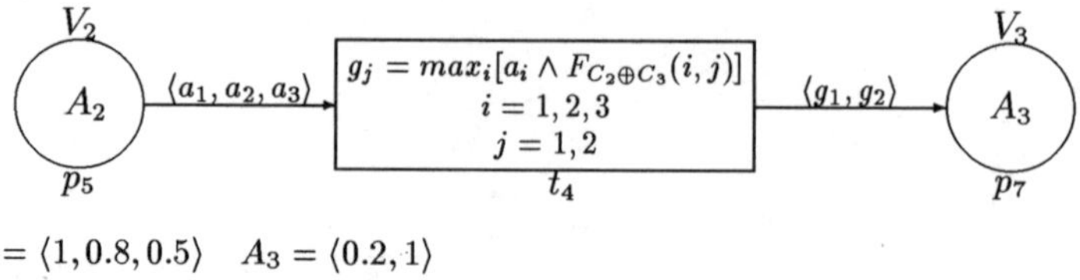

Figure 3: Detailed inscribed formulae in transition t_4.

this by applying the forward chaining algorithm developed in [15] and [18], which consists of following the firing rules of the transitions, until there are no more transitions to fire.

We consider as initial places those places with empty input set. In case we have some initial place without any initial token in it, that is, there are no data propositions concerning the variable associated with this place, we insert a token whose label is the base set of that variable. The meaning of this token is that *we don't know anything about the variable.*

Let us assume the fuzzy subsets involved in the knowledge base K of the example have been defined on the following base sets: $X_1 = \{a, b, c\}$, $X_3 = \{l, m\}$, $X_2 = \{1, 2, 3\}$, $X_4 = \{p, q\}$.

Fuzzy sets of X_1:

	a	b	c
A_1	1	1	0.3
B_1	1	0.6	0.9
D_1	1	0.7	0.5
E_1	1	0.3	0.2

Fuzzy sets of X_2:

	1	2	3
A_2	1	0.8	0.5
C_2	1	0.6	0
D_2	1	1	0.3

Fuzzy sets of X_3:

	l	m
A_3	0.2	1
C_3	0.5	1
E_3	0.1	1

Fuzzy sets of X_4:

	p	q
A_4	1	0.5
D_4	1	0.4

We shall also assume that the implication relation $f_{\rightarrow}(A, B)$ used in this example is the bounded sum, defined as $F_{A \oplus B}(x, y) = 1 \wedge [1 - A(x) + B(y)]$.

Initially, only transition t_3 is enabled. Transition t_3 has no inscribed formulae. Right after its firing, tokens are inserted in places p_5 and p_6, both with the same label as the token in the input place of transition t_3 before the firing operation. In this new state, transition t_4 and t_6 are enabled. Transition t_4 have the inscribed formulae showed in detail in figure 3, where $F_{C_2 \oplus C_3}(i, j)$ denotes the element in row i and column j of the array $F_{C_2 \oplus C_3}$, representing the relation on $X^2 \times X^3$, induced by implication I_1.

The variables a_1, a_2, and a_3 in figure 3 are set to 1, 0.8 and 0.5 respectively. The evaluation of the formulae in t_4 results in $g_1 = 0.8$ and $g_2 = 1$, which are the values of the token inserted in p_7. Place p_7 has now two tokens: one corresponding to the data proposition d_4:V_3 is A_3 and one corresponding to the evaluation of implication I_1. This requires that place p_7 be replaced by a subnet that performs conjunction of the token values. This process will be discussed later.

The inscribed formulae in transition t_6 perform the conjunction of data propositions in its input places, which corresponds to facts about variables V_2 and V_4, appearing in the antecedent of implication I_2. Assuming that conjunction is performed through the *min* operator, the inscribed formula in transition t_6 is $h_{ij} = a_i^1 \wedge a_j^2$, $i = 1, 2, 3$, $j = 1, 2$. The firing of transition t_6 changes the net to a new state where a token with label $\langle 1, 0.8, 0.5, 0.5, 0.5, 0.5\rangle$ is inserted in place p_{10} representing the relation $A_2 \cap A_4$.

Transition t_7 is now enabled. The implication I_2 induces a fuzzy relation on $X^2 \times X^4 \times X^1$ given by

$$F_{(D_2,D_4)\oplus D_1}(x^2, x^4, x^1) = 1 \wedge [1 - H_{D_2,D_4}(x^2, x^4) + D_1(x^1)]$$

where $H_{D_2,D_4}(x^2, x^4) = D_2(x^2) \wedge D_4(x^4)$.

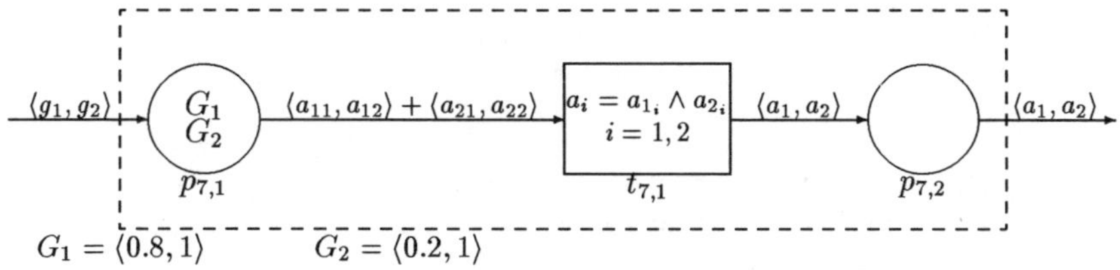

Figure 4: Detailed subnet to be substituted for place p_7.

The result of firing transition t_7 is a net state where a token with label $\langle 1, 0.5, 0.5 \rangle$ is inserted in place p_{11}.

Let us now discuss the case of place p_7. It must be replace by a subnet like the one in figure 4. Transition $t_{7,1}$ performs conjunction of token values in its input place. We note that the arc label $\langle a_{11}, a_{12} \rangle + \langle a_{21}, a_{22} \rangle$ requires that at least two tokens must exist in place $p_{7,1}$ in order to enable transition $t_{7,1}$.

Firing transition $t_{7,1}$ produces the token $\langle 0.2, 1 \rangle$, which is inserted in place $p_{7,2}$ as long as the arc label $\langle a_1, a_2 \rangle$ makes $a_1 = 0.2$ and $a_2 = 1$. Transition t_5 is now enabled.

The firing of transition t_5 is similar to the one of transition t_7, resulting in a token value for p_8 with label $\langle 1, 0.5, 0.5 \rangle$. Transition t_8 finds a single value for V_1, applying conjunction operation on the two partial results for V_1 obtained through I_1 and I_2, and inserts a token in place p_{12}. As place p_{12} already has two other tokens because of the existence of two data propositions with respect to V_1, the same procedure applied to place p_7 must be used to find a final value for V_1, which is $V_1 = \{1/a, 0.3/b, 0.2/c\}$.

5. Conclusions

We have presented an algorithm to answer questions posed to knowledge bases which contain linguistic information. The knowledge base may contain data statements and implication statements and should be represented in a HLFPN model. The procedure introduced here is executed in two phases: one for extracting, from the complete net representing the knowledge base, a subnet with all required information to answer the refered query and one to evaluate the subnet. The proposed algorithm has advantages with respect to the one suggested by Yager in [21] concerning information organization, because it is applied to structured knowledge bases. Thus, the process of succesively look for information in the knowledge base about a variable V in the first phase of the algorithm, consists of directly accessing the entries of array type structures. In a future work we shall expand the HLFPN and its related reasoning algorithms to more general knowledge bases and more complex query types.

Acknowledgments: The first author is grateful to CNPq, the Brazilian National Research Council for grant #521029/94-5. The second author acknowledges CPNq for grant #300729/86-3 and ESPRIT Project ECLA 005.

REFERENCES

[1] Bugarin, A. and Barro, S. (1994) *Fuzzy Reasoning Supported by Petri Nets*, IEEE Trans. on Fuzzy

Systems, 2, 2, pp. 135-150.
[2] Cao, T. and Sanderson, A. (1993) *Variable Reasoning and Analysis about Uncertainty with Fuzzy Petri Nets*, In: Application and Theory of Petri Nets 1993, (Marco A. Marsan ed.), LNCS, Spring-Verlag, pp. 126-145.
[3] Cardoso, J. ; Valette, R. ; Dubois, D. (1990) *Petri Nets with Uncertain Markings*, In: Advances in Petri Nets 1990, (G. Rosenberg ed.), LNCS 483, Springer-Verlag, pp. 64-78.
[4] Chen, S., Ke,J. S. and Chang, J. (1990) *Knowledge Representation using Fuzzy Petri Nets*, IEEE Trans. Knowledge and Data Eng., 2, 3, pp. 311-319.
[5] Chun, M. ; Bien, Z. (1993) *Fuzzy Petri Net Representation and Reasoning Methods for Rule-Based Decision Making Systems*, IEICE Trans. Fundamentals, vol. E76-A, n. 6.
[6] Genrich, H. J. (1986) *Predicate/Transition Nets*, In: Petri Nets: Central Models and Their Properties, (W. Brauer, W. Reisig and G. Rozenberg eds.), LNCS 254, Springer-Verlag, pp.207-247.
[7] Giordana, A. ; Saitta, L. (1985) *Modeling Production Rules by Means of Predicate Transition Networks*, Information Sciences, 35, 1, pp.1-41.
[8] Jensen, K. (1986) *Coloured Petri Nets*, In: Petri Nets: Central Models and Their Properties, (W. Brauer, W. Reisig and G. Rozenberg eds.), Lecture Notes in Computer Science 254, Springer-Verlag, pp.248-299.
[9] Looney, C. (1988) *Fuzzy Petri Nets for Rule-Based Decision Making*, IEEE trans. on Systems, Man and Cybernetics, 18, 1, pp. 178-183.
[10] Murata, T. ; Zhang, D. (1988) *A Predicate-Transition Net Model for Parallel Interpretation of Logic Programs*, IEEE Trans. on Software Engineering, 14, 4, pp. 481-497.
[11] Murata, T. (1989) *Petri Nets: Properties, Analysis and Applications*, Proceedings IEEE,77, 4, pp.541-580.
[12] Murata, T. ; Subrahmanian, V. S. ; Wakayama, T. (1991) *A Petri Net Model for Reasoning in the Presence of Inconsistency*, IEEE Trans. on Knowledge and Data Engineering, 3, 3, pp. 281-292.
[13] Pedrycz, W. (1993)*Fuzzy Control and Fuzzy Systems*, John Wiley & Sons Inc., second extended edition.
[14] Pedrycz, W. ; Gomide, F. (1994) *A generalized Fuzzy Petri Net Model*, IEEE Trans. on Fuzzy Systems, 2, 4, pp. 295-301.
[15] Scarpelli, H. ; Gomide, F.; Pedrycz, W. (1992) *Modeling Fuzzy Reasoning using High Level Fuzzy Petri Nets*, Technical Report RT-DCA 023/92, DCA/ FEE/ UNICAMP. (submitted for publication)
[16] Scarpelli, H.; Gomide, F. and Yager, R. (1993) *A Backward Reasoning Algorithm for High Level Fuzzy Petri Nets*, Technical Report RT-DCA 004/93, DCA/ FEE/ UNICAMP. (submitted for publication)
[17] Scarpelli, H. ; Gomide, F. (1993) *Fuzzy Reasoning and Fuzzy Petri Nets*, Fifth IFSA World Congress, Seoul, Korea, July, 5-9, pp. 1326-1329.
[18] Scarpelli, H. and Gomide, F. (1993) *Fuzzy Reasoning and Fuzzy Petri Nets in Manufacturing Systems Modeling*, Journal of Intelligent and Fuzzy Systems, 1, 3, pp. 225-241.
[19] Scarpelli, H. ; Gomide, F. (1993) *Fuzzy Reasoning and High Level Fuzzy Petri Nets*, First European Congress on Fuzzy and Intelligent Technologies, Aachen, Germany, September, 7-10, pp. 600-605.
[20] Scarpelli, H.; Gomide, F. (1994) *A High Level Fuzzy Petri Net Approach for Discovering Potential Inconsistencies in Fuzzy Knowledge Bases*, Fuzzy Sets and Systems, 64, pp. 175-193.
[21] Yager, R. R. (1993) *Querying Knowledge Base Systems with Linguistic Information via Knowledge Trees.*, Int. J. Man-Machine Studies, 19, pp.73-95.
[22] Zadeh, L. A. (1965) *Fuzzy Sets*, Information and Control, 8, pp.338-353.
[23] Zadeh, L. A. (1979) *A Theory of Approximate Reasoning*, Machine Inteligence 9, (Hayes, Michie & Kulich eds.), 9, pp.149-194.

ON APPROXIMATION OF CONTROLS IN DISTRIBUTED SYSTEMS BY FUZZY CONTROLLERS

Vladik KREINOVICH*, Hung T. NGUYEN**, Ongard SIRISAENGTAKSIN***

*Department of Computer Science
University of Texas at El Paso
El Paso, TX 79968, USA

**Department of Mathematical Sciences
New Mexico State University
Las Cruces, NM 88003, USA

***Department of Computer and Mathematical Sciences
University of Houston-Downtown
Houston, TX 77002, USA

ABSTRACT

We show that fuzzy controllers are good approximations for control strategies in distributed systems, i.e., plants whose states are described by infinitely many parameters. This result generalizes the previous results on the universal approximation property of fuzzy controllers as approximations of functions (i.e., for plants whose states are described by finitely many parameters).

1. Introduction

1.1. Why fuzzy control

The rise of fuzzy control is due to several reasons:

i) The derivation of control laws, say, in standard framework of modern control theory (through optimization) is not always an easy and feasible task. This task is difficult not only because the optimization problem itself is technically difficult, but because the constraints on which this optimization is based (that describe open-loop dynamics) are not always known for large-scale non-linear systems.

ii) In many control problems (e.g., in parking a car) the objective function cannot be specified precisely. This phenomenon occurs for many problems in robotics. In these cases, instead of the precise objective function, we have the experience of human experts.

iii) The design of a good controller should be based upon the maximal use of available knowledge about the plant under consideration. Often, an essential part of this knowledge can be only formulated by experts in imprecise ("fuzzy") terms, so, it is necessary to incorporate fuzzy information into the design of a controller.

Due to these reasons, fuzzy methodology has been placed on the list of possible control tools.

1.2. Fuzzy control and neural network control

The list of possible control tools includes another non-traditional control tool: neural network control. Neural network control has not yet lead to as many practical applications as fuzzy control, but its analysis is important to us because in many cases, neural network control has been more thoroughly analyzed (actually, the results that we present in this paper were first obtained for neural network control, and then transferred to fuzzy control).

Neural network control and fuzzy control methodologies are sometimes viewed as competitors. We believe that they are not competitors, they are *complementary* technologies: namely, they enable us to take into consideration different types of knowledge. Indeed, according to general scientific principles, a successful solution of a real-life must be based on the available data and knowledge, and the type of available data dictates the appropriate solution techniques. For example, in statistical data processing, depending on the available data, we can use different types of estimators, e.g.:

- *Bayesian estimators* are applicable when we have a prior knowledge of the probabilities.
- *Maximum likelihood* estimators are applicable if we know that the probability distribution belongs to a given (finite-parametric) family, so we need to determine its parameters.

The situation in control is similar: the design of a controller depends on the type of available data:

- If this data consists of training samples, then the appropriate tool is *neural network control.*
- If this data consists of expert control rules formulated in a natural language, then *fuzzy control* methodology can be applied.
- In case we have both the samples and the rules, we must apply a combined strategy of *fuzzy neural control.*

1.3. What is a control strategy. The need for universal approximation property

In the majority of real-life control problems, we have a *feedback*, i.e., we have some information about the state of a controlled plant. In this case, the controller must find and apply the control whose value is appropriate for the current state information. The control is thus a *function* of the state. When the plant is controllable, the task of an engineer is to design a controller to do the job.

- In some cases, it is possible to solve the corresponding optimization problem, and to find the function that describes the optimal control strategy. It is not always possible to implement this function exactly, so, the task of an engineer is to *approximate* the ideal control strategy.
- In many real-life cases, we cannot compute the ideal control strategy. In these cases, we can try to implement different control strategies and choose the best one. In these cases, we must be sure that by using a certain control tool, we do not miss the (unknown) control strategy. Since the ideal control strategy is unknown, the only way for us to guarantee that we cannot miss the ideal (optimal) control strategy is to be sure that an *arbitrary* control strategy can be approximated by strategies that can be implemented using the given tool. In other words, we must guarantee that the tool has a *universal approximation* property.

1.4. Universal approximation property for plants with finitely many parameters

In many plants, their states can be described by finitely many parameters. In this case, a state of a system can be represented as a point in some finite-dimensional vector space R^n. Since measurements are in general bounded, the possible states form a bounded subset of R^n, and hence, the state space is a *compact* region of R^n. In this case, a control strategy is simply a continuous (real- or vector-valued) function defined on a compact set of R^n. For such systems, the desired universal approximation property means that for every $\varepsilon > 0$, we can approximate an arbitrary continuous function on a compact with an accuracy ε by a control generated by a given tool. This property has been first proven for neural networks by Hecht-Nielsen [5], Cybenko [3], Funahashi [4], and Hornik, Stinchcombe, and White [6]. These proofs use the Theory of Approximation of Functions, especially Stone-Weierstrass Theorem. It turns out that a similar result can be obtained for fuzzy controllers; see, e.g., Wang [10], Kosko [7], Buckley [1], Buckley and Hayashi [2], Nguyen and Kreinovich [9].

1.5. Distributed systems : formulation of a problem

All these results are about the plants whose states can be described by finitely many parameters. In many real-life situations, we need infinitely many parameters to describe a state: e.g., to describe the state of a chemical reactor, we must describe the values of temperature and concentrations in infinitely many different points. In such cases, the state of a plant can be described as a *function* (e.g., temperature $T(x,y,z)$ as a function of coordinates), and a control strategy is a mapping from the set of all functions to the set of real numbers, i.e., in mathematical terms, a *functional.* For such systems, the optimization problems are even more complicated and less feasible than for plants with finitely many parameters, so, for such plants, it is even more desirable to apply fuzzy or neural network control methodologies. It is therefore important to know whether these methodologies can potentially lead to optimal controls, i.e., to know whether these methodologies have a universal approximation properties for systems with distributed parameters.

1.6. Solution

- For *neural network* control, the positive answer to this question was recently given in [8]: i.e., neural network control is a universal approximation for distributed systems.
- In this paper, we show that a similar universal; approximation result holds also for *fuzzy* control.

2. Distributed systems: the approximation problem

2.1. Fuzzy control can take only finitely many inputs

An expert can only take finitely many parameters into consideration. Fuzzy control methodology formalizes the expert's knowledge. Therefore, a control strategy that stems from fuzzy control methodology can only take finitely many parameters as inputs. Specifically, a fuzzy controller is designed as follows:

- First, we start with the expert rules of the form

$$\text{If } x_1 \text{ is } A_1^j \text{ and ... and } x_n \text{ is } A_n^j, \text{ then } u \text{ is } B_j,\ j=1,2,...,r,$$

  where A_i^j and B^j are words of natural languages (like "large") that are described by membership functions (for simplicity, these membership functions will be denoted by the same letters A_i^j and B^j.
- Modeling logical connectives "and" and "or" as $t-$norms and $t-$conorms in fuzzy logic, we get the following membership function for the variable u:

$$\mu(u) = \vee_{j=1}^r [\wedge_{i=1}^n A_i^j(x_i) \wedge B^j(u)].$$

- The overall output value for u, corresponding to the input $x = (x_1, ..., x_n)$, is $y = f(x) = D(\mu)$, where D is a procedure that maps the function into a value, and that is called *defuzzification.*

Comments.

1. In the above technique, we first combine the rules, and then defuzzify the result of this combination. It is also possible to first defuzzify the result of using each rule, and then combine the results of these defuzzifications.
2. As a result, we get the control u as a function of *finitely many* input variables $x_1, ..., x_n$. How can we use fuzzy control for controlling finite-dimensional systems?

2.2. Reduction of infinite – dimensional cases to finite – dimensional ones : main idea

Let us consider the above example of a chemical reactor. To describe the state of the reactor precisely, we need to know the value $T(x, y, z)$ of the temperature at every point (x, y, z) inside the reactor. In reality, we cannot measure the temperature in infinitely many points. So, in all practical cases, we only have finitely many values $T(x_i, y_i, z_i)$ measured in finitely many points (x_i, y_i, z_i). So, a practical control is based on the values of finitely many parameters.

The more values we measure, the better picture of the temperature distribution we get, and the more accurate control can be thus achieved.

For distributed systems, a state is described by a function $f(z)$. So, a control strategy is a mapping from the set of all functions to the set of real numbers. We will approximate these mappings by mappings that only use the values of the state function in finitely many points $z_1, ..., z_N$. The resulting finite-dimensional control strategy will then be approximated by a fuzzy controller. So, we are approximating the desired control strategy by fuzzy controller that stems from the rules of the type

$$\text{If } f(z_1) \text{ is } A_1^j \text{ and ... and } f(z_n) \text{ is } A_n^j, \text{ then } u \text{ is } B_j,\ j = 1, 2, ..., r.$$

Let us describe this idea formally.

2.3. Distributed systems and their approximations : formal description

The *state space* S of a distributed system is a subset of the set $R^n \times \prod_{j=1}^m C(Z_j)$, where for each j, Z_j is a compact metric space with metric ρ_j, and $C(Z_j)$ denotes the space of all continuous functions from Z_j to R with a metric

$$\|f - g\| = \max_{z \in Z} |f(z) - g(z)|.$$

Thus, an arbitrary element $s \in S$ is of the form $s = (x_1, x_2, ..., x_n, f_1, ..., f_m)$, where $f_j : Z_j \to R$.

On this set S, we can define a metric ρ as follows: if $s = (x_1, x_2, ..., x_n, f_1, ..., f_m)$ and $\tilde{s} = (\tilde{x}_1, \tilde{x}_2, ..., \tilde{x}_n, \tilde{f}_1, ..., \tilde{f}_m)$, then

$$\rho(s, \tilde{s}) = \max\{|x_i - \tilde{x}_i|, \|f_j - \tilde{f}_j\|, i = 1, 2, ..., n; j = 1, ..., m\}.$$

By physical considerations, the set S is compact. So, S is a compact subset of the space $R^n \times \prod_{j=1}^m C(Z_j)$. We are interested in approximating continuous functionals $J : S \to R^p$, where the metric on R^p is defined as

$$\|x - y\| = \max\{|x_k - y_k|, k = 1, 2, ..., p\}.$$

To approximate such a functional, we are going to choose, for every j from 1 to M, a finite set V_j, and approximate a functional $J(s)$ by a value $J_\varepsilon(\pi(s))$ for an appropriately chosen function J_ε, where $\pi_V(s)$ denotes the set of all values of f_j on $z \in V_j$:

$$\pi_V(s) = (x_1, ..., x_n; f_j(v_{ij}), i = 1, ..., N_j, j = 1, ..., m),$$

where $N_j = |V_j|$ is the number of elements in V_j.

To get J_ε closer to J, we select as J_ε, the result of an interpolation of J. To be more precise, we will find a finite set $G \subseteq S$, and define

$$J_\varepsilon(\pi_V(s)) = \sum_{g \in G} \alpha_g(s) J(g) / \sum_{g \in G} \alpha_g(s) \tag{1}$$

for specifically chosen weights $\alpha_g(s)$.

As a result, we get a function J_ε of finitely many variables, such that for every state $s \in S$,

$$\|J(s) - J_\varepsilon(\pi_V(s))\| \leq \varepsilon.$$

This function, in its turn, will be approximated by a fuzzy controller.

3. Reduction to finite-dimensional case: details and the idea of the proof

We will illustrate the main idea of the reduction and of the corresponding proof on the simplest example of a distributed system, when the state is described by a single function (for general case, see [8]). Let us reformulate this case in purely mathematical terms.

3.1. Details of the mathematical construction

Let Z be a compact metric space. By $C(Z)$, we denote the set of all continuous real-valued functions on Z, equipped with the sup-norm.

In the following text, we will deal with several sup-norms on different spaces. To avoid clumsy notations, we will use the same notation $\|\cdot\|$ for all these sup-norms. The context will make it clear which norm we are talking about.

Let F be a compact subset of $C(Z)$, and let $J : F \to R$ be a continuous functional. For every $\varepsilon > 0$, we are going to construct a continuous function J_ε that is defined on a compact subset of some finite-dimensional Euclidean space, and which will be close to J within ε in some reasonable sense (that will be made precise later).

Let $V = \{v_1, v_2, ...\}$ be a finite subset of Z, and let $|V|$ denote the number of elements in V. We will define a mapping $\pi_V : F \to R^{|V|}$ as follows:

$$\pi_V(f) = (f(v_1), ..., f(v_{|V|})).$$

On the range $\pi_V(F)$ of this mapping, we will define the metric as

$$\|x - y\| = \max_{i=1,...,|V|} |x_i - y_i|.$$

So,

$$\|\pi_V(f) - \pi_V(g)\| = \max_{v \in V} |f(v) - g(v)|.$$

Obviously, $\|\pi_V(f) - \pi_V(g)\| \leq \|f - g\|$, so, π_V is continuous, and hence, $\pi_V(F)$ is a compact subset of $R^{|V|}$.

To define J_ε, we will choose a finite set $G \subset F$, and define $J_\varepsilon(\pi_V(f))$ by the formula (1), where the weights $\alpha_g(f)$ are non-negative and non-increasing functions of $\|\pi_V(f) - \pi_V(g)\|$.

Specifically, we will take $\alpha_g(f) = \max\{0, \delta - \|\pi_V(f) - \pi_V(g)\|\}$ for some suitable $\delta > 0$.

Comments.

- Thus defined J_ε depends only on the values of $\pi_V(f)$, i.e., it is indeed a function of finitely many variables (namely, of $|V|$ variables).
- Our "interpolation" is somewhat unusual in the sense that in general, $J_\varepsilon(g) \neq J(g)$ for interpolation values $g \in G$. This is not crucial, and in principle, we could use a more complicated interpolation procedure that has this property, but since our only goal is to prove the approximation property, we decided to use this (imperfect but simpler) interpolation procedure.
- To make sure that the function (1) is defined for all f, we must guarantee that $\sum_{g\in G} \alpha_g(f) > 0$ for all $f \in F$. In other words, we need to guarantee that for every $f \in F$, there exists a $g \in G$ for which $\|\pi_V(f) - \pi_V(g)\| < \delta$. One way to guarantee that is to choose G in such a way that for every $f \in F$, there exists a $g \in G$ for which $\|f - g\| < \delta$; then, we will have $\|\pi_V(f) - \pi_V(g)\| \leq \|f - g\| < \delta$. Such a finite set G (called a $\delta-$*net* for F) exists because F is a compact set.
- For each g, $\alpha_g(f)$ is a continuous function on $\pi_V(F)$, and therefore, the function J_ε is also continuous.

3.2. How to guarantee that the resulting function is an $\varepsilon-$ approximation to J

For a given $\varepsilon > 0$, we would like to construct J_ε so that

$$\forall f \in F, \quad \|J(f) - J_\varepsilon(\pi_V(f))\| \leq \varepsilon. \tag{2}$$

Let us observe that $J_\varepsilon(\pi_V(f))$ is a convex combination of the values $J(g)$, $g \in G$, so

$$J_\varepsilon(\pi_V(f)) - J(f) = \sum_{g\in G} \alpha_g(f)(J(g) - J(f)) / \sum_{g\in G} \alpha_g(f),$$

and hence,

$$|J_\varepsilon(\pi_V(f)) - J(f)| \leq \max_g |J(g) - J(f)|,$$

where max is over all g for which $\|\pi_V(f) - \pi_V(g)\| < \delta$. So, (2) will be satisfied if we have $|J(g) - J(f)| \leq \varepsilon$ whenever $\|\pi_V(f) - \pi_V(g)\| < \delta$. Now, since J is continuous on a compact, it is uniformly continuous, and therefore, there is $\delta(\varepsilon)$ such that $\|f - g\| \leq \delta(\varepsilon) \to |J(f) - J(g)| \leq \varepsilon$. So, if we choose V in such a way that for all $f, g \in F$,

$$\|f - g\| \leq \|\pi_V(f) - \pi_V(g)\| + \alpha, \tag{3}$$

then for $\delta + \alpha \leq \delta(\varepsilon)$, (2) will follow.

Let us show how to choose V so that (3) will be guaranteed.

- If a compact set is covered by a family of open sets, then from this family, we can extract a finite subfamily that will still be a covering. All open balls of radius $\delta(\varepsilon)/2$ form a covering of F, so, there exists a finite subcovering. As G, we will take the set of all centers of the balls from this subcovering. So, for every $f \in F$, there exists a $g \in G$ for which $\|f - g\| < \delta(\varepsilon)/2$.
- Similarly, since Z is a compact, we choose V to be a finite set of centers of open balls of radius β (to be specified shortly), that form a finite covering of Z. Since F is a

compact subset of $C(Z)$, its elements are equicontinuous (Arzela-Ascoli Theorem), thus there exists a β such that for all $f \in F$, if $\|u - v\| \leq \beta$, then $\|f(u) - f(v)\| \leq \delta(\varepsilon)/4$. This is our chosen β.

- Let us take $\delta = \delta(\varepsilon)/2$ in the above formula for α_g. Then, due to our choice of G, for every $f \in F$, there exists a $g \in G$ such that $\|f - g\| < \delta$, hence, $\|\pi_V(f) - \pi_V(g)\| < \delta$, and $\alpha_g(f) > 0$. So, $\sum_{g \in G} \alpha_g(f) > 0$ for all f, and $J_\varepsilon(\pi_V(f))$ is defined for all $f \in F$.

Let us show that for this choice of parameters, we indeed have (3) with $\alpha = \delta(\varepsilon)/2$ (and hence, (2)). Indeed, for an arbitrary $z \in Z$, there is a $v \in V$ for which $\|z - v\| \leq \beta$, implying that $|f(z) - f(v)| \leq \delta(\varepsilon)/4$ and $|g(z) - g(v)| \leq \delta(\varepsilon)/4$. Recalling that $\|\pi_V(f) - \pi_V(g)\| = \max_V |f(v) - g(v)|$, we arrive at:

$$|f(z) - g(z)| \leq |f(z) - f(v)| + |f(v) - g(v)| + |g(z) - g(v)| \leq$$

$$\frac{\delta(\varepsilon)}{4} + \|\pi_V(f) - \pi_V(g)\| + \frac{\delta(\varepsilon)}{4}.$$

Hence, $\|f - g\| = \sup_Z |f(z) - g(z)| \leq \|\pi_V(f) - \pi_V(g)\| + \delta(\varepsilon)/2$.

From this, we see that for each $f \in F$, if $g \in G$ is such that $\|\pi_V(f) - \pi_V(g)\| < \delta = \delta(\varepsilon)/2$, then $\|f - g\| \leq \delta(\varepsilon)$, and hence, $|J(f) - J(g)| \leq \varepsilon$, and $|J(f) - J_\varepsilon(\pi_V(f))| \leq \varepsilon$.

Similar construction can be applied to a more general case, when we need several functions to describe a state. We have thus proven the following result:

Theorem. *Let Z_j, $1 \leq j \leq m$, be compact sets, S be a compact subset of $R^n \times \prod_{j=1}^m C(Z_j)$, and $J : S \to R^p$ be a continuous function. Then, for every ε, there exists a continuous mapping $\pi : S \to E$ of S into a finite-dimensional space E, and a continuous function J_ε that is defined on a compact subset of E such that for every $s \in S$, $\|J(s) - J_\varepsilon(\pi(s))\| \leq \varepsilon$.*

So, to complete the proof that fuzzy controllers are universal approximators for systems with distributed parameters, we can use the known results that fuzzy controllers are universal approximators for functions of several variables: an approximation to a function J_ε of finitely many variables will thus be an approximation to the desired functional $J : S \to R$.

In mathematical terms, these results mean that some sets if functions are *dense* in $C(K)$ for a compact K. So, let us recall these results.

4. Some dense subsets of $C(K)$ related to fuzzy control

Let us describe several classes of designs of fuzzy controllers (or, more general, fuzzy systems) which form dense subsets of $C(K)$:

4.1. Fuzzy systems with Gaussian membership functions

With Gaussian membership functions $\mu_{j,i}(x_i)$ (to describe natural-language terms from the expert's rules), product or minimum inference, and defuzzify-then-combine strategy, we get the class of functions of the form

$$f(x) = \frac{\sum_{j=1}^r u_j \prod_{i=1}^n \mu_{j,i}(x_i)}{\sum_{j=1}^r \prod_{i=1}^n \mu_{j,i}(x_i)}.$$

When restricted to compact domain K of R^n, these functions form a dense subset in $C(K)$ (Wang [10]; Nguyen Kreinovich [9] for min instead of $\prod$). Note that these results were obtained by verifying the sufficient condition for denseness given by the Stone-Weierstrass Theorem in the Theory of Approximation of Functions.

4.2. Sugeno type controllers

One of Sugeno's types of controllers can be described (in the simplest case) as follows (see, e.g., Buckley [1]):

- We start with rules R_j of the type: *"If e (error) is A_j, and change of error (Δe) is B_j, then $y = P_j$"*, where $j = 1, ..., r$, and P_j is some polynomial in e and Δe.
- Using a continuous t–norm T, we compute the weight of each rules as $\lambda_j = T(A_j(e), B_j(\Delta e))$.
- Finally, we compute the control value as a weighted combination of P_j: $y = f(e, \Delta e) = \sum_{j=1}^{r} \lambda_j P_j$.

Using Stone-Weierstrass Theorem, Buckley has shown that Sugeno type controllers are universal controllers, i.e., any process that can be controlled by a continuous control law can be also controlled by Sugeno type controllers.

4.3. Systems with fuzzy outputs

Recently (see, e.g., Buckley and Hayashi [2]), the rules of Sugeno's controllers have been extended to fuzzy inputs and fuzzy outputs. Using a direct proof for ε–approximation, the authors of [2] have shown that these "fuzzy input-output" controllers are also universal approximators.

4.4. Combine – then – defuzzify controllers

If we proceed in the spirit of combine-then-defuzzify, then various classes of fuzzy controllers also form dense subsets of $C(K)$ [9].

5. Conclusions

The results of Wang and Mendel, Wang, Kosko, and ours, show that fuzzy control is a universal control tool in the following sense: for every (potential) control strategy, and for every ε, there exist a fuzzy controller whose control is ε–close to the desired one. In other words, although we do not know what an optimal control is, in principle, by appropriately choosing the rules, we can get a fuzzy control that is as close to the optimal control as possible. So, by restricting ourselves to fuzzy control methodology, we do not lose any possible control strategies.

These results were initially proved for plants whose states can be described by finitely many parameters. We prove a generalization of this result, that shows that fuzzy control can be used in the most general control situations. Namely, we prove that fuzzy control is a universal control tool for distributed systems (i.e., for plants that require infinitely many parameters to describe their current state).

Acknowledgments

This work was supported by a NSF Grant No. CDA-9015006, NSF Research Opportunity Award (for O.S.), NASA Research Grant No. 9-757 and the Institute for Manufacturing and Materials Management grant. The work of Hung Nguyen was partially carried out at the Laboratory for International Fuzzy Engineering Research, Chair of Fuzzy Theory, Japan. We are greatly thankful to B. Bouchon-Meunier for valuable comments.

REFERENCES

[1] Buckley, J. J. (1993) Sugeno type controllers are universal controllers, Fuzzy Sets and Systems, 299–303.

[2] Buckley, J.J., and Hayashi, Y. (1993) Fuzzy input-output controllers are universal approximators, Fuzzy Sets and Systems, 58, 273–278.

[3] Cybenko, G. (1989) Approximation by superpositions of a sigmoidal function, Mathematics of Control, Signals, and System, 2, 303–314.

[4] Funahashi, K. (1989) On the approximate realization of continuous mappings by neural networks, Neural Networks, 2, 183–192.

[5] Hecht-Nielsen, R. (1987) Kolmogorov's mapping neural network existence theorem, IEEE Intl. Conference on Neural Networks, San Diego, SOS Printing, 11–14.

[6] Hornik, K., Stinchcombe, M., and White, H. (1989) Multilayer feedforward networks are universal approximators, Neural Networks, Vol. 2, 359–366.

[7] Kosko, B. (1992) Fuzzy systems as universal approximators, Proceedings of the 1st IEEE International Conference on Fuzzy Systems, San Diego, CA, 1153–1162.

[8] Kreinovich, V., and Sirisaengtaksin, 0. (1993) 3-layer neural networks are universal approximators for functionals and for control strategies, Neural, Parallel, and Scientific Computations, 1, 325–346.

[9] Nguyen, H. T., and Kreinovich, V. (1993) On approximation of controls by fuzzy systems, Proceedings of the Fifth International Fuzzy Systems Association World Congress, Seoul, Korea, 1414–1417.

[10] Wang, L. X. (1992) Fuzzy systems are universal approximators, Proceedings of the IEEE International Conference on Fuzzy Systems, San Diego, CA, 1163–1169.

FUZZY CONTROL WITH NON-PRECISE INPUTS

Sylvie Galichet, Laurent Foulloy

Laboratoire d'Automatique et de MicroInformatique Industrielle
LAMII / CESALP,
Université de Savoie, BP 806
74016 ANNECY Cedex, FRANCE
Tel : (33)50-66-60-40, Fax : (33)50-66-60-20, Email: galichet@univ-savoie.fr

ABSTRACT

Fuzzy control applications usually deal with precise inputs. This paper aims at analyzing the behaviour of a fuzzy controller when the input is also fuzzy. For sake of simplicity, only fuzzy proportional controllers are studied. The input is thus represented by a fuzzy number. An accommodation of the general equations of a Mamdani type fuzzy controller is first derived, leading to some easy software implementation. Results are presented for different inference operators and defuzzification methods. Various shapes of the membership function of the input are also studied.

1• Introduction

Fuzzy logic controllers are usually decomposed into three basic operating procedures: fuzzification, fuzzy inference and defuzzification. Let us clarify each of these sub-operations for a fuzzy controller with one input x and one output u, where x is an element of universe of discourse X and u an element of universe of discourse U. Let E be a set, the set of the fuzzy subsets over E will be denoted $\mathcal{F}(E)$.

The fuzzification interface is a mapping from the input universe of discourse X to $\mathcal{F}(X)$. The fuzzy inference mechanism is a decision making logic which employs fuzzy rules to determine the element of $\mathcal{F}(U)$ corresponding to an input fuzzy subset belonging to $\mathcal{F}(X)$. Finally, the process of obtaining a digital output from a fuzzy subset is called defuzzification. The most commonly used defuzzification strategy is the centroid one, which consists in computing the center of the area under the membership function of the fuzzy subset issued from the fuzzy inference mechanism. Fig. 1 gives a global representation of a fuzzy controller.

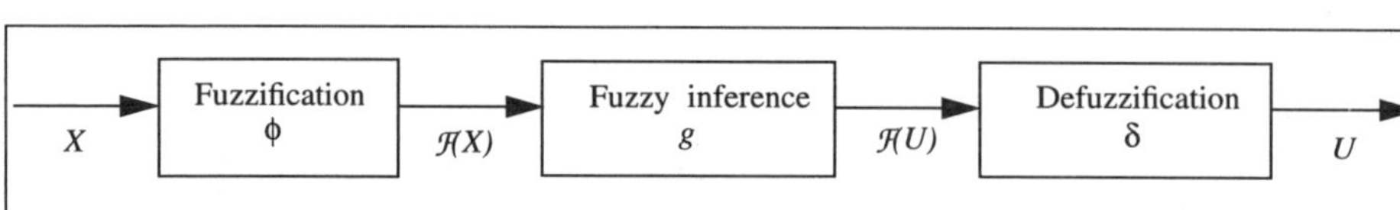

Fig. 1 : Fuzzy control mechanism

If we consider the input x_0 as being precise, then the fuzzy subset associated to x_0 by fuzzification has a membership function equal to zero everywhere in X except at the point x_0 where it takes the value 1. This assumption allows an important simplification in the expression of the fuzzy inference. The projection over U involved in the g function is reduced to some evaluation at the precise input x_0. In most of the fuzzy control applications, precise inputs are implicitly assumed. In that case, the fuzzy inference expression directly includes the fuzzification step, even when not explicitly notified.

Our purpose is to study experimentally the answer of a single-input, single-output Mamdani type fuzzy controller for non-precise inputs. The first step consists of the accommodation of the general equations of a Mamdani type fuzzy controller to some non-precise input represented by a fuzzy number. Section 2 aims at developing this approach. The expressions finally derived have been implemented by software. Section 3 presents the results obtained with different conjunction and disjunction operators and defuzzification methods. Various kinds of non-precise inputs have also been tested. Section 4 is dedicated to the corresponding results.

2• Dealing with non-precise inputs

Without loss of generality, we restrict our attention to a single-input, single-output fuzzy controller in which the rules are expressed in the form:

$$\textbf{If } x \text{ is } A_i \textbf{ then } u \text{ is } B_i \tag{1}$$

In that case, the fuzzy subset F of $\mathcal{F}(U)$ inferred for an input fuzzy subset E of $\mathcal{F}(X)$ is given by its membership function, according to Zadeh's combination/projection principle [Zadeh, 73], [Dubois 91].

$$\mu_F(u) = \sup_{x \in X} (\mu_E(x) \text{ T } \mu_\Gamma(x, u)) \tag{2}$$

where T represents any triangular norm and Γ the graph of the g function, that is a fuzzy subset of the cartesian product X x U.

For Mamdani's type fuzzy controllers, the expression of the membership function of the graph is formulated as:

$$\mu_\Gamma(x, u) = \perp_{i \in Z} (\mu_{\tau_X(A_i)}(x) \text{ T } \mu_{\tau_U(B_i)}(u)) \tag{3}$$

where $\perp$ represents any triangular conorm. $\mu_{\tau_X(A_i)}$ (respectively $\mu_{\tau_U(B_i)}$) is the fuzzy meaning of the symbol A_i (resp. B_i), that is the fuzzy subset over X (resp. U) associated to the symbol A_i (resp. B_i).

Our purpose is now to rewrite expressions (2) and (3) in order to facilitate the software implementation. Let input E be a fuzzy number (see for example [Pedrycz, 89]), that is a fuzzy subset of X such that:

- E is normal, i.e. $\exists\, x \in X$ such that $\mu_E(x) = 1$
- E has a bounded support,
- E is unimodal
- E is upper semi-continuous

Let x_{E-} and x_{E+} be the inferior and superior bounds of the support of E. Fig. 2 gives different possible choices for E. Regularly distributed triangular membership functions are chosen for $\mu_{\tau_X(A_i)}$ and $\mu_{\tau_U(B_i)}$ (see Fig. 3). We also impose an overlap of 1, i.e. $\frac{\sigma a}{\Delta a} = \frac{\sigma b}{\Delta b} = 1$. All these requirements square with fuzzy proportional controllers [Galichet, 93].

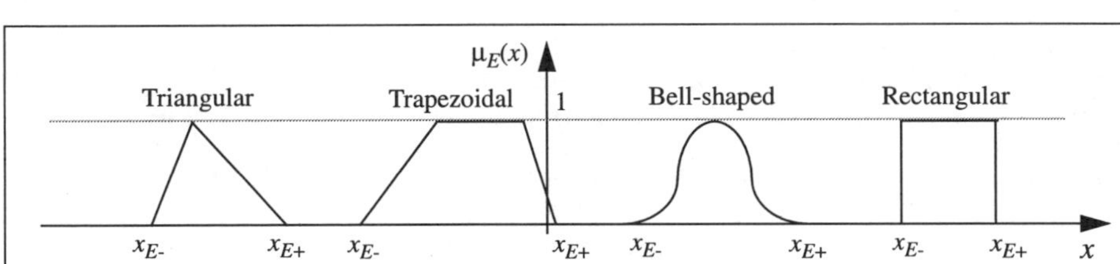

Fig. 2 : Possible fuzzy inputs

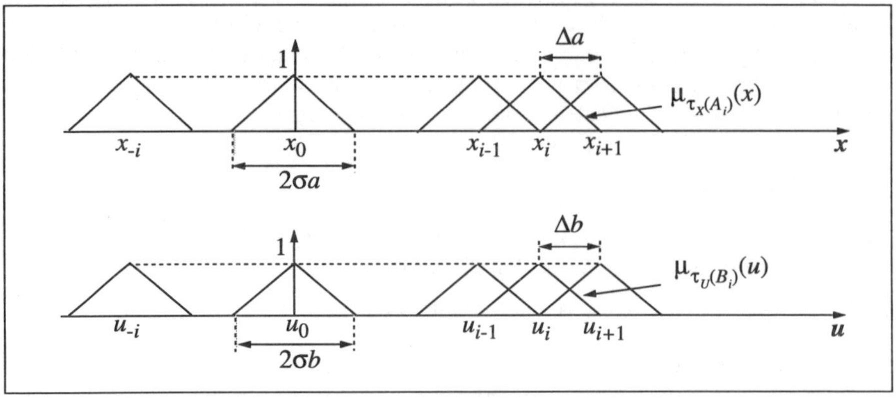

Fig. 3 : Membership functions of the fuzzy meanings of the symbols A_i and B_i

Let us now examine how the equation (2) can be transformed for the given assumptions. Noting that:

$\mu_E(x) = 0$ for $x \notin [x_{E-}, x_{E+}]$

and $a \mathrm{T} 0 = 0$ for any T-norm and any $a \in [0,1]$,

equation (2) can be expressed as:

$$\mu_F(u) = \max_{x \in [x_{E-}, x_{E+}]} (\mu_E(x) \mathrm{T} \mu_\Gamma(x, u)) \tag{4}$$

Let Ind(x) be an integer-valued function which returns the index i of x such that $x_i \le x < x_{i+1}$. For any $x \in [x_{E-}, x_{E+}]$, $\mu_{\tau_X(A_i)}(x) = 0$ if $i \notin \{\mathrm{Ind}(x), \mathrm{Ind}(x)+1\}$. As T-norm and ⊥-conorm verify:

$a \mathrm{T} 0 = 0$

$a \perp 0 = a$ for any $a \in [0,1]$,

only the two rules numbered Ind(x) and Ind(x)+1 are fired for a given x. The equation (3) may be finally rewritten as:

$$\mu_\Gamma(x, u) = \perp_{i \in \{\mathrm{Ind}(x), \mathrm{Ind}(x)+1\}} (\mu_{\tau_X(A_i)}(x) \mathrm{T} \mu_{\tau_U(B_i)}(u)) \tag{5}$$

The ultimate step to obtain the output control consists of the defuzzification process. The centroid method leads to the control signal u expressed as:

$$u = \frac{\int_U \mu_F(v) v dv}{\int_U \mu_F(v) dv} \tag{6}$$

The integration over U can be reduced to an integration over [*umin, umax*] where *umin* and *umax* represent the minimal and maximal values such that μ_F is not zero. For the given fuzzy input E, the minimal and maximal indices of the fired rules are respectively Ind(x_{E-}) and Ind(x_{E+})+1.

The assumptions made on the distribution of the $\mu_{\tau_U(B_i)}$ (see Fig. 3) allow us to write:

$$u_i = i \,.\, \Delta b \tag{7}$$

Expression (6) can be finally rewritten as:

$$u = \frac{\int_{umin}^{umax} \mu_F(v) v dv}{\int_{umin}^{umax} \mu_F(v) dv} \quad \text{where} \quad \begin{cases} umin = u_{\text{Ind}(x_{E-})} - \Delta b = (\text{Ind}(x_{E-}) - 1) \,.\, \Delta b \\ umax = u_{\text{Ind}(x_{E+})+1} + \Delta b = (\text{Ind}(x_{E+}) + 2) \,.\, \Delta b \end{cases} \tag{8}$$

Illustration of equations (8) is given in Fig. 4.

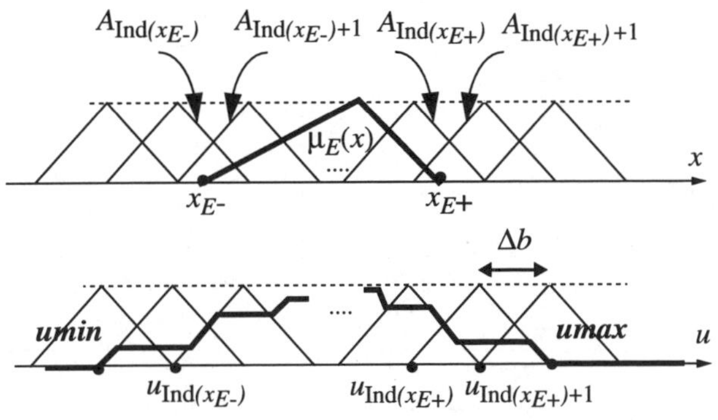

Fig. 4 : Support of the inferred fuzzy subset

Software implementation is finally easily obtained. It only consists of the writing of the functions returning the evaluation of the right hand side of the equations (5), (4) and (8).

The two following sections are dedicated to the presentation of the control law obtained for different kinds of fuzzy input. Regularly distributed triangular membership functions with an overlap of 1 have been assumed for $\mu_{\tau_X(A_i)}$ and $\mu_{\tau_U(B_i)}$ (see Fig. 3) in all the presented results. Furthermore, we have chosen $\Delta a = \Delta b = 1$. So, the limit control law of the fuzzy controller with precise input [Bouslama, 92], [Buckley, 89], [Galichet, 93] is simply: $u = x$, that is that is a proportional controller with a gain $K_p = 1$.

3• Fuzzy triangular inputs

First example consists in examining the behaviour of the fuzzy controller with triangular inputs of different width. The T-norm and ⊥-conorm used in the inference process are the classical min - max operators. The defuzzification is made with the center of gravity method. The fuzzy triangular input is symmetrical with a base of width lE. The different curves in Fig. 5 represent the output control u with respect to the modal value x of the fuzzy triangular input.

Whatever the width of the fuzzy triangular input, the output of the fuzzy controller is similar to the output of the linear controller when the modal value of the fuzzy triangular input coincide with some modal value of the $\mu_{\tau_X(A_i)}$. In our case, it happens when $x = 0$ and $x = 1$.

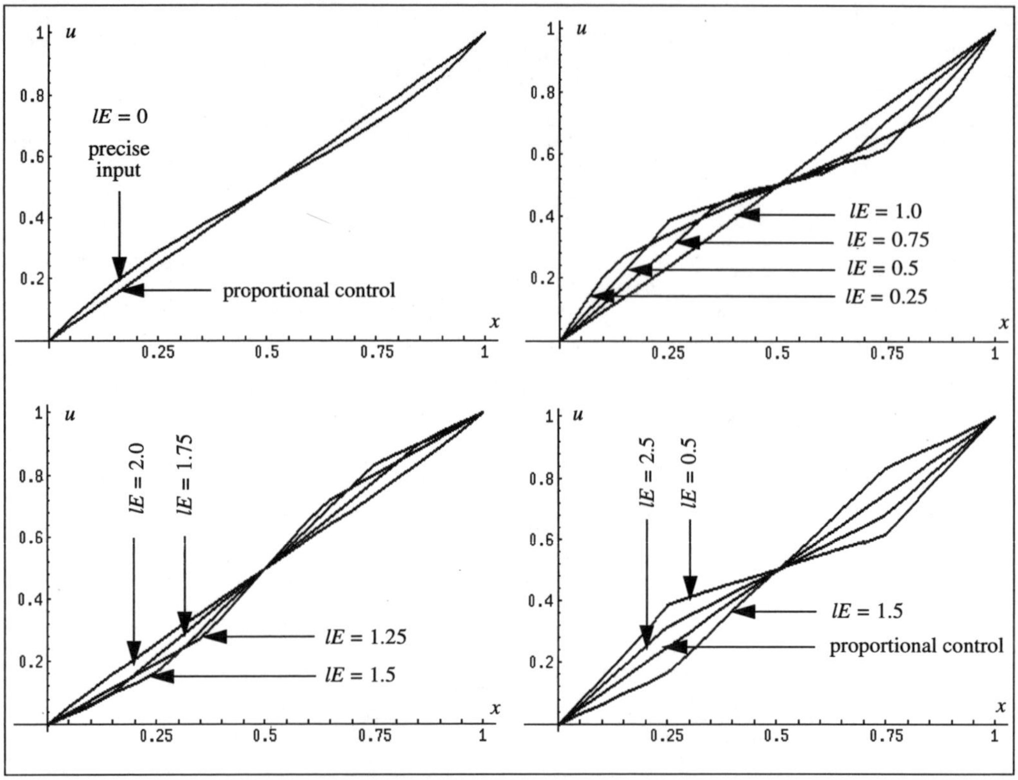

Fig. 5 : Fuzzy triangular inputs with different width

Between the modal values, some non-linear interpolation is obtained. As for fuzzy control with precise input, the middle point, i.e. the point at the same distance to two consecutive modal values, satisfies the linear law. It may also be observed that choosing *lE* as a multiple of Δa, i.e. *lE* = 1, 2, .. , results in approaching the linear control law.

Different origins of the control law non-linearities may be distinguished. Some non-linear effects may appear because of the fuzzy input. But, the inference operators or the defuzzification strategy also introduce non-linearities. In order to correctly evaluate the real consequence of inferring with fuzzy inputs, some results have been produced with linear inference operators and linear defuzzification method.

Linear defuzzification can be obtained by using the "height" defuzzification method [Bouslama, 92]. In that case, the equation (8) becomes:

$$u = \frac{\sum_{i = imin}^{imax} (i \cdot \Delta b)\, \mu_F(i \cdot \Delta b)}{\sum_{i = imin}^{imax} \mu_F(i \cdot \Delta b)} \quad \text{where} \quad imin = \mathrm{Ind}(xE-) \text{ and } imax = \mathrm{Ind}(xE+) + 1 \tag{9}$$

For obtaining linear interpolation it is proper to use the Zadeh AND and the Lukasiewicz OR [Ying, 90], that is:

$$a \, T \, b = a \, . \, b \quad \text{and} \quad a \perp b = \text{Min}\,(1, a{+}b)$$

Fig. 6 shows that linear interpolation is approached by using linear inference operators and defuzzification method. In that case, the fuzzy controller has the same behaviour with precise and fuzzy inputs.

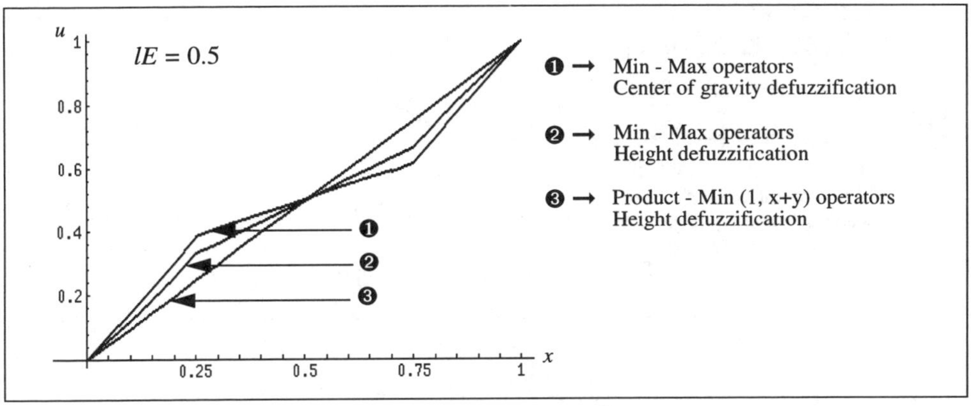

Fig. 6 : Linear inference operators and defuzzification strategy

4• Different shapes of fuzzy input

The same test has been realized for various kinds of fuzzy input. Fig. 7 gives the output control for some fuzzy rectangular input of width *lE*. Results obtained with some fuzzy trapezoidal input are reported in Fig. 8. In that case, *lE* denotes the trapeze support. The trapeze core has been set up to the half. Fuzzy gaussian inputs have also been studied (see Fig. 9). The fuzzy input support has been reduced to $lE = 6\sigma$, where σ denotes the standard deviation of the gaussian membership function. Fig. 10 allows a comparison of the control output with respect to the shape of the fuzzy input.

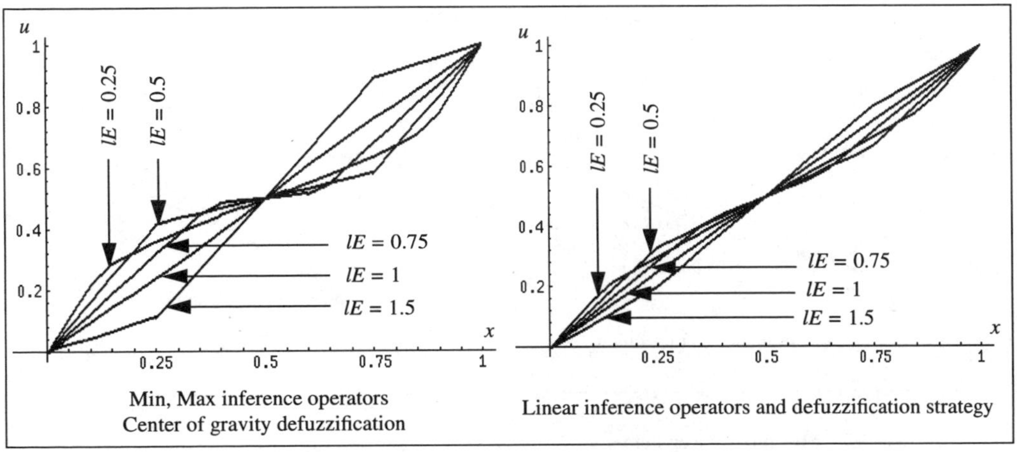

Fig. 7 : Fuzzy rectangular inputs

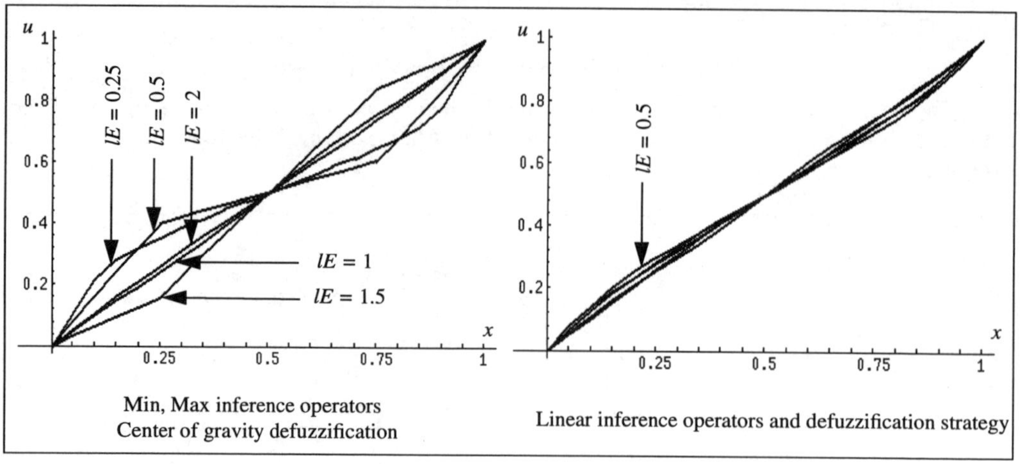

Fig. 8 : Fuzzy trapezoidal inputs

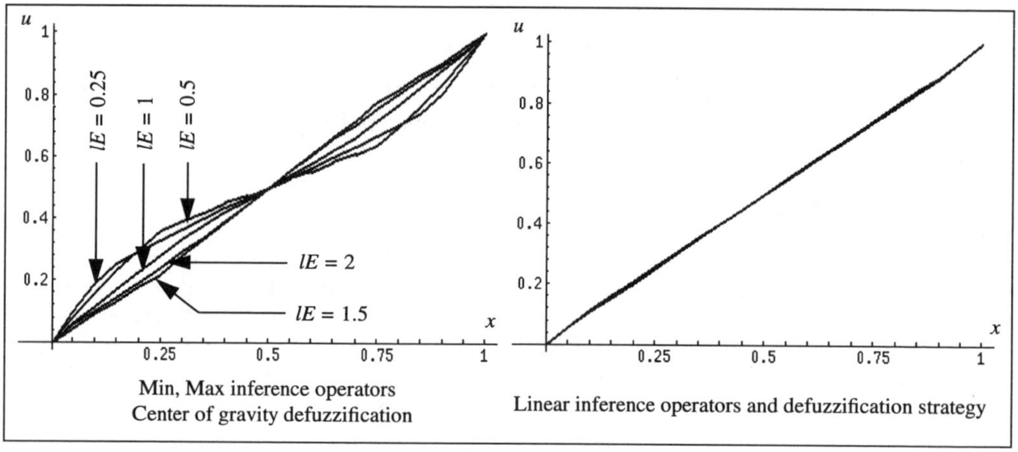

Fig. 9 : Fuzzy gaussian inputs

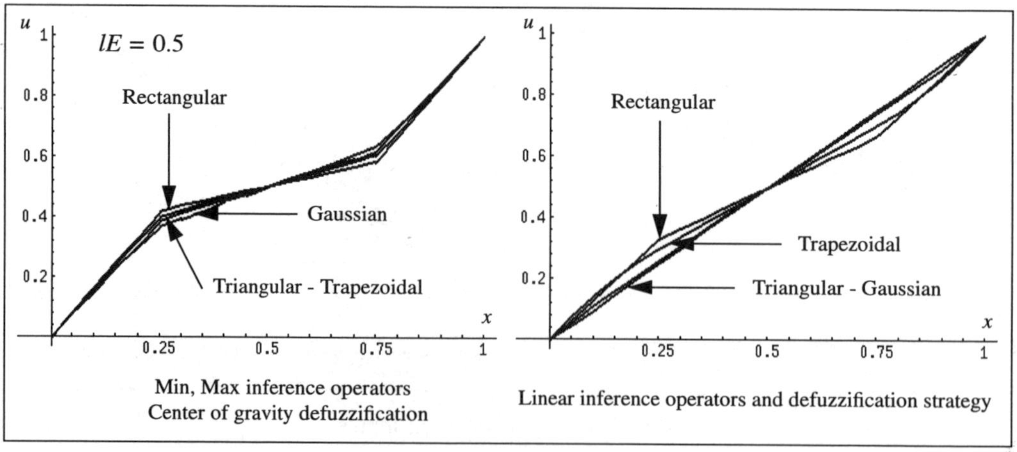

Fig. 10 : Various shapes of fuzzy input

5• Conclusion

The behavior of a Mamdani type fuzzy controller has been experimentally studied when a fuzzy input is assumed. The principal point is that the fuzzy controller delivers the same kind of output control as when dealing with a precise input. In particular, a precise equivalence between the fuzzy controller and a proportional controller is obtained when the fuzzy input is centered on some modal value of the input partition. Furthermore, by using linear inference operators and defuzzification method, the fuzzy controller generates quasi-linear interpolation between the modal values.

Based on these experimental conclusions, further ideas may be stated:

- a fuzzy controller does not require precise information to correctly control a process. In that case, it can be directly interfaced with any system able to produce some information about the process, precise or not.

- the good reaction of a fuzzy controller in the presence of non-precise informations is in accordance with the robustness property observed in many fuzzy control applications.

REFERENCES

[Bouslama, 92] Bouslama F., Ichikawa A., *"Fuzzy control rules and their natural control laws"*, Fuzzy Sets and Systems, 1992, No 48, pp. 65-86.

[Buckley, 89] Buckley J.J., Ying H., "Fuzzy controller Theory ; limit theorems for linear fuzzy control rules", *Automatica*, Vol. 25, No 3, 1989, pp. 469-472.

[Dubois, 91] Dubois D., Prade H., "*Fuzzy sets in approximate reasoning. Part I : Inference with possibility distributions*", Fuzzy Sets and Systems, 25th Anniversary Memorial, Vol 40, No 1, 1991, pp.143-202..

[Galichet, 93] Galichet S., Foulloy L., "*Fuzzy Equivalence of Classical Controllers*", First European Congress on Fuzzy and Intelligent Technologies, EUFIT' 93, Aachen, Germany, September1993.

[Pedrycz, 89] Pedrycz W., "*Fuzzy control and fuzzy systems* -1989, John Wiley & sons inc.

[Ying, 90] Ying H., Siller W., Buckley J., *"Fuzzy control theory : a nonlinear case"*, Automatica, Vol 26, No 3, 1990, pp. 513-520.

[Zadeh, 73] Zadeh L.A., "*Outline of a new approach to the analysis of complex systems and decision processes"*, IEEE trans. on systems, man and cybernetics, 1973, vol. SMC3, no 1, pp. 28-44.

DECISION AND AGGREGATION TECHNIQUES

TOWARD A THEORY OF IMMEDIATE PROBABILITIES

Ronald R. Yager, Kurt J. Engemann & Dimitar P. Filev

Machine Intelligence Institute
Iona College
New Rochelle, NY 10801

ABSTRACT

The concept of immediate probabilities is introduced as a modification of the probabilties of the outcomes with information about the payoffs. We investigate the case where this transformation is mediated through dispositional information (optimism/pessimism). We use the Dempster rule of combination to help obtain an expression for these probabilities. We show how immediate probabilities allows us to explain the Allais paradox.

1 Introduction

The concept of **immediate probability** is based upon the premise that a decision makers perception of probabilities used in decision making are effected by the value of the payoffs associated with the potential outcomes of a decision. Classically decision making environments when no probabilistic information is available involve the introduction of subjective information regarding the decision makers pessimistic/optimistic attitude. The paradigm of decision making under ignorance [1, 2] is based very strongly on the use of this kind of information. However even in environments in which one has probabilistic information pessimistic/optimistic attitudinal information reflects itself into the decision making process. As shown by the Allais [3] paradox the use utility theory doesn't always provide a satisfying way of introducing this kind of information into probabilistic environments. In [4, 7] we have introduced the concept of immediate probabilities by using the pessimistic/optimistic attitudinal (dispositional) information of the decision maker as a method for modifying the probability distributions to reflect the payoffs associated with the possible outcomes of a decision. In this work we look further at the use of the disposition of the decision maker as a method of obtaining immediate probabilities.

2. Decision Making With Dispositions

In decision making under ignorance when probabilistic information is not available, an approach often used [1, 2] is based upon the decision maker's attitude or disposition. Four approaches in this spirit have been often cited:

(1). Pessimistic - The decision maker assumes the worst outcome for each alternative

(2). Optimistic - The decision maker assumes the best outcome for each alternative

(3). Neutral - The decision maker takes the average of the outcomes for each alternative

(4) Hurwicz - The decision maker selects a level of optimism α, between 0 and 1. For a given alternative the decision maker takes α times the optimistic value plus $(1 - \alpha)$ the pessimistic value.

We note that each of these approaches can also be seen as some mapping f from the outcome set for an alternative, A, to real line, f(A) = a,where

(1) **Pessimist**: $f(A) = \text{Min}_i[a_i],\ a_i \in A$

(2) **Optimist:** $f(A) = \text{Max}_i[a_i]$

(3) **Neutral**: $f(A) = \frac{1}{n}\sum_i a_i$

(4) **Hurwicz:**

$$f(A) = \alpha\, \text{Max}_i[a_i] + (1 - \alpha)\, \text{Min}_i[a_i]$$

It is not clear how the decision maker selects his attitude. For example, is the choice of a pessimistic strategy the result of a recent spate of bad luck which is making the decision maker conservative or is it a strategy which allows him to prepare in a conservative way ?

In [5] Yager suggested a generalization of this approach. This generalization is based upon an aggregation using OWA operators [6].

Definition: An OWA operator of dimension n is a mapping f: $R^n \rightarrow R$ that has an associated weighting vector $W = [w_1, \ldots w_n]^T$ such that

(1) $\sum_i w_i = 1$

(2) $w_i \in [0, 1]$

and where

$$f(a_1, \ldots a_n) = \sum_{i=1}^{n} w_j\, b_j$$

with b_j being the jth largest element of the collection of a_i values.

Using this new aggregation, Yager [5] suggested a means for implementing decision making under a disposition. Assume A is an alternative whose potential outcomes are $(a_1, \ldots a_n)$. Let W be a OWA weighting vector reflecting the decision maker disposition. Then the evaluation of the alternative A is

$$f(a_1, \ldots a_n) = \sum_{i=1}^{n} w_j\, b_j.$$

As we indicated, the vector W is used to carry the decision maker's disposition. We note that if W is defined such that $w_1 = 1$ and $w_j = 0$ for all other j then $f(a_1, \ldots a_n) = \text{Max}_i[a_i]$. Thus this vector with one in the top position, which we denote as W^*, portrays the pure optimist.

On the other hand when $w_n = 1$ and $w_j = 0$ for all other j then $f(a_1, \ldots a_n) = \text{Min}_i[a_i]$. Thus this vector with one in the bottom position, which we denote as W_*, portrays the pure pessimist. The neutral case is carried by the weighting vector where $w_i = 1/n$ for all i. The Hurwicz case is represented by a vector in which $w_1 = \alpha$ and $w_n = 1 - \alpha$ and all other weights are zero.

The use of the OWA aggregation operator to implement the evaluation of an alternative in this decision making under disposition allows for a whole family of different types of dispositions, each carried by a different weighting vector W. Thus in this framework the decision maker's attitude is captured by a particular OWA weighting vector W which we call his dispositional vector..

In [6] a measure to capture the degree of optimism inherent in a weighting vector was introduced. Assume we have a weighting vector W used to reflect a decision maker's disposition in a decision environment then the measure of *optimism* associated with that vector is

$$\Omega = \frac{1}{n-1} \sum_{i=1}^{n} w_i (n-i)$$

It can be shown that $\Omega \in [0, 1]$ and that Ω attains the value 1 if and only if W corresponds to the pure optimistic case. Ω attains the value zero if and only if W corresponds to the pure pessimistic case. We note that for the case when $w_i = 1/n$ for all i, then $\Omega = .5$. In the Hurwicz case $\Omega = \alpha$.

In [5] Yager provided a further semantics for this environment which unifies it with the probabilistic approach. We first note that the requirements on the weights in W are such that $w_i \in [0, 1]$ and $\sum_i w_i = 1$, which are exactly the requirements associated with a probability distribution. Consider now the form of the aggregation used in the dispositional method to evaluate the alternative $A = [a_1, \ldots a_n]$,

$$E(A) = \sum_j w_j b_j$$

where b_j is the jth largest element in A. In [5] it was suggested that we can interpret the weight w_j as a probability, in particular *the probability that the jth best thing will occur.* Interpreting the weights in this manner we see that the formulation for E(A) can be viewed as an expected value. We see this as an expected value since w_j may be interpreted as the probability of the jth best outcome and b_j is the jth best outcome.

Using this semantics we see that the classic decision making under ignorance can be viewed in the same framework as decision making with probabilities. Only in this case the probabilities rather than being associated with a particular outcome are associated with an ordered position.

Under this semantics the dispositional vector W can be viewed as a probability distribution, which we shall call the *dispositional probability distribution,* reflecting the decision maker's attitude. Given a dispositional probability distribution we can associate with it a measure of entropy capturing the degree of uncertainty inherent in the disposition,

$$H(W) = -\sum_{i=1}^{n} w_i \ln w_i.$$

In some environments it may be more appropriate to only require the decision maker to indicate a degree α of optimism rather than the whole distribution. Using a method of maximum entropy [8] we can generate the dispositional probability distribution, the weighting vector W. In particular we can generate the maximum entropy probability by solving the following constrained optimization problem for the w_i:

$$\text{Max} -\sum_{i=1}^{n} w_i \ln w_i.$$

<u>s/t</u>:

$$\alpha = -\frac{1}{n-1} \sum_{i=1}^{n} w_i (n-i)$$

$$\sum_{i=1}^{n} w_i = 1$$

$$w_i \in [0, 1]$$

The probabilities generated using this approach are *ME-OWA* probabilities. In [4] an alternative approach to generating the vector W is suggested.

In our discussion of decision making under ignorance we have assumed that the dispositional probabilities are a reflection of the decision maker's attitude on the pessimistic/optimistic scale. In the techniques discussed we have assumed that these probabilities are independent of the actual payoffs. It is interesting to consider the possibility that the dispositional probabilities are a function of the actual payoffs, the a_i's. In the following we briefly suggest an approach to implement this agenda.

One possible model for incorporating the dispositional probabilities is based upon the BADD-OWA aggregation introduced by Yager and Filev [9]. In this environment the weights, dispositional probabilities, are defined as

$$w_i = \frac{(a_i)^\gamma}{\sum_j a_j^\gamma}$$

where $\gamma \in [-\infty, \infty]$. In the case when $\gamma = \infty$, we can easily show that

$$w_i = 1 \qquad \text{for } a_i = \text{Max}_i[a_i]$$
$$w_i = 0 \qquad \text{for all others}$$

Thus in the case when $\gamma = \infty$ we get the pure optimist. When $\gamma = -\infty$ we get that

$$w_i = 1 \qquad \text{for } a_i = \text{Min}_i[a_i]$$
$$w_i = 0 \qquad \text{for all others}$$

Thus this is the pure pessimist.

When $\gamma = 0$ we see that

$$w_i = 1/n$$

and hence we get the neutral disposition.

In general we note that for $\gamma > 0$ we have that if $a_i > a_j$ then $w_i \geq w_j$. Thus this is an optimistic type of attitude where the better outcomes have higher probability. We also see that if $\gamma < 0$ we have that if $a_i > a_j$ then $w_i < w_j$. This can be seen as a pessimistic attitude.

In the case when $\gamma = 1$, here we get $w_i = \frac{a_i}{\sum a_j}$ and the expected value, the OWA aggregation, is $E(V) = \frac{a_i^2}{\sum a_j}$. In the case when $\gamma = -1$ we get $w_i = \frac{\frac{1}{a_i}}{\sum \frac{1}{a_j}}$ and therefore $E(V) = \frac{1}{\sum_{i=1}^{n} \frac{1}{a_i}}$

3. Immediate Probabilities

In decision environments in which we have probabilistic information about the outcomes associated with the actions we use this information to help in the decision process. However even in situations in which we have very good probabilistic information about the outcomes, decision makers still only use a modified version of these probabilities to guide their actions in the current immediate problem.

Consider a fair die with six sides. The probability of occurrence of each of the sides is 1/6. Assume now we associate some payoff with each of the sides. In particular assume that we assign a payoff of $1,000,000 to the face with six on it and a $100 payoff to each of the other sides. Most decision makers will not act as if the probability associated with the side with six dots is the same as all other sides. The probabilities which guide the decision maker's action in this lottery have been distorted from the generally acceptable 1/6 to reflect some information about this immediate decision to be made. In particular some reflection of the payoffs associated with each of the outcomes will be used by the decision maker. Thus even in situations where the probabilities are obtained from the structure of the situation we can find a modification. The situation is more pronounced in cases where the probabilities are drawn from past history.

We feel that one cause of the transformation of what we shall call the **pristine probabilities** to the immediate decision is a reflection of the decision maker's attitude (disposition) reflected through the payoffs. It is important to distinguish this from the utility theory [1] where the transformation of a decision is manifested through a transformation of the value of the outcomes rather than the probabilities themselves. To reflect this transformation of the pristine probabilities to the immediate decision at hand we [4, 7] introduced the idea of **immediate probabilities**. In this section, with the aid of the Dempster rule of combination [10], we provide a formal derivation of the concept of immediate probabilities.

Assume we have a decision making problem involving an alternative under doubt with n possible outcomes, $a_1, \ldots a_n$. In the following we shall assume that the outcomes have been indexed such that a_i is the ith best outcome. We are interested in determining information about the outcome which will occur. Assume we have available some probability distribution based upon either historical observation of similar experiments or upon the structure of the problem. We shall denote these probabilities as $p_1, \ldots p_n$. We shall call these the pristine probabilities. We can view these pristine probabilities as some piece of evidence about what is going to happen in the current situation.

A second piece of evidence which we have regarding the current performance of the action is the *decision maker's attitude* captured in terms of a dispositional probability vector W where w_i is the decision makers probability that the ith best outcome will happen.

In this situation we have two sources of evidence regarding the appropriate probabilities for the current decision, one being the pristine probabilities and the other being the dispositional probabilities. We are faced with the problem of combining these two pieces of evidence. In order to accomplish this task we shall represent these two pieces of information as belief structures [10] and then use Dempster's rule of combination to combine them.

If X is a set, a Dempster-Shafer belief structure on X is a collection of non-null subsets of X, F_i, called focal elements and a mapping

$$m:\ 2^X \rightarrow [0, 1]$$

called the <u>basic assignment function</u> such that

$$1)\ m(F_i) > 0 \text{ and } 2)\ \sum m(F_i) = 1.$$

A special case of a Dempster-Shafer belief structure is one where the focal elements are singletons, these are called Bayesian belief structures.

We consider the first piece of evidence to be the pristine probabilities. In the following we let $X = \{a_1, \ldots a_n\}$ be the set of outcomes associated with a decision which have non-zero pristine probabilities. In this case we represent this information by the belief function m_1, which has singleton focal elements, thus $E_i = \{a_i\}$ and $m_1(E_i) = p_i$, the probability associated with outcome a_i.

The second piece of evidence is the decision maker's dispositional attitude. We represent this also as a belief structure with belief function m_2 which has singleton focal elements F_j where

$$F_j = \{\text{the } j^{th} \text{ best outcome}\}$$

and where $m_2(F_j) = w_j$, the dispositional probability associated with occurrence of the jth best outcome.

Dempster's rule provides a means for combining these two pieces of evidence regarding the occurrence of the appropriate outcome. In particular the resulting belief structure,

$$m = m_1 \oplus m_2,$$

is defined as follows. For all $E_i \cap F_j \neq \Phi$ we have

$$m(E_i \cap F_j) = \frac{m_1(E_i) * m_2(F_j)}{1 - C}$$

where

$$C = \sum_{\substack{i,\, j \\ E_i \cap E_j = \Phi}} m_1(E_i) * m_2(F_j)$$

Let us look at this formulation in more detail. First it can be easily seen that

$$1 - C = \sum_{\substack{i,\, j \\ E_i \cap E_j \neq \Phi}} m_1(E_i) * m_2(F_j) = K$$

thus for $E_i \cap F_j \neq \Phi$ we get

$$m(E_i \cap F_j) = \frac{m_1(E_i) * m_2(E_j)}{\sum\limits_{\substack{i,j \\ E_i \cap E_j \neq 0}} m_1(E_i) * m_2(E_j)}$$

Consider now the intersection $E_i \cap F_j$. First both $E_i = \{a_i\}$ and $F_j = \{j^{th}$ best outcome$\}$ are singleton sets. Since we have previously assumed that the indexing of the payoffs are such that a_j is the jth best outcome we have then $F_j = \{a_j\}$. From this we see that

$$E_i \cap F_j = \Phi \text{ if } i \neq j$$

$$E_i \cap F_j = a_i \text{ if } i = j$$

Thus we get

$$m(E_i \cap F_i) = m(\{a_i\}) = \frac{m_1(E_i) * m_2(F_i)}{\sum\limits_{i=1} m_1(E_i) * m_2(F_i)}$$

$$m(\{a_i\}) = \frac{p_i * w_i}{\sum\limits_{i=1}^{n} p_i * w_i}$$

We call these resulting values the **immediate probabilities** associated with the outcome a_i, and indicate it as

$$\widehat{p}_i = \frac{p_i * w_i}{\sum\limits_{i=1}^{n} p_i * w_i}.$$

We use these probabilities to define the immediate expected value (IEV) defined as

$$IEV(A) = \sum_{i=1}^{n} \widehat{p}_i\, a_i$$

In [4] we indicate that it is this immediate expected value that the decision maker uses to evaluate a particular alternative and hence compare alternatives.

It should be noted that the assumption that $p_i > 0$ for all the outcomes under consideration is very important, it guarantees that the $\hat{p}_i$ are well defined. We see that since $p_i > 0$ for all i this guarantees that $w_i \, p_i > 0$ for at least one i which assures us that the $\hat{p}_i$ are well defined.

4. Immediate Probabilities and Allais Paradox

In introducing the concept of immediate probabilities we have provided a mechanism for allowing a transformation of the pristine probabilities associated with a decision. This transformation is based upon a subjective component involving the decision maker's disposition. It should be carefully noted that this idea is distinct from the concept of utilities often used in the decision literature. In the case of utility theory it is the value of the outcome that is transformed based upon the decision maker's disposition not the value of the probability associated with an outcome.

We now show how the introduction of the immediate probabilities allows to solve an important paradox in the decision literature called the Allais paradox [3]. Also, see Tsukamoto [11] for an alternative solution to this paradox.

In the following we shall consider four lotteries. Each lottery will be described in the format $[p_1, p_2: a_1, a_2]$ where a_i and a_2 are the outcomes and p_1 and p_2 are their respective pristine probabilities. The following are the four lotteries:

L_1: [1, 0: \$30, 0] L_2: [.8, .2: 40, 0]

L_3: [.25, .75: 30, 0] L_4: [.2, .8: 40, 0]

Experimental observations have indicated that in most cases people have the preference $L_1 > L_2$ and $L_4 > L_3$. Since the expected values of these lotteries are $E(L_1) = 30$; $E(L_2) = 32$; $E(L_3) = 7.5$; $E(L_4) = 8$ the preference $L_1 > L_2$ is not justified based upon the use of expected value.

One potential explanation of these preferences is in terms of utility theory. In this case we transform the payoffs to reflect their values or utility to the decision maker. We shall denote these utilities, U(A), where A is some dollar value. The only requirement we shall impose is that if $A > B$ then $U(A) \geq U(B)$. Consider now the use of expected utility, EU. In the case of our lotteries we get $EU(L_1) = U(30)$; $EU(L_2) = .8U(40) + .2U(0)$; $EU(L_3) = .25U(30) + .75U(0)$; $EU(L_4) = .2U(40) + .8U(0)$. From the experimental observations we require that

I) $L_1 > L_2$: $U(30) > .8U(40) + .2U(0)$

II) $L_4 > L_3$:

$$.2U(40) + .8U(0) > .25U(30) + .75U(0)$$

From II we see that $8U(40) + 3.2U(0) > U(30) + 3U(0)$ and hence $.8U(40) + .2U(0) > U(30)$ which contradicts I. Thus I and II are in contradiction and hence utility theory can't explain the preferences.

We now consider an explanation using the idea of immediate probability. We consider the decision maker has a disposition vector

$$W = \begin{bmatrix} \alpha \\ \\ 1-\alpha \end{bmatrix}$$

The first lottery has $\hat{p}_i = p_i$ and therefore $IEV(L_1) = 30$. For lottery two, since $a_1 > a_2$ then we get

$$\hat{p}_1 = \frac{p_1\alpha}{p_1\alpha + p_2(1-\alpha)} = \frac{.8\alpha}{.2 + .6\alpha}$$

and $\hat{p}_2 = 1 - \frac{.8\alpha}{.2 + .6\alpha}$.

In this second lottery

$$IEV(L_2) = \hat{p}_1 40 + p_2 0 = \hat{p}_1 40$$

and thus

$$IEV(L_2) = \frac{(40).8\alpha}{.2 + .6\alpha}.$$

The requirement that $L_1 > L_2$ imposes the condition that $30 > \frac{32\alpha}{.2 + .6\alpha}$ and requires that $\alpha < 3/7$.

In the third lottery we get

$$\hat{p}_1 = \frac{(.25)\alpha}{(.25)\alpha + (1-\alpha)(.75)} = \frac{\alpha}{3 - 2\alpha}$$

and hence $IEV(L_3) = \frac{30\alpha}{3 - 2\alpha}$

In the fourth lottery we get

$$\hat{p}_1 = \frac{(.2)\alpha}{(.2)\alpha + (1-\alpha)(.8)} = \frac{\alpha}{4 - 3\alpha}$$

and hence

$$IEV(L_4) = \frac{40\alpha}{4 - 3\alpha}$$

The requirement that $L_4 > L_3$ implies

$$\frac{40\alpha}{4 - 3\alpha} > \frac{30\alpha}{3 - 2\alpha}$$

$$(4)(3 - 2\alpha) > 3(4 - 3\alpha)$$

$$\alpha > 0$$

Since we have already required $\alpha < 3/7$ we see that the required condition is $\alpha < 3/7$ to satisfy both conditions. It is interesting to note the requirement that $\alpha < 3/7$ implies a pessimistic attitude on the part of the experimental subjects.

Thus we see that this paradox can be explained by using the transformation of the probabilities using the concept of immediate probabilities.

5. References

[1]. Luce, R. D. and Raiffa, H., Games and Decisions: Introduction and Critical Survey, John Wiley & Sons: New York, 1967.

[2]. Render, B. and Stair, R.M., Introduction to Management Science, Allyn and Bacon: Boston, 1992.

[3]. Allais, M., "Le comportement de l'homme rationnel devant le risque: Critique des postulats et axioms de l'ecole Americaine," Econometrica 21, 503-546, 1953.

[4]. Engemann, K. J., Filev, D. P. and Yager, R. R., "Modeling decision making using immediate probabilities," International Journal of General Systems, To Appear.

[5]. Yager, R. R., "Decision making under Dempster-Shafer uncertainties," International Journal of General Systems 20, 233-245, 1992.

[6]. Yager, R. R., "On ordered weighted averaging aggregation operators in multi-criteria decision making," IEEE

Transactions on Systems, Man and Cybernetics 18, 183-190, 1988.
[7]. Yager, R.R., Engemann, K and Filev, D.P. "On the concept of immediate probabilities," International Journal of Intelligent Systems, to Appear.
[8]. O'Hagan, M., "Using maximum entropy-ordered weighted averaging to construct a fuzzy neuron," Proceedings 24th Annual IEEE Asilomar Conf. on Signals, Systems and Computers, Pacific Grove, Ca, 618-623, 1990.
[9]. Yager, R. R. and Filev, D. P., "Parameterized "andlike" and "orlike" OWA operators," International Journal of General Systems, (To Appear).
[10]. Shafer, G., A Mathematical Theory of Evidence, Princeton University Press: Princeton, N.J., 1976.
[11]. Tsukamoto, Y., "A fuzzy measure model of uncertainty averter," Proceedings of the First Asian Conference on Fuzzy Sets and Systems, Singapore, 1993.

THE COMPUTATIONAL PROBLEM OF USING OWA OPERATORS

Vincenzo CUTELLO
Dep. of Mathematics
University of Catania
Catania, Italy

Javier MONTERO
Dep. of Statistics and O.R.
Complutense University
Madrid, Spain

Abstract

In this paper we will analyze some computational problems related with the use of OWA operators as information aggregators. In particular we will concentrate on ordered hierarchical aggregations of OWA operators as defined in [4].

Key words: Aggregation operators, combinatorial optimization.

1 Introduction and preliminaries

Ordered Weighted Averaging (OWA) operators were firstly proposed by Yager in [13] and studied subsequentely by many other authors both from a theoretical and a practical point of view (see for instance [4,6,8,14]). It is well known that T-norms and T-conorms represent aggregation operators that generalize the notion of conjunction and disjunction of classical logic with, in particular, the min operator being the maximal T-norm and the max operator being the minimal T-conorm (see [10].) OWA operators verify the nice property of filling the gap between min and max. Intuitively, then, by means of OWA operators we can go from conjunction (intersection) to disjunction (union) in a continuous way.

T-norms and T-conorms are also associative. Thus, given any T-norm or T-conorm $F(x, y)$ and given any n values $(a_1, \ldots, a_n)$, we can apply the same operative definition of $F(x, y)$ to obtain $F(a_1, \ldots, a_n)$. More in details, we see that we can evaluate each value $F(a_1, \ldots, a_n)$ either as $F(F(a_1, \ldots, a_{n-1}), a_n)$ or $F(a_1, F(a_2, \ldots, a_n))$. Consequentely, since we can evaluate in time $\mathcal{O}(1)$, i.e. in constant time, the value $F(a, b)$ for any pair a, b we can evaluate $F(a_1, \ldots, a_n)$ in time $\mathcal{O}(n)$, i.e. in linear time.

1.1 OWA operators

To simplify the formalization of OWA operators we make us of the notion of *sorting permutation* of a list.

If $L = [a_1, a_2, \ldots, a_n]$ is a list of numbers, a sorting permutation σ for L is any permutation of the elements of L that produces a list $\sigma(L) = [a_{[1]}, \ldots, a_{[n]}]$ verifying

$$a_{[i]} \geq a_{[j]}$$

for all $i \leq j$.

DEFINITION 1.1 *An OWA operator of dimension n is an aggregation operator ϕ that has an associated list of weights $W = [w_1, \ldots, w_n]$ such that*

1. $w_i \in [0, 1]$ *for all* $1 \leq i \leq n$
2. $\sum_{i=1}^{n} w_i = 1$

3. for any $L = [a_1, a_1, \ldots, a_n]$ *and its corresponding* $\sigma(L) = [a_{[1]}, \ldots, a_{[n]}]$

$$\phi(L) = \sum_{i=1}^{n} w_i a_{[i]}.$$

□

In view of the above definition, it can be immediately verified that OWA operators are commutative, monotone and idempotent. Moreover, the value $\phi(a_1, \ldots, a_n)$ needs an $\mathcal{O}(n \log n)$ time to be evaluated since the n input numbers must be sorted.

As previously commented, for any OWA operator ϕ we have

$$\min_i(a_1, \ldots, a_n) \leq \phi(a_1, \ldots, a_n) \leq \max_i(a_1, \ldots, a_n) \tag{1.1}$$

Two significative measures are associated with OWA operators of dimension n.

(m1) The first measure is called *orness* and it estimates how close an OWA operator is to the max operator. In details the degree of orness is defined as

$$orness(\phi) = \frac{1}{n-1} \sum_{i=1}^{n} (n-i) w_i.$$

As shown in [13],

$$orness(\phi) \leq orness(\phi')$$

when their associated weights $(w_1, \ldots, w_n)$ and $(w'_1, \ldots, w'_n)$ verify that

$$\sum_{j=1}^{i} w_j \leq \sum_{j=1}^{i} w'_j \quad \forall i = 1, \ldots, n \tag{1.2}$$

The converse of the above is obviously not true. This property gives in fact an intuitive idea of when an OWA operator must be more an "or" than one other OWA operator.

Dual to the measure of *orness* is the measure of *andness* defined as

$$andness(\phi) = 1 - orness(\phi)$$

which therefore measures how close an OWA operator is to the min operator.

(m2) The second measure called *dispersion* estimates the degree to which all aggregates are used equally, i.e. the degree to which an OWA operator is close to the simple average operator. The degree of dispersion is defined as

$$Disp(\phi) = -\sum_{i=1}^{n} w_i \ln w_i.$$

2 Ordered hierarchies of OWA operators

2.1 Hierarchical aggregations

In some recent scientific investigations the notion of hierarchical aggregation of information has been introduced along with some characterization theorems (see [3,4]). Intuitively, hierarchical aggregations are aggregations of chunks of information which in turn represent aggregated information. The practical consequences of such hierarchical aggregations are quite interesting. If we have aggregation maps of very big dimensions, hierarchical aggregations will allow us to deal with sub-aggregation operators of smaller dimensions, whose computational jobs can be parallelized. Thus, it will be possible to obtain a significant speed up of the whole aggregation process. Moreover, hierarchical aggregations allow to treat input data in different ways according to the underlying semantic of the aggregation process. Input data can be seen as divided into clusters, whose representative is decided according to an ad hoc aggregation operator, and, in turn these resulting representatives can be aggregated by yet another ad hoc aggregation operator.

2.2 Hierarchical aggregations of OWA operators

If the input data are real numbers then by considering the natural ordering on the reals we can talk about "ordered" hierarchical aggregation rules. This type of hierarchical aggregations allows very naturally to use of Yager's OWA operators as aggregation operators.

Formally, let $\phi_0, \phi_1, \ldots, \phi_c$ be $c+1$ OWA operators such that

- ϕ_0 has dimension c;
- ϕ_i has dimension h_i for any $i = 1, 2, \ldots, c$
- $\sum_{i=1}^{c} h_i = n$

Let $w_{0,1}, \ldots, w_{0,c}$ be the weights associated to ϕ_0, and for all $i = 1, \ldots, c$ let $w_{i,1}, \ldots, w_{i,h_i}$ be the weights associated to ϕ_i.

DEFINITION 2.1 *The* ordered hierachical composition *of $\phi_0, \phi_1, \ldots, \phi_c$ is defined by*

$$\phi_0(\phi_1, \ldots, \phi_c)(a_1, \ldots, a_n) = \phi_0(\phi_1(a_{[1]}, \ldots, a_{[h_1]}), \ldots, \phi_c(a_{[n-h_c+1]}, \ldots, a_{[n]}))$$

for all n-uples $(a_1, \ldots, a_n)$. □

DEFINITION 2.2 *If a property P that holds for $\phi_0, \phi_1, \phi_2, \ldots, \phi_c$ holds for the ordered hierarchical aggregation $\phi_0(\phi_1, \ldots, \phi_c)$ as well, we will say that P* propagates under ordered hierarchical aggregations. □

As proven in [4] we have that the property of being an OWA operator propagates under ordered hierarchical aggregations as well as Monotonicity, Idempotency, Commutativity.

3 On the degrees of Orness and Dispersion

Given a fixed ordered hierarchical aggregation $\phi \equiv \phi_0(\phi_1, \ldots, \phi_c)$ based upon $c+1$ OWA operators, Yagers's orness measure takes the expression

$$orness(\phi) = \phi(\frac{n-1}{n-1}, \frac{n-2}{n-1}, \ldots, \frac{1}{n-1}, 0) = \phi_0(z_1, \ldots, z_i, \ldots, z_c)$$

where

$$z_i = \phi_i(\frac{n - (\sum_{j=1}^{i-1} h_j + 1)}{n-1}, \ldots, \frac{n - \sum_{j=1}^{i} h_j}{n-1}) =$$

$$\sum_{m=1}^{h_i} \frac{n - (\sum_{j=1}^{i-1} h_j + m)}{n-1} w_{i,m} = \frac{n - \sum_{j=1}^{i} h_j}{n-1} + \frac{\sum_{m=1}^{h_i} (h_i - m) w_{i,m}}{n-1} =$$

$$\frac{n - \sum_{j=1}^{i} h_j}{n-1} + (h_i - 1)\frac{orness(\phi_i)}{n-1}$$

for all $i = 1, \ldots, c$, with h_i the dimension of each ϕ_i and $(w_{i,1}, \ldots, w_{i,h_i})$ its associated weights. Hence,

$$orness(\phi) = \sum_{i=1}^{c} \frac{(h_i - 1) orness(\phi_i) + n - \sum_{j=1}^{i} h_j}{n-1} w_{0,i}$$

Obviously, changes in ϕ_j's weights for $j = 1, \ldots, c$ in such a way that $orness(\phi_j)$ does not decrease for any $j = 1, \ldots, c$ will never make $orness(\phi)$ decrease. However, increasing $orness(\phi_0)$ does not necessarely increase $orness(\phi)$ as shown by the following example.

Example 3.1 Let ϕ_0, ϕ'_0 be two OWA operators of dimension c such that their associated weights are, respectively, $w_{0,2} = 1, w_{0,i} = 0$ for all $i \neq 2$ and $w'_{0,1} = w'_{0,3} = 1/2, w'_{0,i} = 0$ for all $i \neq 1,3$. Then, $orness(\phi_0) = orness(\phi'_0)$. Let now ϕ_i be the minimum rule (i.e., $orness(\phi_i) = 0$ for all $i = 1, \dots, c$). Then the difference between $orness(\phi_0(\phi_1, \dots, \phi_c))$ and $orness(\phi'_0(\phi_1, \dots, \phi_c))$ will still depend on the relative sizes h_2 and h_3 of OWA operators ϕ_2, ϕ_3. □

The dispersion of the hierarchical aggregation can be computed as follows.

Denote by ϕ an OWA operator obtained from an ordered hierarchical aggregation of OWA operators $\phi_0, \phi_1, \dots, \phi_c$ with dimensions $c, h_1, \dots, h_c$ respectively. Clearly,

$$\begin{aligned} Disp(\phi) &= -\sum_{j=1}^{c}\sum_{i=1}^{h_j} w_{0,j}w_{j,i}\ln[w_{0,j}w_{j,i}] \\ &= -\sum_{j=1}^{c}\sum_{i=1}^{h_j}(w_{0,j}w_{j,i}\ln w_{0,j} + w_{0,j}w_{j,i}\ln w_{j,i}) \\ &= -\sum_{j=1}^{c} w_{0,j}\ln w_{0,j} - \sum_{j=1}^{c} w_{0,j}Disp(\phi_j) \\ &= Disp(\phi_0) + \sum_{j=1}^{c} w_{0,j}Disp(\phi_j) \end{aligned}$$

Therefore, the dispersion of the hierarchical aggregations depends directly upon the dispersions of the given OWA operators.

In both cases it is therefore clear that once the operators ϕ_0 and $\phi_1, \dots, \phi_c$ are fixed then the ornes and dispersion values of the hierarchical aggregation depend upon the particular ordering of the c OWA operators $\phi_1, \dots, \phi_c$.

It is then natural to ask the following questions:

(1) How can one *quickly* find an ordering of those c OWA aggregated operators which maximizes [resp. minimizes] the dispersion value of the hierarchical aggregation ?

(2) How can one *quickly* find an ordering of those c OWA aggregated operators which maximizes [resp. minimizes] the orness value of the hierarchical aggregation ?

The above mentioned computational problems have already been proposed and characterized in [5,7]

4 Maximizing and minimizing dispersion and orness

Let ϕ_0 be an OWA operator of dimension c and let $\Phi = \{\phi_i : 1 \leq i \leq c\}$ be a given set of OWA operators and let $h_1, \dots, h_c$ their respective dimensions.

4.1 Dispersion

We start by providing polynomial algorithms to maximize [resp. minimize] the dispersion of the hierarchical aggregation.

Let π be a permutation of the indices $\{1, 2, \dots, c\}$ such that

$$i \leq j \text{ if and only if } w_{0,\pi(i)} \geq w_{0,\pi(j)}.$$

Intuitively, for $i = 1, 2, \dots, c$, $w_{0,\pi(i)}$ is the i-th weight in non-increasing order.

Let $\phi_1, \dots, \phi_c$ be a given ordering of the OWA operators in Φ.

The following lemmas hold.

LEMMA 4.1 *The value* $Disp(\phi) = Disp(\phi_0) + \sum_{j=1}^{c} w_{0,j} Disp(\phi_j)$ *is maximum if and only if the ordering* $\phi_1, \ldots, \phi_c$ *verifies*

$$i \leq j \text{ if and only if } Disp(\phi_{\pi(i)}) \geq Disp(\phi_{\pi(j)}).$$

■

LEMMA 4.2 *The value* $Disp(\phi) = Disp(\phi_0) + \sum_{j=1}^{c} w_{0,j} Disp(\phi_j)$ *is minimum if and only if the ordering* $\phi_1, \ldots, \phi_c$ *verifies*

$$i \leq j \text{ if and only if } Disp(\phi_{\pi(i)}) \leq Disp(\phi_{\pi(j)})$$

■

In view of the above Lemmas and from the well-known result (see, e.g., [1]) that any given n numbers can be sorted in $\mathcal{O}(n \log n)$ time, we can claim that the following theorem is true.

THEOREM 4.1 *Given an OWA operator* ϕ_0 *of dimension c and a set of c OWA operators* Φ *and given their dispersions, there exists a* $\mathcal{O}(c \log c)$ *algorithm which produces an ordering of* Φ *maximizing (or minimizing) the dispersion of the hierarchical aggregation.* □

4.2 Orness

The problem of maximizing or minimizing the orness of the hierarchical aggregation at a first sight appears to be quite more difficult from a computational point of view and its (polynomial) solution definitely more subtle. However such a problem can be reduced to the classical combinatorial *assignment problem* for which there exists a $\mathcal{O}(c^3)$ algorithm (see [7] for more details).

Briefly, we have that:

- denoted with $\phi_1, \phi_2, \ldots, \phi_c$ the OWA operators in Φ, and
- denoted for simplicity with δ_{ij} the contribution to the orness of ϕ of the operator ϕ_i once it is chosen as the j-th in the hierarchical aggregation, then
- the problem of maximizing [resp. minimizing] the orness of the hierarchical aggregation is equivalent to the problem of finding a permutation π of the rows of the matrix

$$M = \begin{pmatrix} \delta_{11} & \delta_{12} & \ldots & \delta_{1c} \\ \delta_{21} & \delta_{22} & \ldots & \delta_{2c} \\ \vdots & \vdots & \ddots & \vdots \\ \delta_{c1} & \delta_{c2} & \ldots & \delta_{cc} \end{pmatrix}$$

such that the sum of the elements of the main diagonal of

$$M_\pi = \begin{pmatrix} \delta_{\pi(1)1} & \delta_{\pi(1)2} & \ldots & \delta_{\pi(1)c} \\ \delta_{\pi(2)1} & \delta_{\pi(2)2} & \ldots & \delta_{\pi(2)c} \\ \vdots & \vdots & \ddots & \vdots \\ \delta_{\pi(c)1} & \delta_{\pi(c)2} & \ldots & \delta_{\pi(c)c} \end{pmatrix}$$

is maximum [resp. minimum], i.e. such that

$$\sum_{i=1}^{c} \delta_{\pi(i)i}$$

is maximum [resp. minimum].

The above is the classical ***assignment problem*** (also known as the ***bipartite weighted matching problem***) (see [12], chapter 11) and can be solved by the so-called ***hungarian method*** (see [11]) in time $\mathcal{O}(c^3)$. More recently in [9] it has been proven that by using special data structures it is possible to improve the above bound to $\mathcal{O}(c^2 \log c)$.

As proven in [5] we have that $\mathcal{O}(c \log c)$ algorithms can be given for the most significative cases in practice. Since each one of the given c OWA operators represents partial information to be aggregated into only one index, it is common to define them in such a way that either

- they contain the same amount of information. In this case, all OWA operators in Φ will have the same dimension h such that $n = hc$; or
- they treat their inputs with the same degree of optimism (pessimism). In this case, all OWA operators will have the same degree of orness.

5 Final Comments

In this paper, we continue our research work aimed at a full characterization of hierarchical aggregation operators (see [2,3]) by focussing our attention on the hierarchical aggregations of OWA operators (see [4]) and to the related computational problems ([5]). Since the property of being an OWA operator propagates under ordered hierarchical aggregations and since the orness and dispersion of the obtained OWA operator depend upon the given ordering of the aggregated OWA operators, we have studied how this values can be maximized or minimized. In particular, we have provided polynomial algorithms which solve the general problem for both the dispersion and the orness.

Acknowledgment: This research has been partially supported by Dirección General de Investigación Cientifica y Técnica (Spanish national grant PB91-0389).

References

[1] T.H. Cormen, C.E. Leiserson and R.R. Rivest. *Introduction to Algorithms.* MIT Press, Cambridge, MA, 1990.

[2] V. Cutello and J. Montero. A characterization of rational amalgamation operations. *International Journal of Approximate Reasoning*, vol. 8 (4), pp. 325-344, June 1993.

[3] V. Cutello and J. Montero. Hierarchies of intensity preference aggregations. *International Journal of Approximate Reasoning*, vol.10:123-133, 1994.

[4] V. Cutello and J. Montero. Hierarchies of aggregation operators. *International Journal of Intelligent Systems* , 9:1025-1045, 1994.

[5] V. Cutello and J. Montero. Computational problems of the hierarchical aggregation of OWA operators. *Proceedings of the 5-th Internation IPMU Conference* , Paris, July 1994, pg. 407-411.

[6] V. Cutello and J. Montero. Recursive Families of OWA operators *Proceedings of IEEE World Congress on Computational Intelligence, FUZZ-IEEE 94*, Orlando, Florida, June 1994. P. Bonissone Ed, IEEE Press, pp. 2046-2049.

[7] V. Cutello and J. Montero. Hierarchical aggregation of OWA operators: basic measures and related computational problems *International Journal of Uncertainty, Fuzzyness and Knowledge-Bases* , 1995. To appear.

[8] J. Fodor, J.-L. Marichal and M. Roubens Characterization of the ordered weighted averaging operators. *IEEE Transactions on Fuzzy Systems.* To appear.

[9] M.L. Fredman and R.E. Tarjan. Fibonacci heaps and their uses in improved network optimization algorithms. *Journal of A.C.M.*, 34(3):596-615, 1987.

[10] G.J. Klir and T.A. Folger. *Fuzzy sets, uncertainty and information.* Prentice Hall, Englewood Cliffs, NJ, 1988.

[11] H.W Kuhn. The Hungarian Method for the assignment Problem. *Naval Research Logistics Quarterly*, 2:83-97, 1955.

[12] C.H. Papadimitriou and K. Steiglitz. *Combinatorial Optimization. Algorithms and Complexity.* Prentice Hall, Englewood Cliffs, NJ, 1982.

[13] R.R. Yager. On ordered weighted averaging aggregation operators in multi-criteria decision making. *IEEE transactions on systems, man and cybernetics*, 18:183–190, 1988.

[14] R.R. Yager. Application and extension of OWA aggregation. *Int. Journal of Man-Machine Studies*, 37:103-132,1992.

ON GROUP DECISION MAKING UNDER LINGUISTIC PREFERENCES AND FUZZY LINGUISTIC QUANTIFIERS

Francisco HERRERA, J.Luis VERDEGAY

Dept. of Computer Science and Artificial Intelligence
University of Granada,
18071 - Granada, Spain
e-mail: herrera, verdegay@robinson.ugr.es

ABSTRACT

In this paper some results on group decision making under linguistic preferences and fuzzy linguistic quantifiers are presented. Assuming a set of individual linguistic preferences, representing the preferences of the particular individuals, we develop a solution method for the choice process. We define a linguistic ordered weighted averaging operator, and use it for deriving a collective linguistic preference where the weights are defined using a fuzzy linguistic quantifier. Finally, we use the concept of nondominated alternatives for obtaning a set of maximal nondominated alternatives from the collective linguistic preference, that is, the solution to the decision process.

Keywords: Group decision making, linguistic preferences, linguistic quantifiers.

1. Introduction

Since human judgments including preferences are often vague, fuzzy logic plays an important role in decision making. Several authors have provided interesting results on group decision making or social choice theory with the help of fuzzy sets. They have proved that fuzzy sets provided a more flexible framework for discussion group decision making [15, 8, 9, 11].

In a fuzzy environment it is supposed there exists a finite set of alternatives $X = \{x_1, \ldots, x_n\}$ as well as a finite set of individuals $N = \{1, \ldots, m\}$, and each individual $k \in N$ provides his preference relation on X, i.e., $P^k \subset$ XxX, and $\mu_{P^k}(x_i, x_j)$ denotes the degree of preference of alternative x_i over x_j, $\mu_{P^k}(x_i, x_j) \in [0, 1]$.

Sometimes, however, an individual could have a vague information about the preference degree of the alternative x_i over x_j and can not estimate his preference with an exact numerical value. Then a more realistic approach may be to use linguistic assessments instead of numerical values, that is, to suppose that the variables (preference relations) which participate in the problem are assessed by means of linguistic terms [5, 17, 2, 4, 6, 12]. A scale of certainty expressions (linguistically assessed) would be presented to the individual which could then use it to describe his degree of certainty in a preference.

Assuming a set of individual linguistic preferences, we present a solution method for the choice process. We define a collective linguistic ordered weighted averaging (LOWA) operator, and use it for deriving a collective linguistic preference where the weights are defined using a fuzzy linguistic quantifier. Finally, we use de concept of nondominated alternatives for obtaning the set of maximal nondominated alternatives from the collective linguistic preference, that is, the solution to the decision process.

This model seems to be more human consistent with the social choice in an imprecise environment. The linguistic decision process can be summarized in the figure 1.

In order to develop the linguistic decision process, the paper is organized as follows. Next section describes the linguistic approach. Section 3 presents the use of linguistic preference relations in group decision making and the linguistic choice process. Section 4 presents an example of proposed linguistic choice process. Finally some conclusions are pointed out.

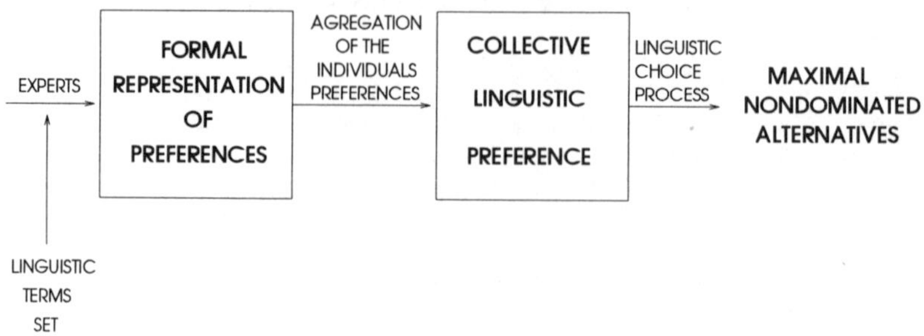

Fig. 1. Linguistic Decision Process

2. The linguistic approach

2.1. Linguistic assessments

The linguistic approach considers the variables which participate in the problem assessed by means of linguistic terms instead of numerical values, [18]. This approach is appropriate for a lot of problems, since it allows a representation of the experts information in a more direct and adequate form whether they are unable of expressing that with precision.

A linguistic variable differs from a numerical variable in that its values are not number s, they are words or sentences in a natural or artificial language. Since words, in general, are less precise that numbers, the concept of a linguistic variable serves the purpose of providing a means of approximated characterization of phenomena which are too complex or too ill-defined to be amenable to description in conventional quantitative terms.

A linguistic variable is associated with two rules:

1. *a syntactic rule*, which may take the form of a grammar for generating the names of the values of the variable, and

2. *a semantic rule*, which defines an algorithmic procedure for computing the meaning of each value.

We need a term set defining the uncertainty granularity, i.e. the finest level of distinction among different quantifications of uncertainty. The elements of the term set will determine the granularity of the uncertainty. In [1] was studied the use of term sets with odd cardinal, representing the middle term a probability of "approximately 0.5", being the rest terms placed symmetrically around it and the limit of granularity 11 or no more than 13.

The semantic of the elements of the term set is given by fuzzy numbers defined on the [0,1] interval, which are described by membership functions. Provided that the linguistic assessments are just approximate ones given by the experts or decision-makers, we can consider that linear trapezoidal membership functions are good enough to capture the vagueness of those linguistic assessments, since to obtain more accurate values may be impossible or unnecessary.

For instance the figure 2 shows an hierarchical structure of linguistic values or labels. As it is clear, the level 1 provides a granularity containing three elements, the level 2 a granularity with nine elements, and different granularity levels could be presented. In fact, in this figure the level 4 presents the finest granularity in a decision process, the numerical values.

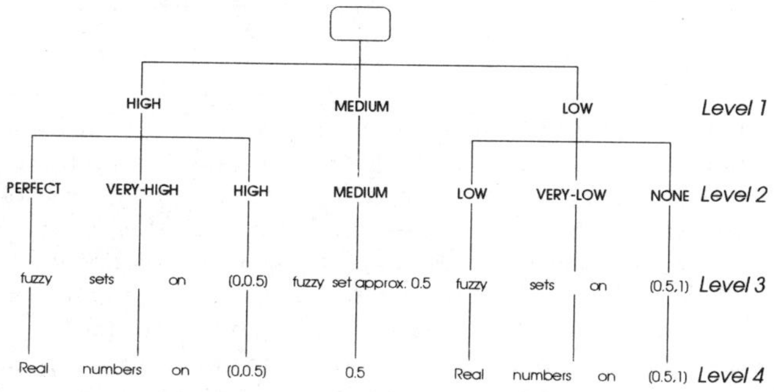

Fig. 2. Hierarchy of labels

Obviously, the kind of label set to be used ought to be established. Then, let $S = \{s_i\}, i \in H = \{0, \ldots, T\}$ be a finite and totally ordered term set on [0,1] in the usual sense [1, 2, 19]. A label s_i represents a possible value for a linguistic real variable, that is, a vague property or constraint on [0,1]. The representation is achieved by the 4-tuple $(a_i, b_i, \alpha_i, \beta_i)$. The first two paremeters indicate de interval in which the membership value is 1.0; the third and fourth parameters indicate the left and right width of the distribution. We consider a term set with odd cardinal, where the middle label represents a uncertainty of "approximately 0.5" an the rest terms are placed symmetrically around it. Moreover, we require the following properties for the term set:

1. The set is ordered:

$$s_i \geq s_j \; if \; i \geq j.$$

2. The negation operator is defined as:

$$Neg(s_i) = s_j \; such \; that \; j = T - i.$$

3. Maximization operator:

$$Max(s_i, s_j) = s_i \; if \; s_i \geq s_j.$$

4. Minimization operator:

$$Min(s_i, s_j) = s_i \; if \; s_i \leq s_j.$$

For example, we can consider the term set of the level 2, *S:*

$$s_6 = P \qquad s_5 = VH$$

$$s_4 = H \qquad s_3 = M$$

$$s_2 = L \qquad s_1 = VL$$

$$s_0 = N$$

where

$$P = Perfect \qquad VH = Very\_High$$

$$H = High \qquad M = Medium$$

$$L = Low \qquad VL = Very\_Low$$

$$N = None$$

As it is obvious this term set verifies each of the above properties.

2.2. Combination of linguistic values

Since aggregation of uncertainty information is a recurrent need in the decision process, combinations of linguistic values are needed. Two main different approaches can be found in order to aggregate and compare linguistic values: The first acts by direct computation on labels [3], and the second uses the associated membership functions. Most of the available techniques belong to the later. However, the final results of those methods are fuzzy sets which do not correspond to any label in the original term set. If one wants finally to have a label, then a "linguistic approximation" is needed [18, 19, 1, 5]. There are neither a general criteria to evaluate the goodness of an approximation nor a general method for associating a label to a fuzzy set, so that specific problems may require to develop taylored methods.

In the following, we present an aggregation operator of linguistic labels by direct computation on labels, based on both the ordered weighted averaging (OWA) operator, [16], and the convex combination of linguistic labels, [3].

A mapping F from

$$I^n \rightarrow I(\text{where } I = [0,1])$$

is called an OWA operator of dimension n if associated with F is a weighting vector $W, W = [w_1, \ldots, w_n]$, such that,

i) $w_i \in [0,1]$,

ii) $\Sigma_i w_i = 1$,

and

$$F(a_1, \ldots, a_n) = w_1 \cdot b_1 + w_2 \cdot b_2 + \ldots + w_n \cdot b_n$$

where b_i is the i-th largest element in the collection $a_1, \ldots, a_n$. Denoting B as the vector consisting of the arguments of F put in descending order,

$$F(a_1, \ldots, a_n) = W \cdot B^T$$

provides an aggregation type operator that always lies between the "and" and the "or" aggregation. Its properties are presented in [16].

This operator can be extended to linguistic arguments, using the convex combination of linguistic labels defined in [3]. In fact, let M be a collection of linguistic labels, $p_k \in M, k = 1, \ldots, m$, and assume $p_m \leq p_{m-1} \leq \ldots \leq p_1$ without loss of generality. For any set of coefficients $\{\lambda_k \in [0,1], k = 1, 2, .., m, \Sigma\lambda_k = 1\}$ the convex combination of these m generalized labels is the one given by

$$\mathbf{C}\{\lambda_k, p_k, k = 1, \ldots, m\} = \lambda_1 \odot p_1 \oplus (1 - \lambda_1) \odot \mathbf{C}\{\beta_h, p_h, h = 2, \ldots, m\}$$

with

$$\beta_h = \lambda_h / \sum_2^n \lambda_k ; h = 2, \ldots, m.$$

In [3] was defined the aggregation of labels by addition, the difference of generalized labels, and the product by a positive real number over a *generalized label space* $\mathcal{S}$, based on S, that is, the cartesian product $\mathcal{S} = Sx\mathbf{Z}^+$, with the basic label set $S = \{(s_i, 1), i \in H\}$. In our context all the operations are realized over the basic set S. Briefly, the result of the expression $\lambda \odot s_j \oplus (1 - \lambda) \odot s_i, j \geq i$, is the s_k such that $k = \min\{T, i + \text{round}(\lambda \cdot (j - i))\}$.

An example to clarify of this operation is the following. Suppose the term set:

$$S = \{s_6 = P, s_5 = VH, s_4 = H, s_3 = M, s_2 = L, s_1 = VL, s_0 = N\}$$

and $\lambda = 0.4$,

		$1-\lambda=0.6$			
		M	**L**	**P**	**VH**
$\lambda=0.4$	**P**	H	H	P	VH
	N	L	VL	H	M
	H	M	M	VH	VH
	L	M	L	VH	VH

where, for example:

$$k_{11} = min\{6, 3 + round(0.4 * (6-3))\} = 4 \text{ (H)}$$
$$k_{21} = min\{6, 0 + round(0.6 * (3-0))\} = 2 \text{ (L)}$$

Therefore the linguistic ordered weighted aggregation (LOWA) operator can be defined as

$$F(a_1,\ldots,a_m) = W \cdot B^T = \mathbf{C}\{w_k, b_k, k=1,\ldots,m\} =$$

$$= w_1 \odot b_1 \oplus (1-w_1) \odot \mathbf{C}\{\beta_h, b_h, h=2,\ldots,m\}$$

where $\beta_h = w_h/\Sigma_2^m w_k, h = 2,\ldots,m$, and B is the associated ordered label vector. Each element $b_i \in B$ is the i-th largest label in the collection $a_1,\ldots,a_m$.

3. Linguistic preference relations in group decision

Suppose we have a set of n alternatives $X = \{x_1,\ldots,x_n\}$ and a set of individuals $N = \{1,\ldots,m\}$. Each individual $k \in N$ provides a preference relation linguistically assessed into the term set S,

$$\phi_{P^k} : XxX \rightarrow S,$$

where $\phi_{P^k}(x_i,x_j) = p_{ij}^k \in S$ represents the linguistically assessed preference degree of the alternative x_i over x_j. We assume that P^k is reciprocal in the sense, $p_{ij}^k = \text{Neg}(p_{ji}^k)$, and by definition $p_{ii}^k = None$ (the minimum label in S).

As it is known, basically two approaches may be considered. A direct approach

$$\{P^1,\ldots,P^m\} \rightarrow \text{ solution}$$

according to which, on the basis of the individual preference relations, a solution is derived, and an indirect approach

$$\{P^1,\ldots,P^m\} \rightarrow P \rightarrow \text{ solution}$$

providing the solution on the basis of a collective preference relation, P, which is a preference relation of the group of individuals as a whole.

Here we consider the indirect derivation, and hence we have two issues to study: First, to derive P from $\{P^1,\ldots,P^m\}$, and second, to obtain the solution from P.

3.1. The collective linguistic preference relation

For the first question it is necessary to aggregate the linguistic preference relations, to obtain $p_{ij} \in S$ from $\{p_{ij}^1,\ldots,p_{ij}^m\}$ for all i,j. We discuss this using the concept of *fuzzy majority*. Fuzzy linguistic quantifiers have provided tools to formally deal with fuzzy majority and can be used to define a weight vector. We then use the LOWA operator to obtaini the collective preference relation P as

$$P = F(P^1,\ldots,P^m)$$

with $p_{ij} = F(p_{ij}^1,\ldots,p_{ij}^m)$ with the weight vector, W, representing the fuzzy majority over the individuals.

The fuzzy linguistic quantifiers were introduced by Zadeh, [20]. Linguistic quantifiers are typified by terms such as *most, at least half, all, as many as possible.* A quantifier Q assumed to be a fuzzy set in [0,1]. Zadeh distinguished between two types of quantifiers, absolute and proportional or relative.

An absolute quantifier

$$Q : R^{+} \rightarrow [0,1],$$

satisfies:

$$Q(0) = 0,$$
$$\exists k \text{ such that } Q(k) = 1.$$

A relative quantifier,

$$Q : [0,1] \rightarrow [0,1],$$

satisfies:

$$Q(0) = 0,$$
$$\exists r \in [0,1] \text{ such that } Q(r) = 1.$$

A nondecreasing quantifier satisfies

$$\forall a, b \text{ if } a > b \text{ then } Q(a) \geq Q(b).$$

being the membership function of a non-decreasing relative quantifier Q, as follows:

$$Q(r) = \begin{cases} 0 & \text{if } r < a \\ \frac{r-a}{b-a} & \text{if } a \leq r \leq b \\ 1 & \text{if } r > b \end{cases}$$

with $a, b, r \in [0,1]$.

Yager computes the weights w_i of the aggregation from the function Q describing the quantifier, [16]. In the case of a nondecreasing absolute quantifier,

$$w_i = Q(i) - Q(i-1), i = 1, \ldots, m,$$

and in the case of a non-decreasing relative quantifier,

$$w_i = Q(i/m) - Q((i-1)/m), i = 1, \ldots, m.$$

3.2. The linguistic choice process

When fuzzy preferences are used, numerous answers (called the solution concepts) derived from a collective fuzzy preference have been presented in the literature: a consensory-like pooling or consensus winner [7], a competitive-like pooling, [10], a fuzzy α-majority uncovered fuzzy set based on the concept of fuzzy tournaments, [13], etc.

Next, we propose a method to derive a solution on the basis of the collective preference relation, method based on the concept of non-dominated alternatives [14].

Let P^s be a *linguistic strict preference relation* $\mu_{P^s}(x_i, x_j) = p^s_{ij}$ such that,

$$p^s_{ij} = None \text{ if } p_{ij} < p_{ji}, \text{ or } p^s_{ij} = s_k \in S \text{ if } p_{ij} \geq p_{ji}$$
$$\text{with } p_{ij} = s_l, p_{ji} = s_t \text{ and } l = t + k.$$

The *linguistic degree of nondominance* of x_i is defined as

$$\mu_{ND}(x_i) = \text{Min}_{x_j \in X}[\text{Neg}(\mu_{P^s}(x_j, x_i))]$$

here the value $\mu_{ND}(x_i)$ is to be meant as a linguistic degree to which the alternative x_i is ominate by no one of the elements in X.

Finally, a set of *maximal nondominated alternatives*, $X^{ND} \subset X$, is obtained as:

$$X^{ND} = \{x \in X / \mu_{ND}(x) = \text{Max}_{y \in X}[\mu_{ND}(y)]\}.$$

Therefore aggregating the knowledge of the experts, X^{ND} is selected as the set of preferred ternatives in our choice process.

. Example

Consider the above linguistic term set S, and suppose a set of four alternatives $X = x_1, x_2, x_3, x_4\}$, and a set of four individuals, whose linguistic preference relations using S are:

$$P^1 = \begin{bmatrix} - & H & H & P \\ L & - & H & VH \\ L & L & - & M \\ N & VL & M & - \end{bmatrix} P^2 = \begin{bmatrix} - & VH & H & M \\ VL & - & VH & L \\ L & VL & - & H \\ M & H & L & - \end{bmatrix}$$

$$P^3 = \begin{bmatrix} - & P & VH & H \\ N & - & VH & L \\ VL & VL & - & VL \\ L & H & VH & - \end{bmatrix} P^4 = \begin{bmatrix} - & P & VH & VH \\ N & - & L & H \\ VL & H & - & VL \\ VL & L & VH & - \end{bmatrix}$$

Using the relative quantifier "As least half" with the pair $(0.0, 0.5)$, and the corresponding)WA operator with $W = (0.5, 0.5, 0, 0)$, the collective linguistic preference relation is:

$$P = \begin{bmatrix} - & P & VH & VH \\ L & - & VH & VH \\ L & M & - & H \\ M & II & VH & - \end{bmatrix}$$

The linguistic strict preference relation is

$$P^s = \begin{bmatrix} - & H & M & L \\ N & - & L & VL \\ N & N & - & N \\ N & N & VL & - \end{bmatrix}$$

d the linguistic degree of nondominance is:

$$\mu_{ND}(x_1, x_2, x_3, x_4) = [P, L, M, H].$$

The set of maximal nondominated alternatives, and therefore the solution, X^{ND}, is

$$X^{ND} = \{x_1\}$$

Conclusion

We have presented a linguistic decision process in group decision making based on an indirect proach, deriving a collective linguistic preference by means of a LOWA operator with a zy linguistic quantifier as a tool to deal with fuzzy majority, obtaining a set of maximal ndominated alternatives based on the nondominated alternative concept.

REFERENCES

[1] Bonissone, P.P., and Decker., K.S. (1986) Selecting Uncertainty Calculi and Granularity: An Experimer in Trading-off Precision and Complexity. In: Uncertainty in Artificial Intelligence, (L.H. Kanal, and J.I Lemmer eds.), North-Holland, p. 217-247.

[2] Delgado, M., Verdegay, J.L., and Vila., M.A. (1992) Linguistic Decision Making Models, Internation Journal of Intelligent Systems, 7, 479-492.

[3] Delgado, M., Verdegay, J.L., and Vila., M.A. (1993) On Aggregation Operations of Linguistic Label International Journal of Intelligent Systems, 8, 351-370.

[4] Delgado, M., Verdegay, J.L., and Vila., M.A. (1994) A Model for Linguistic Partial Information in Decisio Making Problems. To appear in: International Journal of Intelligent Systems.

[5] Fedrizzi, M., and Mich., L. (1992) Rule Based Model for Consensus Reaching in Group Decision Suppor Proc. IPMU92, Spain, 301-304.

[6] Herrera, F., and Verdegay., J.L. (1993) Linguistic Assessments in Group Decision, Proc. of First Europea Congress on Fuzzy and Intelligent Technologies, Aachen, 941-948.

[7] Kacprzyk, J. (1986) Group Decision Making with a Fuzzy Linguistic Majority, Fuzzy Sets and System 18, 105-118.

[8] Kacprzyk, J. (1990) Multiperson Decision Making Models using Fuzzy Sets and Possibility Theory (Kluwe Academic Pub., Dordrecht).

[9] Kacprzyk, J., and Roubens, M. (1988) Non-Conventional Preference Relations in Decision Makin Springer-Verlag, Berlin.

[10] Kacprzyk, J., Fedrizzi, M., and Nurmi., H. (1990) Group decision Making with Fuzzy Majorities Repre sented by Linguistic Quantifiers. In: Approximate Reasoning Tools for Artificial Intelligence, (J.L. Verde gay, and M. Delgado eds.), Verlag TÜV Rheinland, Cologne, p. 267-281.

[11] Kacprzyk, J., Fedrizzi, M., and Nurmi., H. (1992) Fuzzy Logic with Linguistic Quantifiers in Grou Decision Making. In: An Introduction to Fuzzy Logic Applications in Intelligent Systems, (R.R. Yager and L.A. Zadeh eds.), Kluwer Academic, Boston, p. 263-280.

[12] Mich, L., Gaio, L., and Fedrizzi., M. (1993) On Fuzzy Logic-Based Consensus in Group Decision, Proc IFSA93, Seoul, 698-700.

[13] Nurmi, H., and Kacprzyk, J. (1991) On Fuzzy Tournaments and Their Solution Concepts in Group Decisio Making, European Journal of Operational Research, 51, 223-232.

[14] Orlovsky, S.A. (1978) Decision-Making with a Fuzzy Preference Relation, Fuzzy Sets and Systems, 1 155-167.

[15] Spillman, R., and Spillman, B. (1987) A Survey of Some Contributions of Fuzzy Sets to Decision Theory In: Analysis of Fuzzy Information, (J.C. Bezdek ed.), CRC Press, 109-118.

[16] Yager, R.R. (1988) On Ordered Weighted Averaging Aggregation Operators in Multicriteria Decisio Making, IEEE Transactions on Systems, Man and Cybernetics, 18, 183-190.

[17] Yager, R.R. (1992) Fuzzy Screening Systems. In: Fuzzy Logic: State of the Art, (R. Lowen ed.), Kluwe Academic Publisers.

[18] Zadeh, L. (1975) The Concept of a Linguistic Variable and its Applications to Approximate Reasoning Part I, Information Sciences, 8, 199-249, Part II, Information Sciences, 8, 301-357, Part III, Informatior Sciences, 9, 43-80.

[19] Zadeh, L. (1979) Fuzzy Sets and Information Granularity. In: Advances in Fuzzy Sets Theory and Appli cations, (M.M. Gupta et al. eds.), North Holland, 3-18.

[20] Zadeh, L. (1983) A Computational Approach to Fuzzy Quantifiers in Natural Language, Comps. & Maths with Appls, 9, 149-184.

NEGOTIATION AND CONFLICT RESOLUTION IN PRODUCTION ENGINEERING THROUGH SOURCE CONTROL

Luisa MICH*, Mario FEDRIZZI*, Roberto GARIGLIANO**

* Department of Computer Science and Management Science, University of Trento,
Via Inama 5- 38100 Trento, Italy
Tel.: +39.461.882150 - Fax.: +39.461.882124
e-mail: mich@cs.unitn.it

** AI Systems Research Laboratories, Computer Science, SECS, University of Durham,
Durham, DH1 3LE, UK
Tel: +44.91.374 2639 - Fax: +44.91.374 3741
e-mail: Roberto.Garigliano@durham.ac.uk

ABSTRACT

This paper presents a general strategy for the management of decision making under uncertainty in industry, and its application to the specific problem of route generation in an electric engineering company. The general strategy is based on the fusion of several complementary approaches, using both linguistic and numerical information. The overall goal is to produce a system which is highly interactive, allows the use of as much information as possible, and leads the decision making process without forcing it.
Key Words: Simultaneous Engineering, Manufacturing Industry, Source Control System, Consensus, Uncertainty, Negotiation

1. Introduction

A major problem in manufacturing industry is the integration of the design and production engineering processes [11]. There is a strong link between these two phases, since the characteristics of a design determine the manufacturing processes needed and, vice versa, features of the production cycle act as constraints on the acceptable designs.

The design and production phases are together known in manufacturing engineering as "route". The specific problem we are addressing is thus "Route Generation and Selection".

The traditional approach is sequential and is inadequate to cope profitably with the feedback nature of these processes, which are naturally iterative and interactive [16] (Fig. 1).

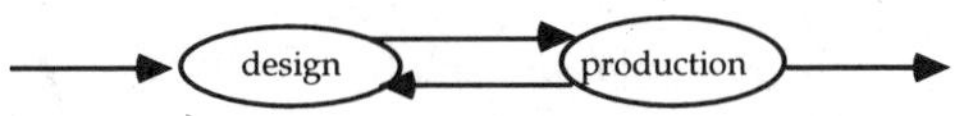

Figure 1

Moreover, currently the route generation and selection process in industry includes a considerable subjective elements, and involves a rather stressing decisional environment [15]. In a Computer Supported Cooperative Work perspective, we need to consider organisational, technological and relational issues [18].

In this paper we are mostly concerned with the support requirements for coordination and decision, as the problem of aiding the experts in the initial generation of the routes is a separate question. It is however necessary to give a brief introduction to the issues involved in the route generation, and to clarify the kind of data structures which can be used in that process.

To reduce lead time and costs and to improve communication it is necessary to adopt a systemic approach to the generation of routes for a discrete component or sub-assembly. From a Simultaneous Engineering point of view, product and process design should be parallel and synchronous [12,14] (designers and production engineers working separately are likely to get solutions that are local "optimum"). Another key issue is the different kinds of uncertainty which are involved. Very schematically, we have to face:

- a changing environment/context
- different teams of people (with different skills, goals, relational-decisional behaviours)
- many sources of information (internal and external to the industry)
- various constraints, such as time, money, etc.

More abstractly, we have:

- many variables
- many processes
- different levels of uncertainty.

A computer-based AI system to support the designers, production engineers and all others involved parties in the process of route generation requires an architecture which encompasses different modules [2].

2. The approach

A schema of the proposed approach is made of the following major phases (details are presented later on):

- Initialisation and definition of the problem
- Selection
- Evaluation of the consensus degree and shifting of positions
- Negotiation.

2.1. Initialisation and definition of the problem

• Identification of the Simultaneous Engineering Team' experts: SET. In this phase, each group is identified, and relevant information about each group and its members is collected. Since this information is for the use by the SCS, the type and quantity of information is controlled by the basic assumptions and needs of the SCS. In general, the SCS is built so as to need information of a linguistic, very general nature: such information is easy to obtain, and can always be defaulted.

• Assignment of the initial weights to the experts: this is done by the SCS, on the basis of the information obtained by the previous process, and of general heuristics in the SCS.

• Individuation of a set of possible solutions and evaluation of their advantages and disadvantages: this is done by SET, using also the Route Production tool (RP).

2.2. Selection

• Routes feasibility: elimination (goals-criterion) and classification (structure). In this phase some routes are eliminated or given a weight in relation to their consistency with the technical, domain dependent knowledge of the RP. The RP also classifies the remaining routes by structure and value.

• Routes acceptability and changing of the experts' weights: this is done by the SCS, by checking the internal consistency of each pair route/experts, according to some domain independent heuristics (e.g., responsibility vs. advantage).

2.3. Evaluation of the consensus degree and shifting of positions

• The CM module builds a matrix representing the judgement of each group on each solution, expressed in fuzzy linguistic terms. The experts are made aware of the weight on each solution put by the RP.

• The distance of each group from the general agreement on the value of each solution is then measured, taking into account the weights on the experts placed by the SCS. Some solutions are then chosen for further discussion, and some groups are invited to shift positions.

• The solutions are then ranked, and the process repeated until one solution passes the RKB conditions, or deadlock is reached.

2.4. Negotiation

If deadlock is reached, a negotiation process is started in order to reach a new solution, which should be a compromise between the best solutions causing the deadlock. The choice of the solutions is made using the RKB, while the RP is instrumental in achieving the compromise by proposing strategies suitable to the situation, using the previous classification.

3. The Modules

The various modules of the system are now presented in more details.

The Simultaneous Engineering Team (SET)

$SET = \{G_1, G_2, \ldots, G_K\}$

Ex.:

G_1 = group of production engineers
G_2 = designers
G_3 = accountants
G_4 = personnel manager
G_5 = marketing manager
etc.

Each group has realistically specific competencies and local goals that they have to be asked in order to define (individuate or complete) the components of the production routes vectors. Afterwards the SCS [7,8,9] assign to each group (constituted by one or more experts) an initial weight (given a common task it is possible to aggregate the internal and external weight obtained by the SCS) on the basis of ability and trustworthiness. The groups' weights are obtained using information of different kind: official documents, interviews to the experts and to external people, e.g., customers, personal experiences, etc. So that weights can be changed at different stages of the "session".

We then have:

$SET = \{<G_1,w_1(t)>,<G_2,w_2(t)>,\ldots,<G_K,w_k(t)>\}.$

The Route Processor Unit (RP)

This is the module that contains the domain specific, technical knowledge about the problem of route generation. It should be used by the teams of experts as an aid in the productions of their initial solutions. It is also used to prune them initially, to give them some weights, and to classify them. It might also be used as an aid in the final negotiation phase.

The Consensus Module (CM)

This is the unit that deals with calculating how far away the positions of the various experts are, and with ranking the solutions according to these global judgements. It uses a system based on fuzzy linguistic variables.

The Relational Knowledge Base (RKB)

This is the unit that contains the relational meta-level knowledge that controls what is acceptable in the particular circumstances in the way of agreement. It is used in the control of the selection process and in the negotiation process.

The Source Control System (SCS)

The main principles of the source control mechanism (S) are:

- the system must know who it is communicating with;
- the system keeps a record of each source, as detailed below;
- this source information is used, when needed, to judge the data;
- external features of the piece of information (such as source, tone, consistency, language etc.) are used rather then the content of the information itself;
- evaluation of the data over time is used to adjust the indices for the source (feed-back).

Example of representation of information: Information A with indices representing Certainty, Source, Time, Responsibility and Action Point.

Representation of source:

- Ability:
  - expertise
  - reasoning
  - other sources
  - personal experience
- Belief and Interests
- Relation to X
  - helpfulness
  - trustworthiness.

The Source Control System is composed of:

- source models (including defaults and classifications);
- importance analyser (to decide how far the analysis should go): this is called at various stages;
- information consistency analyser (this unit checks that the various part of an information listed above can consistently stay together) [17];
- heuristics set (used to extract the various parameters needed from the pieces of information if they are not explicitly present);
- conflict resolution (use to resolve conflicts, based mainly on source models) [1];
- enquiry unit (needed if more information is required);
- source re-evaluation unit (feeds back the result of the analysis onto the source models).

Although the SCS can be used as part of any environment which produces suitable information, it is especially designed to be used as part of an advanced Natural Language Processing Unit (such as the LOLITA system developed at Durham University [10]).

The Source Control Unit of the proposed system is built by adapting the general framework to the task at hand: in particular, this means:

- detailed *a-priori* models for classes of specific sources (e.g., designers, planners etc.);
- detailed track record for individual participants;
- specialist knowledge about terms, forms of arguments etc., used in the domain;
- specialistic set of heuristics to deal with such extra information;
- adaptation of importance and enquiry units to task at hand;
- use of some problem specific knowledge in the conflict resolution phase (possibly thorough the integration of some form of the endorsement system);
- design of a specific, non-Natural Language front-end.

4. Description of a session

Each group proposes one ore more solution as "production routes". Given the complexity of the problem, there are a number of possible solutions. Each route can be defined as a couple of vectors, a structure vector R and a goal vector C:

$PR = \{PR_1, PR_2, ..., PR_N\}$ set of all the given routes

$PR_i = <R_i, C_i>$.

The elements of both R_i and C_i contain linguistic information, quantitative information, processes, etc.

The structure of a route can be defined in terms of the components given from the different groups at the beginning of the process (each group gives initially the set of structure types and goals which it consider should appear in the route definition). In particular we can have a set of types as components of the structure:

$T = \{t_1, t_2, ..., t_m\}$.

For example we can have as routing options (types): machines selected, operator, sequences, etc.

So that each R_i is given as instantiation of these types:

$R_i = <o_1, o_2,...o_l>$.

Remark: not all components have to be assigned from all the groups. A generic type and a set of default values for all the t_i are used for these purposes. The goal vector C_i should contain the estimated level of satisfaction of the goals (criterion) of the group, e.g.:

$C_i =$ {cost, personnel, relational stress, quality of the final product, marketability etc.}

Note that the goal vector should at least contain the specific goal of each expert group (these can be requested directly from the experts, or the SCS can infer them from its classification).

Some of these solutions would be unfeasible on the basis of overall limitation on technical grounds (e.g., if there is a deadline that all solution have to respect): the RP would then eliminate them.

It should be remarked that at each step the systems communicate to the experts as much useful information as possible on the session. In this case, it would be clarified why some solutions are deemed unfeasible.

Other solution could be "similar", according to an *ad-hoc* metric that must be agreed with the customer or the experts before the session starts. Because of the representation of the solutions, we can have similar structure and/or similar goal's satisfaction. There are thus four main classes of routes:

- similar structure and similar goals
- similar structure and different goals
- different structure and similar goals
- different structure and different goals.

On the basis of structure and goals, the RP classify the routes in order to reduce their number. This process gives:

$PR = \{PR_1, PR_2, ..., PR_{N'}\}$ with $N' \leq N$.

Then the RP (routes processor) evaluates the PR and assigns to each of them a global "index of goodness" (W) to discount the fact that the groups goals evaluation are (partially) subjective, by checking against domain dependent technical knowledge. It is important to note that this form of discount of subjectivity is implicit, i.e., at this stage there is no explicit model of subjectivity. This is then introduced by using the SCS later on.

The SCS will have assigned weights to the expert groups, using the standard four SCS techniques for model building: class defaults, reported information, direct information and system feed-back. Note that the model built by the SCS is quite complex, and it is possible to produce a single weight only because a very precise task with a simple overall goal is at hand. Once the routes are received, however, the SCS is able to modify these weights using some information about the solutions.

Examples of these (domain independent) heuristics are:

If all of a group's solutions have been eliminated or had received a low RP index, the weight should be decremented.

If the solution proposed by young experts are similar to those of senior engineers, their weight should be incremented.

If a solution is such that it maximises only the goal specific to the group's class, and low responsibility is accepted, the weight should be decremented.

Note that there is a danger here that the implicit discount of subjectivity by the RP and the explicit one by the SCS might overlap, causing a lowering of a solution's value below what is warranted. To remove this danger, a process of tuning the system in the actual job conditions would be necessary.

The next step is a group decisional process. The CM implements a fuzzy-logic based formalism to support interactively the experts in achieving a common solution [4,5,6,13]. We have as input for the CM:

$$SET = \{<G_1,w_1(t)>,<G_2,w_2(t)>,....,<G_K,w_k(t)>\}$$

$$PR' = \{(<R_1,C_1>,W_1),(<R_2,C_2>,W_2),...,(<R_{N'},C_{N'}>,W_{N'})\}$$

where the lower case weights are the experts' weight given by the SCS, and the upper case ones are the route weight given by the RP.

Then, all the components of the SET are asked to give their evaluation on the PR elements. According to the model proposed [3], groups' preferences can be expressed in linguistic terms [19], (that is, choosing a linguistic label represented by "fuzzy numbers" in a term set V, the range of which is pre-established).

From a formal point of view, the results of this phase of consultation can be expressed in a matrix form. The elements of these matrixes represent the "linguistic performances" that groups have assigned to each PR.

The CM identifies some consensus strategy using a "cost for changing opinion" that should take into account the weights assigned to the groups by the SCS. There are two stages: 1. Identify candidate solutions for the discussion (eliminate those whose total value - has judged by the SET - does not pass a fixed threshold). 2. The remaining solutions are evaluated again, after a discussion based on the advice of the CM , which in turn is based on the "cost for changing opinion": this process may be repeated a number of times depending on various elements.

In reality we observe different behaviours depending on the available time (when the deadline approaches there is usually an extra effort to settle differences) and on the liveliness of the meeting (often, if some participants have changed opinion, the discussion is carried on beyond what is needed for the minimum consensus, since there is a chance of reaching a wider agreement). By monitoring the above effects it is possible to tailor the heuristics to the particular stage of the discussion.

To take into account some relational rules (meta-rules or heuristic on the negotiation process), and to increment the flexibility of the CM, potential common solutions have to be examined by the RKB module. We can have for example in the RKB rules as:

The department manager can not be ignored. A PR that leaves any group totally unsatisfied can not be accepted.

Elements from each group have to be contained in the proposed solution. A qualified majority is required.

If after some round of consultation there isn't a PR that had passed all those tests and that can be chosen as the final one, it is necessary to change level of negotiation and change the PR themselves.

The group has to be told why the consensus hasn't been reached and how to use this information as starting point for the next steps. The identification of the routes involved in the deadlock is as follows: the first candidate is the route that achieved the maximum level of agreement (either actual or measured by the cost of changing opinion). If there is a deadlock, there must be at least one group whose disagreement causes this solution to fail because of some relational rule from the RKB. The second candidate is then the solution closest to the first one (e.g., in terms of structure/goal classification) among those preferred by the "dissident" group.

The interaction among the groups now moves to the level of "content".

The candidate solutions are compared according to their existing classification (by structure and value). For each of the four possible situations (similar structure, similar goals; similar structures, different goals etc.), suitable strategies of negotiations are proposed. For example, in the situation "similar structure - different goals", it is advisable to use a strategy of "localising the problem"; in the situation "different structure, similar goals" either a strategy of diversification or one of exchange are possible. It is more difficult to find a suitable strategy in the case "different structure, different goals".

5. Conclusions

In this paper we have presented a general strategy for group decision under uncertainty, based on the fusion of several different AI techniques: SCS, consensus support metrixes, technical knowledge bases and relational heuristics. It has also been shown how this approach can be applied to a real industrial domain, namely the generation and selection of production routes by simultaneous engineering teams.

The Source Control System provides a way of explicitly discounting subjective elements in the decision making; the Route Processing Module uses both mathematical and knowledge based techniques to aid and constrain the generation of options from a technical point of view, and to cluster them; the Consensus Module provides a way of calculating the cost of solutions which takes into account the opinions of the experts and the results from the Source Control System and the Route Processing Module; finally, a set of meta-rules - defining what kind of consensus is acceptable in the specific situation - is used, together with the results from the Consensus Module phase, to modify the original options when necessary. This is done by identifying the likely candidates, by recognising their relation using the clustering results, and by applying a set of modification heuristics.

The main points of this approach are: 1. The complexity of the decision making process is recognised, and no attempt is made to reduce it to simple *formulae*; 2. the role of uncertainty in the decision making process is stressed, which implies a need to use techniques from the Management of Uncertainty field, such as the Source Control System; 3. human factors and relations play a crucial role even in technical decisions: this role must be made explicit, in order to be able to discount it from the decisions, while reducing the emotive impact on the participants; 4. the approach is highly flexible, because it is largely domain independent and because of the presence, iteration and interaction of several techniques.

REFERENCES

[1] Cohen, P.R. (1985) Heuristic Reasoning about Uncertainty: an Artificial Intelligence Approach, Pitman, London.
[2] Dewan, P., Riedl, J. (1993) Toward Computer-Supported Concurrent Software Engineering, IEEE Computer, January, 17-27.
[3] Fedrizzi, M., Kacprzyk, J., Zadrozny, S. (1988) An interactive multi-user decision support system for consensus reaching processes using fuzzy logic with linguistic quantifiers, Decision Support System, 4, 313-327.
[4] Fedrizzi, M., Mich, L. (1991) Consensus Reaching in Group Decision Using Production Rules, Proc. Annual Conference of the Operational Research Society of Italy (AIRO'91), September 18-20, Riva del Garda, Italy, 1991, 118-121.
[5] Fedrizzi, M., Mich, L., Gaio, L. (1992) A Fuzzy Logic-Based Model for Consensus Reaching in Group Decision Support. In: Advanced Methods in AI (B. Bouchon-Meunier, L. Valverde, R.R.Yager eds.), LNCS 682, Springer-Verlag, Proc. Int. Conf. on Information Processing and Management of Uncertainty (IPMU'92), July 6-10 1992, Mallorca, Spain, 301-304.
[6] Fullér, R., Gaio, L. Mich, L., (1994) OCA functions for consensus reaching in group decisions in fuzzy environment, Proc. 3rd Int. Conf. on Fuzzy Logic, Neural Nets and Soft Computing (IIUZUKA'94), August 3-7, Iizuka, Japan, 1994, 101-102.
[7] Garigliano, R., Bokma A. (1990) Towards a Conflict Resolution Mechanism for Multiple Sources, LNCS 521, Springer-Verlag.
[8] Garigliano, R., Bokma, A. (1992) Heuristics for Knowledge Extraction in Source Control. In: Advanced Methods in AI (B. Bouchon-Meunier, L. Valverde, R.R.Yager eds.), LNCS 682, Springer-Verlag, Proc. Int. Conf. on Information Processing and Management of Uncertainty IPMU'92, July 6-10 1992, Mallorca, Spain.
[9] Garigliano, R., Long, D., Bokma, A. (1988) A Model for Learning by Source Control. In: Uncertainty and Intelligent Systems (B. Bouchon-Meunier, Saitta, R.R. Yager eds.), LNCS 313, Springer-Verlag.
[10] Garigliano, R., Morgan, R.G., Smith, M. (1992) LOLITA - Progress Report 1, Computer Science Technical Report 12/92, Durham University.
[11] Garigliano, R., Maroupoulos, P., Bennett, K.H., (1993) Werner Electric: A Computer Aided Route Processing System, Proposal to the Science and Engineering Research Council, UK.
[12] Klein, M. (1993) Capturing Design Rationale in concurrent Engineering Teams, IEEE Computer, January, 39-47.
[13] Mich, L., Gaio, L., Fedrizzi, M. (1993) Approximate Reasoning in Modelling of Consensus in Group Decision, 8th Austrian AI Conf. on Fuzzy Logic in Artificial Intelligence (FLAI'93), June 28-30 1993, Linz, Austria. In: Fuzzy Logic in AI (E.P. Klement, W. Slany eds.), Springer-Verlag, LNAI 695, 91-102.
[14] Migliarese, P., Polucci, E., (1993) A System for Group Production Planning in Manufacturing, Interfaces, May-June, 23, 29-40.
[15] Ng Keung-CHI, Abramson, B. (1992) Consensus Diagnosis: A simulation study, IEEE transactions on Systems, Man and Cybernetics, 5, 22: 916-928.
[16] Reddy, Y.V.R., Srinivas, K., Jagannathan, V., Karinthi, R. (1993) Computer Support for Concurrent Engineering, IEEE Computer, January, 12-16.
[17] Smith, B., Kelleher, G. (1988) Reason Maintenance Systems and Their Applications, Ellis Horwood, Chichester, UK.
[18] Winograd, T., Flores, F. (1986) Understanding Computers and Cognition. Norwood, NJ: Ablex Publishing.
[19] Zadeh, L. (1975) The concept of a linguistic variable and its application to approximate reasoning, Information Sciences, (Part I) 8, 199-249, (Part II) 8, 301-357, (Part III) 9, 43-80.

MEANINGFUL MEANS ON ORDERED SETS

Yakov Kamen*, Sergei Ovchinnikov**

* Sun Microsystems, Inc.
2550 Garcia Avenue
Mountain View, CA 94043, USA

** San Francisco State University
1600 Holloway Avenue
San Francisco, CA 94132, USA

ABSTRACT

It is shown that, under some rather general conditions, any comparison meaningful mean on a strict simple order is an order statistic.

1. Introduction

In a very informal way, the problem under consideration can be formulated as follows. Suppose M is a rule which assigns to each sample taken from a set X, its mean value which is an element of the same set. In the framework of ordinal measurement X is supposed to be an ordered set. Suppose now that we compare mean values taken for two samples of the same size n and

$$M(x_1, \ldots, x_n) < M(y_1, \ldots, y_n).$$

Let ϕ be a strictly increasing bijection of X onto itself. In other words, ϕ is an isomorphism of the ordered set X to itself. In order to call this comparison meaningful in the spirit of measurement theory, we want

$$M(\phi(x_1), \ldots, \phi(x_n)) < M(\phi(y_1), \ldots, \phi(y_n))$$

to hold for any such ϕ.

Our results generalize those obtained in [5, Chapter 3] and [3].

2. Definitions

We use notations from [4, chapter 1] and [6, section 3.3] and begin with some definitions.

Let (A, R) be a binary relation. A mapping ϕ: $A \to A$ is an *isomorphism* if it is a bijection and $aRb \Leftrightarrow \phi(a)R\phi(b)$. (A, R) is called *transitive* if, for any two elements a and b in A, there is an isomorphism ϕ such that $\phi(a) = b$.

Let $(X, <)$ be a *strict simple order*, i.e., $<$ is an asymmetric, complete, and transitive binary relation on X. An *open ray* in X is a subset in the form $(\leftarrow, a) = \{x \mid x < a\}$ or $(a, \to) = \{x \mid a < x\}$, $a \in X$. An *open interval* in $(X, <)$ is either a ray or a subset in the form $(a, b) = \{x \mid a < x < b\}$, $a, b \in X$. Similarly, *closed intervals* $(\leftarrow, a]$, $[a, \to)$, and $[a, b]$ are defined. Obviously, any interval I defines a simple strict order $(I, <)$. We assume that $(X, <)$ satisfies the following conditions

(i) $(X, <)$ has no gaps, i.e., all open intervals (a, b) are nonempty sets.

(ii) all open intervals $(I, <)$ are transitive.

It is readily verified that any (infinite or not) interval of the real line satisfies these two condition. The same is true for intervals in the set of all rational numbers.

Definition 1. Let $(X, <)$ be a simple strict order. A symmetric function $M(x_1, \ldots, x_n)$ on X^n with values in X is a *Cauchy mean* if

$$\min\{x_1, \ldots, x_n\} \leq M(x_1, \ldots, x_n) \leq \max\{x_1, \ldots, x_n\}$$

It looks like Cauchy first introduce this notion in [2]. Cauchy means are *internal* functions in the context of functional equations theory ([1], p.243).

Example 1. Suppose $X = Re$ (real line) endowed with the natural order. We define n functions $x^{(i)}$ by arranging any given n-tuple $(x_1, \ldots, x_n)$ in the increasing order

$$x^{(1)} \leq x^{(2)} \leq \ldots \leq x^{(n)}.$$

These functions are called order statistics. Obviously, any of these functions is a Cauchy mean. In particular, min, max, and median of a sample are Cauchy means.

Definition 2. A Cauchy mean M on a simple strict order $(X, <)$ is said to be *comparison meaningful* if, for any isomorfism ϕ of $(X, <)$,

$$M(x_1, \ldots, x_n) < M(y_1, \ldots, y_n)$$

implies

$$M(\phi(x_1), \ldots, \phi(x_n)) < M(\phi(y_1), \ldots, \phi(y_n))$$

for all n-tuples $(x_1, \ldots, x_n)$ and $(y_1, \ldots, y_n)$.

3. Main results

For the proof of the following theorem we define $x \leq y$ iff ($x < y$ or $x = y$).

Theorem 1. Let $(X, <)$ be a simple strict order satisfying conditions (i) and (ii). For any comparison meaningful Cauchy mean M on $(X, <)$,

$$M(x_1, \ldots, x_n) \in \{x_1, \ldots, x_n\}.$$

Proof. Since M is a symmetric function, we may assume that $x_1 \leq \ldots \leq x_n$. Then $x_1 \leq M(x_1, \ldots, x_n) \leq x_n$. Suppose that $M(x_1, \ldots, x_n) = x_0$ where $x_i < x_0 < x_{i+1}$ for some i. By conditions (i) and (ii), there are elements a and b in X and an isomorphism ϕ of the interval (x_i, x_{i+1}) such that $x_i < a < x_0 < b < x_{i+1}$ and $\phi(a) = b$. Let ϕ^* be equal to ϕ on the interval (x_i, x_{i+1}) and be the identity mapping for all other elements in X. Then ϕ^* is an isomorphism of $(X, <)$. Obviously,

$$a = M(a, \ldots, a) < M(x_1, \ldots, x_n) = x_0.$$

Since M is comparison meaningful and $\phi^*(a) = b$, $\phi^*(x_i) = x_i$ for all i, we have

$$b = M(b, \ldots, b) < M(x_1, \ldots, x_n) = x_0 ,$$

a contradiction. Thus $M(x_1, \ldots, x_n)$ must be equal to one of the x_i's.

□

Obviously, any ordered statistics is a comparison meaningful Cauchy mean on the simple strict order $(Re, <)$ where $<$ is the natural ordering. The following example shows that there are other comparison meaningful Cauchy means on $(Re, <)$.

Example 2. Let $X = [0,1]$. We define $M(x_1, x_2) = \max\{x_1, x_2\}$ if $x_1 \neq 0$, and $M(x_1, x_2) = 0$ otherwise. This function is a comparison meaningful Cauchy mean but different from any order statistic on X.

Example 3. For any real $x_1 \leq x_2 \leq x_3$ we define $M(x_1, x_2, x_3) = x_1$, if $x_1 = x_2$, and $M(x_1, x_2, x_3) = x_3$, if $x_1 < x_2$. For all other orderings of arguments, M is defined by symmetry. This function is again a comparison meaningful Cauchy mean but different from any order statistic.

Note, that all order statistics are continuous functions on Re while the means from the previous examples are not.

Definition 3. Let $(X, <)$ be a simple strict order. As in the case of real numbers, we define *order statistic* $x^{(i)}$ by arranging any given n-tuple $(x_1, \dots, x_n)$ in the increasing order

$$x^{(1)} \leq x^{(2)} \leq \dots \leq x^{(n)}.$$

Our goal is to establish conditions under which order statistics are the only comparison meaningful Cauchy means on $(X, <)$. Two previous examples suggest that we probably need continuity and connectedness. In what follows, we consider only simple strict orders satisfying conditions (i) and (ii) and endowed with interval topology. X^n is a topological space with the product topology on it. We shall use notation $\overline{x} = (x_1, \dots, x_n)$ for n-tuples.

The following proposition belong to the general theory of topological ordered sets.

We will use notations $x \wedge y$ and $x \vee y$ for the minimum and maximum of two elements x and y in $(X, <)$, respectively.

Proposition 1. Let $\mathcal{O}^n$ be a subset of X^n defined by $\mathcal{O}^n = \{\overline{x} \mid x_1 < \dots < x_n\}$. Then $\mathcal{O}^n$ is an open subset in X^n and its closure is $\overline{\mathcal{O}^n} = \{\overline{x} \mid x_1 \leq \dots \leq x_n\}$, provided that condition (i) is satisfied.

Proof. Consider sets

$$X_i^n = \{\overline{x} \mid x_i < x_{i+1}\}$$

for $1 \leq i \leq n-1$. Obviously, $\mathcal{O}^n$ is the intersection of all these sets. By condition (i), for any $\overline{x} \in X_i^n$, we can find elements $a, b \in X$ such that $x_i < a < b < x_{i+1}$. Consider a direct product $\prod_k X_k$ where $X_i = (\leftarrow, a)$, $X_{i+1} = (b, \rightarrow)$, and $X_k = X$ for all other k. It is an open subset of X_i^n containing $\overline{x}$. Thus, X_i^n, and therefore $\mathcal{O}^n$, are open sets in X^n.

Consider sets $\overline{X_i^n} = \{\overline{x} \mid x_i \leq x_{i+1}\}$ for $1 \leq i \leq n-1$. By using the previous arguments, it is easy to show that complements of these sets are open. Therefore, each $\overline{X_i^n}$ is open and so is their intersection $\overline{\mathcal{O}^n}$. It suffices to prove now that any $\overline{x} \in \overline{\mathcal{O}^n} \setminus \mathcal{O}^n$ belongs to the closure of $\mathcal{O}^n$. *We prove it here only in the case $n = 2$ leaving the general case for the full-length paper.* Let $\overline{x} = (x, x)$ and U be a neighborhood of $\overline{x}$. Without loosing generality, we may assume that $U = I \times I$ where I is an open interval in X containing x. Then, by condition (i), there is an element $y \neq x$ in I and the pair $(x \wedge y, x \vee y) \in U \cap \mathcal{O}^2$. Thus $\overline{x}$ belongs to the closure of $\mathcal{O}^2$.

□

The second part of the previous theorem is not true if condition (i) is not satisfied. Indeed, consider $X = [0,1] \cup \{2\}$ endowed with the interval topology. Then $(2,2)$ does not belong to the closure of

$$\mathcal{O}^2 = \{(x,y) \mid 0 \leq x < y \leq 1 \text{ or } (y = 2 \text{ and } x < 2)\}.$$

Definition 4. Suppose $\overline{x}'$ and $\overline{x}''$ are two points in X^n such that

$$\overline{x}' = (x_1, \ldots, x_{k-1}, x'_k, x_{k+1}, \ldots, x_n)$$

and

$$\overline{x}'' = (x_1, \ldots, x_{k-1}, x''_k, x_{k+1}, \ldots, x_n)$$

where $x'_k \neq x''_k$. The set

$$[\overline{x}', \overline{x}''] = \{(x_1, \ldots, x_{k-1}, x, x_{k+1}, \ldots, x_n)\},$$
$$x'_k \wedge x''_k \leq x \leq x'_k \vee x''_k$$

is called a *line segment*.

Lemma 1. Let $(X, <)$ be a simple strict order connected in the interval topology and satisfying condition (ii). The restriction of a continuous comparison meaningful Cauchy mean M on any line segment in $\mathcal{O}^n$ is equal to an order statistic.

Proof. First, note that the restriction of an order statistic on $\mathcal{O}^n$ is a coordinate function, i.e. $x^{(i)} = x_i$ for all i. Since $(X, <)$ is connected, it has no gaps, i.e. empty open intervals ([2], lemma 3.5.9.) Thus condition (i) is satisfied and, according to Theorem 1, $M(x_1, \ldots, x_n) \in \{x_1, \ldots, x_n\}$. Let $[\overline{a}', \overline{a}'']$ be a line segment. This line segment is homeomorphic to the closed interval $[a'_k \wedge a''_k, a'_k \vee a''_k]$ which is a connected subset of X. The image of $[\overline{a}', \overline{a}'']$ under M is a connected subset of

$$\{a_1\} \cup \ldots \cup \{a_{k-1}\} \cup [a'_k \wedge a''_k, a'_k \vee a''_k] \cup \{a_k\} \cup \ldots \cup \{a_n\}$$

Since $[a'_k, a''_k] \in \mathcal{O}^n$, $M(x_1, \ldots, x_n) = x_i$ for some i on the line segment under consideration.

□

Theorem 2. Suppose simple strict order $(X, <)$ is connected in the interval topology and satisfies condition (ii). Any continuous comparison meaningful Cauchy mean M on $(X, <)$ is an order statistic.

Proof. It suffices to prove that the restriction of M on $\mathcal{O}^n$ is an order statistic. Indeed, then, by continuity of M and any order statistic, M is an order statistic on the closer of $\mathcal{O}^n$, and, by Proposition 1 and symmetry of M, it is the same order statistic on the entire X^n.

Let $\overline{a} = (a_1, \ldots, a_n)$ and $\overline{b} = (b_1, \ldots, b_n)$ be any two points in $\mathcal{O}^n$. For any $1 \leq i \leq n$, we define $\overline{a}_0 = \overline{a}$ and

$$\overline{a}_i = (a_1 \wedge b_1, \ldots, a_i \wedge b_i, a_{i+1}, \ldots, a_n).$$

Similarly, $\overline{b}_0 = \overline{b}$ and

$$\overline{b}_i = (a_1 \wedge b_1, \ldots, a_i \wedge b_i, b_{i+1}, \ldots, b_n).$$

We say that the sequence of points $\{\overline{a}_0, \ldots, \overline{a}_n = \overline{b}_n, \ldots, \overline{b}_0\}$ *connects* points $\overline{a}$ and $\overline{b}$ in $\mathcal{O}^n$ and call two consecutive elements of this sequence *adjacent points*. Note, that all points in this sequence indeed belong to $\mathcal{O}^n$ and two adjacent points are not necessarily distinct. It follows from our definition, that, for any two distinct adjacent points $\overline{x}$ and $\overline{y}$ in the above defined sequence, there is an index i such that $x_k = y_k$ for all $k \neq i$ and $x_i \neq y_i$. Thus, any two distinct adjacent points define a line segment in $\mathcal{O}^n$. We constructed a sequence

$$[\overline{z}_1, \overline{z}_2], [\overline{z}_2, \overline{z}_3], \ldots, [\overline{z}_{m-1}, \overline{z}_m],$$

of line segments in $\mathcal{O}^n$ such that $\overline{z}_1 = \overline{a}$ and $\overline{z}_m = \overline{b}$. By Lemma 1, M is an order statistic on each of these line segments. Since all $\overline{z}_i$'s belong to $\mathcal{O}^n$, it is the same order statistic on all these line segments and

$$M(a_1, \ldots, a_n) = M(b_1, \ldots, b_n)$$

for any two points in $\mathcal{O}^n$.

□

REFERENCES

[1] J. Aczel and J. Dhombres, *Functional Equations in Several Variables* (Cambridge University Press, 1989)
[2] A.L. Cauchy, *Cours d'analyse de l'École Royale Polytechnique,* Ire partie, Analyse algébraique (Paris, 1821)
[3] J.-L. Marichal and M. Roubens,Characterization of some stable aggregation functions, Proc. Int. Conf. on Indust. Eng. and Product Management, Mons (1993), 187-196.
[4] J. Pfanzagl, *Theory of Measurement* (Physica-Verlag, 1971)
[5] A. Orlov, *Stability in Social Economics Systems*, (Nauka, Moscow, 1979) in Russian
[6] F.S. Roberts, *Measurement Theory* (Addison-Wesley, 1979)

CHARACTERIZATION OF SOME AGGREGATION FUNCTIONS ARISING FROM MCDM PROBLEMS

János C. FODOR*, Jean-Luc MARICHAL** and Marc ROUBENS**

* Department of Computer Science
Eötvös Loránd University
Múzeum krt. 6–8
H-1088 Budapest, Hungary

** Institute of Mathematics
University of Liège
15, avenue des Tilleuls - D1
B-4000 Liège, Belgium

ABSTRACT

We investigate the aggregation phase of multicriteria decision making procedures. Characterizations of some classes of nonconventional aggregation operators are established. The first class consists of the ordered weighted averaging operators (OWA) introduced by Yager. The second class corresponds to the weighted maximum defined by Dubois and Prade. The dual class (weighted minimum) and some ordered versions are also characterized. Results are obtained via solutions of functional equations.

1. Introduction

In fuzzy (valued) multicriteria decision making problems it is typical that we have quantitative judgments on the pairs of alternatives concerning each criterion. These judgments are very often expressed by the help of fuzzy preference relations.

More formally, let A be a given set of alternatives and $R_1, R_2, \ldots, R_m$ be valued binary relations on A representing m criteria. That is, each R_i is a function from $A \times A$ to D ($i = 1, \ldots, m$), where $D \subseteq \mathbb{R}$ (usually, D is either $[0, 1]$ or $\mathbb{R}$).

As it is well-known, multicriteria decision making procedures consist of three main steps (phases) as follows.

1. *Modelling phase*

   In this phase we look for appropriate models for valued monocriterion relations R_i ($i = 1, \ldots, m$) and also for determining the importance of each criterion (i.e., the weights).

2. *Aggregation phase*

   In this step we try to find a unified (global) relation R on A, on the basis of monocriterion relations and the weights.

*Supported in part by OTKA (National Scientific Research Fund, Hungary) I/6-14144, and by the Foundation for Hungarian Higher Education and Research 615/94. J. Fodor is also with the Department of Mathematics, University of Agricultural Sciences, Gödöllő, Hungary.

3. *Exploitation phase*

   In this phase we transform the global information about the alternatives either into a (partial or complete) ranking of the elements in A, or into a global choice of the best actions in A.

In this paper we are dealing with the aggregation phase only. That is, we are looking for an *aggregation function* $M : D^m \rightarrow D$ which satisfies a number of "desirable" properties so that the global relation R, expressed by

$$R(a,b) = M(R_1(a,b), R_2(a,b), \ldots, R_m(a,b))$$

for all $a, b \in A$, reflects an overall opinion on the pairs of alternatives. For more details see Fodor and Roubens [6].

In addition to the classical aggregation operations (e.g. weighted arithmetic means, geometric means, root-power means, quasi-arithmetic means, etc), two new classes have been introduced in the eighties.

Dubois and Prade [4] defined and investigated the weighted maximum (and its dual: weighted minimum) operators in 1986. The formal analogy with the weighted arithmetic mean is obvious.

Yager [16] introduced the ordered weighted averaging operators (OWA) in 1988. The basic idea of OWA is to associate weights with a particular ordered position rather than a particular element.

The same idea was used by Dubois *et al.* [5] to introduce ordered weighted maximum (OWMAX) and minimum for modelling soft partial matching.

The main difference between OWA and OWMAX (resp. OWMIN) is in the underlying non-ordered aggregation operation. OWA uses arithmetic mean while OWMAX (resp. OMIN) applies weighted maximum (resp. weighted minimum). At first glance, this does not seem to be an essential difference. However, Dubois and Prade [4] proved that OWMAX is equivalent to the median of the ordered values and some appropriately choosen additional numbers used instead of the original weights.

Although several papers have dealt with different aspects of these operations, their characterizations have not been known yet. The main aim of the present paper is to deliver these missing descriptions.

First we study the ordered weighted averaging operators in details. We formulate some natural properties which are obviously possessed by the OWA operators. Then we show that those conditions are sufficient to characterize the OWA family. Quasi-OWA aggregators are also introduced and a particular class is characterized.

Then we investigate the weighted maximum and minimum operators in the same spirit as in case of OWA. Finally, ordered weighted maximum and minimum are characterized. For more details and proofs see [8] and [7].

2. Ordered weighted averaging aggregation operators (OWA)

The ordered weighted averaging aggregation operator (OWA) was proposed by Yager [16] in 1988. Since its introduction, it has been applied to many fields as neural networks (Yager [14]), data base systems (Yager [15]), fuzzy logic controlers (Yager [17]) and group decision making (Yager [16], Cutello and Montero [2]). Its structural properties (Skala [13]) and its links with fuzzy integrals (Grabisch [9]) were also investigated.

We consider a vector $(x_1, \ldots, x_m) \in \mathbb{R}^m$, $m > 1$, and we are willing to substitute to that vector a single value $M^{(m)}(x_1, \ldots, x_m) \in \mathbb{R}$, using the aggregation operator (aggregator) M.

An OWA aggregator $M^{(m)}$ associated to the m non negative weights $(\omega_1^{(m)}, \ldots, \omega_m^{(m)})$ such that $\sum_{k=1}^m \omega_k^{(m)} = 1$ corresponds to

$$M^{(m)}(x_1, \ldots, x_m) = \sum_{i=1}^m \omega_i^{(m)} x_{(i)}, \qquad x_{(1)} \leq \cdots \leq x_{(i)} \leq \cdots \leq x_{(m)},$$

where numbers $x_1, \ldots, x_m \in \mathbb{R}$ are rearranged increasingly and are denoted as $x_{(1)} \leq \cdots \leq x_{(i)} \leq \cdots \leq x_{(m)}$. $\omega_1^{(m)}$ is linked to the lowest value $x_{(1)}$, ..., $\omega_m^{(m)}$ is linked to the greatest value $x_{(m)}$.

This class of operators includes

- $\min(x_1, \ldots, x_m)$ if $\omega_1^{(m)} = 1$.
- $\max(x_1, \ldots, x_m)$ if $\omega_m^{(m)} = 1$.
- any order statistics $x_{(k)}$ if $\omega_k^{(m)} = 1$, $k = 1, \ldots, m$.
- the arithmetic mean if $\omega_1^{(m)} = \cdots = \omega_m^{(m)} = \frac{1}{m}$.
- the median $(x_{(m/2)} + x_{(m/2)+1})/2$ if $\omega_{(m/2)}^{(m)} = \omega_{(m/2)+1}^{(m)} = \frac{1}{2}$ and m is even.
- the median $x_{(m+1)/2}$ if $\omega_{(m+1)/2}^{(m)} = 1$ and m is odd.
- the arithmetic mean excluding the two extremes if $\omega_1^{(m)} = \omega_m^{(m)} = 0$ and $\omega_i^{(m)} = \frac{1}{m-2}$, $i \neq 1, m$.

Well-known and easy to prove properties of the OWA aggregators are summarized as follows (see also Yager [16], Cutello and Montero [2]).

Any OWA aggregator is

- *neutral* (or symmetric or commutative): $M(x_1, \ldots, x_m) = M(x_{i_1}, \ldots, x_{i_m})$ holds for all $(x_1, \ldots, x_m) \in \mathbb{R}^m$, when $(i_1, \ldots, i_m) = \sigma(1, \ldots, m)$, where σ represents a permutation operation;
- *monotonic*: $x_i' > x_i$ implies $M(x_1, \ldots, x_i', \ldots, x_m) \geq M(x_1, \ldots, x_i, \ldots, x_m)$;
- *idempotent*: $M(x, \ldots, x) = x$ holds for all $x \in \mathbb{R}$;
- *compensative*: $\min_{i=1,m} x_i \leq M^{(m)}(x_1, \ldots, x_m) \leq \max_{i=1,m} x_i$.

Moreover, the following conditions, which are nonusual in the literature of MCDM, are also satisfied by any OWA aggregator.

- *ordered linkage property* (Marichal and Roubens [11]):

  For any given real numbers $\{x_1, \ldots, x_{2m}\}$, ordered as $x_{(1)} \leq x_{(2)} \leq \cdots \leq x_{(2m)}$, we have

  $$M^{(m+1)}(y_1, \ldots, y_i, \ldots, y_{m+1}) = M^{(m)}(z_1, \ldots, z_j, \ldots, z_m)$$

  where $y_i = M^{(m)}(x_{(i)}, \ldots, x_{(m+i-1)})$ $(i = 1, \ldots, m+1)$ and $z_j = M^{(m+1)}(x_{(j)}, \ldots, x_{(m+j)})$ $(j = 1, \ldots, m)$;

- *stability for the same positive linear transformation:*
$$M^{(m)}(rx_1+t,\ldots,rx_m+t)=rM(x_1,\ldots,x_m)+t,$$
for all $(x_1,\ldots,x_m)\in\mathbb{R}^m$, all $r>0$, all $t\in\mathbb{R}$;
- *ordered stability for positive linear transformations with the same unit and independent zeroes:*
$$M^{(m)}(rx_1+t_1,\ldots,rx_m+t_m)=rM(x_1,\ldots,x_m)+T(t_1,\ldots,t_m)$$
holds for all $(x_1,\ldots,x_m)\in\mathbb{R}^m$, all $r>0$, all $(t_1,\ldots,t_m)\in\mathbb{R}^m$ and for the ordered values $x_{(1)}\le x_{(2)}\le\ldots\le x_{(m)}$, $t_{(1)}\le t_{(2)}\le\ldots\le t_{(m)}$.

Notice that, in general, OWA aggregators fail to satisfy

- *associativity:*
$$\begin{aligned}M(M(x_1,x_2),x_3) &= M(x_1,M(x_2,x_3)),\\ &\vdots\\ M(M(x_1,\ldots,x_{m-1}),x_m) &= M(x_1,M(x_2,\ldots,x_m))\end{aligned}$$
for all $(x_1,\ldots,x_m)\in\mathbb{R}^m$;
- *decomposability:* (Kolmogorov [10], Nagumo [12])
$$M^{(m)}(x_1,\ldots,x_k,x_{k+1},\ldots,x_m)=M^{(m)}(x,\ldots,x,x_{k+1},\ldots,x_m)$$
when $x=M^{(k)}(x_1,\ldots,x_k)$, for all $(x_1,\ldots,x_m)\in\mathbb{R}^m$.

In addition, associativity and decomposability together imply the ordered linkage property, while the converse is not true in general.

Now we turn to the characterization of OWA operators, i.e., we choose two sets of sufficient conditions from the above list of necessary conditions.

3. Characterization of the OWA aggregator

A foundational paper of Aczél, Roberts and Rosenbaum [1] shows that the general solution of a functional equation related to the stability for positive linear transformations (case ♯5 in [1]) :
$$M(rx_1+t,\ldots,rx_m+t)=rM(x_1,\ldots,x_m)+t,\quad r>0,$$
where M is a mapping from $\mathbb{R}^m\to\mathbb{R}$ given by
$$M(x_1,\ldots,x_m)=\begin{cases}S(\underline{x})f\left(\frac{x_1-A(\underline{x})}{S(\underline{x})},\ldots,\frac{x_m-A(\underline{x})}{S(\underline{x})}\right)+A(\underline{x}) & \text{if } S(\underline{x})\neq 0\\ x & \text{if } S(\underline{x})=0,\end{cases}$$
where $\underline{x}=(x_1,\ldots,x_m)$, $S^2(\underline{x})=\sum_i(x_i-A(\underline{x}))^2$ and $A(\underline{x})$ represents the arithmetic mean (whence $S(\underline{x})=0\Leftrightarrow x_1=x_2=\cdots=x_m=x$); f is an arbitrary function from $\mathbb{R}^m$ to $\mathbb{R}$.

It is also true that the weighted mean corresponds to monotonic and idempotent aggregators which satisfy the (SPLU)-property (see [1], case ♯ 9).

From results obtained by Marichal and Roubens [11], we know that neutral, continuous, stable for the same positive linear transformations and associative (resp. decomposable) operators are characterized by the min or max operators (resp. min or max or $A(\underline{x})$).

Weaker property than associativity or decomposability is needed to be able to characterize the OWA operators which include min, max and the arithmetic means. This intermediate property is related to the ordered linkage property.

Theorem 1 *The class of ordered weighted averaging aggregators corresponds to the operators which satisfy the properties of neutrality, monotonicity, stability for the same positive linear transformations and ordered linkage.*

Another characterization of OWA operators corresponds to the following proposition.

Theorem 2 *The class of ordered weighted averaging operators corresponds to the aggregators which satisfy the properties of neutrality, monotonicity, idempotency and stability for positive linear transformations with the same unit, independent zeroes and ordered values.*

Proofs of these results and some related issues can be found in [8].

4. Decomposable quasi-OWA aggregators

The quasi-arithmetic mean was first considered and characterized by Kolmogorov [10] and Nagumo [12]. It corresponds to the aggregator

$$M(x_1, \dots, x_m) = f^{-1}\left[\frac{1}{m}\sum_i f(x_i)\right]$$

where f is a continuous strictly monotonic function.

It is natural to consider the quasi-OWA operators

$$M(x_1, \dots, x_m) = f^{-1}\left[\sum_i \omega_i^{(m)} f(x_{(i)})\right].$$

These aggregators have still to be characterized but one can prove the following proposition.

Theorem 3 *Any decomposable quasi-OWA operator corresponds to the* min *or* max *or the quasi-arithmetic mean.*

5. Weighted maximum and minimum

Using the concept of possibility and necessity of fuzzy events [18, 3], one can evaluate the possibility that a relevant goal is attained, and the necessity that all the relevant goals are attained by the help of the following formulas (see [4] for more details) *weighted maximum:*

$$\max_{i=1,m}\{\min(w_i, x_i)\}, \quad w_i \in [0,1], \quad \max_{i=1,m} w_i = 1 \tag{1}$$

and

weighted minimum:

$$\min_{i=1,m}\{\max(w_i, x_i)\}, \quad w_i \in [0,1], \quad \min_{i=1,m} w_i = 0. \tag{2}$$

The analogy between the weighted arithmetic mean and the weighted maximum is obvious: product corresponds to minimum, sum does to maximum. It is emphasized in [4] that weighted maximum and minimum operators can be calculated as medians, i.e., the qualitative counterparts of means.

It is easy to see that weighted maximum satisfies idempotency and monotonicity. Moreover, it fulfils also (with $T^{(m)} = M^{(m)}$)

- *stability for maximum* (SMAX):

$$M^{(m)}(x_1 \vee t_1, \ldots, x_m \vee t_m) = M^{(m)}(x_1, \ldots, x_m) \vee T^{(m)}(t_1, \ldots, t_m)$$

  for all $(x_1, \ldots, x_m) \in [0,1]^m$, $(t_1, \ldots, t_m) \in [0,1]^m$.

- *stability for minimum with the same unit* (SMINU):

$$M^{(m)}(r \wedge x_1, \ldots, r \wedge x_m) = r \wedge M^{(m)}(x_1, \ldots, x_m)$$

  for all $(x_1, \ldots, x_m) \in [0,1]^m$, $r \in [0,1]$.

In a sense, the converse is also true, as we state in the following theorem.

Theorem 4 *Suppose that M is a nondecreasing function from $[0,1]^m$ to $[0,1]$ such that $M(0,\ldots,0) = 0$ and $M(1,\ldots,1) = 1$. Then M satisfies SMAX and SMINU if and only if there exist weights $w_1, \ldots, w_m \geq 0$ with* $\max w_i = 1$ *such that*

$$M(x_1, \ldots, x_m) = \max_{i=1,\ldots,m} \{\min(w_i, x_i)\}.$$

By duality, we can introduce the corresponding stability conditions in the case of the weighted minimum as follows:

- *stability for minimum* (SMIN):

$$M^{(m)}(x_1 \wedge t_1, \ldots, x_m \wedge t_m) = M^{(m)}(x_1, \ldots, x_m) \wedge T^{(m)}(t_1, \ldots, t_m)$$

  for all $(x_1, \ldots, x_m) \in [0,1]^m$, $(t_1, \ldots, t_m) \in [0,1]^m$.

- *stability for maximum with the same unit* (SMAXU):

$$M^{(m)}(r \vee x_1, \ldots, r \vee x_m) = r \vee M^{(m)}(x_1, \ldots, x_m)$$

  for all $(x_1, \ldots, x_m) \in [0,1]^m$, $r \in [0,1]$.

Obviously, the weighted minimum (2) satisfies both conditions. We state that the converse is also true in the following sense.

Theorem 5 *Suppose that M is a nondecreasing function from $[0,1]^m$ to $[0,1]$ such that $M(0,\ldots,0) = 0$ and $M(1,\ldots,1) = 1$. Then M satisfies SMIN and SMAXU if and only if there exist weights $w_1, \ldots, w_m \geq 0$ with* $\max w_i = 1$ *such that*

$$M(x_1, \ldots, x_m) = \min_{i=1,\ldots,m} \{\max(w_i, x_i)\}.$$

6. Ordered weighted minimum and maximum

Suppose that $(x_1, \ldots, x_m) \in [0,1]^m$ and order these numbers increasingly: $x_{(1)} \leq x_{(2)} \leq \ldots \leq x_{(m)}$. The ordered weighted maximum (OWMAX) operator associated to the m nonnegative weights $(w_1, \ldots, w_m)$ with $\max w_i = 1$ corresponds to

$$M(x_1, \ldots, x_m) = \max_{i=1\ldots,m} \{\min(w_i, x_{(i)})\}, \tag{3}$$

see [5].

The weight w_1 is linked to the lowest value $x_{(1)}$, $\ldots$, w_m is linked to the greatest value $x_{(m)}$. This class of operators includes

- $\min(x_1, \ldots, x_m)$ if $w_1 = 1$ and $w_i = 0$ for $i \geq 2$;
- $\max(x_1, \ldots, x_m)$ if $w_m = 1$;
- any order statistics x_k if $w_k = 1$ and $w_i = 0$ for $i > k$;

Obviously, any OWMAX operator is neutral, nondecreasing, idempotent and compensative. In addition, SMAX and SMINU are also satisfied for the ordered values $x_{(1)} \leq \ldots \leq x_{(m)}$, $t_{(1)} \leq \ldots \leq t_{(m)}$ as follows:

$$\begin{aligned} M(x_{(1)} \vee t_{(1)}, \ldots, x_{(m)} \vee t_{(m)}) &= M(x_{(1)}, \ldots, x_{(m)}) \vee T(t_{(1)}, \ldots, t_{(m)}), \\ M(r \wedge x_{(1)}, \ldots, r \wedge x_{(m)}) &= r \wedge M(x_{(1)}, \ldots, x_{(m)}). \end{aligned}$$

Fortunately, the converse is also true in the following form.

Theorem 6 *A nondecreasing function $M : [0,1]^m \to [0,1]$ with $M(0, \ldots, 0) = 0$ and $M(1, \ldots, 1) = 1$ satisfies SMAX and SMINU for ordered elements if and only if there exist weights $1 = w_1 \geq \ldots \geq w_m \geq 0$ such that*

$$M(x_1, \ldots, x_m) = \max_{i=1\ldots,m} \{\min(w_i, x_{(i)})\}.$$

Notice that we can obtain similar characterization for OWMIN operators when using SMIN and SMAXU for ordered values. We formulate the statement without proof as follows.

Theorem 7 *A nondecreasing function $M : [0,1]^m \to [0,1]$ with $M(0, \ldots, 0) = 0$ and $M(1, \ldots, 1) = 1$ satisfies SMIN and SMAXU for ordered elements if and only if there exist weights $1 \geq w_1 \geq \ldots \geq w_m = 0$ such that*

$$M(x_1, \ldots, x_m) = \min_{i=1\ldots,m} \{\max(w_i, x_{(i)})\}.$$

7. Conclusion

We have characterized some aggregation functions which can be useful in multicriteria decision making procedures. The results contribute to the theory of MCDM, but at the same time can help the reader in choosing at least a particular family of aggregators on the basis of some properties expected in advance from an aggregation function.

REFERENCES

[1] J. Aczél, F.S. Roberts and Z. Rosenbaum, On the scientific laws without dimensional constants, *Journal of Math. Analysis and Appl.*, **119** (1986), 389–416.

[2] V. Cutello and J. Montero, Hierarchies of aggregation operators, Working paper (1993), unpublished.

[3] D. Dubois and H. Prade, *Théorie des Possibilités. Application à la Représentation des Connaissances en Informatique*, Masson, Paris, 1985.

[4] D. Dubois and H. Prade, Weighted minimum and maximum operations in fuzzy set theory, *Inf. Sci.* **39** (1986) 205–210.

[5] D. Dubois, H. Prade and C. Testemale, Weighted fuzzy pattern-matching, *Fuzzy Sets and Systems* **28** (1988) 313–331.

[6] J. Fodor and M. Roubens, *Fuzzy Preference Modelling and Multicriteria Decision Support,* Kluwer, Dordrecht, 1994.

[7] J. Fodor and M. Roubens, Characterization of weighted maximum and some related operations, *Inf. Sci.* (to appear).

[8] J. Fodor, J.-L. Marichal and M. Roubens, Characterization of the ordered weighted averaging operators, *IEEE Trans. on Fuzzy Systems* (to appear).

[9] M. Grabisch, On the use of fuzzy integral as a fuzzy connective, Proceedings of the Second IEEE International Conference on Fuzzy Systems, San Francisco (1993) 213-218.

[10] A.N. Kolmogoroff, Sur la notion de la moyenne, *Accad. Naz. Lincei Mem. Cl. Sci. Fis. Mat. Natur. Sez.,* **12** (1930) 388-391.

[11] J.L. Marichal and M. Roubens, Characterization of some stable aggregation functions, in : Proc. Intern. Conf. on Industrial Engineering and Production Management, Mons, Belgium (1993) 187–196.

[12] M. Nagumo, Über eine Klasse der Mittelwerte, *Japanese Journal of Mathematics*, **6** (1930) 71-79.

[13] H.J. Skala, Concerning ordered weighted averaging aggregation operators, Statistical Papers **32** (1991), 35–44.

[14] R.R. Yager, On the aggregation of processing units in neural networks, Proc. 1st IEEE Int. Conference on Neural Networks, San Diego, Vol. II (1987) 927–933.

[15] R.R. Yager, A note on weighted queries in information retrieval systems, *J. Amer. Soc. Information Sciences*, **28** (1987) 23–24.

[16] R.R. Yager, On ordered weighted averaging aggregation operators in multicriteria decision making, *IEEE Trans. on Systems, Man and Cybernetics*, **18** (1988), 183–190.

[17] R. R. Yager, Connectives and quantifiers in fuzzy sets, *Fuzzy Sets and Systems* **40** (1991) 39–75.

[18] L. A. Zadeh, Fuzzy sets as a basis for a theory of possibility, *Fuzzy Sets and Systems*, **1** (1978), 3–28.

A GENERAL FRAMEWORK FOR THE UTILITY-BASED AND OUTRANKING METHODS

J. Dombi

Department of Applied Informatics
Jozsef Attila University[1]
H-6720 Szeged, Arpad t. 1. Hungary.

ABSTRACT

In this paper on the utility and preference based decision process is presented in unified way. We first deal with the utility based decision making comparing with the preference model. We define an universal preference function with useful properties and we show that the utility based decision making gives the same result as the preference based using the universal preference function. The preference based decision making procedures (the outranking approaches) can be get also with the help of the universal preference function i.e. we proved that there exist a unary function applying to the universal preference function getting the preference relations of the different outranking approaches.

1. Utility and preference based decision making

The most widespread methods used in multicriteria decision analysis are mainly of two types: the ones based on utility theory or on preference relations. Naturally, different methods provide different results even if the same information is used. The immediate question arises under which condition do the two approaches give the same result. More precisely, let a and b two actions (alternatives) and c_1, c_2, ... ,c_n different criteria, x_i and y_i are the utilities (evaluations) of a and b from the perspective of c_i, respectively.

	c_1	c_2	...	c_n
a	x_1	x_2	...	x_n
b	y_1	y_2	...	y_n
	w_1	w_2	...	w_n

(1)

Let w_1, w_2, ... ,w_n denote the nonnegative weights of the criteria such that they are normalised, i.e.:

$$\sum_{i=1}^{n} w_i = 1 \tag{2}$$

The aggregated utility of actions a and b are

[1] This work has been partially supported by OTKA

$$A = U(a) = \sum_{i=1}^{n} w_i x_i \ , \quad B = U(b) = \sum_{i=1}^{n} w_i y_i \tag{3}$$

The usual ordering of the numbers A and B induces an ordering of actions a and b (generally an ordering on the set of the actions).

On the other hand, in the preference-based methods first a and b are compared from the perspective of each c_i. This is usually done by calculating a preference index p_i

$$p_i = p(x_i, y_i) \tag{4}$$

with the help some function p. The next step is the aggregation of the p_i-s:

$$p(a,b) = \sum_{i=1}^{n} w_i p(x_i, y_i) \tag{5}$$

To be able to compare the two approaches let us use the same function p and the aggregated utility values A and B to obtain another preference index $p(A,B)$ of actions a and b.

1. Theorem:

$$p(a,b) = p(A,B) \ \text{ iff } \ p(x,y) = \frac{1}{2}(y - x + 1). \tag{6}$$

where p(x,y) is the linear form of the universal preference function.

Proof:

$$p(a,b) = \frac{1}{2}(b - a + 1) = \frac{1}{2}\left(\sum_{i=1}^{n} w_i y_i - \sum_{i=1}^{n} w_i x_i + \sum_{i=1}^{n} w_i\right) =$$

$$= \frac{1}{2}\left(\sum_{i=1}^{n} w_i (y_i - x_i + 1)\right) = \frac{1}{2}\left(\sum_{i=1}^{n} w_i p(x_i, y_i)\right). \ \square \tag{7}$$

Another way of comparing the two approaches is by calculating for each action the value

$$V(a) = \frac{1}{n}\sum_{b} p(a,b) \tag{8}$$

where n is the number of actions, and p is the preference function given in the above theorem.

2. Theorem:

$$V(a) > V(b) \ \text{ iff } \ U(a) > U(b). \tag{9}$$

Proof:

Let $a_1, a_2, \ldots, a_m$ alternatives, $c_1, c_2, \ldots, c_n$ criteria and e_{ki} are utilities of a_k from the perspective of c_i, respectively. For $p(a_k)$ we get:

$$p(a_k)=\frac{1}{n}\sum_{x=1}^{n}p(a_k,a_x)=\frac{1}{n}\sum_{x=1}^{n}\sum_{i=1}^{m}w_i\frac{e_{ki}-e_{xi}+1}{2}=\frac{1}{2n}\sum_{x=1}^{n}\sum_{i=1}^{m}w_i(e_{ki}-e_{xi}+1) \quad (10)$$

and for

$$p(a_k)-p(a_l)=\frac{1}{2n}\sum_{x=1}^{n}\left(\sum_{i=1}^{m}w_i((e_{ki}-e_{xi}+1)-(e_{li}-e_{xi}+1))\right)=$$

$$=\frac{1}{2n}\sum_{x=1}^{n}\left(\sum_{i=1}^{m}w_ie_{ki}-\sum_{i=1}^{m}w_ie_{li}\right)=\frac{1}{2}(U(a_k)-U(a_l)).\ \square \quad (11)$$

The function in (6) has some useful properties:

$$p(1,0)=0,\quad p(0,1)=1,\quad p(x,x)=\frac{1}{2} \quad (12)$$

$$p(x,y)+p(y,x)=1 \quad (13)$$

Let the negation function $n(x)=1-x$. Then

$$p(x,y)+p(n(x),n(y))=1. \quad (14)$$

The next result gives the generalisation of the theorem (7), using the aggregation operator

$$A=f^{-1}\left(\sum_{i=1}^{n}w_if(x_i)\right) \quad (15)$$

axiomatised by Dyckhoff [1] and Fodor [2], where f is the generator function of the aggregation, $f:[0,1]\to[0,1]$ continuous and strictly monotone increasing. The generalised negation function is:

$$n(x)=f^{-1}(1-f(x)) \quad (16)$$

3. Theorem:

If $p^*(x,y)=f^{-1}\left(\frac{1}{2}(f(y)-f(x)+1)\right)$ then $p^*(a,b)=p^*(A,B)$ (17)

where $p^*(x,y)$ is the universal preference function.

In other words, the preference and the aggregation are interchangeable.

$$p(U(a),U(b))=U(p(a,b)).$$

Proof:

$$p^*(a,b)=f^{-1}\left(\frac{1}{2}(f(a)-f(b)+1)\right)=f^{-1}\left(\frac{1}{2}\left(f\left(f^{-1}\left(\sum_{i=1}^{m}w_i f(x_i)\right)\right)-f\left(f^{-1}\left(\sum_{i=1}^{m}w_i f(y_i)\right)\right)+1\right)\right)=$$

$$=f^{-1}\left(\frac{1}{2}\left(\sum_{i=1}^{m}w_i f(x_i)-\sum_{i=1}^{m}w_i f(y_i)+\sum_{i=1}^{m}w_i\right)\right)=f^{-1}\left(\frac{1}{2}\sum_{i=1}^{m}w_i(f(x_i)-f(y_i)+1)\right). \quad \square$$

The properties (9),(10),(11) hold for p^* as well with the modification

$$p^*(x,x)=v, \tag{18}$$

where v is the neutral value of $n(x)$, i.e. $n(v)=v$.

2. Outranking approach based on the universal preference function

In "The Outranking approach and the Foundation of ELECTRE Methods" [3], Bernard Roy presents.the fundamental aspects of this approach: the basic principles in the construction of the outranking relation, the concepts of the concordance and discordance. In this paper we are concentrating only to the concordance relation. For validating a comprehensive outranking relation S, it is necessary take into account the fact that the role which has to be devoted to each criterion in the aggregation procedure is not necessarily the same. In other words, we need to characterize what is usually referred to as "the greater or less importance" give to each criterion. In methods, the importance of the j-th criterion is taken into accaunt. By definition, the concordance index c(a,b) characterizes the strength of the positive arguments able to validate the assertion aSb. By definition in ELECTRE I method

$$c_1(a,b)=\sum_{j\in aSb}w_j$$

and in ELECTRE II method

$$c_2(a,b)=\sum_{j\in aSb}w_j\frac{p_j+q_j(a)-g_j(b)}{p_j-q_j}$$

where

$$\sum_{i=1}^{n}w_i=1$$

The outranking method of the PROMETHEE method is quite similar to the concordance index in the ELECTRE method. In order to estimate the preferences the decision maker is offered a choice, for each criterion between six forms of curves presented in [4]. In function of the way his preference increase with difference b-a, the decision-maker sets, for each criterion, the form of the function and associated parameter(s). The parameters to be estimated have simple interpretation since they are indifference and preference thresholds.

The outranking methods are not preference-based methods in the above sense (i.e. universal preference function), because in the ELECTRE methods [3]

$$p(x,x) = 1 \tag{19}$$

and in the PROMETHEE methods [4]

$$p(x,x) = 0. \tag{20}$$

According to (16) and (17) PROMETHEE and ELECTRE seem to be dual approaches. Furthermore, $p(x,y)$ and $p(y,x)$ are not related. The main reason is that $p(x,y)$ differs from the form (6) and the function $p(x,y)$ varies from criterion to criterion, i.e.

$$p(a,b) = \sum_{i=1}^{n} w_i p_i(x_i, y_i). \tag{21}$$

We define the general form of the outranking and the utility approach:

$$p(a,b) = \sum_{i=1}^{n} w_i \tau_i \big(p_i(x_i, y_i)\big) \tag{22}$$

where τ_i:[0,1]→[0,1] are unary functions.

The next result presents the two outranking methods in a unified way.

4. Theorem:

For the preference function p^{EL}, p^{PR} of the ELECTRE and PROMETHEE methods there exist unary functions τ_i^{EL} and τ_i^{PR} such that

$$p^{\mathrm{EL}}(a,b) = \sum_{i=1}^{n} w_i \tau_i^{\mathrm{EL}}(p(x_i, y_i)) \tag{23}$$

$$p^{\mathrm{PR}}(a,b) = \sum_{i=1}^{n} w_i \tau_i^{\mathrm{PR}}(p(x_i, y_i)) \tag{24}$$

and moreover in the sense of 2. theorem we can get the utility approach from the general form (22).

Proof:

Let $\tau^{\mathrm{EL}}(x)$ and $\tau^{\mathrm{PR}}(x)$ the following functions:

$$\tau_i^{\mathrm{EL}}(x) = \begin{cases} 0 & x \le p_i \\ \dfrac{x}{q_i - p_i} & p_i < x < q_i \\ 1 & x \ge q_i \end{cases} \tag{25}$$

for some $0 \le p_i \le q_i \le \frac{1}{2}$, and

$$\tau_i^{\mathrm{PR}}(x) = \begin{cases} 0 & x \le p_i \\ \dfrac{x}{q_i - p_i} = \varphi(x) & p_i < x < q_i \\ 1 & x \ge q_i \end{cases} \tag{25$'$}$$

for some $\frac{1}{2} \le p \le q \le 1$.

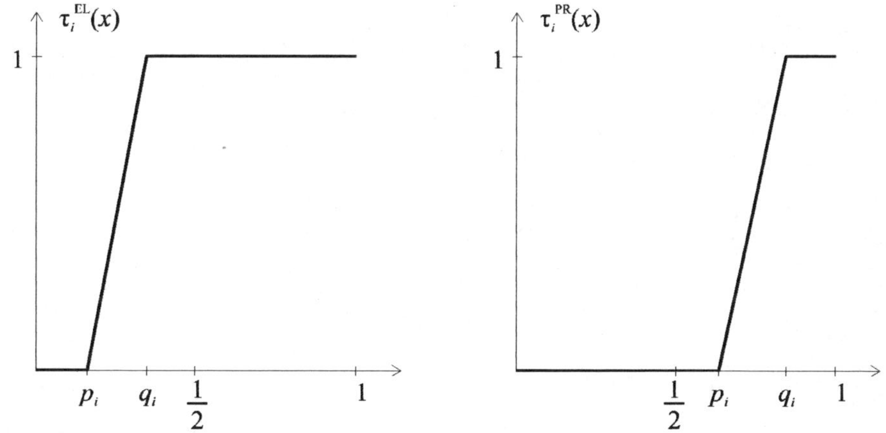

Figure.
τ^{EL} and τ^{PR} functions.

The function $\varphi(x)$ might have different forms, see PROMETHEE [4].

The utility methods can be obtained by

$$\tau_i(x) = x.$$

We remark that if the τ function commute with the negation function, i.e.

$$\tau(n(x)) = n(\tau(x)) \tag{26}$$

then the properties (9),(10),(11) remain valid for p^*.

These constructions can be used to build a more general decision method choosing less restrictive functions for $\tau(x)$. It is incorporated into the Dec.Art method.

REFERENCES:

[1] Dyckhoff, H., Basic concepts for a theory of evaluation: Hierarchical aggregation via autodistributive connectives in fuzzy set theory, Eur J. of Oper Res 20 (1985) 221-233.

[2] Fodor, J. C. and Roubens, M., Valued preference structures, Eur J. of Oper Res

[3] Roy, B., The outranking approach and the foundations of ELECTRE methods, in Readings in Multiple Criteria Decision Aid, Ed. C. A. Bana e Costa, Springerverlag (1990), 155-183.

[4] Brans, J. P. and Mareshal, B., The PROMETHEE methods for MCDM; the PROMCALC, GAIA and bankadviser software, in Readings in Multiple Criteria Decision Aid, Ed. C. A. Bana e Costa, Springerverlag (1990), 253-276.

DEFINING FUZZY COVERING RELATIONS FOR DECISION AID

Patrice PERNY

LAFORIA - IBP
Université Paris 6
4 Place Jussieu
75252 Paris cedex 05

ABSTRACT

The aim of this paper is to show how multivalued logics underlying fuzzy sets theories allow some concepts and properties used in decision aid procedures to be extended. In particular, some procedures based on the use of a conventional covering relation are presented and generalized to the fuzzy case. The main interest of these procedures is their ability to build transitive preference structures from intransitive pairwise comparisons of objects. This typical feature should be useful not only to rank fuzzy numbers or to discriminate between the alternatives of a decision problem, but also to classify multidimensional objects.

1. Introduction

In the field of Decision Aid, the analyst is faced with a complex environment, where multiple objectives and point of views are to be considered. Within this context, imprecise knowledge and necessary uncertain hypothesis are always present. This is the reason why preference analysis requires the use of non-conventional models able to reflect the imperfect nature of preference information.

In many cases, the decision problem consist in analysing the elements of a discrete finite set A of objects or *alternatives*, knowing their partial evaluations according to each relevant dimension or criterion. The so-called *"Relational Approach"* is an interesting attempt to structure such a problem and to systematize the analysis of preferences. The typical feature of the relational approach is to aggregate ordinal comparisons (*i.e.* preference relations) instead of numerical evaluations of objects. This approach close to some democratic voting procedures allows a fine description of preferences, both at the one-dimentional level and at the comprehensive level. Let us recall the basic 3-stages process which characterise this approach:

Stage 1: Evaluation of objects on each dimension or criterion. This leads to build a performance vector for each object.

Stage 2: Pairwise comparisons of objects on each dimension. This leads to build a one-dimensional fuzzy preference relation aiming at catching with nuances and realism the confidence we have in statements about preferences or similarities.

Stage 3: Aggregation of one-dimensional relations into a comprehensive fuzzy relation reflecting a compromise attitude between the various relevant dimensions.

However, it should be noted that the comprehensive relation obtained at the end of the third stage mostly lacks of transitivity property (see [9]). The possible presence of oriented cycles in the graph of this relation makes the analysis more difficult, and deriving a clear conclusion becomes a complex and artificial task. This is the major difficulty of this natural approach, known for its simplest occurences as the so-called *Condorcet Paradox* or later as the impossibility theorem of Arrow [1]. Results obtained in Social Choice theory (see for instance [13]) show that pairwise *context free* comparisons of objects do not allow to build transitive multidimensional comparisons. A fine analysis as shown the transitivity requierment and the independance of context property to be incompatible (see [1, 9]). Hence, an attempt to avoid this difficulty is to adopt context dependant comparisons in order to get some transitivity property. Among the various technics obtained in this direction, procedures based on the *covering* concept are realy interesting (for a synthesis, see [3]). They allow transitive information to be extracted from conventional cyclic relations. Roughly speaking, an alternative *a* is said to cover another alternative *b* if *a* beats *b* when they are compared vis-à-vis other alternatives respectively. Our aim in this paper is to investigate some possible extension of this useful concept when preferences are fuzzy. To this end, after some basic definitions introduced in section 2, we propose a systematic way of generalizing some concepts linked with binary relations (section 3). Then fuzzy covering relations and associated similarity relations are introduced (section 4). Finally their interest for Multiple-Criteria comparisons of alternatives is briefly discussed in the conclusion (section 5).

2. Fuzzy Binary relations and fundamental definitions

When comparing objects or candidates, when analysing similarities or preferences, the concept of binary relation appears as a key notion. Indeed, binary relations provide us with a natural model favouring pairwise comparisons of alternatives. Moreover, fuzzy binary relations seem to be an appropriate concept to compare and evaluate alternatives from ill-known or ill-defined preference information. Let us introduce some basic definitions :

Definition 1. A *fuzzy preference logic* is a formal system *L* characterised by:

• a conventionnal syntax

It is characterized by the following language:

- the variables : $t, u, v, w, \ldots$
- the constants : $a, b, c, \ldots, m, \ldots$
- the functional symbols $f, g,$, ...
- the predicative symbols $H, I, P, R, S\ldots$
- the logical connectives ($\neg$: *negation*, $\wedge$: *conjunction*, $\supset$: *implication*)
- the quantifiers $\exists$, $\forall$,
- the brackets (,).

Remark 1. In order to simplify formulae, additionnal connectives may be introduced according the following rules:

disjunction: $(A \vee B)$ stands for $\neg(\neg A \wedge \neg B)$
equivalence: $(A \equiv B)$ stands for $((A \supset B) \wedge (B \supset A))$

The set of *well formed formulae* (*WFF*) is defined recursively by :

- the atomic propositions belong to WFF
- if $A \in WFF$ then $(A) \in WFF$, $(\neg A) \in WFF$, $(\exists\, u\, A) \in WFF$ et $(\forall\, u\, A) \in WFF$,
- if $A \in WFF$ and $B \in WFF$ then $(A \wedge B) \in WFF$, $(A \vee B) \in WFF$, $(A \supset B) \in WFF$ et $(A \equiv B) \in WFF$.

• a multivalued semantic

An L-interpretation is a function υ defined from WFF into [0, 1] characterised by its values on atomic propositions. Let $\upsilon(P)$ be the truth value of the proposition P for all $P \in WFF$. $\upsilon(P) \geq \upsilon(Q)$ means that P is at least as true than Q. P is said to be true if and only if $\upsilon(P) = 1$ and false if and only if $\upsilon(P) = 0$. This interpretation may be extended from atomic propositions to more complex formulae using the following rules:

$$\upsilon(\neg A) = N(\upsilon(A)) \quad (a)$$
$$\upsilon(A \wedge B) = T(\upsilon(A), \upsilon(B)) \quad (b)$$
$$\upsilon(A \vee B) = V(\upsilon(A), \upsilon(B)) \quad (c)$$
$$\upsilon(A \supset B) = I(\upsilon(A), \upsilon(B)) \quad (d)$$
$$\upsilon(A \equiv B) = E(\upsilon(A), \upsilon(B)) \quad (e)$$

where A and B are elements of WFF and (N, T, V, I, E) is a set of functions completely defined by the given of N and T using the following construction :

T is a t-norm
N is a negation $N(x) = \phi^{-1}(1 - \phi(x))$ where ϕ is an automorphism on [0, 1] (see [4]).
V is the co-norm of T for the negation N: $V(x, y) = N(T(N(x), N(y)))$ for all x, y in [0, 1],
I is the quasi-inverse of T: $I(x, y) = \sup\{\varepsilon \in [0, 1] \,/\, T(x, \varepsilon) \leq y\}$
E is the symetric part of I: $E(x, y) = T(I(x, y), I(y, x))$.

This provides a truth-functionnal extension of classical logical operations. Moreover, if the u variable is defined on a countable set $\{u_1, ..., u_m\}$ we set :

$$\upsilon(\forall\, u\, A(u)) = T^m(\upsilon(A(u_1)), ..., \upsilon(A(u_m))) = T(\upsilon(A(u_1)), T^{m-1}(\upsilon(A(u_2)) ..., \upsilon(A(u_m))) \quad (f)$$
$$\upsilon(\exists\, u\, A(u)) = V^m(\upsilon(A(u_1)), ..., \upsilon(A(u_m))) = V(\upsilon(A(u_1)), V^{m-1}(\upsilon(A(u_2)) ..., \upsilon(A(u_m))) \quad (g)$$

Remark 2. With the above definition the following property holds:

$$\forall\, x, y \in [0, 1], \qquad x \leq y \Leftrightarrow I(x, y) = 1$$
$$x = y \Leftrightarrow E(x, y) = 1$$

Definition 2. Let us denote B_A the set of all the fuzzy binary relations on a finite discrete set A. Let us consider $\sigma_{N,T}$ and $\alpha_{N,T}$, the transformations defined from B_A onto itself as follows:

$\forall\, H \in B_A$, $\forall\, (a, b) \in A^2$,

$$\sigma_{N,T}(H)(a, b) = \upsilon(H(a, b) \wedge H(b, a)) = T(H(a, b), H(b, a))$$
$$\alpha_{N,T}(H)(a, b) = \upsilon(\neg\,(\neg\, H(b, a) \supset \neg\, H(a, b))) = N(I[N(H(b, a)), N(H(a, b))])$$

The last formula may appear as complex and somewhat artificial. Is is worth noting that the formula $\neg\,(\neg\, H(b, a) \supset \neg\, H(a, b))$ is equivalent to $(H(a, b) \wedge \neg\, H(a, b))$ in classical logic. Here, the first expression is preferred to the second because it provides a way of building an ordinal extension of the asymetric part concept in the fuzzy case. The resulting relations $\sigma_{N,T}(H)$, $\alpha_{N,T}(H)$ associated with H are natural extensions of the symmetric and asymmetric part of H respectively. The α operator is useful to define a strict-preference relation from a fuzzy weak-preference relation whereas σ may be used to extract a similarity relation from the weak preference relation.

3. Fuzzy extension of properties for binary relations

The above definition provides an interpretation allowing a systematic extension of classical definitions used in preference modelling and Social Choice Theory. Let H denote a binary relation defined on a set A of alternatives. For simplicity the notation $H(u, v)$ stands either for the syntaxic representation of the proposition *"the relation H hold for the ordered pair (u, v)"* or for the semantic evaluation of this proposition that should be denoted $\upsilon(H(u, v))$. Here is an illustration of our extension mecanism applied to completeness and transitivity properties.

3.1. Extension of the completeness property

A conventionnal binary relation H defined on a set A is said to be *complete* if and only if the following relation hold :

$$\forall u\ \forall v\ \ (H(u, v) \vee H(v, u))$$

If H is a fuzzy relation we get: $\upsilon(\forall u\ \forall v\ (H(u, v) \vee H(v, u))) = 1$

therefore by (f) we get: $T^{m2}\{\upsilon(H(a, b) \vee H(b, a)) \,/\, (a, b) \in A^2\} = 1$

hence : $inf\{\upsilon(H(a, b) \vee H(b, a)) \,/\, (a, b) \in A^2\} = 1$

This implies : $\forall\ a, b \in A^2,\ \ V(H(a, b), H(b, a)) = 1$

Thus we get infinite possibilities to derive a condition of completeness depending on the choice of a particular conorm V.

If $V_1(x, y) = \max\{x, y\}$ we get the standard condition of *Completeness (version 1)*:
$\forall\ a, b \in A,\ \ H(a, b) = 1$ or $H(b, a) = 1$

If $V_2(x, y) = \min\{x + y, 1\}$ we get the following condition of *Completeness (version 2)*:
$\forall\ a, b \in A,\ \ H(a, b) + H(b, a) = 1$

The first version is the strongest we can imagine while V_1 is the smallest conorm. More generally, if $(V_j)_{j=1,\dots,k}$ is an ordered sequence of co-norms ($V_j \geq V_{j+1}$, $\forall\ j = 1, \dots, k\text{-}1$), we get a nested family of completeness conditions (V_{j+1}-complete $\Rightarrow$ V_j-complete, $\forall\ j = 1, \dots, k\text{-}1$).

3.2. Extension of the transitivity property

A conventionnal binary relation H defined on a set A is said to be *transitive* if and only if the following relation hold :

$$\forall u\ \forall v\ \forall w\ \ ((H(u, w) \wedge H(w, v)) \supset H(u, v))$$

that is to say : $\upsilon(\forall u\ \forall v\ \forall w\ ((H(u, w) \wedge H(w, v)) \supset H(u, v))) = 1$

therefore by (f) we get : $T^{m3}\{\upsilon((H(a, b) \wedge H(b, c)) \supset H(a, c)) \,/\, (a, b, c) \in A^3\} = 1$

this implies : $\forall\ a, b, c \in A^3,\ I(\upsilon(H(a, b) \wedge H(b, c)), H(a, c)\ ) = 1$

therefore we have : $\forall\ a, b, c, \in A^3,\ \ H(a, c) \geq T(H(a, b), H(b, c))$

Thus we get infinite possibilities to derive a condition of transitivity depending on the choice of a particular t-norm T.

If $T_1(x, y) = \min\{x, y\}$ we get the standard condition of *Transitivity (version 1)*:
$\forall\ a, b, c, \in A^3,\ \ H(a, c) \geq \min\{H(a, b), H(b, c)\}$

This first version is the strongest we can imagine while T_1 is the greatest conorm.

If $T_2(x, y) = \max\{x + y - 1, 0\}$ we get the following condition of *Transitivity (version 2)*:
$\forall\ a, b, c, \in A^3,\ \ H(a, c) \geq H(a, b) + H(b, c) - 1$

This version is weaker than the first one.

Definition 3. Let L be a given multivalued logic and H a fuzzy binary relation defined on a set A. We will say that H is

- *L-reflexive* iff $\upsilon(\forall u\ \ H(u, u)) = 1$
- *L-irreflexive* iff $\upsilon(\forall u\ \ \rceil H(u, u)) = 1$
- *L-symmetric* iff $\upsilon(\forall u\ \forall v\ \ (H(u, v) \equiv H(v, u))\ ) = 1$
- *L-asymmetric* iff $\upsilon(\forall u\ \forall v\ \ \rceil (H(u, v) \wedge H(v, u))\ ) = 1$
- *L-complete* iff $\upsilon(\forall u\ \forall v\ \ (H(u, v) \vee H(v, u))\ ) = 1$
- *L-transitive* iff $\upsilon(\forall u\ \forall v\ \forall w\ \ ((H(u, w) \wedge H(w, v)) \supset H(u, v))\ ) = 1$

In practical applications we suggest interpreting this formulae using the fuzzy preference logic characterized by $N(x) = 1 - x$ and $T(x, y) = \min\{x, y\}$. Thus we get the strongest system of conditions. This system is the only one that is compatible with the notion of α-cut. That is to say that an *L-property* hold for a given fuzzy binary relation H if and only it holds for every α-cut of H. Bearing in mind that a fuzzy binary relation may be seen as a collection of nested crisp binary relations, the resulting definitions may be seen as natural extensions of classical definitions used in the non-fuzzy case. Choosing these pair of operators we get:

Definition 4. Let us consider a fuzzy binary relation on A, that is a mapping H associating with each ordered pair of alternative $(a, b) \in A \times A$ an element $H(a, b)$ of $[0, 1]$. H is said to be:

- *reflexive* if $\forall\ a \in A,\ \ H(a, a) = 1$,
- *irreflexive* if $\forall\ a \in A,\ \ H(a, a) = 0$,
- *symmetric* if $\forall\ a, b \in A,\ \ H(a, b) = H(b, a)$,
- *asymmetric* if $\forall\ a, b \in A,\ \ H(a, b) > 0\ \ \Rightarrow\ \ H(b, a) = 0$,
- *complete* if $\forall\ a, b \in A,\ \ H(a, b) = 1$ or $H(b, a) = 1$,
- *transitive* if $\forall\ a, b, c \in A,\ \ H(a, b) \geq \min\{H(a, c), H(c, b)\}$.

In order to make definition 2 more precise while keeping consistency with the above notion of transitivity, we will use the following definition (for more details see [9]).

Definition 5. Let us denote B_A the set of all fuzzy binary relations on a finite discrete set A. For any relation $H \in B_A$, the symmetric and asymetric part of H are respectively given by the σ and α operators defined from B_A onto itself by setting :

$$\sigma(H)(a, b) = min\{H(a, b), H(b, a)\}$$

$$\alpha(H)(a, b) = \begin{cases} H(a, b) & if\ \ H(a, b) > H(b, a) \\ 0\ otherwise & \end{cases}$$

for all H in B_A, for all ordered pairs (a, b) in A^2.

We are now in position to define fuzzy covering relations and to present the main result.

4. Fuzzy covering relations and their properties

Covering relations have been extensively used in Social Choice Theory. They have been introduced in order to define non-trivial choice functions from intransitive preference relations (for a introduction to choice functions see [12]). We are going to define three type of covering relation. For any crisp preference relation H, a covering relation hold for any ordered pair (a, b) if and only if a takes systematically the advantage over b when compared to other alternatives respectively. For example a is said to cover b if b beats c implies a beats c for every c in A (*forward covering*). It is a way of establishing that a is at least as good as b, taking into account the context of the comparison represented by $A\backslash\{a, b\}$. Considering the oriented graph associated with H, only the forward arcs leaving respectively a and b are used to decide whether a covers b forward or not. A complementary notion can be added, taking into account the backward arcs entering in a and b respectively. Thus the *backward covering* relation hold for any ordered pair (a, b) if and only if a takes systematically the advantage over b when compared respectively to other alternatives on the basis of backward arcs. Thus a covers b backward if and only if c beats a implies c beats b for every c in A. Finally a comprehensive covering relation may be defined by intersecting the forward and backward covering relations. Other covering relations could be considered as well but they are not use in the following. In this paper, three covering relations will be associated with any given preference relation H.

The *forward covering* relation FC:	$FC(H)(a, b) \equiv (\forall\ c\ (H(b, c)\ \supset H(a, c)))$
The *backward covering* relation BC:	$BC(H)(a, b) \equiv (\forall\ c\ (H(c, a)\ \supset H(c, b)))$
The *comprehensive covering* relation CC:	$CC(H)(a, b) \equiv (\forall\ c \in A, (FC(H)(a, b) \wedge BC(H)(a, b)))$

This last relation hold for an ordered pair (a, b) if and only if a behave better than b when they are compared vis-à-vis the other alternatives respectively. When H is a conventional relation, the relations $FC(H)$, $BC(H)$ and $CC(H)$ are proved to be transitive. Therefore, selecting uncovered elements in some sense should lead to a non-empty choice. For any set A of alternatives and any weak-preference relation H, the choice function will select a non-empty subset $UC(A, H)$ of uncovered elements defined by:

$$UC(A, H) = \{a \in A\ /\ \forall\ b \in A,\ \ CC(H)(b, a) \supset CC(H)(a, b)\}$$

The $UC(A, H)$ set is never empty but it may contain all the alternatives belonging to A. In most cases, the covering relation is rather poor and the resulting selection turns out to be almost trivial. In our opinion, this is the only drawback of the procedure.

From this observation, one interesting idea is to extend this technic to the case of fuzzy binary relations. For instance if the $H(a, c)$ degree is almost as great as the $H(b, c)$ degree for all c in A, the covering degree $FC(H)(a, b)$ should be close to one. On the contrary, if there exists at least one d such that $H(b, d)$ is great whereas $H(a, d)$ is small, the covering degree $FC(H)(a, b)$ should be close to zero. Such a covering relation should be more discriminating and could provide more selective choice functions.

Using the extension mecanism introduced in section 3, and assuming that the truth value of the proposition $(FC(H)(a, b) \equiv (\forall\ c\ (H(b, c)\ \supset H(a, c))))$ equals one, we get the following general definition of the fuzzy forward covering relation, for a given cunjunction operator T.

$$FC_T\,(H)(a, b) = \underset{c \in A}{T}\,\{\mathrm{Sup}\{\varepsilon \in [0, 1]\ /\ T(H(b, c), \varepsilon) \leq H(a, c))\}$$

Applying the same mecanism to the other covering relations leads to the following results:

$$BC_T\,(H)(a,\,b) = \underset{c\,\in\,A}{T}\,\{\mathrm{Sup}\{\varepsilon \in [0,\,1]\;/\;T(H(c,\,a),\,\varepsilon) \le H(c,\,b)\}\}$$

$$CC_T\,(H)(a,\,b) = T\{FC_T\,(H)(a,\,b),\,BC_T\,(H)(a,\,b)\}$$

Let us consider any set of alternatives $A = \{a_1, \ldots a_n\}$ we get the following

Theorem 1. There exists one unique *continous* t-norm T such that the relation $FC_T(H)$ is *transitive* on A, whatever the relation H is. This is the *min* t-norm, $T(x, y) = \min\{x, y\}$ for all $(x, y) \in [0, 1]^2$. Moreover, the same result hold for the relation $FC_T(H)$.

Proof. The continuity of the *min* t-norm is obvious. Let us show that this t-norm make the relation $FC_T(H)$ transitive. If T is the *min* t-norm the we get:

$$I(x,\,y) = \begin{cases} 1 & \textit{if } x \le y \\ y & \textit{else} \end{cases}$$

and the fuzzy forward covering relation is defined as follows:

$$FC(H)(a,b) = \begin{cases} 1 & \textit{if } \forall c \in A,\ H(b,c) \le H(a,c) \\ \underset{\{c\in A/H(b,c)>H(a,c)\}}{\textit{inf}} \{H(a,c)\} & \textit{otherwise} \end{cases} \tag{1}$$

Let us consider an arbitrary triple (a, b, c) in A^3 and let us prove the following inequality:

$$FC(H)(a,\,c) \ge \min\{FC(H)(a,\,b),\,FC(H)(b,\,c)\} \tag{2}$$

To this end, let us suppose that $FC(H)(a, c) < 1$ (if this is not the case the relation (2) is obvious). By (1) we get:

$$\exists d \in A\ /\ \begin{cases} H(a,\,d) = FC(H)(a,\,c) \\ H(c,\,d) > FC(H)(a,\,c) \end{cases} \tag{3}$$

1st Case: $FC(H)(a, c) \ge H(b, d)$

By (3) we get $H(c, d) > H(b, d)$. This implies by (1) that $H(b, d) \ge FC(H)(b, c)$. The hypothesis allow to conclude: $FC(H)(a, c) \ge FC(H)(b, c)$.

2nd Case: $FC(H)(a, c) < H(b, d)$

By (3) we get $H(b, d) > H(a, d)$. This implies by (1) that $H(a, d) \ge FC(H)(a, b)$. By (3) we get $FC(H)(a, c) \ge FC(H)(a, b)$.

Therefore, in any case, $FC(H)(a, c) \ge \min\{FC(H)(a, b), FC(H)(b, c)\}$

Let us show now that the *min* is the only continous solution. To this end, let us consider a t-norm T distinct of the *min*. By hypothesis, there exists an ordered pair (x, y) such that $T(x, y) < \min\{x, y\}$. In the following of the proof, we will suppose that $x > y$. This assumption is not restrictive because if $x < y$ we have still $T(y, x) < \min\{y, x\}$, and if $x = y$ then we have $T(x, y) < y$ and $T(1, y) = y$; by continuity $\exists\ z \in\ ]x, 1[\ /\ T(z, y) = (y + T(x, y))/2$; by construction $z > x$ and thus $z > y$; moreover by definition $T(z, y) < y$ and therefore $T(z, y) < z$. Hence we can see that $T(z, y) < \min\{z, y\}$ with $z > y$, z being substituting x.

Now let us consider $A = \{a, b, c\}$ a set of 3 alternatives and H the following fuzzy preference relation: $H(a, b) = 1$, $H(b, c) = x$, $H(a, c) = y$, $H(b, a) = 0$, $H(c, b) = 0$, $H(c, a) = 0$, $H(a, a) = 1$, $H(b, b) = 1$, $H(c, c) = 1$.

We have:

ε_1 = Sup{$\varepsilon \in$ [0, 1] / $T(H(b, c), \varepsilon) \leq H(a, c))$} = Sup{$\varepsilon \in$ [0, 1] / $T(x, \varepsilon) \leq y$}
ε_2 = Sup{$\varepsilon \in$ [0, 1] / $T(H(b, a), \varepsilon) \leq H(a, a))$} = Sup{$\varepsilon \in$ [0, 1] / $T(0, \varepsilon) \leq 1$} = 1
ε_3 = Sup{$\varepsilon \in$ [0, 1] / $T(H(b, b), \varepsilon) \leq H(a, b))$} = Sup{$\varepsilon \in$ [0, 1] / $T(1, \varepsilon) \leq 1$} = 1

Since $T(x, y) < y$ and $T(x, 1) > y$, the continuity of T implies $\exists\, t \in\,]y, 1[\, / \,T(x, t) = y$. Hence $\varepsilon_1 = t$ and $\varepsilon_1 > y$. Hence $FC_T\,(H)(a, b) = T^3(\varepsilon_1, \varepsilon_2, \varepsilon_3) = T^3(\varepsilon_1, 1, 1) = \varepsilon_1$. Therefore $FC_T\,(H)(a, b) > y$. Moreover, it is easy to see that $FC_T\,(H)(a, c) = y$ and $FC_T\,(H)(b, c) = x$ since $H(c, a) = H(c, b) = 0$. Thus $FC_T\,(H)(a, c) < min\{FC_T\,(H)(a, b), FC_T\,(H)(b, c)\}$, this contradict transitivity of $FC_T\,(H)$. A similar proof may be used for $BC_T\,(H)$ since $BC_T\,(H) = (FC_T\,(H^{-1}))^{-1}$ □

Theorem 2. There exists one unique *continous* t-norm T such that the relation $CC_T(H)$ is *transitive* on A, whatever the relation H is. This is the *min* t-norm, $T(x, y) = \min\{x, y\}$ for all $(x, y) \in [0, 1]^2$.

Proof. If T is the *min* t-norm, then $FC_T\,(H)$ and $BC_T\,(H)$ are transitive (theorem 1) and thus $CC_T(H)$ is transitive. Conversely, suppose there exists an ordered pair (x, y) such that $T(x, y) < min\{x, y\}$. Let us consider the following relation : $H(a, b) = 1$, $H(b, c) = 1$, $H(a, c) = y$, $H(b, a) = 0$, $H(c, b) = 0$, $H(c, a) = 0$, $H(a, a) = 1$, $H(b, b) = 1$, $H(c, c) = 1$. Hence we get:
$FC_T\,(H)(a, b) = y$, $FC_T\,(H)(b, c) = 1$, $FC_T\,(H)(a, c) = y$, $FC_T\,(H)(b, a) = 0$, $FC_T\,(H)(c, b) = 0$, $FC_T\,(H)(c, a) = 0$, $BC_T\,(H)(a, b) = 1$, $BC_T\,(H)(b, c) = y$, $BC_T\,(H)(a, c) = y$, $BC_T\,(H)(b, a) = 0$, $BC_T\,(H)(c, b) = 0$, $BC_T\,(H)(c, a) = 0$.
Therefore the comprehensive covering relation is: $CC_T\,(H)(a, b) = y$, $CC_T\,(H)(b, c) = 1$, $CC_T\,(H)(a, c) = T(y, y) \leq T(x, y) < y$, $FC_T\,(H)(b, a) = 0$, $FC_T\,(H)(c, b) = 0$, $FC_T\,(H)(c, a) = 0$. This show that $CC_T\,(H)$ is not transitive since $CC_T\,(H)(a, c) < min\{CC_T\,(H)(a, b), CC_T\,(H)(b, c)\}$
□

These results show the only possible way for the covering concept to be extended to the fuzzy case, if both continuity and transitivity properties are required. Thus, we get the following definition:

Definition 6. Let A be a finite discrete set of alternatives and H a fuzzy preference relation on A. The relation $CC(H)$ will be said to be the *fuzzy covering relation of H on A* if the following equalities hold:

$$\forall\ (a, b) \in A^2,\ \ CC(H)(a, b) = \mathrm{Min}\{FC(H)(a, b), BC(H)(a, b)\}$$

with

$$FC(H)(a,b) = \begin{cases} 1 & \text{if } \forall c \in A,\ H(b,c) \leq H(a,c) \\ \inf\limits_{\{c \in A / H(b,c) > H(a,c)\}} \{H(a,c)\} & \text{otherwise} \end{cases}$$

$$BC(H)(a,b) = \begin{cases} 1 & \text{if } \forall c \in A,\ H(c,a) \leq H(c,b) \\ \inf\limits_{\{c \in A / H(c,a) > H(c,b)\}} \{H(c,b)\} & \text{otherwise} \end{cases}$$

Remarks. Actually the relations $FC(H)$ and $BC(H)$ are the left and right traces of the relation H (for more details see [6]). If H is reflexive then $CC(H) \subset H$. This shows that computing $CC(H)$ is a natural way of building a transitive relation without adding artificial information. This would not be the case if we would have used a transitive closure operation. Moreover, since the covering relation is reflexive and transitive, the symmetric part of $CC(H)$ is a fuzzy similarity relation (*i.e.*

reflexive symmetric and transitive). The index $\sigma(C(H))(a, b) = \min\{C(H)(a, b), C(H)(b, a)\}$ reflects the degree to which a and b have the same behaviour when compared to other alternatives respectively. This relation may be suitable to perform a hierachical classification of alternatives.

It is also possible to extend the choice function introduced above to fit with fuzzy relations. The extension consist in defining a fuzzy subset of uncovered alternatives as follows:

$$UC(A, H) = \{(a, \mu_{UC}(a)), a \in A\}$$

with $$\begin{aligned}\mu_{UC}(a) &= \upsilon(\forall\, b \in A, CC(H)(b, a) \supset CC(H)(a, b)\} \\ &= \inf_b \text{Sup}\{\varepsilon \in [0, 1] \,/\, T(CC(H)(b, a), \varepsilon) \leq CC(H)(a, b)\}\end{aligned}$$

A crisp selection may be derived from the fuzzy set by considering the support of $UC(A, H)$, that is to say the set of maximal elements for the relation $C(H)$. This set is defined as follows:

$$\begin{aligned}M(A, CC(H)) &= \{a \in A \,/\, \mu_{UC}(a) = 1\} \\ &= \{a \in A \,/\, \forall\, b \in A, CC(H)(b, a) \leq CC(H)(a, b)\}\end{aligned}$$

Using theorem 2, it is easy to prove that the set of maximal elements $M(A, CC(H))$ is never empty. Therefore the associated choice function is well-defined. For any subset B of A the choice function returns the non-trivial selection $M(B, CC(H))$.

5. Conclusion

The new type of choice functions introduced here is not far from classical choice functions designed to select the subset of *undominated alternatives* (see [8, 7]). However it is worth noting that the first Orlowski's choice function is ill-defined when all the fuzzy binary relations on A are admissible. In order to get a well-defined choice function, two independant solutions are usually considered:

1 - to restrict the domain of admissible relations (avoiding cyclic preferences).
2 - to select the less dominated elements on the basis of a non-domination score.

In most cases, authors introducing a choice function adopt the first or the second solution to ensure that the selected subset be not empty (see [8], [5] and [11]). Our procedure offers a third solution to the selection problem. Using covering relations we dont need restriction to get a well defined choice function selecting a *maximal set*. Strictly uncovered elements do exist since the relation $C(H)$ is transitive. Thus, knowing the family of classical choice functions, mainly defined from scoring functions (see [2]), the use of covering relations appear as a complementary approach to derive exact choices when preference are fuzzy.

Moreover, any covering relation contains a nested family of *partial preorders* (*i.e.* reflexive, transitive but not necessary complete relations). This allows to get more or less discriminating rankings of alternatives. It is also possible to get a complete ranking of alternatives by implementing an iterated choice mecanism (see [9]).

Finally, let us remark that the covering index $CC(H)(a, b)$ is a non-decreasing function of the quantities $H(a, c)$ and $H(c, b)$, and a non-increasing function of the quantities $H(c, a)$ and $H(b, c)$, for all $c \in A$. This is a suitable garanty to preserve monotonicity properties in the exploitation of intransitive relations. Moreover, the use of the *min* operator recommended by theorems 1 and 2 implies that monotonic increasing transformations applied to H should not modify the selection, the classification or the ranking derived from the associated covering relation. Therefore, fuzzy covering relations appear as a natural tool to derive robust conclusions from preference relations resulting from ordinal aggregation procedures.

For these reasons, we believe that the covering concept could be used to derive a prescription from pairwise comparisons, in the domain of MCDA. For instance, it may be a useful notion to exploit fuzzy outranking relations introduced in [10]. Unfortunately practical tests have shown the difficulty to found a significant discrimination from covering relations, even if their fuzzy extension is considered. A possibility to avoid this drawback is to choose a smaller t-norm than the *min*. It is indeed easy to show that, for any pair of t-norms, if T_1 is smaller than T_2 then the relations $FC_{T2}(H)$, $BC_{T2}(H)$ and $CC_{T2}(H)$ are included in the relations $FC_{T1}(H)$, $BC_{T1}(H)$ and $CC_{T1}(H)$ respectively. Thus, choosing the *min* t-norm leads to the smallest fuzzy covering relation. On the contrary, choosing a smaller t-norm T (for instance the Lukasiewicz t-norm) leads to more discriminating covering relations, but theorem 1 show us the resulting loss of transitivity. Actually, relaxing the t-norm means loosing the transivity in its strongest version. This last solution may be interesting in some applications as shown in [9].

REFERENCES

[1] Arrow, K.J., (1951), *Social Choice and Individual Values*, Cowles Foundations and Wiley, New York.

[2] Barrett, C.R., Pattanaik, P.K., Salles, M., (1990), "On choosing rationally when Preferences are fuzzy", *Fuzzy Sets and Systems*, **34**, 197-212.

[3] Bordes, G., (1983), "On the possibility of reasonable consistent majoritarian choice: some positive results", Journal of Economic Theory, **31**, 122-132.

[4] Dombi, J., (1982), "A general class of fuzzy operators, the De-Morgan class of fuzzy operators and fuzziness measures induced by fuzzy operators", *Fuzzy Sets and Systems*, 150-163.

[5] Dutta, B., Panda, S., Pattanaik, P.K., (1986), "Exact choice and fuzzy preferences", *Mathematical Social Sciences*, **11**, 53-68.

[6] Fodor, J., (1992), "Traces of fuzzy binary relations", *Fuzzy sets and systems*, **50**, 331-341.

[7] Ovchinnikov, S.V., Ozernoy, V.R., (1988), "Using fuzzy binary relations for identifying noninferior decision alternatives", Fuzzy Sets and systems, **25**, 21-32.

[8] Orlowski, S.A., (1978), Decision-making with a fuzzy preference relation, *Fuzzy sets and Systems*, **1**, 155-167.

[9] Perny, P., (1992), "Modelisation, agregation et exploitation de preferences floues dans une problématique de rangement : bases axiomatiques, procedures et logiciels", Doctoral disertation and LAFORIA document n°9.

[10] Perny, P., Roy, B., (1992), "The Use of fuzzy Outranking Relations in Preference Modelling", *Fuzzy Sets and Systems*, **49**, 33-53.

[11] Roubens, M., (1989), "Some properties of choice functions based on valued binary relations", *European Journal of Operational Research*, **40**, 309-321.

[12] Sen, A.K., (1971), "Choice functions and revealed preferences", *Review of Economic Studies*, **38**, 307-313.

[13] Sen, A., K., (1986), "Social Choice Theory", dans *Handbook of Mathematical Economics vol. III*, K. J. Arrow, M. D. Intriligator (Eds.), Elsevier Sciences Publishers B. V., North-Holland, Chap. 22, 1073-1181.

FUZZY SENSOR AGGREGATION: APPLICATION TO COMFORT MEASUREMENT

Eric BENOIT, Gilles MAURIS, Laurent FOULLOY

LAMII/CESALP
Université de Savoie
41, avenue de la Plaine
B.P. 806, F-74016 Annecy cedex

ABSTRACT

This paper focuses on the acquisition of abstract information, i.e. information which are not analytically related to conventional physical quantities as for example the comfort. In these complex cases, we propose to use fuzzy sensors which compute and report linguistic assessment of numerical acquired values. Two methods are proposed to realize the aggregation from basic measurements. The first one performs the combination of the relevant features by means of a rule based description of the relations between them. With the second one, the aggregation is realised through an interpolation mechanism that make a fuzzy partition of the numeric multi-dimensional space of the basic features.

1. Introduction

When attempting to qualify complex phenomena, especially those related to human perception, one is often led to use words of the natural language [1], [2]. This linguistic description is less precise than the numerical one obtained through usual sensors. This description is also subjective because it depends on the observer.

Several advantages may be found in such an approach. In particular, the linguistic description is easily understood by human beings even if the concepts are abstract or if the context is changing. For example, everybody can figure out what is the concept of danger and can qualify it even if the context is unknown. Everybody is also able to classify colours without having an explicit knowledge of the origin of colours and of the perception mechanisms. Even for simple problems such as temperature measurements, human can easily integrate a priory knowledge. Imagine two people, one living close to the north pole and the other living close to the equator, speaking of a comfortable outdoor temperature in their respective countries. they are both able to abstract the concept of comfort even if the related temperatures are not the same. The representation of measurements by means of linguistic scales provides abstract information which can be integrated in decision, diagnosis or control especially with systems using symbolic coding.

Integrating capabilities of symbolic representation directly at the sensor level has led to the concept of symbolic sensor [3], [4]. The numeric-symbolic conversion has been studied in particular when the symbols of the linguistic scale have their meaning represented by means of fuzzy subsets. We have proposed to call these new sensors *fuzzy symbolic sensors* or simply *fuzzy sensors* [5], [6].

Two ways could be considered in order to obtain a linguistic description of properties or attributes from the features linked to them: defining the linguistic description by means of a rule formalism that involves a linguistic description of every feature returned by the corresponding fuzzy sensor or building the linguistic description directly on the numerical product space of the independent features by means of an interpolation method. After having recalled our formalism of the numeric-linguistic conversion, these two strategies will be investigated in this paper and then applied to a linguistic description of comfort.

2. The numeric to linguistic conversion

To perform a symbolic measurement, it is necessary to clearly specify the relation between symbols and numbers. Let $\mathcal{E}$ be the set of all possible measurements. Let $\mathcal{L}(\mathcal{E})$ be the set of symbols associated to this universe. Denote $\mathcal{P}(\mathcal{E})$ the set of subsets of $\mathcal{E}$. An injective mapping, called a *meaning* and denoted $\tau : \mathcal{L}(\mathcal{E}) \to \mathcal{P}(\mathcal{E})$ associates any symbol with a subset of measurements. Injectivity means that two symbols with the same meaning should be considered as identical. Symbolic measurement is now obtained from a mapping, called a *description* and denoted ι: $\mathcal{E} \to \mathcal{P}(\mathcal{L}(\mathcal{E}))$. It associates any measurement with a subset of the symbolic set $\mathcal{L}(\mathcal{E})$.

There is a fundamental relation between description and meaning. If a symbol belongs to the description of a measurement, then the measurement belongs to the meaning of the symbol, i.e:

$$L_1 \in \iota(x) \Leftrightarrow x \in \tau(L_1)$$

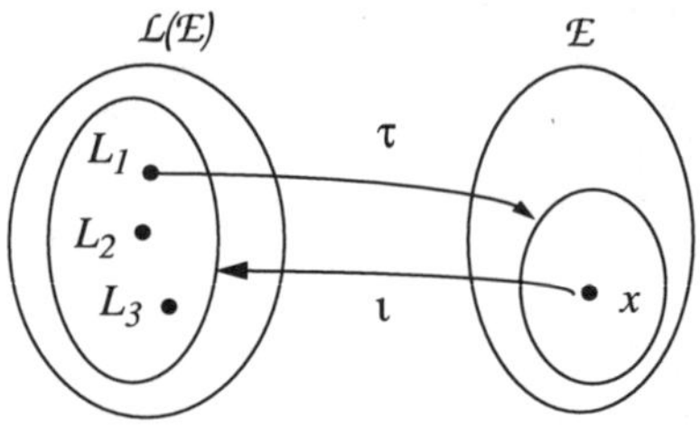

Fig. 1 : Relation between the description and the meaning

Here is an example for a sensor that returns the ambient temperature. Let the measurement set be $T = [0\ ^{o}C, 40\ ^{o}C]$ and the symbolic set be $\mathcal{L}(T) = \{$***cold***, ***cool***, ***mild***, ***warm***, ***hot***$\}$ The following figure shows an example of symbol meanings:

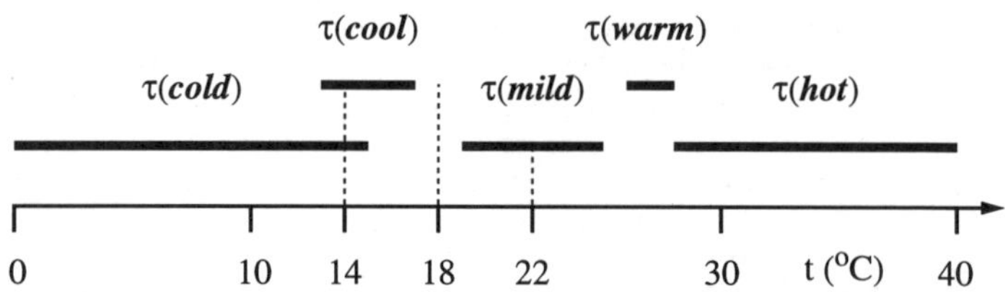

Fig. 2 : Meanings of items of $\mathcal{L}(T)$

The descriptions of measurements comes directly from the definitions. For example:

$$\iota(22\ ^{o}C) = \{\textbf{\textit{mild}}\}$$

$$\iota(14\ ^{o}C) = \{\textbf{\textit{cold}}, \textbf{\textit{cool}}\}$$

$$\iota(18\ ^{o}C) = \varnothing$$

In the general case, the description of a measure can contain any number of symbols. In a natural language, a measure is usually described by only one symbol. To apply this condition we impose the set of the meanings to be a partition of the measurement set (fig. 3).

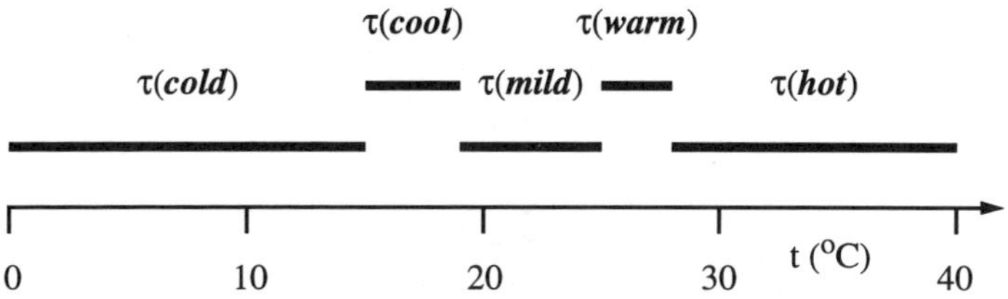

Fig. 3 : Meanings satisfying the partition

3. The fuzzy numeric-symbolic conversion

The previous approach, based on subsets, obviously leads to sharp transitions in the sensor response. Fuzzy subset theory, developed by Zadeh, provides a nice solution to this problem [7], [8]. The extension of the characteristic function of a *crisp* subset (values in {0,1}) to the membership function of a fuzzy subset (values in [0,1]) can be used to model gradual transitions between symbols.

The fuzzy numeric to symbolic interface depends on the extension to the fuzzy case of the definitions of the previous section. So fuzzy meaning and fuzzy description have to be defined [9]. The fuzzy meaning is a mapping from the symbolic set $\mathcal{L}$ to the set of the fuzzy subsets of measurements. The fuzzy meaning of a symbol L is characterized by its membership function denoted $\mu_{\tau(L)}(x)$.

In a same manner, a fuzzy description can be defined as a mapping from the measurement set into the set of the fuzzy subsets of symbols, so the fuzzy description is characterized by its membership function denoted $\mu_{\iota(x)}(L)$.

The relation between the membership functions of a fuzzy description and the corresponding fuzzy meaning comes directly from the fundamental relation between the meaning and the description, i.e:

$$\mu_{\iota(x)}(L) = \mu_{\tau(L)}(x)$$

In order to conserve a link between the sensor description and the human feeling, we impose that the set of symbols meaning is a fuzzy partition of the measurement set in the sense of Bezdek [10]. Then, the triplet $<E, \mathcal{L}(E), \iota>$ is a fuzzy nominal scale [6].

$$\Sigma_{a \in \mathcal{L}(T)} \mu_{\iota(t)}(\boldsymbol{a}) = 1$$

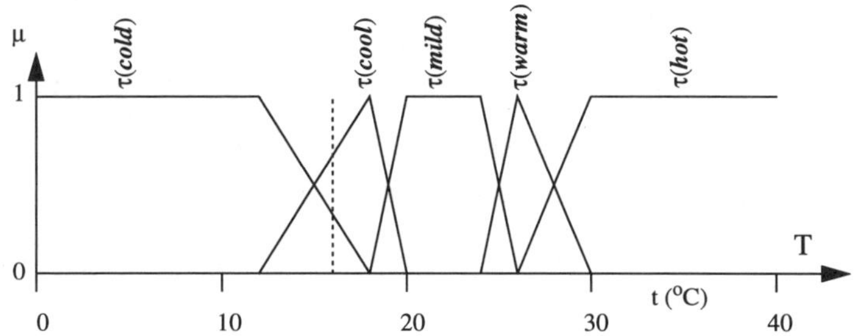

Fig. 4 : Meanings of the temperature lexical set

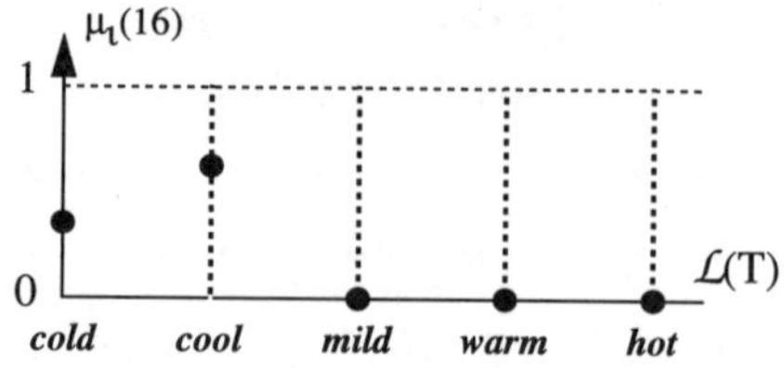

Fig. 5 : Description of the measure 16 °C

4. Rule based symbolic aggregation

With multi-dimensional measurements, the definition of symbolic meanings leads to a combination between basic measurements. For example, the description of comfort needs the knowledge of the relative humidity and of the temperature. Then a linguistic definition of the comfort can be modelized by a set of rules. Let us describe by this example the mechanism used to define the semantic of comfort.

The temperature and humidity measurements take their respective values in the sets denoted T and H. The temperature is described by symbols in the set $\mathcal{L}(T)$ = {***cold***, ***cool***, ***mild***, ***warm***, ***hot***}. The humidity is described by symbols in the set $\mathcal{L}(H)$ = {***very_low***, ***low***, ***medium***, ***high***}. It also assumed that the meanings of the symbols generate a partition of the respective numerical sets.

The problem is now to aggregate both measurements to obtain information about the feeling of comfort. Such a feeling could be symbolically defined as follows.

atmosphere is ***comfortable*** when : the temperature is ***mild*** **and** the humidity is ***medium***

atmosphere is ***acceptable*** when : the temperature is ***cool*** **or** ***mild*** **or** ***warm*** **and** the humidity is ***low*** **or**
the temperature is ***cool*** **or** ***warm*** **and** the humidity is ***medium***

atmosphere is ***uncomfortable*** when :
atmosphere is **not** (***comfortable*** **or** ***acceptable***)

One solution is to consider that two symbols connected by the **and** operator are in fact one symbol whose meaning is defined on the cartesian product of the numerical sets. For example, "the temperature is ***mild*** **and** humidity is ***medium***" should be no more considered as the two expressions connected by the operator **and**, but as one expression such as "the temperature_humidity is ***mild_medium***". The new symbolic variable temperature_humidity takes its value in a new set of symbols, for example denoted $\mathcal{L}(TH)$. The symbol ***mild_medium*** belongs to $\mathcal{L}(TH)$, and its meaning is defined on the cartesian product TxH. Then the meaning of ***comfortable*** is defined by

$$\tau(\textbf{\textit{comfortable}}) = \tau(\textbf{\textit{mild\_medium}})$$
$$= \tau(\textbf{\textit{mild}}) \times \tau(\textbf{\textit{medium}})$$

The meanings of the **or** and the **not** operators are respectively defined by the union operator and by the negation operator.

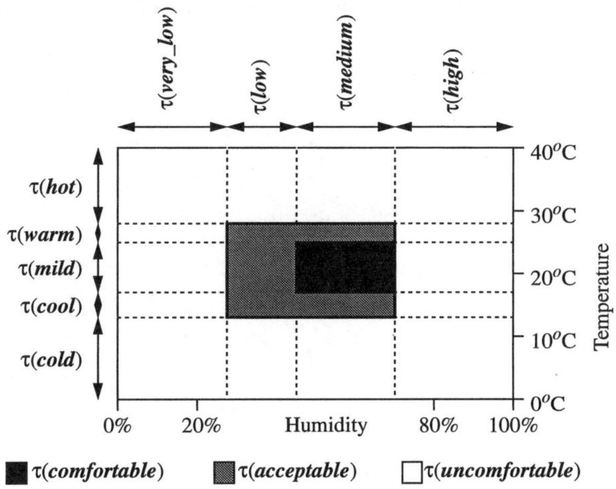

Fig. 6 : Graph representing the meanings of ***comfortable***, ***acceptable*** and ***uncomfortable***

5. Rule based fuzzy sensor aggregation

Let t and h be the temperature and humidity measurement. The fuzzy description of the couple (t, h) can be expressed in terms of the fuzzy descriptions of t and h by means of a triangular norm T.

$$\mu_{\iota(t,\, h)}(\boldsymbol{a\_b}) = \mu_{\iota(t)}(\boldsymbol{a}) \top \mu_{\iota(h)}(\boldsymbol{b})$$

As a fuzzy partitioning, in the sense of Bezdek, was chosen for the temperature and the humidity, we have :

$$\Sigma_{\mathrm{a} \in \mathcal{L}(T)} \mu_{\iota(t)}(\boldsymbol{a}) = 1 \text{ and } \Sigma_{\mathrm{b} \in \mathcal{L}(H)} \mu_{\iota(h)}(\boldsymbol{b}) = 1$$

Therefore, if we want to obtain a fuzzy partition for the new symbolic set $\mathcal{L}(TH)$, the triangular norm has to be distributive with respect to the addition (for example the product).

$$\begin{aligned}
&\Sigma_{\mathrm{a_b} \in \mathcal{L}(HT)} \mu_{\iota(t,\, h)}(\boldsymbol{a_b}) \\
&= \Sigma_{\mathrm{a} \in \mathcal{L}(T)} \Sigma_{\mathrm{b} \in \mathcal{L}(H)} \mu_{\iota(t)}(\boldsymbol{a}) \top \mu_{\iota(h)}(\boldsymbol{b}) \\
&= \Sigma_{\mathrm{a} \in \mathcal{L}(T)} \mu_{\iota(t)}(\boldsymbol{a}) \top \Sigma_{\mathrm{b} \in \mathcal{L}(H)} \mu_{\iota(h)}(\boldsymbol{b}) \\
&= \Sigma_{\mathrm{a} \in \mathcal{L}(T)} \mu_{\iota(t)}(\boldsymbol{a}) \top 1 \\
&= \Sigma_{\mathrm{a} \in \mathcal{L}(T)} \mu_{\iota(t)}(\boldsymbol{a}) = 1
\end{aligned}$$

Now, we have to define the fuzzy meaning of the **or** operator. Let E be a set of measurements and $\mathcal{L}(E)$ its associated set of symbols. Let L_1 and L_2 be two symbols of $\mathcal{L}(E)$. Let $\perp$ be the triangular conorm which defines the fuzzy meaning of **or**:

$$\begin{aligned}
&\mu_{\tau(L1_\mathbf{or}_L2)}(x) \\
&= \mu_{\tau(L1)}(x) \perp \mu_{\tau(L2)}(x) \\
&= \mu_{\iota(x)}(L_1) \perp \mu_{\iota(x)}(L_2)
\end{aligned}$$

Denote $\mathcal{L}^*(E) = \mathcal{L}(E) - \{L_1, L_2\}$, and $\mathcal{L}^+(E) = \mathcal{L}(E) \cup \{L_1\_\mathbf{or}\_L_2\}$

The triangular conorm must be chosen such that the meaning of the symbols of $\mathcal{L}^{+}(E)$ forms also a fuzzy partition:

$$\sum_{L \in \mathcal{L}(E)} \mu_{\iota(x)}(L)$$
$$= \sum_{L \in \mathcal{L}^{*}(E)} \mu_{\iota(x)}(L) + \mu_{\iota(x)}(L1) + \mu_{\iota(x)}(L2)$$
$$=1$$

$$\sum_{L \in \mathcal{L}^{+}(E)} \mu_{\iota(x)}(L)$$
$$= \sum_{L \in \mathcal{L}^{*}(E)} \mu_{\iota(x)}(L) + \mu_{\iota(x)}(L1\_\mathbf{or}\_L2)$$
$$=1$$

Therfore we have

$$\mu_{\iota(x)}(L_1) \perp \mu_{\iota(x)}(L_2) = \mu_{\iota(x)}(L1) + \mu_{\iota(x)}(L2) \leq 1$$

Thus, the triangular conorm should verify

$$x \perp y = x+y \text{ if } x+y \leq 1$$

for example : $x \perp y = \min(x+y, 1)$. This is the triangular conorm chosen to define the **or** operator.

The meaning of the negation operator **not** is defined by: $\mu_{\tau(\mathbf{not}\ L)}(x) = 1 - \mu_{\tau(L)}(x)$

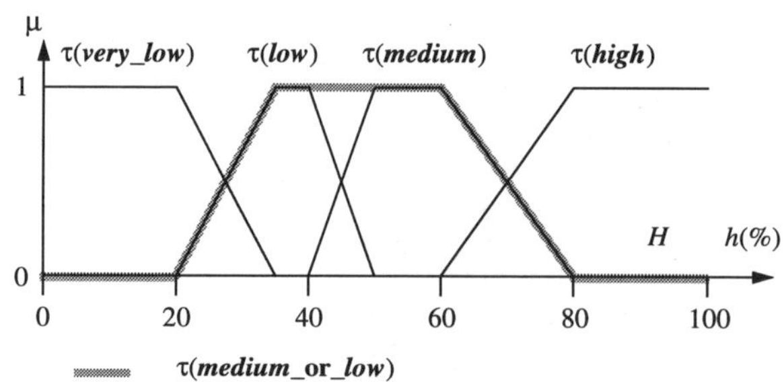

Fig. 7 : Meanings of the humidity lexical set, and meaning of a compound symbol

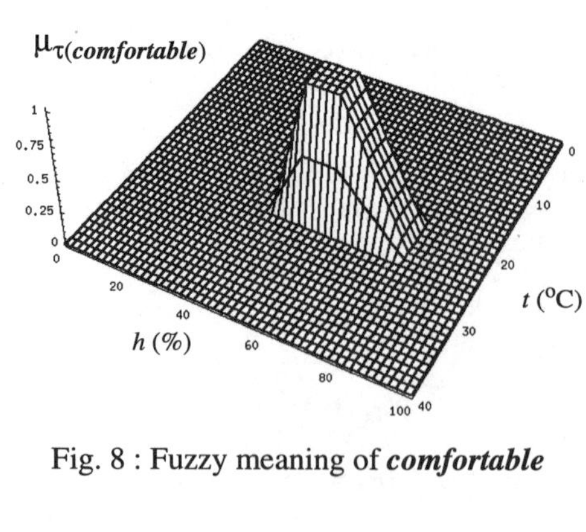

Fig. 8 : Fuzzy meaning of ***comfortable***

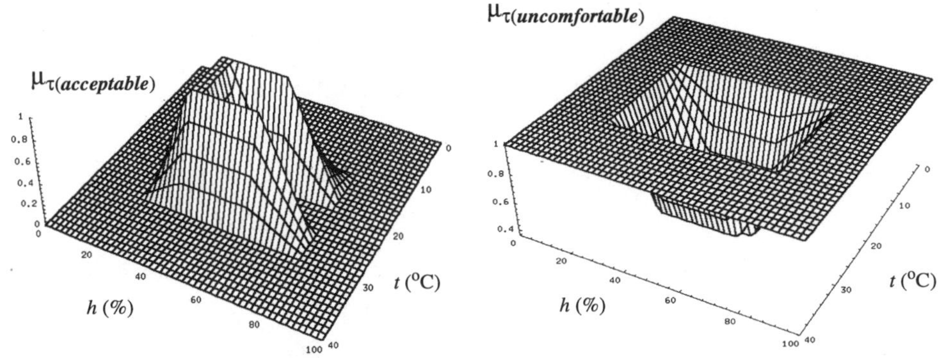

Fig. 9 : Fuzzy meaning of ***acceptable*** and of ***uncomfortable***.

The fuzzy sensor is now able to describe the temperature and humidity for the symbol comfortable by a grade of membership which qualify the comfort feeling. Three results are given below for three different values of the temperature in °C and humidity in %.

$$\mu_{\iota(23,\ 50)}(\textbf{\textit{comfortable}}) = 1$$

$$\mu_{\iota(25,\ 60)}(\textbf{\textit{comfortable}}) = 0.5$$

$$\mu_{\iota(5,\ 80)}(\textbf{\textit{comfortable}}) = 0$$

6. Multi-component numeric-symbolic conversion

When a sensor uses several transducers, the measure is a vector of numerical values, and the measurement set is a multi-dimensional volume. An alternative way to the preceding method is to define directly the meaning of each symbol on this multi-dimensional volume. Furthermore the set of meanings has to be a fuzzy partition of the measurement set.

In this section, we consider an initial knowledge about the measurements. This knowledge is materialized by the meaning of symbols on a small subset *V* of the measurement set. Then the measurement set is partitioned in n-simplexes with the Delaunay triangulation method. A n-simplex in a n-dimensional space is a polyhedra with n+1 vertices. For example, a 2-simplex is a triangle and a 3-simplex is a tetrahedron. The points used to perform the triangulation are the elements of the subset *V*.

The membership function of the meaning of each symbol is defined by piece on the n-simplexes. A multi-linear interpolation is used to define this function on each n-simplex. We suppose the restriction on a n-simplex of the membership function of the meaning of a symbol ***s*** is :

$$\mu_{\tau_E(s)}(v) = \mu_{\tau_E(s)}(x_1, \ldots, x_n) = a_1x_1 + \ldots + a_nx_n + a_{n+1}$$

The value of this function is known for the n+1 vertices of the n-simplex. Indeed, the vertices are elements of the subset *V*. So the n+1 factors a_i can be computed by resolving the system of n+1 equations with n+1 unknowns.

$$A = M^{-1}B \qquad B = \begin{bmatrix} \mu_{\tau_V(s)}(v_1) \\ \mu_{\tau_V(s)}(v_2) \\ \dots \\ \mu_{\tau_V(s)}(v_{n+1}) \end{bmatrix}$$

$$M = \begin{bmatrix} x_{1_1} & \dots & x_{1_n} & 1 \\ \dots & \dots & \dots & \dots \\ x_{n+1_1} & \dots & x_{n+1_n} & 1 \end{bmatrix} \qquad A = \begin{bmatrix} a_1 \\ \dots \\ a_{n+1} \end{bmatrix}$$

Where v_i is the ith vertex of the n-simplex, and x_{i_j} is its jth component.

This process is performed on each n-simplex and for each symbol. Then we have a fuzzy nominal scale defined on $\mathcal{E}$. This scale is an extension of the fuzzy nominal scale on *V*.

With this method, the knowledge needed to configure the sensor is very compact. It can be acquired during a learning phase by a communication with a system called teacher which can be a man or an expert system. During the learning phase, the teacher and the sensor analyse the same phenomenon, and the teacher gives its description to the sensor. The sensor increases its knowledge with its measure associated to the teacher description. Then it owns a crips meaning of the symbols on the subset of the measurement set. The sensor can now build the fuzzy nominal scale on the measurement set. This technique was succefully used to implement a fuzzy color sensor [11].

If we consider the example of comfort measurement, the meaning of ***comfortable***, ***acceptable***, ***uncomfortable*** under a subset *V* of *T*x*H* are defined as follow:

τ_V(***comfortable***) = {(20,50)}

τ_V(***acceptable***) = {(26,35), (16,35), (24,65), (18,65), (28,50), (15,50), (20,25), (20,80)}

τ_V(***uncomfortable***) = {(0,0), (20,0), (40,0), (0,100), (20,100), (40,100), (0,50), (40,50)}

The following figure shows the meaning of ***comfortable*** and of ***acceptable*** under *T*x*H*.

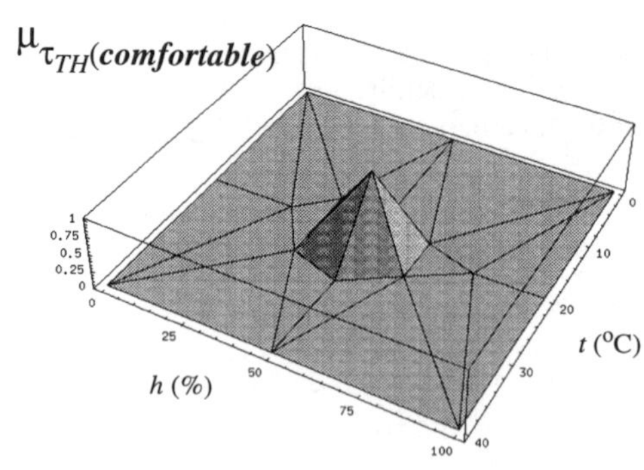

Fig. 10 : Meaning of ***comfortable***

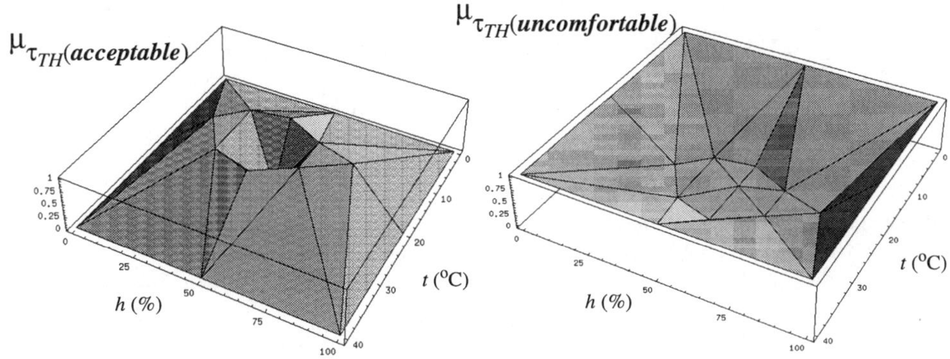

Fig. 11 : Meaning of ***acceptable*** and of ***uncomfortable.***

7. Conclusion

This paper has been concerned by the aggregation of complementary information using fuzzy sensors which compute and report linguistic assessments of numerical acquired values. When the aggregation cannot be represented by a numerical model, we have proposed to use a linguistic model of the aggregation or to use a set of characteristic examples associated with a fuzzy interpolative method. The rule based method is particularly efficient when the basic features are of a linguistic type as it is often the case at the reasoning level. In this approach, all the numerical operations are made at the low level (i.e. inside the sensor or the actuator). So the central unit has a reduced computation load, because it works on compact information. With the interpolative method, the computation load is higher because it works on a multidimensional numerical space. But it provides a richer structure for the output information. The first proposed method seems more adapted to complex systems that handle a lot of information and in which the perception is only one of the different considered tasks. The second proposed method seems more adapted to cases where the perception is the principal objective.

REFERENCES

[1] Finkelstein L. *"Representation by symbol systems as an extension of the concept of measurement"*, Kybernetes, Vol. 4, pp. 215-223, 1975.

[2] Zingales G., Narduzzi C., *"The role of artificial intelligence in measurement."*, Proc. of IMEKO TC7 int. symp. on AIMaC'91, Kyoto, Japan, 1991, pp. 2-12.

[3] Benoit E., Foulloy L., *"Symbolic sensors : one solution to the numerical-symbolic interface"*, Proc. of the IMACS DSS&QR workshop, Toulouse, France, march 1991.

[4] Benoit E., Foulloy L., *"Symbolic sensors"*, 8th International Symposium on Artificial Inteligence based Measurement and Control (AIMaC'91), september 1991, Ritsumeikan University, Kyoto, Japan.

[5] Foulloy L., Benoit E., Mauris G., *"Applications of fuzzy sensors"*, European Workshop on Industrial Fuzzy Control and Applications, Barcelona, Spain, April 93.

[6] Mauris G., Benoit E., Foulloy L., *"Fuzzy symbolic sensors : From concept to applications"*, Int. Journal of Measurement, N° 12, 1994, pp. 357-384.

[7] Zadeh L. A., *"Fuzzy sets"*, Information and control, Vol. 8, pp.338 - 353, 1965.

[8] Zadeh L. A., *"The concept of a linguistic variable and its application to approximate reasoning"*, Information Sciences, part 1:Vol 8, No 3, pp. 199-249,part 2: Vol 8,pp. 301-357, part 3:Vol 9, pp.43-80, 1975.

[9] Zadeh L.A., *"Quantitative fuzzy semantics"*, Information Sciences, Vol. 3, 1971, pp. 159-176

[10] Bezdek J., *"Pattern recognition with fuzzy objective functions"*, Plenum, 1981, New-York.

[11] Benoit E., *"Capteurs symboliques et capteurs flous: un nouveau pas vers l'intelligence"*, Thèse de doctorat de l'Université Joseph Fourier Grenoble I, January 1993.

EXTENDED FUZZY CLUSTERING MODELS FOR ASYMMETRIC SIMILARITY

Mika SATO*, Yoshiharu SATO**

* Hokkaido Musashi Women's Junior College, Kita 22, Nishi 13, Kita-ku, Sapporo 001, Japan,

** Division of Information Engineering, Hokkaido University Kita 13, Nishi 8, Kita-ku, Sapporo 060, Japan

ABSTRACT

As a generalization of the additive clustering model [6], we discuss the following three additive fuzzy clustering models: a simple additive fuzzy clustering model, an overlapping fuzzy clustering model and a fuzzy clustering model for ordinal scaled similarity [4]. The essential merits of fuzzy clustering models are 1) the amount of computations for the identification of the models are much fewer than a hard clustering model and 2) fewer number of clusters are needed to get a suitable fitness.

These fuzzy clustering models are extended to the model for asymmetric similarity. In this model, the concept of the similarity among clusters is introduced. The crucial assumption of this model is that the asymmetry of the similarity between the pair of objects is caused by the asymmetric similarity among clusters. The validity of this model is shown by some examples.

1. Introduction

The additive clustering model in hard cluster analysis is intended to find the structure of the similarity between the pair of objects. However, a large number of cluster is required by the constraint of the model. Usually, the model is denoted by the following:

$$s_{ij} \approx \sum_{k=1}^{K} w_k p_{ik} p_{jk}, \tag{1.1}$$

where s_{ij} $(i, j = 1, 2, \cdots, n)$ is the similarity between object i and j, and K is the number of clusters, w_k is a weight representing the salience of the property corresponding to cluster k.If object i has the property of cluster k then $p_{ik} = 1$, otherwise is 0. Notice that the product $p_{ik}p_{jk}$ is unity if and only if both objects i and j belong to cluster k and in this model, the similarity between the pair of objects is defined to be the common property of objects. Moreover, if the pair of objects shares some common properties, the grade which the pair of objects contribute to the similarities is assumed to be mutually independent (Shepard, R.N. and Arabie, P. [6]).

On the other hand, in the case of the fuzzy clustering, a fuzzy cluster is defined to be a fuzzy subset on the set of objects and a fuzzy grade of each object represents the degree of belongingness. If we assume that a cluster is the group whose element share common properties, then the fuzzy grade shows the degree which the object has the common properties of each cluster. By introducing the concept of the fuzzy cluster into the additive clustering model, we can construct a natural clustering model which is possible to interpret as the structure of similarity by using fewer clusters.

In this paper, we propose the simple additive fuzzy clustering model, the overlapping additive fuzzy clustering model, and the additive fuzzy clustering model for ordinal scaled similarity. Furthermore, we discuss the case that the observed similarity data is asymmetric, and extend the model to asymmetric similarity.

2. Additive fuzzy clustering model for symmetric similarity

2.1. Simple additive fuzzy clustering model

We define the simple additive fuzzy clustering model as follows:

$$s_{ij} \approx \alpha \sum_{k=1}^{K} u_{ik} u_{jk}. \tag{2.1}$$

where α is non-negative constant, the similarity s_{ij} has a ratio scale and $0 \leq s_{ij} \leq 1$. u_{ik} is a fuzzy grade which represents the degree of belongingness of object i to cluster k. Generally, u_{ik} are denoted by using the matrix representation $U = (u_{ik})$ called a partition matrix, which satisfy the following condition:

$$u_{ik} \geq 0, \quad \sum_{k=1}^{K} u_{ik} = 1. \tag{2.2}$$

In this case, the product $u_{ik}u_{jk}$ is the degree of simultaneous belongingness of objects i and j to cluster k. That is, the product denotes the degree of the sharing common properties. And we introduce the positive constant α to control the goodness of fit between s_{ij} and $\sum_{k=1}^{K} u_{ik}u_{jk}$, because the constraint of $u_{ik}u_{jk}$ exists (by (2.2)). Namely, we assume that s_{ij} is proportional to $\sum_{k=1}^{K} u_{ik}u_{jk}$. Moreover, in the case that s_{ij} has an interval scale, we define the following model:

$$s_{ij} \approx \alpha \sum_{k=1}^{K} u_{ik} u_{jk} + \beta. \tag{2.3}$$

where β is an additive constant. If the similarity s_{ij} has an interval scale, then the origin of s_{ij} is arbitrary, and s_{ij} needs not satisfy $s_{ij} \geq 0$. Assuming that ζ_{ij} has an absolute scale which determines the unit and the origin uniquely, and s_{ij} is observed in a certain unit, then we may adjust the unit by the following:

$$s_{ij} = \alpha \zeta_{ij} \quad (\alpha > 0).$$

And if s_{ij} has an interval scale, then it is represented by the following:

$$s_{ij} = \alpha' \zeta_{ij} + \beta \quad (\alpha' > 0, \beta \neq 0).$$

Consequently, we propose the model (2.3).

The method of the fuzzy clustering based on this model is to find the partition matrix $U = (u_{ik})$ which satisfies the condition (2.2) and has the best fitness for the model (2.1) or (2.3). Then we find U and α which minimize the following sum of square error η_1^2 under the condition (2.2),

$$\eta_1^2 = \frac{\sum_{i \neq j=1}^{n} (s_{ij} - \alpha \sum_{k=1}^{K} u_{ik} u_{jk})^2}{\sum_{i \neq j=1}^{n} (s_{ij} - \bar{s})^2}, \quad \bar{s} = \frac{1}{n(n-1)} \sum_{i \neq j=1}^{n} s_{ij}.$$

In this case, we introduce the following transformation, to avoid the constrained optimization problem.

$$\begin{pmatrix} u_{i1} \\ u_{i2} \\ u_{i3} \\ \vdots \\ u_{iK-1} \\ u_{iK} \end{pmatrix} = \begin{pmatrix} \cos^2\theta_{i1} \\ \sin^2\theta_{i1}\cos^2\theta_{i2} \\ \sin^2\theta_{i1}\sin^2\theta_{i2}\cos^2\theta_{i3} \\ \vdots \\ \sin^2\theta_{i1}\sin^2\theta_{i2}\cdots\cos^2\theta_{iK-1} \\ \sin^2\theta_{i1}\sin^2\theta_{i2}\cdots\sin^2\theta_{iK-1} \end{pmatrix}. \tag{2.4}$$

From this, we know that the condition (2.2) is satisfied for any $\theta_{ik} \in [0, 2\pi]$, that is,

$$\sum_{k=1}^{K} u_{ik}(\theta_{i1}, \theta_{i2}, \cdots, \theta_{iK-1}) = 1, \quad u_{ik}(\theta_{i1}, \theta_{i2}, \cdots, \theta_{iK-1}) \geq 0.$$

2.2. Overlapping additive fuzzy clustering model

This model is defined by the following:

$$s_{ij} \approx \tilde{\alpha} \sum_{k=1}^{K} u_{ik} u_{jk}. \tag{2.5}$$

We loosen the constraint of condition (2.2) and assume

$$0 \leq u_{ik} \leq 1.$$

The looseness of condition (2.2) makes us consider the concept of overlapping flexibly. That is, it allows that objects belong to all clusters or no clusters. In this case, the condition $0 < \tilde{\alpha} \leq 1$ is needed, because when $\tilde{\alpha} \sum_{k=1}^{K} u_{ik} u_{jk} = \text{const.}$, if $\tilde{\alpha} \to \infty$ then $\sum_{k=1}^{K} u_{ik} u_{jk} \to 0$. In order to avoid the constrained problem, we use the following transformations:

$$\tilde{\alpha} = \frac{1}{1 + e^{-a}}, \quad u_{ik} = sin^2\theta_{ik} \tag{2.6}$$

where $-\infty < a < \infty$, $\theta_{ik} \in [0, \frac{\pi}{2}]$, and notice that $\tilde{\alpha}$ does not attain 1, but only approximately.

2.3. Ordinal additive fuzzy clustering model

The model is defined as follows:

$$s_{ij} \overset{m}{=} \sum_{k=1}^{K} u_{ik} u_{jk}, \tag{2.7}$$

where $\overset{m}{=}$ denotes the monotone relation. If we denote $w_{ij} = \sum_{k=1}^{K} u_{ik} u_{jk}$, then we assume the following relation:

$$\begin{cases} s_{ij} < s_{hl} \Rightarrow w_{ij} \leq w_{hl} \\ s_{ij} = s_{hl} \Rightarrow w_{ij} = w_{hl} \end{cases}$$

$$i, j, h, l = 1, \cdots, n.$$

To simplify this, we rewrite $W = f(U)$ and $S = (s_{ij})$, then (2.7) is replaced by

$$S \overset{m}{=} W = f(U).$$

The purpose is to find such a solution U, using this model (2.7), satisfying the monotone relation between S and W. That is, when we introduce the $\tilde{W} = T(S)$, where T denotes a monotone transformation, we may fit $\tilde{W}$ to $W = f(U)$ by using the least square method (Kruskal, J.B. [2],[3]) . This process is represented by

$$S \stackrel{m}{=} \tilde{W} \cong W = f(U)$$

where $\cong$ means the approximation solved by the least square method. The practical calculation is to determine the $\tilde{W}$ by using S and W based on the monotone regression principle (Kruskal, J.B. [2]), and to find W, which minimizes the following η_2^2:

$$\eta_2^2 = \frac{\sum\limits_{i \neq j=1}^{n} (w_{ij} - \tilde{w}_{ij})^2}{\sum\limits_{i \neq j=1}^{n} (w_{ij} - \bar{w})^2}, \qquad \bar{w} = \frac{1}{n(n-1)} \sum_{i \neq j=1}^{n} w_{ij}.$$

under the condition (2.2).

3.Reproduction and robustness

It is an important problem whether the algorithm of the fitting is able to reproduce the structure or not when the observed data have the structure of the model actually. And whether the small change of the data has an extreme influence on the result or not is also important problem. The former is known as a reproduction property and the latter is a robustness of the model. By the Monte Carlo simulation, these problems have been investigated in [4]. And it has shown that these algorithms have a reproduction property and the models are sufficiently robust for moderate errors.

4. Additive fuzzy clustering model for asymmetric similarity

4.1 Simple additive fuzzy clustering model for asymmetric similarity

If the observed similarity data is asymmetric, then the proposed additive fuzzy clustering models in the foregoing sections are not available. Then we extend the model (2.1) as follows:

$$\tilde{s}_{ij} \approx \sum_{k=1}^{K} \sum_{l=1}^{K} w_{kl} u_{ik} u_{jl}. \tag{4.1}$$

Where the observed asymmetric similarity $\tilde{s}_{ij}$ has a ratio scale and $0 \leq \tilde{s}_{ij} \leq 1$. u_{ik} is a fuzzy grade (the degree of belonging) which object i belongs to cluster k. The condition (2.2) is also assumed:

$$u_{ik} \geq 0, \quad \sum_{k=1}^{K} u_{ik} = 1.$$

In this model, the weight w_{kl} is considered to be a quantity which shows the asymmetric similarity between the pair of clusters. That is, we assume that the asymmetry of the similarity between the objects is caused by the asymmetry of the similarity between the clusters. If $k = l$, $u_{ik}, u_{jl} = 1$ and $w_{kl} > 1$, then the right hand side of (4.1) clearly exceed 1.0. Hence we need at least the following condition

$$0 \leq w_{kl} \leq 1,$$

because $0 \leq \tilde{s}_{ij} \leq 1$. If the model (4.1) has the following condition

$$w_{kl} = 0 \ \ (k \neq l), \quad \forall k \ w_{kk} = \alpha,$$

where α =const., then the model (4.1) identifies the model (2.1). That is, the model (2.1) is the special case of model (4.1).

We avoid the constrained problem by the transformation (2.4) and

$$w_{kl} = \frac{1}{1+e^{-b}}, \tag{4.2}$$

where $-\infty < b < \infty$, and notice that w_{kl} does not attain 0 or 1, but only approximately.

4.2 Overlapping additive fuzzy clustering model for asymmetric similarity

The model is defined by the following:

$$\tilde{s}_{ij} \approx \sum_{k=1}^{K}\sum_{l=1}^{K} w_{kl} u_{ik} u_{jk}. \tag{4.3}$$

This model is the extended model of (2.5). In this model, we also assume that

$$0 \leq u_{ik} \leq 1, \ \ 0 \leq w_{kl} \leq 1.$$

And we use the transformation (2.6) and (4.2) for the optimization process.

4.3 Ordinal additive fuzzy clustering model for asymmetric similarity

The model is defined as follows:

$$s_{ij} \overset{m}{=} \sum_{k=1}^{K}\sum_{l=1}^{K} w_{kl} u_{ik} u_{jl}, \tag{4.4}$$

where $\overset{m}{=}$ denotes the monotone relation. This model is the extended model of (2.7).

5. Numerical example

We shall show the application of the fuzzy clustering model for a dissimilarity data among terms of a kinship [5]. Table 5.1 shows the matrix of the similarity. In the optimization algorithm, using (2.4), 20 sets of initial values for u_{ik} are given by using uniform pseudorandom numbers for the interval $[0, \frac{\pi}{2}]$, and finally we select the best result. The number of clusters is determined based on the value of fitness. Figure 5.1 and Figure 5.2 show the variation of the value of fitness with respect to the number of clusters. Increasing the number of clusters, the value of fitness decreases, but even if the number of clusters is greater than 3, the decrease of the fitness is not so much. From the principle of parsimony, it should be considered that the number of clusters is determined to be 3. Figure 5.3~Figure 5.5 are visualizations of these results. In the abscissa, the terms of the kinship are located, and the ordinate shows the degree of belonging of clusters. And each part of drawing represents the degree of belonging for each cluster. From this, the terms represented by using a oblique line upward to the right is large, mean female. And the terms whose degree of belonging is represented by the large black part, mean male. The cluster which is characterized by the oblique line downward to the right, means the neuter term like cousin. Figure 5.4 is the result from the overlapping clustering model. The result is almost the same as Figure 5.3. But the similarity between aunt and niece and between uncle and nephew is very conspicuous and the feature of cousin is clear comparing Figure 5.3. And also the terms which are characterized by the oblique line downward to the right, means the collateral family. Figure 5.5 is the result of the rank ordered similarity, and this means that the value of degree of belonging is $\{0, 1\}$. This method seems to classify roughly, comparing the above simple additive clustering or the overlapping additive clustering, because this method is based on the information of the rank order of the similarity. Then in the case of this data, the hard clustering result is obtained.

Table 5.2 shows the result when we apply the symmetric similarity data (shown in Table 5.1) into the model (4.1). From this, we can find that the values of nondiagonal elements of similarity matrix (shown in Table 5.2) are approximately 0, that is, the clusters are not similar each other. Then this shows that the validity of the result in Figure 5.3, because we need not consider the similarity between the clusters in this data. To show the applications of the models (4.1) ,(4.3), (4.4), we use the asymmetric similarity data shown in Table 5.3(Yoshiharu Sato [7]). This data show the human relations among 16 children and the value of this data shows the degree of likes and dislikes between the children. The results are shown in Table 5.4 , Table 5.5 and Table 5.6. The number of clusters is determined as 5 based on the value of fitness (shown in Figure 5.6, Figure 5.7). In Table 5.4 , Table 5.5 and Table 5.6, C_1, C_2, C_3, C_4, C_5 represent the five clusters, and the children who consist of a same cluster are good friends with each other. Especially, C_1 and C_2 are clusters which consist of only boys, and C_3, C_4, C_5 are only girls. Moreover, w_{kl} means the relationship between a pair of clusters, for example, in both cases, Table 5.4, Table 5.5 and Table 5.6, the boys who belong to cluster C_2 have an interest in the girls who belong to cluster C_5, because the value of w_{25} is large in these cases. On the other hand, the girls of C_5 are not interested in boys of C_2, because the value of w_{52} is small.

Table 5.1 Similarity between terms of kinship

language	a	b	c	d	f	gd	gf	gm	gs	m	ne	ni	si	so	u
a	*														
b	.01	*													
c	.30	.12	*												
d	.55	.18	.11	*											
f	.05	.72	.03	.21	*					(s_{ij})					
gd	.57	.09	.12	.69	.04	*									
gf	.05	.56	.06	.05	.60	.24	*								
gm	.55	.03	.06	.57	.05	.79	.40	*							
gs	.04	.66	.11	.11	.61	.39	.79	.21	*						
m	.59	.15	.05	.81	.38	.61	.05	.62	.04	*					
ne	.29	.59	.31	.07	.53	.07	.54	.03	.62	.00	*				
ni	.84	.06	.32	.60	.01	.61	.04	.53	.09	.51	.44	*			
si	.53	.40	.12	.75	.16	.65	.04	.57	.07	.74	.04	.56	*		
so	.04	.75	.10	.40	.80	.12	.56	.04	.68	.22	.60	.10	.18	*	
u	.41	.53	.26	.01	.60	.03	.54	.04	.56	.06	.84	.29	.01	.53	*

a:	aunt	b:	brother	c:	cousin	d:	daughter
f:	father	gd:	granddaughter	gf:	grandfather	gm:	grandmother
gs:	grandson	m:	mother	ne:	nephew	ni:	niece
si:	sister	so:	son	u:	uncle		

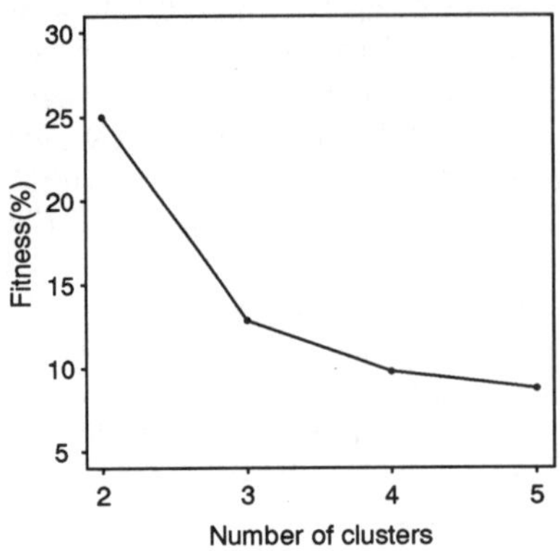

Figure 5.1 Number of clusters and Fitness (Simple additive fuzzy clustering)

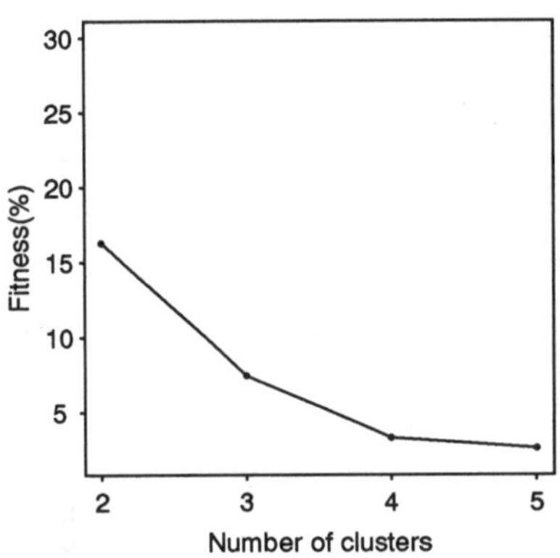

Figure 5.2 Number of clusters and Fitness (Overlapping additive fuzzy clustering)

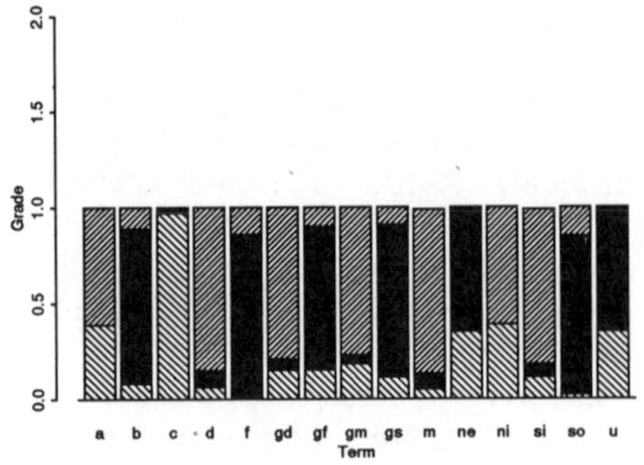

Figure 5.3 Simple clustering of language

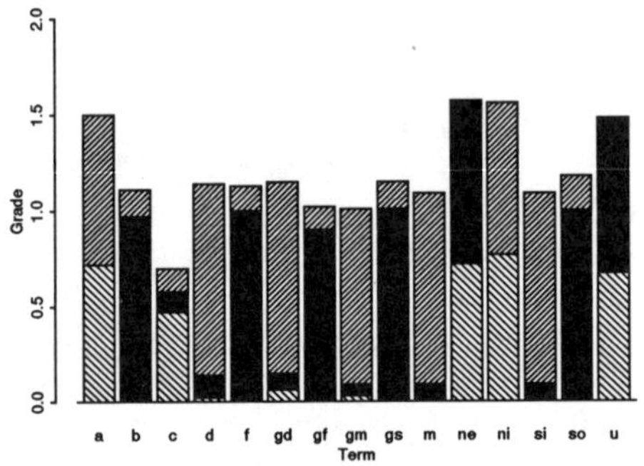

Figure 5.4 Overlapping clustering of language

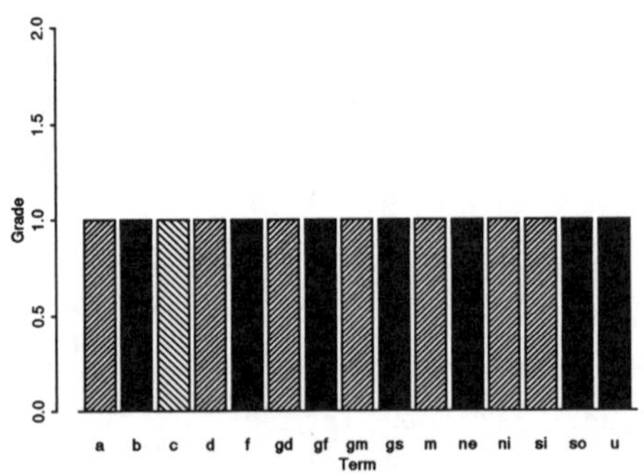

Figure 5.5 Ordered clustering of language

Table 5.3 Dissimilarity matrix of children

child	1	2	3	4	5	6	7	8	9	10	11	12	13	14	15	16
1	0	2	3	3	1	1	2	1	3	6	2	3	6	4	6	4
2	6	0	1	1	6	6	6	6	1	6	6	2	6	2	6	2
3	6	1	0	2	6	6	6	6	1	6	6	1	6	2	6	2
4	6	1	2	0	6	6	6	6	1	6	6	2	6	1	6	2
5	1	3	3	4	0	1	1	2	4	6	3	3	6	4	6	4
6	1	3	2	4	1	0	2	2	3	6	3	2	6	3	6	3
7	1	3	3	4	1	1	0	2	4	6	3	3	6	4	6	4
8	6	1	2	2	6	6	6	0	2	6	1	3	6	3	6	3
9	6	6	6	6	6	6	6	6	0	6	6	1	6	1	6	1
10	6	2	3	3	6	6	6	2	3	0	1	4	1	4	1	4
11	6	1	2	2	6	6	6	1	2	6	0	3	6	3	6	3
12	6	6	6	6	6	6	6	6	1	6	6	0	6	1	6	1
13	6	2	3	3	6	6	6	2	3	1	1	4	0	4	1	4
14	6	6	6	6	6	6	6	6	1	6	6	1	6	0	6	1
15	6	2	3	3	6	6	6	2	3	1	1	4	1	4	0	4
16	6	6	6	6	6	6	6	6	1	6	6	1	6	1	6	0

Table 5.2 Simple clustering for asymmetric similarity

language	C1	C2	C3
a	0.38	0.00	0.62
b	0.07	0.82	0.10
c	1.00	0.00	0.00
d	0.03	0.11	0.87
f	0.00	0.87	0.13
gd	0.13	0.06	0.80
gf	0.15	0.76	0.09
gm	0.16	0.06	0.78
gs	0.12	0.81	0.07
m	0.01	0.11	0.88
ne	0.36	0.64	0.00
ni	0.39	0.00	0.61
si	0.09	0.09	0.83
so	0.01	0.85	0.14
u	0.36	0.64	0.00

$\eta^2 = 0.1321$

Similarity matrix between clusters

cluster	C1	C2	C3
C1	1.00	0.05	0.05
C2	0.05	1.00	0.00
C3	0.05	0.00	1.00

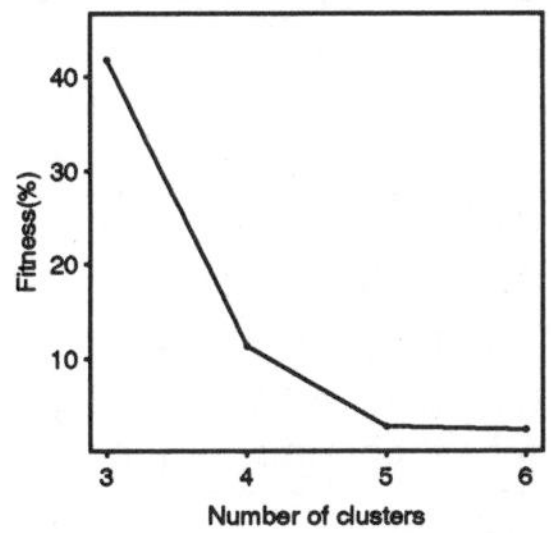

Figure 5.6 Number of clusters and Fitness (Simple additive fuzzy clustering for asymmetric similarity)

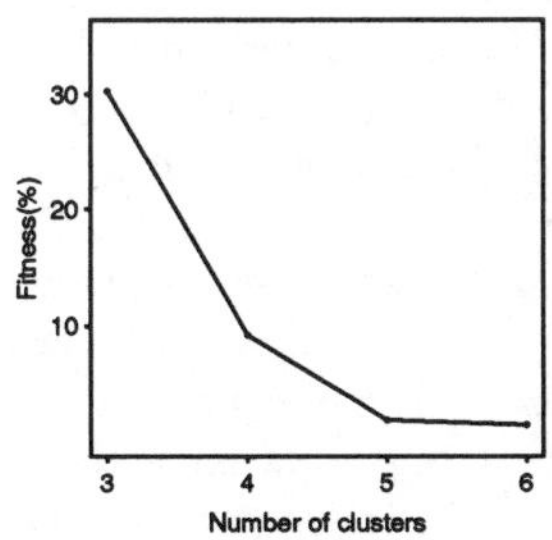

Figure 5.7 Number of clusters and Fitness (Overlapping additive fuzzy clustering for asymmetric similarity)

Table 5.4 Simple fuzzy clustering for asymmetric similarity

child	C1	C2	C3	C4	C5
1	0.94	0.00	0.04	0.02	0.00
2	0.00	0.91	0.09	0.00	0.00
3	0.06	0.87	0.04	0.00	0.03
4	0.00	0.88	0.00	0.08	0.04
5	0.98	0.00	0.00	0.02	0.00
6	0.94	0.04	0.00	0.01	0.01
7	0.95	0.00	0.00	0.05	0.00
8	0.06	0.00	0.94	0.00	0.00
9	0.00	0.06	0.00	0.00	0.94
10	0.00	0.00	0.00	1.00	0.00
11	0.00	0.00	0.91	0.09	0.00
12	0.05	0.00	0.00	0.00	0.95
13	0.00	0.00	0.00	1.00	0.00
14	0.00	0.00	0.00	0.03	0.97
15	0.00	0.00	0.00	1.00	0.00
16	0.02	0.00	0.00	0.03	0.95

$\eta^2 = 0.0292$

Similarity matrix between clusters

cluster	C1	C2	C3	C4	C5
C1	0.92	0.47	0.66	0.00	0.42
C2	0.01	0.98	0.02	0.01	0.77
C3	0.00	0.76	0.98	0.00	0.56
C4	0.00	0.56	0.77	0.88	0.37
C5	0.01	0.00	0.00	0.01	0.92

Table 5.5 Overlapping fuzzy clustering for asymmetric similarity

child	C1	C2	C3	C4	C5
1	0.93	0.00	0.02	0.01	0.00
2	0.00	0.93	0.04	0.00	0.00
3	0.04	0.86	0.00	0.00	0.00
4	0.00	0.83	0.00	0.01	0.00
5	0.89	0.00	0.00	0.00	0.00
6	0.91	0.02	0.00	0.00	0.01
7	0.84	0.00	0.00	0.00	0.00
8	0.04	0.00	0.90	0.00	0.00
9	0.00	0.05	0.00	0.01	0.95
10	0.00	0.00	0.00	0.91	0.00
11	0.00	0.00	0.88	0.06	0.00
12	0.05	0.00	0.00	0.00	0.91
13	0.00	0.00	0.00	0.91	0.00
14	0.00	0.00	0.00	0.00	0.89
15	0.00	0.00	0.00	0.91	0.00
16	0.00	0.00	0.00	0.00	0.87

$\eta^2 = 0.0183$

Similarity matrix between clusters

cluster	C1	C2	C3	C4	C5
C1	1.00	0.59	0.75	0.01	0.49
C2	0.00	1.00	0.01	0.01	0.90
C3	0.01	0.88	1.00	0.01	0.63
C4	0.00	0.69	0.89	1.00	0.44
C5	0.00	0.00	0.00	0.01	1.00

6. Conclusion

As the method of the fuzzy clustering which the observations are given by similarity or dissimilarity, we proposed three additive fuzzy clustering models and three extended additive fuzzy clustering models for asymmetric data. These methods are regarded as an extended method of the method proposed by Shepard, R.N. et al, but there is a difference of the concept of an overlapping and the consideration of the order of a similarity.

As the method of fuzzy clustering whose an object is a similarity data, RFCM(relational fuzzy clustering method) (Hathaway, R.J, et al. [1]) has been proposed. This method is to calculate centroids by using a partition matrix $U = (u_{ik})$, minimizing an extended within dispersion, when a dissimilarity given as a data is defined as an inner product. But in this proposed method, a similarity does not have to be an inner product, and the ratio of the object of this analysis can be a number of cases, namely a ratio scale, an interval scale, or an ordinal scale. The structure of the similarity seems to be extracted clearly by loosening the condition $\sum_{k=1}^{K} u_{ik} = 1$. And the reproduction and the robustness of this method is confirmed by simulating. Furthermore, we extend the model to the asymmetric similarity data. Using this extended model, the structure of an asymmetric similarity between objects can be summarized by the asymmetric similarity between clusters.

Table 5.6 Ordinal fuzzy clustering for asymmetric similarity

child	C1	C2	C3	C4	C5
1	0.98	0.00	0.00	0.02	0.00
2	0.00	0.98	0.02	0.00	0.00
3	0.02	0.95	0.01	0.00	0.02
4	0.02	0.96	0.00	0.01	0.02
5	0.98	0.00	0.00	0.02	0.00
6	0.98	0.01	0.00	0.01	0.00
7	0.98	0.00	0.00	0.02	0.00
8	0.03	0.00	0.97	0.00	0.00
9	0.00	0.00	0.00	0.00	1.00
10	0.00	0.00	0.00	1.00	0.00
11	0.00	0.00	0.97	0.03	0.00
12	0.00	0.00	0.00	0.00	1.00
13	0.00	0.00	0.00	1.00	0.00
14	0.00	0.00	0.00	0.00	0.99
15	0.00	0.00	0.00	1.00	0.00
16	0.00	0.00	0.00	0.00	0.99

$\eta^2 = 0.105069e - 03$

Similarity matrix between clusters

cluster	C1	C2	C3	C4	C5
C1	0.84	0.78	0.80	0.01	0.78
C2	0.01	0.84	0.00	0.02	0.81
C3	0.00	0.81	0.87	0.01	0.78
C4	0.00	0.79	0.82	0.81	0.78
C5	0.02	0.00	0.02	0.02	0.82

References

[1] Hathaway, R.J., Davenport, J.W. and Bezdek, J.C. (1989) Relational Duals of The c-Means Clustering Algorithms. *Pattern Recognition, 22*, pp. 205-212.

[2] Kruskal, J.B. (1964) Multidimensional scaling by optimizing goodness of fit to a nonmetric hypothesis. *Psychometrica, 29*, pp. 1-27.

[3] Kruskal, J.B. (1964) Nonmetric multidimensional scaling: A numerical method. *Psychometrica, 29*, pp. 115-129.

[4] Mika Sato and Yoshiharu Sato (1994) An Additive Fuzzy Clustering Model. *Journal of Japan Society for Fuzzy Theory and Systems, 6*, pp. 319-332.

[5] Rosenberg, S. and Kim, M.P. (1975) The Method of sorting as a data gathering procedure in multivariate research. *Multivariate Behavioral Research, 10*, pp. 489-502.

[6] Shepard, R.N. and Arabie, P. (1979) Additive Clustering: Representation of Similarities as Combinations of Discrete Overlapping Properties. *Psychological Review, 86, 2*, pp. 87-123.

[7] Yoshiharu Sato (1992) Multidimensional Scaling in Minkowski space. *Hokkaido Behavioral Science Report, Series M, No.20*, pp. 69-99.

FUZZY LOGIC IN DATABASES

THE GENERALIZED SELECTION: AN ALTERNATIVE WAY FOR THE QUOTIENT OPERATIONS IN FUZZY RELATIONAL DATABASES

J. C. CUBERO, J. M. MEDINA, O. PONS and M. A. VILA
Department of Computer Sciences and Artificial Intelligence
University of Granada
Adv Andalucia n.38
10872 Granada Spain

ABSTRACT

In this paper the quotient operation in fuzzy relational database is analyzed. With the basis on the classical case a fuzzy quotient definition is given in a general fuzzy database context. Furthermore this concept in generalized by introducing linguistic quantifiers different of $\forall$ and developing the quotient operations by using them.
Keywords: Fuzzy database, fuzzy relational algebra, fuzzy quotient, linguistic quantifier.

1 Introduction

In recent years several authors have studied the problem of how to soft the relational data model so that it can deal with imprecise data and queries. In this connection fuzzy logic and fuzzy set theory have appeared as useful tools to represent and manage the lack of precision, and the fuzzy relational database concept have been developed. Two main approaches to this concept have appeared: The homogeneous model which basically assumes the domains attributes to be linguistic labels and the values themselves to be crisp set of them, and the possibilistic model, the most widely used, were the attribute domains are any kind of sets, but the values of attributes are given by means of possibility distributions defined in these domains.

Both approaches have developed associated query languages based mainly in a generalization of relational algebra. But, in the case of possibilistic model, these generalizations have covered all of classical relational operations with the exception of the quotient. Only the Yager's work (Yager 1992) deals with this operation but in a particular context of fuzzy database.

However, in the fuzzy querying process it is necessary to have available a general definition of quotient. Notice that, the classical quotient were defined to model queries which agree with following general structure:

"To get those objects which connect through a relation with *all* of elements of some set described by means of other relation" and this structure may be made more flexible in two ways:

1. to consider the involved set is the representation of some imprecise property and consequently a fuzzy set.

2. to soft the quantifier *all* by imposing the connection with *almost all, most* etc. of elements of some set.

the use of any way or both two generates an important class of imprecise queries which have to be solved in the fuzzy database context.

Therefore we are concerned in this paper with to give a quotient definition as general as possible for a very general model of fuzzy database. In order to do it, we will are founded in the following ideas:

- In the classical case, the quotient may be considered as a special form of selection if the condition of first normal form is avoided. Thus, the fuzzy quotient must be defined by means of a fuzzy selection since the relations of a fuzzy database are unnormalized by definition.

- Any set of linguistic quantifiers involves some order relation and includes the $\exists$ and $\forall$ as limit elements. Thus, the generalization of the quotient according to a linguistic quantifier must cover both particular cases and reproduce this order relation.

The paper begins by introducing some initial definitions and the used notation. Next the relation between quotient and selection in classical database are analyzed. The conclusions of such analysis are used to give a definition of fuzzy quotient in the following section. Finally this definition is generalized in the last section by using a set of linguistic quantifiers.

2 Initial definitions and notation

In this paper, the considered approach of fuzzy database will be the Possibility-based one (Prade and Testemale 1984, Zemankova-Leech and Kandel 1984). Therefore we will assume that a Fuzzy Relational Database (FRDB) is a set of relations $R, S, \ldots$, where each relation R is defined in the following way:

Definition 2.1 *Let us consider a sequence of attributes $A_1, \ldots, A_n$, $\forall i \in \{1, \ldots, n\}$ the basic domain of A_i will be denoted by D_i. We will assume some of A_is attributes are allowed to take imprecise values which will be modeled by means of possibility distribution on the basic domains. The set $\mathcal{I}$ will be the index set of such attributes and the set of possibility distributions defined on $D_i; i \in \mathcal{I}$ will be denoted by $P(D_i)$. In these conditions we will define the relation R as a subset of the cartesian product:*

$$\prod_{i \in \mathcal{I}} \mathcal{P}(D_i) \times \prod_{i \in \{1 \ldots n\} - \mathcal{I}} D_i$$

The set of attributes $\{A_1, \ldots, A_n\}$ will be called schema of R, as in the classical case. In the following, the attribute sets, schemes or subeschemes of some relation, will be denoted by $\mathcal{A}$, $\mathcal{B}$, ...

With this definition crisp and fuzzy values may appear in the same relation, we will denote by $x, y \ldots$ the first ones whereas the second ones by $X, Y, \ldots$. For the sake of simplicity, the membership function of each fuzzy value will be denoted with the same symbol as it. That is: if we consider a fuzzy attribute with basic domain D, to be in the schema of a relation R then any tuple $\bar{t} \in R$ will have the form $\bar{t} = (\ldots, x, \ldots)$ or $\bar{t} = (\ldots, X, \ldots)$ where $x \in D$ and $X \in \mathcal{P}(D)$ with membership function $X : D \longrightarrow [0,1]$. In any case we shall denote by $\bar{t}[A]$ the value of the attribute A in the tuple $\bar{t}$.

The query management for the possibilistic model of FRDB may carried out by using two different ways: the first one is relational algebra-based and it has been mainly developed in Prade and Testemale (1984) and Zemankova-Leech and Kandel (1984); the second way lies on the use of the domains calculus-based query language developed for FRDB in Vila et al. 1992, with the special features of the possibilistic model which were given in Vila et al (1993).

Since our goal is to develop an way to carry out operations like to the quotient, we will be focused in the algebra-based approach. Concretely, we are concerned in this paper with the fuzzy $\simeq$-selection with atomic conditions and the fuzzy projection, they are defined as follow:

Definition 2.2 *Let us consider a relation R which includes at least one fuzzy attribute A in its schema, let D be the A basic domain and $P \in \mathcal{P}(D)$ the possibility distribution of any fuzzy property to be fulfilled by the A values of the R tuples. We shall denote by $\sigma\Gamma_{A \simeq P}(R)$ the $\simeq$-selection of R according to P and by using the measure Γ, and the result of this operation will be the following one. If*

$$T = \sigma\Gamma_{A \simeq P}(R)$$

, then any tuple of T has the form $(\bar{t}, \alpha)$ where $\bar{t} \in R$ and $\alpha \in [0,1]$ representing the Γ-measure of how $\bar{t}[A]$ and P match. We will denote

$$\alpha = \Gamma(\bar{t}[A] \mid P)$$

Obviously there are versions of the selection with atomic conditions being more general than the above presented one. They are those that consider fuzzy operators different of the $\simeq$, such that "much greater than", "not much smaller than" etc. which are modeled by fuzzy relations. These kinds of relations are used by the operations defined in Prade and Testemale (1984) or Zemankova-Leech and Kandel (1984), however the definition 2.2 covers both two approaches in the case of equality.

The form of the projection depends on if the considered relation belongs to the initial FRDB or not. In the first case we have a classical relation with fuzzy values and the projection will be done as in the classical way. In the second one the relation comes from several operations and it is a fuzzy relation, therefore after a classical projection tuples being equals with different membership levels could appear, this redundancy must be avoided by considering a single tuple with maximum membership level. All these ideas are formulated in the following definitions.

Definition 2.3 *Let R be a relation with schema $\mathcal{A}$ belonging to a FRDB and the attribute set $\mathcal{B}$ included in the schema of R. We will call projection of R on $\mathcal{B}$, and we will denote by $\pi_{\mathcal{B}}(R)$, the relation T whose schema is $\mathcal{B}$ and which verifies:*

$$\forall \bar{t} \in T \Rightarrow \exists \bar{r} \in R \;/\; \forall A \in \mathcal{B}\; \bar{t}[A] = \bar{r}[A]$$

Definition 2.4 *Let S be a fuzzy relation whose schema has the form $\mathcal{A} \cup [0,1]$ and the attribute set $\mathcal{B} \subset \mathcal{A}$. The projection of S on $\mathcal{B}$, $\pi_{\mathcal{B}}(R)$, will be a fuzzy relation T with schema $\mathcal{B} \cup [0,1]$ which verifies:*

$$\forall (\bar{t}, \alpha) \in T \Rightarrow \exists (\bar{r}, \alpha) \in R \;/\; \forall A \in \mathcal{B}\; \bar{t}[A] = \bar{r}[A]$$

and

$$\alpha = max\{\beta \;/\; (\bar{s}, \beta) \in R\,;\; \bar{s}[A] = \bar{t}[A]\; \forall A \in \mathcal{B}\}$$

3 Some reflections about the quotient in classical relational databases

We are now concerned in to clarify the concept of quotient in classical relational databases, therefore all relations are assumed to be crisp and with crisp attributes. According with the classical relational database theory the quotient operation is defined in the following way:

Definition 3.1 *Let R and S be two relations with schemes $\mathcal{A}$ and $\mathcal{B}$ respectively, where $\mathcal{B} \subset \mathcal{A}$. The quotient $R \div S$ is defined as a relation T with schema $\mathcal{A} - \mathcal{B}$, which verifies:*

$$\forall \bar{t} \in T \Rightarrow \forall \bar{s} \in S\; \exists \bar{r} \in R \;/\; \pi_{\mathcal{B}}(\bar{r}) = \bar{s}\;\; \pi_{\mathcal{A}-\mathcal{B}}(\bar{r}) = \bar{t}$$

I we assume the R attributes are ordered in such way that those belonging to $\mathcal{B}$ are the latest, this definition has a more intuitive version.

$$\forall \bar{t} \in T \Rightarrow \forall \bar{s} \in S\; \exists \bar{r} \in R \;/\; \bar{r} = (\bar{t}, \bar{s})$$

The semantic interpretation of the quotient is quite straightforward if we consider the tuples in the relation R to have two parts: the first one corresponds with the subeschema $\mathcal{A} - \mathcal{B}$ and this subeschema must include the primary key of another relation of the database which describes some object set, the second one corresponds with the subeschema $\mathcal{B}$ and it includes the primary key of at least the relation S, which in turn describes another object set. Thus R may be viewed as the representation of a relationship between both two object sets, and the quotient selects those objects of the first set which have the property of to connect with *all* of objects which are described by S. The following example explains these ideas:

Example 1

Let us consider the classical example of suppliers and parts given by:

SUPPLIERS		
S#	*Sname*	*City*
S1	Smith	London
S2	Jones	Paris
S3	Blake	Paris
S4	Clark	London

PARTS		
P#	*Pname*	*Color*
P1	Nut	Red
P2	Bolt	Green
P3	Screw	Blue
P4	Screw	Red
P5	Cam	Blue

SHIPMENTS	
S#	*P#*
S1	P1
S1	P2
S1	P3
S1	P4
S1	P5
S2	P1
S2	P3
S2	P4
S3	P4
S4	P1
S4	P2
S4	P4

The property "To supply all red parts" can be applied to the suppliers set, and we can construct the set of red parts as

$$INT = \pi_{S\#}(\sigma_{Color=red}(PARTS))$$

and we can obtain the suppliers which connect with all of elements of this set by means of

$$SOL = SHIPMENTS \div INT$$

The solutions in this concrete example are:

$$INT = \{P1, P4\} \text{ and } SOL = \{S1, S2, S4\}$$

With this interpretation, we can deduce that the quotient has a similar meaning than the selection, as it selects those elements of any set which verify a concrete property. The difference between both operations lies on the property itself, since it can not be expressed by a logical formula relating attribute values and constants. This is because of the property imposed by the quotient involves the equality with a set of objects and the attributes of a classical relation can only take atomic values.

However if the first normal form condition is avoided the quotient could be reduced to a selection. Let us return to the above example to explain this idea:

If we allow the relation SHIPMENTS to take non-atomic values in $P\#$ we can "compress" it on this attribute and obtain:

C-SHIPMENTS	
S#	*P#*
S1	P1,P2,P3,P4,P5
S2	P1,P3,P4
S3	P4
S4	P1,P2,P4

On the other hand, the relation INT can be viewed as an element of the domain where the second attribute of C-SHIPMENTS is now defined. In these conditions we can obtain the same result than $SHIPMENT \div INT$ by using:

$$\pi_{S\#}(\sigma_{P\# \subseteq INT}(C - SHIPMENTS))$$

Obviously all these reflections can be generalized to the case of more than one attribute are involved, suffice it to consider them as a composite attribute whose domain is a cartesian product.

In the next section these ideas will be used to define the quotient and others related operations in FRDBs. They are specially suitable for this purpose since the relations of FRDBs are unnormalized by definition.

4 The fuzzy quotient in FRDBs

According to the above ideas we will establish the quotient as a selection done over a "compressed" relation, thus we have first to formalize this later concept.

Definition 4.1 *Let R be a relation of an FRDB with schema $\mathcal{A}$, let $A \in \mathcal{A}$, we define the compression of R on A, $\delta_A(R)$ as a relation T which verifies:*

1. *T has as schema $\mathcal{A}$ and its primary key is $\mathcal{A} - A$, that is:*

$$\forall \bar{t}_1, \bar{t}_2 \in T \Rightarrow \bar{t}_1[\mathcal{A} - A] \neq \bar{t}_2[\mathcal{A} - A]$$

where $\bar{t}[\mathcal{A} - A]$ stands for the $\bar{t}$ subtuple which is formed by the values of the attributes of $\mathcal{A} - A$.

2. *$T[A]$ collects the $R[A]$ values of all those tuples of R being equal in the rest of attributes,that is:*

$$\forall \bar{t} \in T \;\; \bar{t}[A] = \{\bar{r}[A] \,/\, \bar{r} \in R \, \bar{r}[\mathcal{A} - A] = \bar{t}[\mathcal{A} - A]\}$$

Obviously, to carry out this operation we must to consider a tuple $\bar{t}$ in T for every different subtuple $\bar{r}[\mathcal{A} - A] \in R$ with $\bar{r}[\mathcal{A} - A] = \bar{t}[\mathcal{A} - A]$ and to take $\bar{t}[A]$ as the union of the A values of those tuples belonging to R which also have $\bar{r}[\mathcal{A} - A]$ as subtuple. That is, we realize a true "compression" of the relation.

It should be remarked that the type of attribute A has not been taken into account in the above definition. In principle, A may be fuzzy or crisp in the relation R, however it must be considered as fuzzy in $\delta_A(R)$, as it takes set values in this relation. But if A is a crisp attribute of R it will only have crisp sets values in $\delta_A(R)$

This definition can be easily generalized to the case where more than one attribute are involved.

Definition 4.2 *If we have to compress the relation R on the attribute set $\mathcal{B} \in \mathcal{A}$, where $\mathcal{B}$ can include crisp and fuzzy attributes, we will consider $\mathcal{B}$ as a single composite crisp attribute B, whose domain is the cartesian product of the all of $\mathcal{B}$ attributes domains and we will carry out $\delta_B(R)$.*

Notice that with this definition $\mathcal{B}$ becomes a single attribute B of the compression, and its values will be sets of tuples which in turn can involve fuzzy and crisp values. That is, the values of B really are complete relations of FRDB.

Once we have defined the compression we are able to cope with the quotient concept. First we will give the definition for the single attribute case and the generalization to the multiple attribute case will be after offered. According to the presented ideas in 3, the quotient will be done by using the composition of a compression with a selection, thus a initial definition of the quotient will be the following

Definition 4.3 *Let R be a relation of an FRDB, with schema $\mathcal{A}$, let $A \in \mathcal{A}$ be a fuzzy attribute of R, and let P be a possibility distribution on the A domain which represents some imprecise property by means a fuzzy set of objects. Let us assume that $\mathcal{A}$ includes the primary key $\mathcal{B}$ of other relation of the FRDB which describes other object set, and $A \notin \mathcal{B}$. In these conditions the set of objects, described by $\mathcal{B}$, which connect through the relation R with* all *of the objects belonging to the fuzzy set P will be given by:*

$$\pi_{\mathcal{B}}(\sigma\Gamma_{A \simeq P}(\delta_A(\pi_{\mathcal{B},A}(R))))$$

This definition provides us a formal expression for the fuzzy quotient in FRDBs but we need a more concrete version of the $\simeq$-selection to completely solving the problem, as the selection depends on a fuzzy measure Γ.

We will choose this measure in such way that it to represent the accomplishment degree of the property of to match with *all* of objects belonging to a fuzzy set. To do it, we will use the following fuzzy logic-based criteria:

Let X and P be two fuzzy sets defined on any domain D, the logical representation of the sentence " all of P are X" is given by:

$$\forall d \in D \, (P(d) \rightarrow X(d))$$

and, if we take $I(P(d), X(d))$ as truth value of $(P(d) \rightarrow X(d)$ where $I(.,.)$ is an implication function (Trillas and Valverde 1985), the accomplishment degree of the above sentence is given by:

$$N(X|P) = min_{d \in D} I(P(d), X(d))$$

thus, if the $\simeq$-selection imposes to match with all of elements of a fuzzy set we can use N as fuzzy measure. Therefore, a more concrete expression for the fuzzy quotient is given by

$$\pi_B(\sigma N_{A \simeq P}(\delta_A(\pi_{B,A}(R))))$$

The measure N has some properties which have important consequences in the quotient and even in the selection.

Property 4.1

N is actually a fuzzy measure, that is: $\forall X_1, X_2, P \in \mathcal{P}(D) \;\; X_1 \subseteq X_2 \Rightarrow N(X_1|P) \leq N(X_2|P)$
Proof
It can been directly deduced from the following property of the implication functions:

$$\forall x, y, z \in [0, 1] \;\; y \leq z \Rightarrow I(x, y) \leq I(x, z)$$

Property 4.2

If X and P are crisp sets then: $N(X|P) = \begin{cases} 1 & \text{if } P \subseteq X \\ 0 & \text{otherwise} \end{cases}$

Proof
Since X and P are crisp sets, X(.) and P(.) becomes characteristic functions. Furthermore the implication functions verifies:

$$\forall x \in [0, 1] \; I(0, x) = 1 \; I(1, x) = x$$

thus, the expression of $N(X|P)$ will be:

$$N(X|P) = min_{d \in P}(X(d))$$

which is equal to zero if $\exists d \in P$ such that $p \notin X$ and equal to 1 otherwise.

From this property we can deduce that $\sigma N_{A \simeq P}(R)$ becomes $\sigma_{A \subseteq P}(R)$ if both A values in R and P are assumed to be crisp sets. This conclusion is important because it assures us that the fuzzy quotient reproduces the classical quotient if the involved attributes and sets are crisp.

We must now to make more specific the expression of N in order to obtain additional properties.Therefore we will assume the function I to be founded in the material implication, and it may be represented by using a continuous t-conorm $\oplus$ by means of:

$$N(X|P) = min_{d \in D}((1 - P(d)) \oplus X(d))$$

Furthermore, by considering the maximum as t-conorm, we have:

$$N(X|P) = min_{d \in D}((1 - P(d)) \vee X(d))$$

thus N becomes the necessity measure of X with respect to P. With this later expression N verifies:

Property 4.3

$N(X|P) = 1$ if and only if $\{d \in D/P(d) > 0\} \subseteq \{d \in D/X(d) = 1\}$
Proof
$N(X|P) = 1 \Rightarrow \forall d \in D \;\; (1 - P(d)) \vee X(d) = 1$ thus $1 - P(d) = 1 \Leftrightarrow P(d) = 0$ if $X(d) < 1$.

Reciprocally if $\{d \in D \; P(d) > 0\} \subseteq \{d \in D \; X(d) = 1\}$, for every $d \in D$ we have either $X(d) = 1$ or $P(d) = 0$ which implies $1 - P(d) = 1$, in any case

$$\forall d \in D \;\; (1 - P(d)) \vee X(d) = 1 \Rightarrow N(X|P) = 1$$

The above property suggests us that the necessity may be a too strict measure, since it does not take value 1 even if $P \equiv X$. We think this is because of the maximum is the lower bound of t-conorms, and the use of another one such as the Lukasiewicz's one could be more meaningful. However the necessity has been used as matching measure by some authors, concretely: Prade and Testemale (Prade and Testemale 1984) have defined the selection with two accomplishment degrees given by means of necessity and possibility measures, and the quotient defined by Yager (Yager 1992) corresponds with our definition when the N measure is assumed to be a necessity. This author also generalizes the operation by considering different linguistic quantifiers such that, *most, a few* etc., instead of *all.* We will deal with the same problem in the next section.

In the following we will assume that the quotient is given by using a N measure based on the material implication but with a general t-conorm in its expression.

We will finish this section by generalizing the quotient definition to the case where more than one attribute are involved. As in the compression case, the way to do it is to considering the attribute set as a single composite attribute whose domain is given by a cartesian product. This leads to the following definition:

Definition 4.4 *Let R be a relation of an FRDB, with schema $\mathcal{A}$, let $\mathcal{B} \subset \mathcal{A}$ be which can include both crisp and fuzzy attributes, we will denote by D the cartesian product of the domains of the $\mathcal{B}$ attributes. Let P be a possibility distribution on the D domain which represents some imprecise property by means a fuzzy set of objects. Let us assume that $\mathcal{A}$ includes the primary key $\mathcal{D}$ of other relation of the FRDB which describes other object set, and $\mathcal{B} \cap \mathcal{D} = \emptyset$. In these conditions the set of objects, described by $\mathcal{D}$, which connect through the relation R with* all *of the objects belonging to the fuzzy set P will be given by:*

$$\pi_{\mathcal{D}}(\sigma N_{B \simeq P}(\delta_B(\pi_{\mathcal{D},\mathcal{B}}(R))))$$

where B is the composite attribute formed by all of $\mathcal{B}$ ones.

It should be remarked that, with these assumptions, P becomes a fuzzy set of tuples which can have both fuzzy and crisp attributes, thus P actually is a complete fuzzy relation of the FRDB which can arise from other previous operations. In consequence, the above defined operation has so much meaning as the classical relational quotient have and is so general as it is.

5 The use of linguistic quantifiers

In the above section, the quotient have been defined for FRDBs by using the property of to connect with *all of* elements being in a fuzzy set. This is , in fact, a "classical" version of the quotient so that in a imprecise context , other different of *exists* and *for all* quantifiers should be able to be used. Therefore, in this section, we will cope with the property of to connect with *Q of* elements of a fuzzy set, where Q is a linguistic proportion of elements such as *most, a few etc.*.

The linguistic quantifier concept was introduced by Zadeh in 1986 (Zadeh 1986) as a fuzzy set either of positive real numbers or of [0,1] interval. None special hypothesis was formulated by this author about the shape of the membership function of these sets, but some previous conditions must be imposed, to use them in the quotient generalization :

1. We will assume that the membership function $Q(.)$ of any quantifier Q is non-decreasing. This is because of we will actually express by means of Q the "minimal" linguistic proportion of objects which verify a concrete property. For example, when we state "give those objects of A which connect with *most* of B objects" we implicitly include in the assertion those A objects connected with all of B objects. Other authors such as Yager (1992) made the same hypothesis.

2. There must by some kind of order between the quantifiers, which, in some way represents the amount of proportion . For example, "a few" should go before "most" and "exists" and "for all" will be respectively the first and the later of all of them.

Thus in the following we will consider a set of linguistic quantifiers according to the next definition:

Definition 5.1 *Let $\mathcal{Q} = \{Q_1, \ldots, Q_l\}$ be a linguistic quantifier set, we will say it is coherent if it verifies:*

i *The membership functions of $\mathcal{Q}$ elements are non-decreasing functions.*

ii *A partial order relation $\succeq$ is defined in $\mathcal{Q}$. It has as maximal element $Q_1 = \exists$ and as minimal one $Q_l = \forall$. Furthermore $\forall Q_i, Q_j \in \mathcal{Q}\ Q_i \subseteq Q_j \Rightarrow Q_j \succeq Q_i$*

iii *The membership function of the modifier $\exists$ is given by $Q_1(x) = 1$ if $x \neq 0$ and $Q_1(0) = 0$, whereas the membership functions of $\forall$ will be $Q_l(x) = 0$ if $x \neq 1$ and $Q_l(1) = 1$*

Once we have established the characteristics of quantifier set, we can generalize the quotient concept. For the sake of simplicity, a single attribute will be considered; however all of definitions and conclusions may be easily generalized to the multiple attribute case by using of the same reasoning as in the definition 4.4. Moreover, to give the quotient generalization, the definition 4.3 can be used with a few modifications.

Definition 5.2 *Let us consider $R, \mathcal{A}, A, \mathcal{B}$ and P with the same assumptions than the definition 4.3 ones, let also Q be a quantifier belonging to a coherent set. In these conditions, the set of objects, described by $\mathcal{B}$, which connect through the relation R with Q of the objects belonging to the fuzzy set P will be given by:*

$$\pi_{\mathcal{B}}(\sigma\Gamma^{Q}_{A\simeq P}(\delta_A(\pi_{\mathcal{B},A}(R))))$$

where Γ^Q represents the accomplishment degree of the property of to match with Q of objects described by P

We have again to specify the measure Γ^Q. To do it, we could use the necessity measure expression, by changing only the aggregation way of the implication truth values. That is, if we have two fuzzy set X and P defined on the same domain D, we take the accomplishment degree of the sentence "Q of P are X" as:

$$\Gamma^Q(X|P) = \mathcal{O}^Q_{d\in D}((1 - P(d) \oplus X(d))$$

$\mathcal{O}^Q$ stands for some aggregation operator defined by using the quantifier Q, such as Yagers OWA which appear in Yager 1992. This is the way used by this last author for the quotient generalizations, however it has the following drawback:

If Q is the $\exists$ quantifier $\mathcal{O}^Q$ becomes the maximum and Γ^Q is given by:

$$\Gamma^{\exists}(X|P) = max_{d\in D}((1 - P(d) \oplus X(d))$$

It can be easily proved that this measure is too big for represent anything, since it can be equal to 1 even when X and P have empty intersection.

On the other hand, an usual way to represent the accomplishment degree of the quantifier $\exists$ is to consider the generalized possibility measure and to give:

$$\Gamma^{\exists}(X|P) = \Pi(X|P) = max_{d\in D}(P(d) \odot X(d))$$

were $\odot$ stands for any continuous t-norm. Therefore, if any quantifier is assumed to be between $\exists$ and $\forall$, a good way to give its associated measure to consider it as a mixture of the generalized possibility and necessity which correspond to both two limit quantifiers. This idea leads us to the following definition:

Definition 5.3 *Let X and P be two fuzzy set with the same domain. Let also $\mathcal{Q}$ be a coherent set of quantifiers, and let us assume it is established a set of [0,1] values associated to $\mathcal{Q}$ elements such that:*

i $\alpha_{\exists} = 1\ \alpha_{\forall} = 0$ **ii** $\forall Q, Q' \in \mathcal{Q}\ Q \succeq Q' \Rightarrow \alpha_Q \geq \alpha_{Q'}$

In these conditions we define:

$$\forall Q \in \mathcal{Q}\ \ \Gamma^Q(X|P) = \alpha_Q \Pi(X|P) + (1 - \alpha_Q) N(X|P)$$

were the generalized possibility and necessity are done by using a pair of dual t-norm and t-conorm.

The following properties of Γ_Q have a direct proof, by using the condition iii of above definition and the properties of Π and N.

Property 5.1

Γ_Q is actually a fuzzy measure, that is: $\forall X_1, X_2, P \in \mathcal{P}(D)\ \ X_1 \subseteq X_2 \Rightarrow \Gamma_Q(X_1|P) \leq \Gamma_Q(X_2|P)$

Property 5.2

Γ_Q keeps up the order relation being in $\mathcal{Q}$. That is: $\forall Q, Q' \in \mathcal{Q}\ \ Q \succeq Q' \Rightarrow \Gamma_Q(X|P) \geq \Gamma_{Q'}(X|P)$

Example 2

The following example shows the use of all these concepts in a practical case. Let us consider a coherent set $\mathcal{Q}$ of quantifiers which have the following membership function:

$$Q(x) = \begin{cases} 0 & \text{if } 0 \leq x < a_Q \\ (x - a_Q)/(b_Q - a_Q) & \text{if } a_Q \leq x < b_Q \\ 1 & \text{if } b_Q \leq x \leq 1 \end{cases}$$

where a_Q, b_Q are [0,1] values such that $a_Q \leq b_Q$. With this definition, the $\mathcal{Q}$ elements are characterized by a pair of real values. Concretely, $\exists$ corresponds to (0,0) and $\forall$ to (1,1). It can easily see that:

$$Q \succeq Q' \Leftrightarrow a_Q \leq a_{Q'} \text{ and } b_Q \leq b_{Q'}$$

gives us an adequate partial order relation on $\mathcal{Q}$, and that:

$$\alpha_Q = (b_Q - a_Q)/2 + 1 - b_Q$$

verifies all of conditions which were imposed in the definition 5.2 for these values.

Now let us return to the example 1 context by softening some of its hypotheses. Concretely we will assume that it is possible to have imprecise information about of what are the suppliers of a shipment, this implies to consider the attribute *P#* of the relation SHIPMENTS as fuzzy. Thus we will keep the example 1 tables corresponding to SUPPLIERS and PARTS and we will replace SHIPMENTS by the following one:

SHIPMENTS	
S#	*P#*
S1/0.8,S2/0.5	P1
S1	P2
S1/0.6,S3/0.9	P3
S1/0.7,S3/0.2	P4
S1	P5
S2/0.6,S1/0.7	P1
S2	P3
S2	P4
S3	P4
S4/0.7,S3/0.8	P1
S4	P2
S4/0.5,S1/0.6	P4

Let us consider the property "to be good" is applied to the supplier set and represented by the following fuzzy set:

$$P = \{S1/0.5, S2/0.7, S3/1\}$$

We want now to get those part which are supplied by: "all", "almost all" , "most", "a few", and "at least one" good suppliers, thus we will carry out the operations described in the definitions. For the sake of simplicity we have chosen the maximum and minimum as t-conorm and tnorm and the following values associated with the quantifiers have been considered

$\forall$ is defined by (1,1) and $\alpha = 0$, "almost all" by (0.7,0.9) and $\alpha = 0.2$, "most" by (0,1) and $\alpha = 0.5$, "a few" by (0.1,0.3) and $\alpha = 0.8$ and finally "at least one" is equivalent to $\exists$ and it is associated with (1,1) and $\alpha = 1$.

The results of these operations are the following ones:

Compression of SHIPMENTS	
$S\#$	$P\#$
S1/0.8 S2/0.6 S3/0.8 S4/0.7	P1
S1/1 S4/1	P2
S1/0.6 S2/1 S3/0.9	P3
S1/0.7 S2/1 S3/1 S4/0.4	P4
S1	P5

Final Results					
$P\#$	$\Gamma^{\exists}$	Γ^{a-few}	Γ^{most}	$\Gamma^{almost-all}$	$\Gamma^{\forall}$
P1	0.80	0.76	0.70	0.64	0.60
P2	0.50	0.40	0.25	0.10	0.00
P3	0.90	0.84	0.75	0.66	0.60
P4	1.00	0.94	0.85	0.74	0.70
P5	0.50	0.40	0.25	0.10	0.00

REFERENCES

[1] Prade H. and Testemale C. (1984)*Generalizing database relational algebra for the treatment of incomplete and uncertain information and vague queries* Information Science Vol. 34 pp. 115-143

[2] Trillas E. and Valverde L. (1985) *On implications and indistinguishability in the setting of Fuzzy Logic.* In R.R. Yager and J. Kacprzyk (eds.) Management Decision and Support Systems using Fuzzy Sets and Possibility Theory. North Holland

[3] Vila M.A., Cubero J.C., Medina J.M., Pons O. (1992) *A logic approach to fuzzy relational databases* To appears in International Journal on Intelligent Systems.

[4] Vila M.A., Cubero J.C., Medina J.M., Pons O. (1993) *On the use of a logical definition of fuzzy relational database* Proc. of The Second IEEE International Conference on Fuzzy Systems pp. 489-499.

[5] Yager R.R. (1988) *On ordered weighted averaging aggregation operators in multicriteria decision-making* IEEE Transc. on System, Man and Cybernetics Vol.18 (1) pp. 183-190

[6] Yager R.R. (1992) *Fuzzy Quotient Operators* Proc. of IPMU'92 pp. 317-322

[7] Zadeh L.A. (1983) *A Computational approach to fuzzy quantifiers in natural languages* Computing and Mathematics with Applications Vol. 9 pp. 149-184

[8] Zemankova-Leech M., Kandel A. (1984) Fuzzy Relational Databases: A key for Expert Systems Verlag TUV Rheiland

ON THE EFFICIENCY OF THE ALPHA-CUT DISTRIBUTION METHOD TO EVALUATE SIMPLE FUZZY RELATIONAL QUERIES

Patrick BOSC, Olivier PIVERT

IRISA-ENSSAT
BP 447 22305 LANNION Cédex
FRANCE
e-mail : bosc@enssat.fr, pivert@enssat.fr

ABSTRACT

In this paper, we consider the evaluation of queries addressed to usual relations and involving imprecise predicates. The queries considered here are modelled as alpha-cuts of complex fuzzy predicates and we study to what extent query processing can be at least partly performed by a regular database management system. In this respect, we show that any tuple must comply with some Boolean conditions in order to belong to the alpha-cut of a compound predicate. The transformation process looks like a rewriting mechanism based on rules. As a matter of fact, some selected tuples do not really belong to the desired alpha-cut and measures, reported here, were made in order to estimate the difference depending on the connectors and the predicates appearing in the query.

1. Introduction

In the database domain, one important issue presently concerns the extension of the facilities and capacities of querying the user is provided with. As a matter of fact, database management systems currently available suffer several limitations regarding user or application needs. In particular, it is assumed that data are precisely known (or fully unknown) and queries are based on crisp conditions. As far as this second aspect is concerned, we have been suggesting to allow for imprecision inside user queries [2]. More specifically, we propose a language including imprecise conditions that are interpreted according to the fuzzy set theory. An SQL-like language has been designed that permits conditions to apply to both individual tuples and sets of tuples [3].

In [4], this powerful language was shown to be as expressive as most of the Boolean based systems aiming at imprecise querying [9, 11] but an important point concerns the evaluation of such queries. In conventional DBMS's, the problem of query evaluation remains somewhat open since given a query, in general the optimal evaluation way cannot be reached. We do not intend to solve the problem in general for fuzzy queries which would be unrealistic since the process cannot become easier for two reasons : i) the available access paths cannot be directly used, and ii) a larger number of tuples is selected by fuzzy conditions with respect to Boolean ones. We will rather consider a subset of queries for which an "efficient" processing can be found out in order to design a method that is expected to afford acceptable performances.

In this context, it appears interesting and useful to understand the connections which exist between properties tied to regular (Boolean) conditions and fuzzy ones, so that fuzzy query processing can come down to Boolean query processing (at least partly). In this study, we will assume that the elements stored in the database are precisely known and can be seen as tuples whose field take their values in ordinary universes. In the context of a language supporting imprecise querying such as SQLf [3], queries are viewed as fuzzy predicates. A query is associated to a threshold α and the retrieved data are the elements of the α-cut. The basic idea is to discard by means of a Boolean condition as many tuples as possible provided they do not compete to acceptable resulting tuples. The ideal situation is reached if we succeed in the determination of exactly those tuples that are the desired α-cut. Then, fuzzy query processing can be naïve without damage for performances. But generally, this strategy produces extra tuples with respect to the desired α-cut. So, we had to perform measures in order to predict the difference.

The condition for the validity of the process lies in the fact that the fuzzy query must lead to the same result as if it were processed on the original set of tuples and this is similar to equivalences between queries [1]. We consider relational fuzzy queries, which are called "simple", i.e. involving only one projection and a number of selections and joins, known as PSJ queries. Since selection and join are basic algebraic operations, Boolean PSJ queries have received attention and it seems interesting to consider their extension to the fuzzy case.

The paper is structured as follows. Firstly, we present the principle used to derive a Boolean query starting with a fuzzy one. The next section contains the Boolean expressions obtained for different atomic and compound predicates. Then, we present a theoretical study about the performances of the derivation rules associated with a particular class of aggregation operators, i.e. the means. In the last section, we set out the practical results obtained with a benchmark, and the influence of several parameters on the global efficiency of the derivation method is brought to the fore.

2. The derivation principle

The idea which will be developed in the rest of the paper is situated in the scope of the evaluation of (possibly compound) fuzzy predicates for a given threshold α. We will show that any considered fuzzy predicate P can be associated with a Boolean predicate that defines a superset of the elements which satisfy P at a degree over α (sometimes this set can be reached exactly). If one considers a fuzzy predicate P applying to a set X of tuples, we suggest the construction of a Boolean query called envelope $E = P [d \geq \alpha]$ which delivers a set X' involving at least all the tuples of X whose grade d is over α. It should be noted that due to the fact that predicates can be negated, the case $P [d \leq \alpha]$ has to be considered too. We can design a set of transformation (derivation) rules to construct the envelope. Basically, the objective is to express the envelope for P, which may be a fuzzy selection or fuzzy join predicate, as a Boolean combination of those of the components of P.

Let us consider a fuzzy query Q applying to a base B made of relations $\{R_1, \ldots, R_n\}$ whose extensions are denoted $\{S_1, \ldots, S_n\}$ and form a database state S. We can define a partial order (≤) over the database states in the following way : $S \leq S' \Leftrightarrow \forall i \; card(S_i) \leq card(S'_i)$. The idea is to change the initial database state S into S' by means of a Boolean predicate BP such that : i) $S \leq S'$, ii) $Q(S) = Q(S')$ i.e the result of the query Q is preserved, and iii) BP is a condition derived from a fuzzy predicate FP applying to a given relation R_i (or two relations in case of a fuzzy join). Successive changes can be performed using predicates connected to the various relations implied in a query Q.

A derivation rule tells that we can replace the expression A by the expression B and two cases occur depending on whether the set of tuples selected by A is the same as (or a superset of) those selected by B. In the first situation, we have a **strong rule** denoted **A ↔ B** (for any tuple t, $A(t) \Leftrightarrow B(t)$, i.e B is a condition **equivalent** to A), otherwise the rule is said to be **weak** and is written **A → B** (for any tuple t, $A(t) \Rightarrow B(t)$, i.e B is a **necessary condition** for A). When only strong rules are used, a minimal state is reached, i.e a state in which every tuple participates in the construction of the answer. On the contrary, if at least one weak rule is applied, the state that is built is potentially reduced but not minimal. This point is obviously important since in the worst case, step by step, there is a risk for the method to deliver the initial set if only weak rules are applied. The situations likely to happen for weak rules are detailed in the following.

3. Some derived expressions

Various kinds of predicates can appear in an extended relational query and we successively describe the Boolean expressions tied to atomic, modified and complex fuzzy predicates.

3.1. Atomic and modified predicates

Let us recall that according to our assumption, we want to characterize the x's of a set X whose degree is over a threshold α with respect to a predicate P. First of all, we consider the case where P is an elementary predicate represented by a one parameter increasing-decreasing membership function (trapezium or triangle for instance). In this context, the condition $P(x) \geq \alpha$ is equivalent to "x belongs to an interval" and then we have the rule :

$$P\,[d \geq \alpha] \leftrightarrow x \in [v_1, v_2] \quad (1).$$

For a negation, we can apply the following rule :

$$((\text{not } P)(x))[d \geq \alpha] \leftrightarrow x \notin [v_1, v_2].$$

We now define the rules which may apply for usual modified predicates. Consider the predicate "mod P" where mod is a modifier interpreted as an exponential operator [10, 15]; the condition $((\text{mod } P)(x)) \geq \alpha$ is equivalent to $(P(x))^h \geq \alpha$ and then we have the two rules :

$$(\text{mod } P)[d \geq \alpha] \leftrightarrow P\,[d \geq \alpha^{1/h}];$$
$$(\text{mod } P)[d \leq \alpha] \leftrightarrow P\,[d \leq \alpha^{1/h}] \quad (2).$$

Other modifiers like powers of fuzzy sets [14], translations [6] and antonyms [13] will not be detailed here but the associated rules can be found in [5]. It can be noted that in all cases, the rules obtained are strong.

3.2. Complex predicates

First, we present the derived expressions tied to the usual connectors (AND/OR), then we will deal with mean operators. We specify the envelopes related to both $P\,[d \geq \alpha]$ and $P\,[d \leq \alpha]$.

If a predicate P is a conjunction $\text{AND}(P_1,\ldots,P_n)$, we have $P(x) \geq \alpha \Leftrightarrow \min(P_1(x),\ldots,P_n(x)) \geq \alpha$ and the rule is :

$$P\,[d \geq \alpha] \leftrightarrow P_1\,[d \geq \alpha] \textbf{ and } \ldots \textbf{ and } P_n\,[d \geq \alpha] \qquad (3).$$

Similarly, we can get the rule :

$$P\,[d \leq \alpha] \leftrightarrow P_1\,[d \leq \alpha] \textbf{ or } \ldots \textbf{ or } P_n\,[d \leq \alpha].$$

In case of a disjunction $P = \text{OR}(P_1, \ldots, P_n)$:

$$P(x) \geq \alpha \Leftrightarrow \max(P_1(x), \ldots, P_n(x)) \geq \alpha$$

and the rules are :

$$P\,[d \geq \alpha] \leftrightarrow P_1\,[d \geq \alpha] \textbf{ or } \ldots \textbf{ or } P_n\,[d \geq \alpha];$$
$$P\,[d \leq \alpha] \leftrightarrow P_1\,[d \leq \alpha] \textbf{ and } \ldots \textbf{ and } P_n\,[d \leq \alpha].$$

Although not developed here, strong rules exist for the weighted-AND/OR [8].

Example.

Let us consider the condition "not young AND very well-paid" with $\alpha = .6$, where very P is interpreted as P^2 and the membership functions for young and well-paid are drawn below.

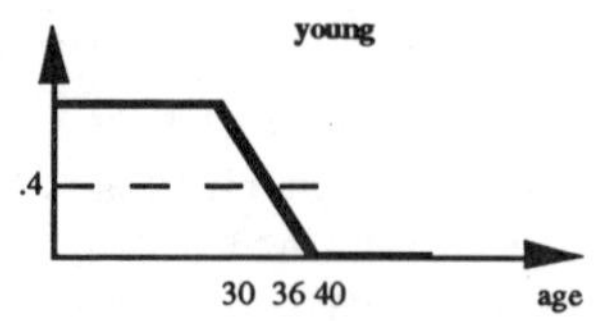

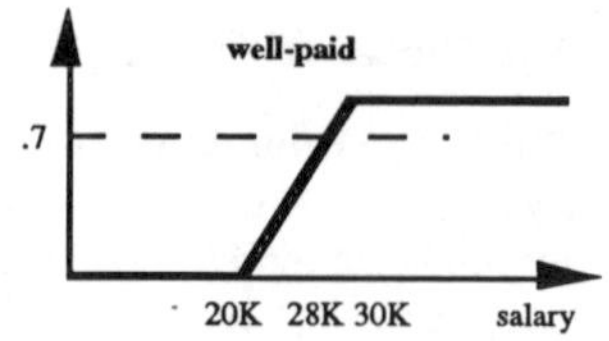

We have to derive a Boolean condition for : (not young AND very well-paid) [d ≥ .6].

According to rule (3), we have :

(not young AND very well-paid)[d ≥ .6] ↔ (not young)[d ≥ .6] **and** (very well-paid)[d ≥ .6].

We can derive : (not young)[d ≥ .6] ↔ young [d ≤ .4]

and according to rule (2) :

(very well-paid)[d ≥ .6] ↔ well-paid [d ≥ .77].

Finally, using rules of type (1), we have a Boolean expression which is equivalent to the initial α-cut:

age≥ 36 **and** salary ≥ 28.

For the operators belonging to the class of "means", that allow to modelize different kinds of compromise between predicates, only weak rules are available. Nevertheless, some specific properties tied to each different mean allow the definition of more or less restrictive rules. In this case, it is an important matter to study the performances induced by these rules i.e. the proportion of extra tuples that risk to be obtained. In the next section, we present a theoretical study of these rules, which is based on a geometrical interpretation.

4. Derivation of mean operators : a theoretical study

4.1. Introduction

In order to study the efficiency of the derivation rules for the means, we propose to represent their behaviour in the particular case where each mean is used to aggregate two fuzzy predicates. This special case is considered only for the sake of clarity of the diagrams. The idea is to evaluate the proportion of selected extra tuples by means of a geometrical representation of the rules. A tuple will be represented by a point of a plane. Each predicate defines one of the coordinate of the point ($x = \mu_{P_1}(t)$, $y = \mu_{P_2}(t)$). The points corresponding to the tuples all belong to the unit square ($\mu_{P_i}(t) \in [0, 1]$). A projection of the α-cuts in the unit square allows to distinguish three interesting areas: i) the restriction area, where the tuples are not selected by the rule, ii) the satisfaction area where the tuples reach the initial threshold, iii) the error area where the tuples do not reach the initial threshold but are selected by the rule. The satisfaction area and the error area are separated by the boundary of the α-cut. In order to simplify the interpretation, we will assume here that the distributions of values are uniform. In this case, the intrinsic efficiency of a rule is directly tied to the size of the error area. It is then interesting to evaluate this size for each weak derivation rule.

4.2. General rules for the means

The operators belonging to the class of "means", denoted mc, have two general properties. The first one is : $\min(P_1(x), \dots, P_n(x)) \leq mc(P_1, \dots, P_n)(x) \leq \max(P_1(x), \dots, P_n(x))$. As a consequence, the following two weak rules hold for any mean :

$$mc(P_1, \dots, P_n)[d \geq \alpha] \rightarrow P_1[d \geq \alpha] \textbf{ or } \dots \textbf{ or } P_n[d \geq \alpha] \quad (4)$$

$mc(P_1, \dots, P_n)[d \leq \alpha] \rightarrow P_1[d \leq \alpha]$ **or** ... **or** $P_n[d \leq \alpha]$.

In this case, we can only determine the restriction area, because the boundary of the α-cut obviously depends of the kind of mean which is used.

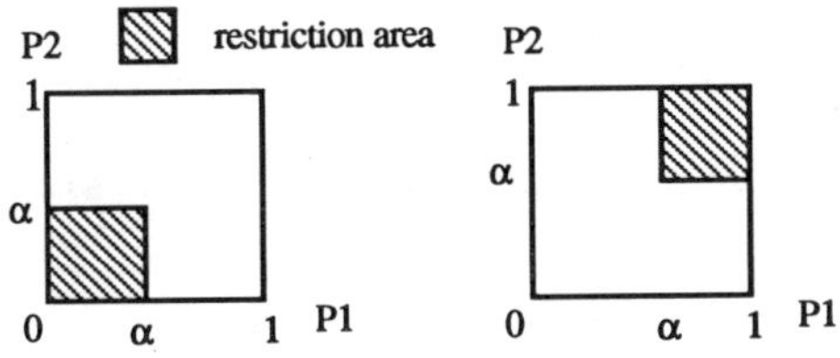

$P_1[d \geq \alpha]$ **or** $P_2[d \geq \alpha]$ $P_1[d \leq \alpha]$ **or** $P_2[d \leq \alpha]$

Fig. 1 : General derivation rules for the means.

The size of the restriction area is a function of α which is increasing in the first case, decreasing in the second one. With more than two predicates, this area becomes a cube or an hypercube. Its size relatively to the global volume will then decrease when the number of predicates will increase. So, one can foresee that these rules will not be very efficient in case of many predicates to be aggregated.

4.3. <u>Specific rules for the means</u>

The second property of the means is monotonicity : $mc(x_1, \dots, x_n) \geq mc(x'_1, \dots, x'_n)$ iff $\forall i\ x_i \geq x'_i$. According to this, we will point out some more specific rules for the geometric mean, the weighted mean which subsumes the arithmetic mean and then the ordered weighted averaging aggregation [12].

4.3.1. The geometric mean

We have : $gm(P_1, \dots, P_n)(x) \geq \alpha \Leftrightarrow (P_1(x) * \dots * P_n(x))^{1/n} \geq \alpha \Rightarrow \forall i\ P_i(c) \geq \alpha^n$.

The rule is : $gm(P_1, \dots, P_n)[d \geq \alpha] \rightarrow P_1[d \geq \alpha^n]$ **and** ... **and** $P_n[d \geq \alpha^n]$.

No specific rule was found for $gm(P_1, \dots, P_n)[d \leq \alpha]$.

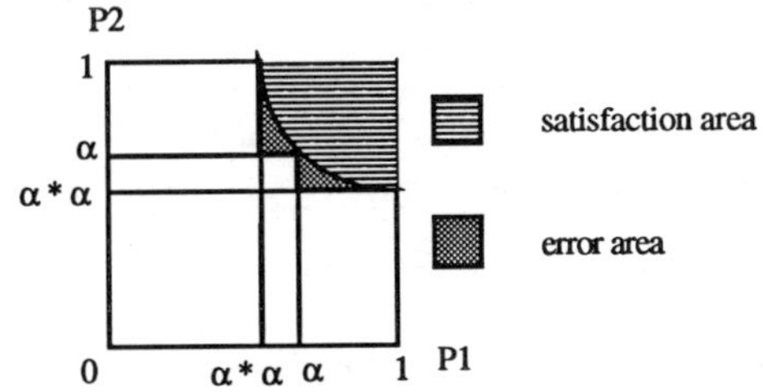

($P_1[d \geq \alpha^2]$ **and** $P_2[d \geq \alpha^2]$) **and** ($P_1[d \geq \alpha]$ **or** $P_2[d \geq \alpha]$)

Fig. 2 : Specific derivation rule for the geometric mean.

It can be noted that the error area is relatively small which means that this specific rule is quite restrictive. This result will be verified in the experimental study.

4.3.2. The weighted mean

We have : $wm(w_1, \dots, w_n, P_1, \dots, P_n)(x) \geq \alpha \quad \Leftrightarrow (w_1 * P_1(x) + \dots + w_n * P_n(x)) \geq \alpha$
$\Rightarrow \forall i \ \ w_i * P_i(x) + (1 - w_i) \geq \alpha.$

Then, the rule (5) is :

$wm(w_1, \dots, w_n, P_1, \dots, P_n)[d \geq \alpha] \rightarrow P_1[d \geq (\alpha+w_1-1)/w_1]$ **and** ... **and** $P_n[d \geq (\alpha+w_n-1)/w_n]$.

Similarly, we have the rule :

$wm(w_1, \dots, w_n, P_1, \dots, P_n)[d \leq \alpha] \rightarrow P_1[d \leq \alpha/w_1]$ **and** ... **and** $P_n[d \leq \alpha/w_n]$.

These rules will be useful only if ($\alpha > 1 - w_1$ or ... or $\alpha > 1 - w_n$) for the first one, and if ($\alpha <$ w1 or ... or $\alpha <$ wn) for the second one. Otherwise, all the derived conditions would be trivially satisfied.

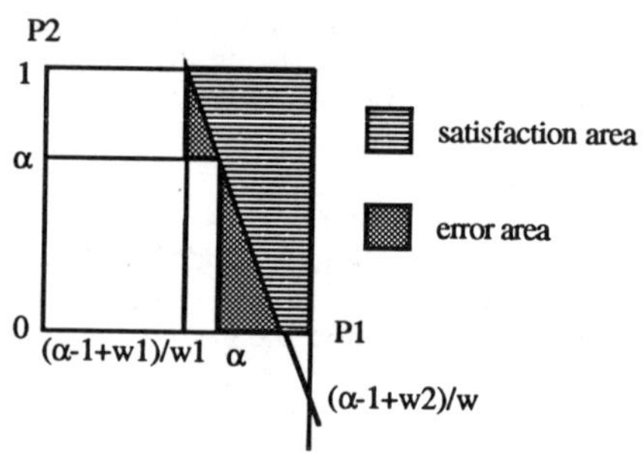

$P_1[d \geq (\alpha+w_1-1)/w_1]$ **and** $P_2[d \geq (\alpha+w_2-1)/w_2]$ **and** ($P_1[d \geq \alpha]$ **or** $P_2[d \geq \alpha]$)

Fig. 3 : A specific derivation rule for the weighted mean.

The ratio w_1/w_2 gives the slope of the limit of the α-cut.

The rules valid for the arithmetic mean are obtained with the previous formulae with $w_i = 1/n \ \forall i$. So, we have :

$am(P_1, \dots, P_n)[d \geq \alpha] \rightarrow P_1[d \geq (n*\alpha + 1 - n)]$ **and** ... **and** $P_n[d \geq (n*\alpha + 1 - n)]$
$am(P_1, \dots, P_n)[d \leq \alpha] \rightarrow P_1[d \leq n*\alpha]$ **and** ... **and** $P_n[d \leq n*\alpha]$.

4.3.3. The Ordered Weighted Averaging operator (OWA)

$owam(w_1, \dots, w_n, P_1, \dots, P_n)(x) \geq \alpha \quad \Leftrightarrow (w_1 * P_{k_1}(x) + \dots + wn * P_{k_n}(x) \geq \alpha$
$\Rightarrow \forall i \ \ w_1 + \dots + w_{n-1} + w_n * P_{k_i}(x) \geq \alpha;$
$\Rightarrow \forall i \ \ w_n * P_{k_i}(x) + (1 - w_n) \geq \alpha$

where $P_{k_i}(x)$ denotes the ith greatest value among the $P_i(x)$'s and the sum of the w_i's equals 1. Then, the rule is :

$owam(w_1, \dots, w_n, P_1, \dots, P_n)[d \geq \alpha] \rightarrow P_1[d \geq (\alpha + w_n - 1)/wn]$ **and** ...
and $P_n[d \geq (\alpha + w_n - 1)/w_n]$.

Similarly, we get : $owam(w_1, \ldots, w_n, P_1, \ldots, P_n)[d \leq \alpha] \rightarrow P_1\ [d \leq \alpha/w_1]$ **and** ... **and** $P_n\ [d \leq \alpha/w_1]$.

The first rule is useful only if $\alpha > 1 - w_n$, and the second one only if $\alpha < w_1$.

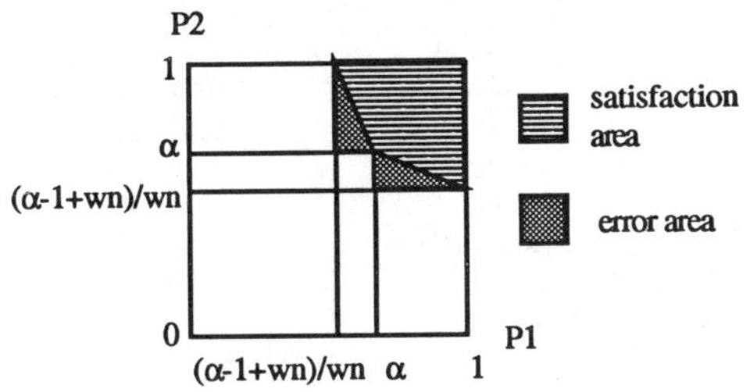

$P_1\ [d \geq (\alpha + w_2 - 1)/w_2]$ **and** $P_2\ [d \geq (\alpha + w_2 - 1)/w_2]$ **and** $(P_1\ [d \geq \alpha]$ **or** $P_2\ [d \geq \alpha])$

Fig. 4 : A specific derivation rule for the OWA operator.

Here again, the efficiency depends on the distribution of the weights. When w1 equals 0, the OWA operator is the min and the specific rule becomes strong. When w1 equals 1, the OWA is the max, and the specific rule is no more valid but the general rule becomes strong. These two cases correspond to optimal efficiency. In-between, we reach at worst the efficiency corresponding to the derivation of the arithmetic mean.

5. Simulation and measures

5.1. Experimental protocol

In order to have a more precise idea about the difference between the result (issued from the envelope) and the pure α-cut, we performed some measurements based on a simulation. The parameters considered here are : i) the type of fuzzy connectors, ii) the value of the threshold α, iii) the number of fuzzy predicates, iv) the selectivity of the fuzzy predicates, v) the distribution of the data. We worked with a 5000 tuples relation EMPLOYEE(#emp, #dep, age, salary, commission, sales) very close to that of DeWitt's benchmark [7]. The values were first assumed to be uniformly distributed for each attribute which take respectively, #dep : 10 values, age : 40 values, salary : 105 values, commission : 45 values and sales : 70 values. Let us recall that the selectivity of a predicate P measures the proportion of tuples that satisfy P i.e. for an attribute A with a uniform value distribution, we have $sel(P) = \Sigma_{x \in domain(A)}\ \mu_P(x)/|domain(A)|$. In this experiment performed using the RdB database management system, different operators (arithmetic, geometric, weighted mean and OWA) connecting several fuzzy predicates (young, well-paid, high sales, high commission) were considered. The ratio r = (n2 - n1) / n where n2 is the number of tuples selected by the derived condition, n1 is the number of elements of the pure α-cut, and n is the cardinality of the relation (5000) is taken as representing the behaviour of a query.

5.2. Variations according to the connector and the number of predicates

The results obtained are drawn below :

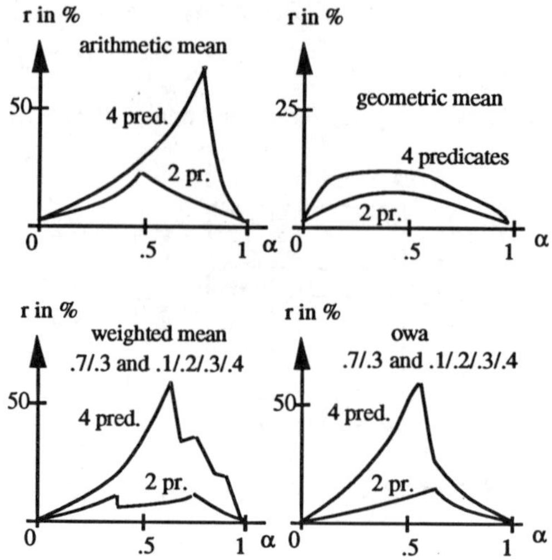

Fig. 5 : Influence of the number of predicates on the performances.

One can note that r remains small for the geometric mean even with four predicates. This is because the specific derivation rule is valid whatever the value of the threshold. The performances concerning the arithmetic mean, the weighted mean and the OWA operator are close to each other. The curves clearly show the different points where the specific rules become available. For example, in the case of the arithmetic mean with two predicates, the specific condition applies for $\alpha \geq .5$, point where the curve decreases significantly. An interesting fact is that the specific rules are not very sensitive to the number of arguments, while the general rule is. Globally, it appears that an increase of the number of parameters implies a significant degradation of the performances. Nevertheless, this bad result can be mitigated by noting that the most frequent queries involve a high threshold, and that in the area close to 1, the rules all have a very good efficiency.

5.3. <u>Influence of the selectivity</u>

In order to evaluate the influence of the selectivity, we used the condition am(young, well-paid) with different selectivity values for the fuzzy predicate well-paid. The effect of the selectivity is especially important on the general rule (cf 4.2). This result can be explained with the help of a geometrical interpretation

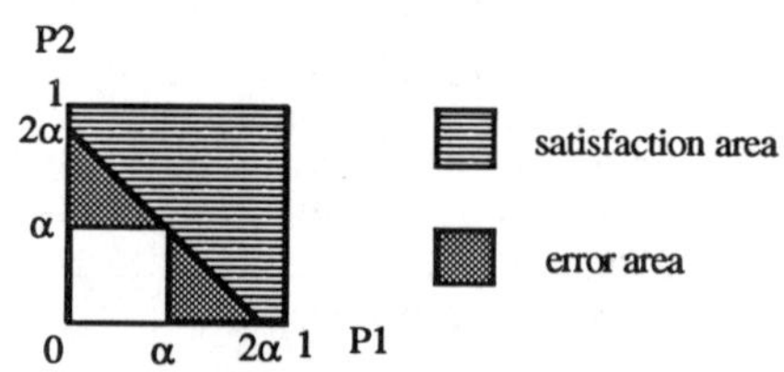

Fig. 6 : General derivation rule for the arithmetic mean.

The general rule is more efficient near the axis joining (0, 0) and (1, 1) than it is near the coordinate axis. When a lot of tuples have a zero degree for at least one predicate, the density in the error area is high. The resulting curve is drawn below :

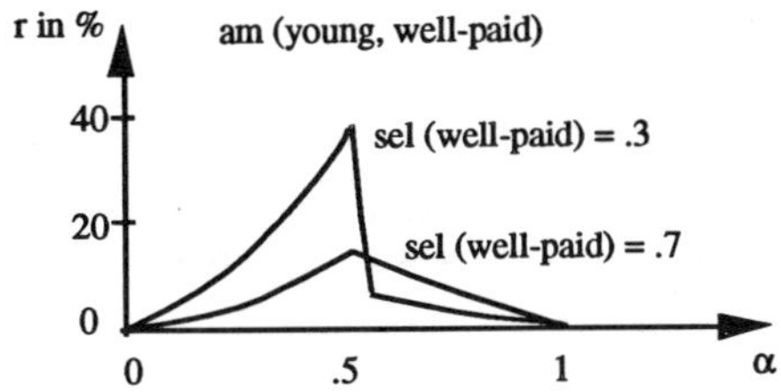

Fig. 7 : Influence of the selectivity (case of the arithmetic mean).

The conclusions are :
i) the shape of the curves does not change significantly,
ii) the value of the extremum increases (from 20 to 40% in this case) and
iii) the breaking when a specific condition becomes valid is more stressed.
For other connectors, one can expect the same kind of behaviour. In each case, if the proportion of tuples is high in the area where the rule is not particularly efficient, then the performances will be low.

5.4. Influence of the data distribution

To perform the preceding measures, we used data whose values where uniformly distributed on the related domains. Unfortunately, this case is not very realistic. The most common distribution actually are exponential or normal. So, we decided to perform other measures in order to delimit the influence of this parameter. We evaluated again the condition am(young, well-paid) with a uniform distribution for the attribute salary and different kinds of distributions for the attribute age. Each curve obtained was then compared to the one corresponding to a uniform distribution for the two attributes.The effect observed is the same in all cases. The performances deteriorate when the density of the tuples increases in the error area. Globally, this parameter has a slight influence on the performances. Its influence depends on the zone where the data is concentrated. If most of the data have a membership degree close to 0 (resp. 1), the performances will be better (resp. worst) than in the uniform distribution case (cf geometrical interpretation).

This result can be generalized to the other mean operators. In each case, the performances vary according to the distribution of the tuples in the zones where the rule is more or less efficient.

6. Conclusion

In this paper, we have been dealing with the evaluation of complex fuzzy predicates in the scope of a database management system. One interesting objective is to use a regular DBMS to process at least partly imprecise queries. We presented a mechanism called derivation, based on the distribution of the α-cut of a complex predicate over its components, thus providing Boolean conditions. If this approach works for a wide range of complex predicates, we have shown that very often the initial α-cut and the obtained Boolean condition were not equivalent, since the latter represents only a necessary condition with respect to the former. In order to its effective and efficient use, it is important to estimate the difference between the desired result (the α-cut) and the result of the derived Boolean condition. To this end, some measures concerning different mean operators were performed which show that this difference remains small when only two predicates are aggregated, but can be over 70% for four predicates. In the near future, these measures should be extended to a wider range of operators (t-norms/t-conorms, quantifiers, etc) for which weak derivation rules exist.

REFERENCES

[1] Bernstein P. A. and Chiu D. M. W. (1981) Using semi-joins to solve relational queries, Journal of the ACM, 28, 1, 25-40.
[2] Bosc P., Galibourg M., Hamon G. (1988) Fuzzy querying with SQL : extensions and implementation aspects, Fuzzy Sets and Systems, 28, 333-349.

[3] Bosc P., Pivert O. (1991) About equivalences in SQLf, a relational language supporting imprecise querying, Proc. IFES, Yokohama (Japan), 309-320.
[4] Bosc P., Pivert O. (1992) Discriminated answers and databases : fuzzy sets as a unified expression means, Proc. 1st International IEEE Conference on Fuzzy Systems FUZZ-IEEE'92, San Diego (USA), 745-752.
[5] Bosc P., Pivert O. (1993) On the evaluation of simple fuzzy relational queries : principles and measures, in: Fuzzy Logic: State of the Art (R. Lowen & M. Roubens eds.), Kluwer Academic Publishers, 355-364.
[6] Bouchon-Meunier B., J. Yao (1992) Linguistic modifiers and imprecise categories, Journal of Intelligent Systems, 7, 25-36.
[7] DeWitt D.J., Bitton D., Turbyfill C. (1983) Benchmarking database systems - A systematic approach, Proc. VLDB Conference, 8-19.
[8] Dubois D., Prade H. (1986) Weighted minimum and maximum operations in fuzzy set theory, Information Sciences, 39, 205-210.
[9] Ichikawa T., Hirakawa M. (1986) ARES : a relational database with the capability of performing flexible interpretation of queries, IEEE Transactions on Software Engineering, 12, 5, 624-634.
[10] Lakoff G. (1973) Hedges : a study in meaning criteria and the logic of fuzzy concepts, Journal of Philosophical Logic, 2, 458-508.
[11] Motro A. (1988) VAGUE : a user interface to relational databases that permits vague queries, ACM Trans. on Off. Inf. Syst., 6, 3, 187-214.
[12] Yager R.R. (1988) On ordered weighted averaging aggregation operators in multicriteria decisionmaking, IEEE Trans. on Syst., Man and Cyber., 18, 183-190.
[13] Yager R.R. (1991) Connectives and quantifiers in fuzzy sets, Fuzzy Sets and Systems, 40, 39-75.
[14] Yong-Yi C. (1981) An approach to fuzzy operators, Busefal, 9, 59-65.
[15] Zadeh L.A. (1972) A fuzzy-set-theoretic interpretation of linguistic hedges, Journal of Cybernetics, 2, 4-34.

AGGREGATE ATTRIBUTES WITH FUZZY VALUES FOR MORE FLEXIBILITY IN INFORMATION RETRIEVAL

Pascal SUBTIL, Noureddine MOUADDIB, Odile FOUCAUT

CRIN / CNRS URA 262
B.P. 239
54506 Vandœuvre-lès-Nancy Cedex - FRANCE
Tel : (33) 83 59 20 00
Fax : (33) 83 41 30 79
e-mail : {foucaut,mouaddib,subtil}@loria.fr

Abstract

In the present work we will attempt an extension of an object oriented data model:

- to describe fuzzy objects with a new attribute type that we termed fuzzy complex attribute. A complex attribute can be either calculated (i.e., its value is calculated from values of other attributes) or aggregate (i.e., it is composed by a set of attributes). All manipulated values can be fuzzy.
- to extend the retrieval process with the new attribute type in order to allow more flexibility in queries.

The approach we propose is illustrated by a real example in human resources management.

Introduction

It is known that the main lack in the actually databases management system (DBMS) is their flexibility. For instance, the user must know the data model used to describe objects and he must express his research profile (or query) according to this model.

Well, in some cases, it is not possible to describe the objects in database with the same structure because attributes aren't known with the same precision from an object to another. In opposite, two users express their research profile with different precision because they haven't the same perception of reality.

This problem is not new and a number of techniques exist for resolve it [1][2][3] and softwares allowing more flexible queries too.

In our precedent studies [4][5][6], we were taking an interest in flexible query from a database of nuanced objects. We proposed :

- a model to represent nuanced information;
- a progressive and interactive process to assist the user to identify a nuanced object (profile) among a set of nuanced known objects.

In these studies, profile and objects are described with the same structure, i.e. with the same set of attributes, as in the most of DBMS. We used these solutions to realise a Fuzzy Information Retrieval and Management System (FIRMS) [7] in Smalltalk.

In the present work, we provide an extension to the FIRMS data model to take into account complex attributes (calculted and aggregate). This extension is also valid in an object oriented DBMS (OODBMS).

So, the identification process is extended with the new attribute types to allow more flexibility in profiles (queries).

In the following paragraphs of this article, we firstly give our data model where we present our definition of complex attributes and their domain, also we compare our approach with linguistic variables [8][9][10]. We secondly show how to retrieve fuzzy objects describing with complex attributes. In the last paragraph, we illustrate our approach by a real example in human resources management.

This work is based on possibility theory [11][12] and fuzzy sets theory [13].

1 Data models

1.1 Attributes Types

An attribute (name, age, ...) can have three different forms : basic attribute, calculated attribute and aggregate attribute. Briefly, we explain these three different kinds of attributes.

- Basic attribute:

  A basic attribute is an attribute which can't be calculated from the values of other attributes or composed by other attributes. It's the case when user must provide the attribute value. For instance, name of somebody, turnover for a firms, ... Number of feminine strength can be view as a basic attribute too, but if we consider its value is calculated, for instance from total strength and masculine strength, then it isn't a basic attribute.

  A basic attribute is represented by a couple $(a, Dom(a))$ where a is the name of attribute and $Dom(a)$ is its domain.

- Calculated attribute

  An attribute a is called calculated attribute if there exists a function v_a permitting to calculate its value. We represent such attributes by a triplet $(a, Dom(a), v_a)$.

  Example : feminine strength = total strength - masculine strength.

- Aggregate attribute

  An aggregate attribute a [14] is an attribute composed with a set of attributes. This set is noted $Agg(a)$. To each aggregate attribute a we associate a function, note v_a, to aggregate the attributes values of $Agg(a)$ for giving the value of a.

  The representation of an aggregate attribute a is a quadruple $(a, Dom(a), v_a, Agg(a))$.

  Obviously, the elements of set $Agg(a)$ can be either basic attributes, calculated attributes or aggregate attributes.

  Example : the knowledge level of a language can be decomposed with speak level, read level and write level. The global level can be the result of the average from these three levels.

1.2 Domains

The domain of an attribute is the set of its possible nuanced values. A nuanced value is a couple (v, n) where v is a crisp, imprecise or vague value and n is a nuance which either expresses uncertainty (e.g., probably, ...) or linguistic quantifiers/modifiers [15] (e.g., very, often, ...). We explain these notions in the following.

First, we remind of fuzzy set concept. A fuzzy set A in X is characterized by a membership (characteristic) function $\mu_a(x)$ which associates with each element in X a real number in the interval $[0, 1]$, with the value of $\mu_a(x)$ at x representing the "grade of membership" of x in A [13].

A crisp value v of an universe U can be viewed as a classical set reduced to one element and characterized by:

$$\mu_v(x) = \begin{cases} 1 & \text{si } x = v \\ 0 & \text{sinon} \end{cases}$$

An imprecise value v of an universe U can be viewed as a classical set characterized by the following function : $\mu_v(x) = 0$ or 1 and $\exists (x_1, x_2) \in U^2$, $x_1 \neq x_2$ $\mu_v(x_1) = \mu_v(x_2) = 1$. In other words, the set contains two elements at least.

A vague value v of an universe U is defined by a fuzzy set and characterized by the following function: $\mu_v(x) \in [0, 1]$ such that:

- $\exists x \in U \quad \mu_v(x) = 1$;
- $\exists x \in U \quad 0 < \mu_v(x) < 1$.

The above definitions are valid in a query. In an object description, we characterize the values by possibility distributions [12]. A possibility distribution allows to indicate the degree of possibility to each element, but only one of these element is exact for the value. We note π_v the possibility distribution associated with the value v.

A nuance n is characterized by a function f_n defined on the set of characteristic functions, noted CF. If a (crisp or imprecise or vague) value v is characterized by μ_v then the characteristic function of n on v is $f_n(\mu_v)$.

See four examples about these notions :

- example 1 : "John is 175cm tall"

  175cm is a precise value represented by the function (figure 1):

$$\pi_{175}(x) = \begin{cases} 1 & \text{if } x = 175 \\ 0 & \text{else} \end{cases}$$

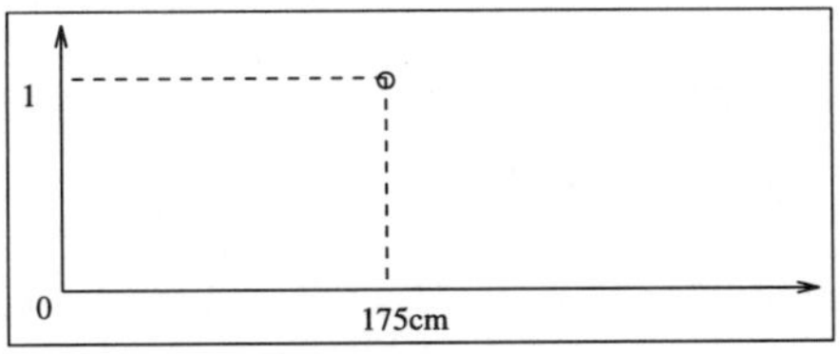

Figure 1: 175 cm

- example 2 : "Height of John is between 172 and 178cm"

  "Between 172 and 178cm" (or [172,178]) is an imprecise value represented by the following function (figure 2):

$$\pi_{[172,178]}(x) = \begin{cases} 1 & \text{if } 172 \leq x \leq 178 \\ 0 & \text{else} \end{cases}$$

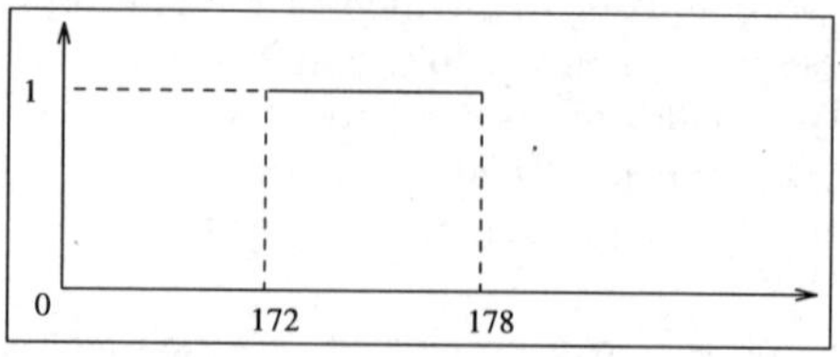

Figure 2: Between 172 and 178 cm

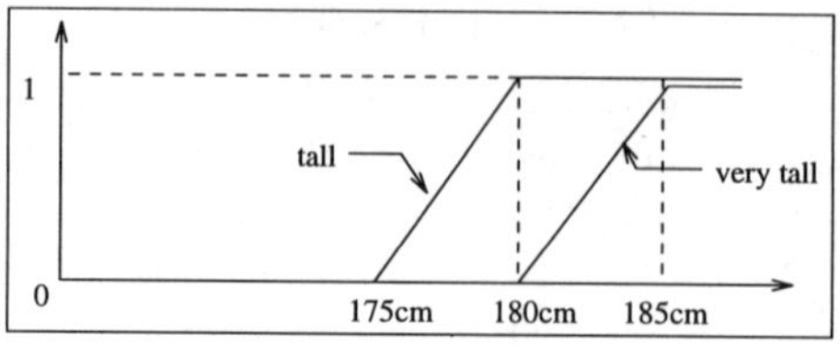

Figure 3: Tall

- example 3 : "John is tall"

  "Tall" is a vague value represented by the following function: (figure 3).

- example 4 : "John is very tall"

  "Very tall" is a nuanced value, (tall, very), defined by (figure 3):

$$(f_{very}(\pi_{tall}))(x) = \pi_{tall}(x - 5)$$

Let us now present our main result of this study which provides a general definition to represent domain of a nuanced attribute a:

$$Dom(a) = \{(na, \eta)\} \cup Refbase \times Nbase[\cup_i E_i \times N_i]$$

where

- na is a term meaning "not applicable" closes to the concept of "extra element" used by Dubois and Prade [12];
- η is a particular nuance meaning "no applied nuance" ($f_\eta(x) = x$);
- $Refbase$ is the basic universe of domain, i.e. a set of crisp values;
- $Nbase$ is a set of possible nuances on the elements of $Refbase$, this set contains η at least;
- E_i are sets of vague values;
- N_i are sets of possible nuances on vague values of E_i, these sets contain η at least;
- brackets indicate an optional part.

Every element e of domain $Dom(a)$ is characterized by a function defined as :

$$\mu_e \; : \; \{na\} \cup Refbase \longrightarrow [0, 1]$$

The set $\{na\} \cup Refbase$ is the universe of attribute a, we note it $ref(a)$.

Example : Let $Dom(height)$ be the domain of the height of person. It is defined as :

$$\begin{array}{l}\{(na,\eta)\}\\ \cup[0,220cm]\times\{\eta,\text{about, more than, less than}\}\\ \cup\{\text{tall, small}\}\times\{\eta,\text{ very, enough}\}\\ \cup\{\text{medium}\}\times\{\eta,\text{ very}\}\end{array}$$

In the next, $\{(na,\eta)\}$ will omit in the domain extension.

1.3 Comparison with Linguistic Variable

A linguistic variable [8][9][10] is a tuple (V, X, T_V) where V is the name of variable (*height* for example), X is the universe of variable V ($[0, 220cm]$ for example), and T_V is the set of possible terms for V described by fuzzy set of X (165, 170 ... *tall*, *small*, ... for example).

Let M be the set of linguistic modifiers (*very*, *about* ... for example) such as $M(T_V)$ gives the fuzzy terms created by M from T_V (*very tall*, *very small* ... for example).

This approach allows to combine every linguistic variable of M with each possible terms of T_V, well, sometimes this association has not a semantic in reality. For example, very 165 for the variable attribute "height of person".

In our model, we can explicitly associate different nuances sets with different values sets allowing to express an exact semantic. For example, we can say the nuance "very" is applicable for the value "tall" but not for the numeric values as 175.

1.4 Object Representation

An object I in the database is described by a set of quadruples : $I = \{< a, nv, rel, agg >\}$ where a is an attribute, nv is a nuanced value, rel is a relevant degree in $[0, 1]$ showing the relevant of the attribute with regard the others, agg is a set of valued attributes in $Agg(a)$ as I. The notion of relevance is very important because, in the reality, an attribute can be more discriminating or more stable in time and, then, more relevant than an other.

1.5 Profile Representation

An profile description is the same as an object. Let PR be a profile. Its description is a set of quadruples : $PR = \{< a, nv, imp, agg >\}$ where a, nv, agg as in object description and imp is the importance degree of the criteria a with regard the others in the profile.

2 Identification Process in FIRMS

The aim of identification process is to provide a list of objects in database which satisfy, correspond to, a profile established by an user. This list is ordered by satisfaction degree. Now, see the mechanism to estimate this degree using possibility and necessity measures.

Consider a profile $PR = \{< a_i, nv_i, imp_i, agg_i >\}$ and an object in database $I = \{< a_j, nv_j, rel_j, agg_j >\}$.

Let n_i be a quadruple $< a_i, nv_i, imp_i, agg_i >$ and let n_j be a quadruple $< a_j, nv_j, rel_j, agg_j >$.

We note :

- $Att(n_i) = a_i$ and $Att(n_j) = a_j$
- $imp(n_i) = imp_i$ and $rel(n_j) = rel_j$
- $v(n_i) = nv_i$ and $v(n_j) = nv_j$

- $agg(n_i) = agg_i$ and $agg(n_j) = agg_j$
- $ref(n_i) = ref(Att(n_i))$ and $ref(n_j) = ref(Att(n_j))$

Let Γ be the set of attributes. Let Γ_0 be the set of attributes of first level, i.e. the set of attributes not in the composition of another attribute. $\Gamma_0 = \{a \in \Gamma \mid \nexists a' \in \Gamma \text{ such that } a \in Agg(a')\}$.

If N is a set of n_i or n_j then $\Upsilon(N)$ is the set of first level valued attributes in N, i.e. $\Upsilon(N) = \{Att(n) \mid n \in N\}$.

Let $select_N(a) = n$ such that $n \in N$, $Att(n) = a$ and $a \in \Gamma$, be the selection of element showing the value of a in N.

The compatibility of a profile with an object in database is measured in terms of possibility and necessity measures.

In the first step, we calculate proximities between the values of attributes in PR with the values of these same attributes in I. Next, we keep the minimum of these compatibilities. Then,

$$\Pi(PR, I) = \min_{a \in \Upsilon(PR)} \Pi_c(select_{PR}(c), select_I(c))$$
$$N(PR, I) = \min_{a \in \Upsilon(PR)} N_c(select_{PR}(c), select_I(c))$$

The compatibility for attribute level is measured, firstly, from the compatibility of its own value (Π^{own} and N^{own}) and, secondly, from the compatibility of the values of aggregate attributes (Π^{agg} and N^{agg}). We keep the minimum of these two compatibilities. Then, in general case,

$$\Pi_c(A, B) = Min(\Pi^{own}(A, B), \Pi^{agg}(agg(A), agg(B)))$$
$$N_c(A, B) = Min(N^{own}(A, B), N^{agg}(agg(A), agg(B)))$$

If $agg(A)$ is empty then $\Pi_c = \Pi^{own}$ and $N_c = N^{own}$.

The compatibilities between own values, Π^{own} and N^{own}, are calculated taking into account relevance and importance degrees:

$$\Pi^{own}(A, B) = Max(1 - imp(A), 1 - rel(B), \Pi^v(A, B))$$
$$N^{own}(A, B) = Max(1 - imp(A), 1 - rel(B), N^v(A, B))$$

where

$$\Pi^v(A, B) = \sup_{x \in ref(A)} Min(\mu_{v(A)}(x), \pi_{v(B)}(x))$$

and

$$N^v(A, B) = \inf_{x \in ref(A)} Max(\mu_{v(A)}(x), 1 - \pi_{v(B)}(x))$$

If B is empty then $\Pi^{own}(A, B) = N^{own}(A, B) = 1 - imp(A)$.

The compatibility of the aggregate attributes, Π^{agg} and N^{agg}, is calculated firstly keeping the maximun compatibility between each element of A and B and, next, keeping the minimum of these maxima. Then,

$$\Pi^{agg}(A, B) = \inf_{a \in A}(\sup_{b \in B}(\Pi(a, b)))$$
$$N^{agg}(A, B) = \inf_{a \in A}(\sup_{b \in B}(N(a, b)))$$

3 Example

Take an example about linguistic competences of somebody in a curriculum vitae.

Only we consider the attributes *linguistic competences (lc), language, level, write, read, speak* such as:

Agg(lc)={language, level}
Agg(level)={write,read,speak}
Dom(language)= a set of languages
The domain of *lc, level, write, read and speak* is:

$$
\begin{array}{ll}
 & [0,20] \times \{\eta\} \\
\cup & \{poor, medium, rich\} \times \{very, enough, \eta\} \\
\cup & \{any\} \times \{\eta\}
\end{array}
$$

So, we have:

$$\Gamma = \{lc, language, level, write, read, speak\}$$

The manager of database gives the aggregation functions to find the values of attributes *lc* and *level.* For example:

- v_{lc} is the average of every languages levels
- v_{level} is the average of write, read and speak levels.

The default value for write, read and speak level in an object description is *any* with the possibility distribution :

$$\mu_{any}(x) = \begin{cases} 0 & \text{si } x = na \\ 1 & \text{else} \end{cases}$$

Now consider the following object I:

lc			
language	Italian	Spanish	Polish
level	rich		very rich
write		rich enough	
read		very rich	
speak		medium	

Then we have $I = \{< lc, -, 1, \{a_1, a_2, a_3\} >\}$ with:

$$
\begin{array}{ll}
a_1 = \{ & < language, \text{ Italian}, 1, - > \\
 & < level, \text{ rich}, 1, - >\}
\end{array}
$$

$$
\begin{array}{ll}
a_2 = \{ & < language, \text{ Spanish}, 1, - > \\
 & < level, -, 1, \\
 & \ \{ \ \ < write, \text{ very rich}, 1, - > \\
 & \ \{ \ \ < read, \text{ rich enough}, 1, - > \\
 & \ \{ \ \ < speak, \text{ medium}, 1, - >\} >\}
\end{array}
$$

$$
\begin{array}{ll}
a_3 = \{ & < language, \text{ Polish}, 1, - > \\
 & < level, \text{ very rich}, 1, - >\}
\end{array}
$$

3.1 First profile

Let PR_1 be the profile: $PR_1 = \{< lc, \text{rich enough}, 1, - >\}$

PR_1 is described with only one attribute, then $\Upsilon(PR_1) = \{lc\}$ and we have:

$$\begin{aligned} \Pi(PR_1, I) &= \Pi_{lc}(\overbrace{select_{PR_1}(lc)}^{A}, \overbrace{select_I(lc)}^{B}) \\ N(PR_1, I) &= N_c(A, B) \end{aligned}$$

More,

$$\begin{aligned} \Pi_{lc}(A, B) &= \Pi^{own}(A, B) \\ N_{lc}(A, B) &= N^{own}(A, B) \end{aligned}$$

because $agg(A)$ is empty.

We have $imp(A) = pert(B) = 1$, then

$$\begin{aligned} \Pi^{own}(A, B) &= \Pi^{v}(A, B) \\ N^{own}(A, B) &= N^{v}(A, B) \end{aligned}$$

We can conclude:

$$\Pi(PR_1, I) = \sup_{x \in ref(A)} Min(\mu_{v(A)}(x), \pi_{v(B)}(x))$$

$$N(PR_1, I) = \inf_{x \in ref(A)} Max(\mu_{v(A)}(x), 1 - \pi_{v(B)}(x))$$

We know the value of A, $v(A)$ = rich enough. Consider the membership function of rich enough is the trapezoid (10, 12, 20, 0) in figure 4.

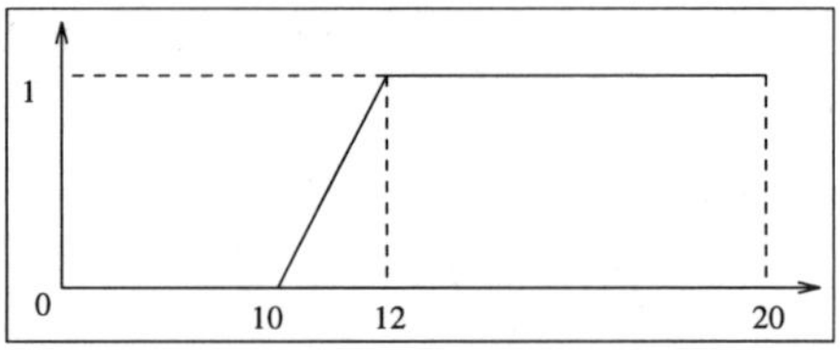

Figure 4: trapezoid of *rich enough*

However, the value of B isn't assigned by the user. Consequently, the value of B is calculated with the aggragation function above defined. $v(B)$ is the average of level values for Italian, Spanish and Polish.

Suppose the level of Spanish is calculated by (rich enough + very rich + medium)/3, then the value of lc can be calculated by the average of languages levels.

The exact function associated to the calculated value is not significant for the understanding, so we consider the possibility distribution for example:

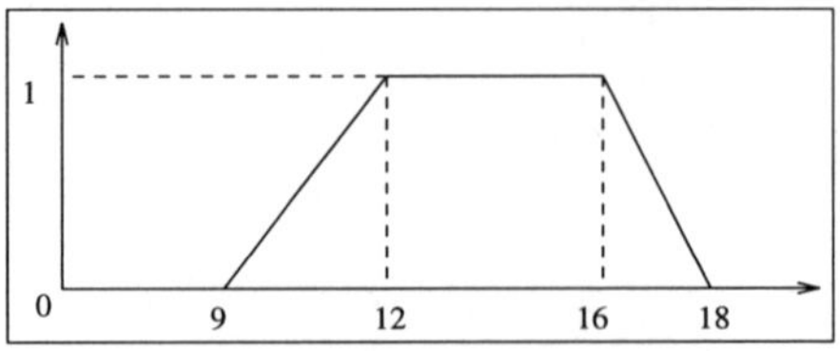

Figure 5: trapezoid of v(B)

Then we have $\Pi(PR_1, I) = 1$ and $N(PR_1, I) = 0.33$.

3.2 Second profile

Let PR_2 be a more detailed profile:

$$PR_2 = \{< lc, \text{any}, 1, \{b\} >\}$$

with
$$\begin{aligned} b = \{ \ & < language, \text{Italian}, 1, - > \\ & < level, richenough, 1, \\ & \{ \ < write, \text{very rich}, 0.8, - >\} >\} \end{aligned}$$

It's easy to see the compatibility between PR_2 and I is abstract to the compatibility between b and a_1, a_2, a_3.

So, we can say $\Pi(b, a_2) = N(b, a_2) = \Pi(b, a_3) = N(b, a_3) = 0$ because a_2 and a_3 haven't the same language as b.

Then we must compare b with a_1. The comparison is exact for the language in b and a_1. Consequently compare the level in b and a_1.

The write level in a_1 is non-existent, then its value is the default value, *any*. So the compatibility is reduced to $\Pi^{own}(A', B')$ and $N^{own}(A', B')$ with:

$$\begin{aligned} A' &= < level, \text{rich enough}, 0.8, - > \\ B' &= < level, any, 1, - > \end{aligned}$$

We have $\Pi^v(A', B') = 1$ and $N^v(A', B') = 0$ comparing the value *rich enoug* in A' with the value *any* in B'. Then,

$$\begin{aligned} \Pi^{own}(A', B') &= Max(1 - imp(A'), 1 - rel(B'), 1) \\ &= 1 \\ N^{own}(A', B') &= Max(1 - imp(A'), 1 - rel(B'), 0) \\ &= Max(1 - 0.8, 1 - 1) \\ &= 0.2 \end{aligned}$$

We can conclude:

$$\begin{aligned} \Pi(PR_2, I) &= 1 \\ N(PR_2, I) &= 0.2 \end{aligned}$$

Conclusion

We introduce the concept of aggregate attribute used in FIRMS. This permits to describe objects in database with not the same precision level. We develop a new identification process to take this concept into account.

Contrary to linguistic variables, our approach allows explicitly to take into account the semantic in associations values-nuances.

This approach is used in a concrete application concerning a firms panel.

References

[1] E. Kerre, R. Zenner, and R. D. Caluwe, "The use of fuzzy set theory in information retrieval and databases: A survey," *J. of the American Society For Information Science*, vol. 37, no. 5, pp. 341–345, 1986.

[2] J. Morrissey, "Imprecise information and uncertainty in information systems," *ACM Transactions on Information Systems*, vol. 8, no. 2, pp. 159–180, 1990.

[3] A. Yazici, R. George, B. P. Buckles, and F. E. Petry, *A survey of conceptual and logical data models for uncertainty management*, ch. 31, pp. 607–643. Lotfi A. Zadeh and Janusz Kacprzyk, John Wiley and Sons ed., 1992.

[4] N. Mouaddib and O. Foucaut, "A progressive and interactive process for the nuanced identification," *Analyse de systèmes*, vol. 16, pp. 23–29, Décembre 1990.

[5] N. Mouaddib and O. Foucaut, "Nuanced object identification. modeling & processing," in *IEEE ISUMA*, pp. 225–232, 1990.

[6] N. Mouaddib, "Fuzzy Identification in Fuzzy Databases. The Nuanced Relational Division," *International Journal of Intelligent Systems*, vol. 9, pp. 461–473, MAY 1994.

[7] N. Mouaddib and P. Subtil, "Tools for Assisting the Design of Fuzzy Applications and the Retrieval of Fuzzy Information from (FIRMS) an Object Oriented Fuzzy System," in *Uncertainty modelling and analysis : theory and applications* (B. Ayyub and M. Gupta, eds.), Machine Intelligence and Pattern Recognition 17, Elsevier (Amsterdam - Lausanne - New York - Oxford - Shannon - Tokyo), 1994.

[8] L. Zadeh, "Concept of a linguistic variable and its application to approximate reasoning-I," *Information Systems*, vol. 8, pp. 199–249, 1975.

[9] L. Zadeh, "Concept of a linguistic variable and its application to approximate reasoning-II," *Information Systems*, vol. 8, pp. 301–357, 1975.

[10] L. Zadeh, "Concept of a linguistic variable and its application to approximate reasoning-III," *Information Systems*, vol. 9, pp. 43–80, 1975.

[11] L. Zadeh, "Fuzzy sets as a basis for a theory of possibility," *Fuzzy Sets and Systems*, vol. 1, pp. 3–28, 1978.

[12] D. Dubois and H. Prade, *Théorie des possibilités - Applications à la représentation des connaissances en informatique.* Masson, 1985.

[13] L. Zadeh, "Fuzzy sets," *Information and Control*, vol. 8, pp. 338–353, 1965.

[14] J. Smith and D. Smith, "Database abstractions: Aggregation and generalization," *ACM Transactions on Database Systems*, vol. 2, pp. 105–133, Juin 1977.

[15] B. Bouchon-Meunier and Y. Jia, "Linguistic modifiers and imprecise categories," *Int. Journal of Intelligent systems*, vol. 7, pp. 25–36, 1992.

A FUZZY SPATIO-TEMPORAL DATA MODEL FOR CIS

Yann Pollet, Eric Ricard, Sébastien Robidou

MATRA CAP Systèmes
Division Recherches et Développement
Parc d'Affaires des Portes
BP n°613
27106 Val de Reuil Cedex
Tel : (16) 32 59 32 32
EMail: pollet/ricard/robidou@mcs-vdr.fr

ABSTRACT

In future CIS systems, the knowledge representation of the information issued from the outside world, will require to have at our disposal some DBMS including the notions of time, space, but also imprecision, uncertainty and incompleteness of information. In this article, we present a data model adapted to the representation and the manipulation of imperfect data in the context of CIS. We also present some concepts in order to describe conceptual schemata featuring the notion of fuzziness on its constitutive elements. The proposed concepts are an extension of existing standards with which a compatibility has been studied (ODMG 93, SQL2)

1. Introduction

The works described here are in keeping with the general pattern of our researchs in the range of **Communication and Information Systems** (CIS). Command aid in CIS is now a domain in full development, that we only begin to handle the potential contribution for the command aid process. So it is reasonable to think that the future operational systems will feature several assistances, based on synthetical reasoning for the intelligence officer. However, data processed by the system are various as far as their nature and their sources are concerned : conjonctural data, received messages, images from sensors, etc. These data getting to the system are generally imprecise (numerical values issued from a sensor), or also uncertain (contradictory information, or not confirmed yet ...). Moreover, human observations arriving at the system by means of messages, combine both imprecision and uncertainty, that is, describing the real world in a more or less vague and uncomplete way. An essential problematic in operational systems design consists in building databases able to take into account all these imperfect information issued from the real world at every moment, but also, more or less complete description that human operator has elaborated (situations).

However, whatever the concepts that we have at our disposal, as far as the data definition is concerned, these have little practical interest if we do not know easily and naturally how to map a partial knowledge from the real worl into a data of the the database, and also, how to **precise, enrich, confirm or infirm** some already existing information. An important aspect of this study was to define a group of operators enabling the creation and the manipulation of imperfect information elements, as data of a base. Here, we developed an original approach generalizing the one of the standard DBMS

2. Need for data management

The CIS context raises up many requirements, as far as the nature of the **real world to model** is concerned. More precisely, the analysis of the necessary data for CIS, clearly reveals several problematics :

- **data structuration**: a first point consists in representing the static aspects of the data schemata. From this angle, the modelling of CIS data implements complicated schemata, requiring strong design concepts based on the object approach [Pollet 92], [Lustman 93]. The need for preserving the data consistency requires, on the other hand, strong mechanisms in order to ensure data integrity, either concerning the relationships (*referential integrity*), or concerning the membership of an attribute to a range of values (*domain integrity*) ;
- **space modelling**: the entities, modelled within a CIS, generally represent physical entities of the operation theatre, and which position is significant (*georeferenced data*). These can be defined as quasi "pinpoint" entities (anti-aircraft battery, radars, ...), or also, fill in a well-defined area of the ground (military units, responsability areas, ...). In this context, a vector map is seen as a collection of georeferenced objects, provided with topologic relationships. The access to such objects must be based on "spatial" criteria (proximity, intersection, ...), or "*topologic*" criteria, requiring specific access operators ;
- **time modelling**: the conjonctural features of these objects (positions, available potentials, ...) evolve in time. So, the "time" concept must take place in their representation, the time considered here, being the "*operational time*" of the real world (and not the internal time of the machine). On the other hand, the access to the past values must be possible thanks to specific operators. To the problematic of representation of the "*evolutionary*" entities of the world, is added the need to take into account purely temporal notions, such as events or durations with their associated characteristics (modelling of the different stages of an action) ;
- **uncertainty and incompleteness modelling**: the imprecise feature of numeric information in entry requires, that we account of these, not as precise values, but as sets of more or less *possible* values. The uncertainty is, of course, also present in geographic positions of the entities (spatial component), as well as in the knowledge of the moments when a description of an entity is made (temporal component). Another form of incompleteness of data concerns the knowledge related to a discrete universe description: by their nature (for example : number of vehicles), or by convention (for example : type of a vehicle, state of a track). The knowledge, that we will term *incomplete*, is an intermediary case between complete unknowledge and perfect knowledge. It requires to store n possible values for a quantity (for example : 40 or 60 tanks regiment) ;
- **data dynamic** : the incomplete knowledge, that we have at our disposal at a given moment about such or such numeric or symbolic attribute, must be easily translated into data of the database by means of specific operations. The simple notion of assignement of a value to an attribute is then no more sufficient, and must be extended by proper manipulation operators to possible values. In the same way, any complementary knowledge subsequently gained must be able to come and enrich the already present information.

3. Chosen approach

The existing relational or object oriented models, supplied by the actual DBMS softwares, only answer partially to the different problems raised up above. If the Object Oriented DBMS, offer strong concepts of modelling, the control of time and space dimensions, necessary for CIS, are not taken into account in a specific way, what leads to search for an extension of the current models. Several approaches have been proposed for time modelling in, but also space in a general context, independent of the CIS context. The approach of the **Operational Data Base,** described in [Pollet 92] and [Pollet 93 a], offers a data model, adapted to CIS, synthetizing these different approaches. This model integrates, in a homogeneous approach, "object" concepts, allowing to take into account the complexity of the universe of interest of the CIS, and on the other hand, the control of temporal and geographic information.

In **Data Fusion systems**, [Pollet 93 b] suggests some concepts in order to take into account the incompleteness of the data in the **possibility theory** context. The *data definition* aspect is dealt by integrating within the model the concepts of attributes, fuzzy semantic relationships, of version and temporal types. These concepts are linked to the existing approaches, such as [Prade 86], [Dubois 87], [Dubois 88], [Chen 93], [Kelman 93].

4. Data definition concepts

4.1. Basic principles

A ***Data Definition Language*** (DDL) must enable the modelling of the structure of the entities and the relationships existing between them. The complexity of the data schemata requires an Object approach [Pollet 92]. As far as the DDL is concerned, an important constraint must be the search of a compatibility with the existing or future standards (SQL2, SQL3, ODMG 93), but also with the already suggested approaches, in other contexts (ESQL2 [Gardarin 93]).

So, our approach provides some modelling operators inspired by the object approach : type notions (data structure without persistency notion), *class* notions (persistent set of objects based on the same model), implicit derivation of properties by *inheritance, multivalued* attributes (set of values), and lastly "reference" links, reflecting the semantic of the *relationships* between classes.

As far as the DDL is concerned, a class "airport" could be defined as :

```
CREATE CLASS airport ISA civil_site
(
        name            CHAR (20) ,
        altitude        INTEGER ,
        track_length    POSITIVE_REAL ,
        include         SET_OF REFERENCE aircraft ,
        shelter_nb      POSITIVE_INTEGER ,
        potential       POSITIVE_REAL ,
        track_state     FUZZY_ENUMERATION (good, average, bad) ,
        ...
);
```

in which "civil_site" and "aircraft" are classes already defined.

4.2. Imperfection of values

The approach presented in [Dubois 88] provides an approach in order to represent different forms of data imperfection in the context of the possibility theory, on the basis of the relational model. Our approach consists in transposing and then extending this approach in the context of an object oriented model. So, two important types of imperfection can be considered :

- the **imprecise real values.** The imprecision, considered here, is reduced to the case of *fuzzy intervals* (trapezoidal possibility distributions). One can also consider more specific forms of distributions (fuzzy real, real interval), as well as the *certain* values as an extreme case. In this form of representation, the extreme cases of unlimited supports are also included, which, for a real attribute A, accounts the case: "A unknown" (*Null* value of the DBMS), "A > a", "A < a", etc.).

  The "date" domain, like generally, any continuous domain of values, provided with an order relation, is integrated in this approach.

- the **discrete quantities** (enumerations, integers, ...) which value is badly known and on which the available knowledge is an enumeration (not necessarly finite) of possible values, assigned with a possibility degree. The *Null* value of the DBMS and the classic assignment of a determined value can be interpreted as two extreme cases of this notion.

  The characters strings, not considered in [Dubois 88], can be easily reduced to the case of a discrete domain. These discrete domains, if they have common basic features, distinguish themselves according to two criteria (finite / infinite, ordinated / non-ordinated), that will not be neutral as far as the data manipulation is concerned.

In these two cases, the conventional value "Unapplicable" is implicitly added to the domain of the considered values by means of an extra-element [Dubois 88].

In order to avoid a too important penalizing of an extended DBMS' performances as far as the queries (select) are concerned, the model suggests the **explicit** statement of the fuzzy feature assigned to an attribute. So, new types (fuzzy_integer, fuzzy_real, ...) are introduced, and the classic domains (integer, real, ...) and the fuzzy types could be both used in a schema. For example, enemy airports, could be expressed as :

```
CREATE CLASS airport ISA civil_site
(
        name        FUZZY_CHAR (20) ,
        altitude    INTEGER ,
        ....
        track_state FUZZY_ENUMERATION (good, average, destroyed) ,
        ......
);
```

4.3. <u>Multi valued attributes</u>

The attributes of type : "set of *atomic values*" (set of integers, reals, symbols enumeration, ...), called *multi valued attributes*, allowed in the basic object model, must be extended to the case of badly known sets. It concerns the modelling of possibility distributions, not on a domain of values D, but on P(D) (set of D's parts). Such entities are named *fuzzy collections*.. An example of such a type of attribute is given by the "languages spoken by a person", which values (not mutually exclusive) are supposed to be "more or less possible" according to the knowledge about this person.

We introduce in a data schema such attributes by statements such as **<u>fuzzy set of</u> real, <u>fuzzy set of</u> integer, <u>fuzzy set of enumeration</u>, ...** We must note that, in the case of a set of real numbers, the associated semantic is not the one of a multi-modal distribution (not taken into account in the present model, we do not know the possible "overlappings" between each fuzzy value of the set).

4.4. <u>Semantical links</u>

The conceptual schemata used in CIS require the use of typed semantical relationships between objects. We make a difference for these relationships, between attributes of type **<u>reference</u> <classe y>** in the definition of a class x (1-1 or n-1 cardinality), and also **<u>set of</u> <u>reference</u> <classe y>** (1-n or n-m cardinality). The notion of reverse link can also be defined. Of course, the relationships themselves must assume a certain degree of fuzziness as the referenced units can be incompletely determined. One can state partially possible links, only.

In order to strictly extend the concept of *reference*, we must distinguish between the mono-valued references (*fuzzy reference*) and the multi-valued references (*fuzzy_set_of reference*). The **fuzzy mono-valued references** are interpreted as a list of possible links with other objects, each weighted by a possibility degree. An extra element meaning a "lack of reference" can also be weighted by a possibility degree. Such a reference must be implemented as an abstract type, enabling to add/remove *possible* values The processing of the reference links is identical to those of other atomic attributes. In the same way, the **fuzzy multi-valued references** are processed as multi-valued attributes.

An example is given by an identified enemy unit, from the regiment level, which membership. division (level n+1) and/or the attached squadrons (level n-1) can be incompletely determined :

```
CREATE CLASS ennemy_regiment ISA enemy_unit
(
        name                 FUZZY_CHAR (20) ,
        membership_division  FUZZY REFERENCE enemy_division ,
        squadrons            FUZZY_SET_OF REFERENCE enemy_squadron ,
        .........
);
```

4.5. Spatial information

The spatial representation of the data is one of the essential requirements in CIS. It is the matter of representing entities of the real world filling a certain position of the space (point, line, area). We account for this notion in [Pollet 92] by means of a predefined basic class "geographic". In the above example, a friendly civil site can be defined by inheritance from this class as :

```
CREATE CLASS friendly_civil_site ISA geographic
(
        name            CHAR (20) ,
        included_sites  SET_OF REFERENCE friendly_civil_site ,
        including_site  REFERENCE friendly_civil_site ,
        ....
);
```

The "geographic" class does not have any directly visible attributes. As far as the data manipulation is concerned, it will be an *abstract type* (Abstract Data Type, or ADT), describing the nature of the spatial position (point, surface, area), and its description in terms of points, vertexes, or also topologic links of connectivity. The *methods* : *get_type*, will allow to ask the type of the spatial information (point, surface,...), of the considered object. However, data about positions are not necessarily precisely known (data about the ennemy universe). It is important to generalize this approach to the case of an imperfect knowledge. Three principal cases of imperfection are met in practice :

- pinpoint object with imprecise position ;
- "polyline" with imprecise vertexes (but in which number of vertexes is known) ;
- enumeration of a finite number of *possible* "shapes" (which can be reduced to the above cases).

These three forms of imprecision are particular cases of a more important problematic, that lies in defining possibility distributions on a certain set of shapes of the plane. In the present model, we are limited to the here above mentioned forms, modelled thanks to the fuzzy_geographic class. The doubt is not allowed, in the model, for the nature (point, "polyline", polygon) of the spatial position. Concerning the representation of an imprecise pinpoint position, we follow the approach presented in [Kelman 93], in which, such a position is represented by a pair of polygons (Kernel, Support). For example, the following statement :

```
CREATE CLASS enemy_civil_site ISA fuzzy_geographic
(
        name            FUZZY_CHAR (20) ,
        included_sites  FUZZY_SET_OF REFERENCE enemy_civil_site ,
        including_sites FUZZY REFERENCE enemy_civil_site ,
        ....
);
```

will correctly represent a civil site, which knowledge can be, at a given moment, imperfect as much as for its imprecise position, as for its description.

4.6. Temporal and historic data

Two notions are necessary in the model :

- the "**intervals**" and "**event**" concepts: they account for particular facts of the real world (and not about physical entities), located more and less precisely on the time axis. These notions have either the status of a temporal area (for example, time interval of an action, of a task, ...) or a particular "time-point" event ;
- the notion of **evolving entity**: such an entity (for example, a military unit) is continually present on the time axis during its whole life-time. However, this one can only be known (more or less precisely) at particular discrete moments of the real world, which are themselves more or less precise. The anterior states of an entity must be stored.

An essential distinction, more judicious than the distinction: physical /logical time, must be carried out :

- the ***technical time*** linked to the machine ; this is the time where the transactions are committed in the database ;
- the ***operational time*** linked to the described real world, and continuous by nature ;
- the ***perceived time*** linked to the (more or less precise) knowledge of the operational time. Discrete by nature, the perceived time scan be viewed as more or less possible sets of operational moments.

The **temporal** class will be used to build, by means of the inheritance principle, new applicative classes, provided with all the temporal features. The chronical account of events is carried out by means of particular "*historic*" attributes. The **fuzzy_temporal** class includes the concepts of imprecise moments. For example, we will have :

```
CREATE CLASS squadron ISA fuzzy_temporal
(
        tanks_number      HISTORIC FUZZY_INTEGER ,
        included_tanks    FUZZY_SET_OF REFERENCE vehicle ,
        ...
);
```

On the other hand, we account for punctual events and duration notions thanks to the **Fuzzy_event** and **Fuzzy_duration** classes. Two logics are suggested in order to manipulate these kind of attributes : the logic of moments and the logic of intervals applied to fuzzy temporal data.

5. Concepts for data manipulation

5.1. Principles

It is interesting to define the necessary operators for the manipulation of imperfect knowledge, that is : insert a new information in the database, enrich or reinforce the certainty about an already present information. The manipulation of the data of a database is, in a CIS, classically based on **two important access types**:

- the **associative access** : it consists in selecting a certain subset of data from the database on the basis of a certain selection criterion. The access is expressed as a SQL clause, and can be applied to more general criteria : alphanumeric data, but also "spatial" and "temporal" criteria as well as combinations of different types of criteria. In the case of a fuzzy database [Dubois88], the result is a subset which elements are weighted by a pair (possibility, necessity).
- the **navigational access** : once a data has been accessed, the problematic is to access another data referenced by the first one. These links can be various: semantic, topologic (for some geographical data), or temporal. The navigational functions of the object model must be extended to the fuzzy data.

5.2. Interface classes

In order to propose a satisfactory approach from the conceptual angle, the manipulation formalism must apply to the objects of the application and to the objects of the database. These last appear as language objects featuring the "persistency" notion. All the operations on the database (selections, insertions, updates, ...) are "persistent" classes methods. So, the interface proposes :

- **bases classes**, independent of the data schema, in particular the "**Db_object**" class, which represents any persistent object, the "**Db_list**" and "**Db_fuzzy_list**" classes, (generic classes allowing the manipulation of fuzzy objects, and returned by selection requests) (the *selection* operation is based on a SQL-like "where" clause, with new keywords for the new operators) ;
- **pre defined classes** of the database model (geographic, temporal classes, ...). **All the presented *fuzzy types*** exist as part of the manipulation interface as ***C++ classes***, the manipulation operators defined for these types appearing as class ***methods*** ;
- **"applicative" classes** which are linked to the data schema and respecting the inheritance hierarchy. The inheritance is possible from geographic classes (resp. fuzzy_geographic) and temporal (resp. fuzzy_temporal) or from both. Except this case, the *inheritance* is reduced to the *simple inheritance*.

5.3. Manipulation of discrete values

The **incomplete** knowledge is related to data with discrete values (integers, enumerations, characters strings). The principle lies in **storing the possibility distributions** related to the available knowledge at a certain moment, under the form of a list of possible values, weighted by a possibility.degree. All the attributes A of a discrete enumerated domain D = $\{a_1, ..., a_n\}$ can be processed by common primitives. Several cases must be considered :

- **complete ignorance** as far the value is concerned. It is the *Null* value of the DBMS. It can be interpreted as a possibility distribution $\mu(a_i) = 1$ whatever i may be. This is the default state of any attributes ;
- **complete knowledge** of the value. It is the usual assignment operation: $A = a_k$. It can be interpreted as a possibility distribution $\mu(a_k) = 1$ and $\mu(a_i) = 0$ for $i \neq k$;
- **imperfect knowledge** leading to store an information of type: "A is certainly equal to a_k" with a "certainty coefficient" c (between 0 and 1). That involves not to completely exclude the a_i, $i \neq k$ from the possible values. If we interprete c as the necessity that A be equal to a_k, then $\mu(a_k) = 1$ and the $\mu(a_i)$, $i \neq k$, are limited to 1-c.

  The **weak assignment operator**, defined as "A.f_set (a_k, c)" (invoked as a method of the Fuzzy_D class, related to the domain D), which effect is to assign all the a_i, $i \neq k$ to the possibility value 1-c. The weak assignment is similar to the classic assignment when c=1. It "assigns" the *Null* value when c = 0 ;
- **alternative between several values.** Such knowledge leads to affirm that A is certainly equal to a_k or a_l or.... (p values) with a given certainty coefficient c (which can be interpreted as a necessity). The $\mu(a_k)$, $\mu(a_l)$, ... cannot be here individually known (we only have $\text{Sup}(\mu(a_k))=1$). The $\mu(a_i)$, $i \neq k$, are limited to 1-c.

  We define the **operator of weak alternative** A.f_alter(c, a_k, a_l, ...) with the convention: $\mu(a_k) = 1$, $\mu(a_l) = 1$, ..., $\mu(a_i)= 1-c$, $i \neq k$. The case c = 1 is similar to a classic disjunction between values.

For instance, let's consider a domain D={1, 2, 3, 4, 5}, and A, a discrete attribute taking its values in the D domain. When stating A, its values are completely unknow. The assignment A = 2 (the value of A is completely known), is made thanks to the usual assignement operator. If A is necessarily (0.7) equals to 3 we use then A.f_set(3,0.7), and if we want to reinforce the trust in A=3 (N=0.8) one can use A.f_set(3,0.8), or else inflect this trust by using A.f_set(3,0.6) (the f_set method remove the existing value and replace it by the new one), and finally if we want to represent the fact that A is certainly (0.8) equals to 2 or 3 then we can use A.f_alter (0.8, 2, 3).

5.4. Manipulation of integers

Integers are discrete values and so, they can be manipulated as such, the infinite cardinal of the set only having an influence on the *implementation* aspect. However, in some cases, one have to search for the record of an approximative value. Here, we rely on the **total order** with which is provided the set of integers. We took into account this specificity by associating the fuzzy_integer type with a **set_fvalue *method*** allowing to initialize the condition of the considered instance with a possibility distribution, reflecting the idea that the integer attribute A is equal to "about n". The arguments gave to this method correspond to the typical form of the distribution that we intend to give to A, and also the spreading factor reflecting the imprecision concerning the value.

5.5. Manipulation of real numbers

The generic case of an imprecise real number is, in our model, a *fuzzy interval* (trapezoidal possibility distribution). The extreme case of this type of distribution is the distribution in which $\mu(x)= 1$ whatever x is. It corresponds to the Null value of real numbers (default value). To the fuzzy_real type of the extended DBMS, corresponds a class of the language provided with methods: "set" (resp. "get") allowing to assign (resp. retrieve) the current value of kernels and supports. A method is added in order to retrieve the possibility and necessity degrees of a precise value. In order to reduce the assignments operations during the program execution, the fuzzy_real class of the language, is derived (in the meaning of the inheritance in C++) into **more specialized classes** reflecting some forms of more specific distributions (fuzzy real, classic interval, pure real), but compatible with a fuzzy_real statement.

5.6. Manipulation of fuzzy collections

The definition of particular values of *fuzzy collections* during the program execution raises up an important problem. The strict definition of an instance of a *fuzzy collection* would require that the application coul list, at least, all the partally possible subsets of the D domain, each weighted by a possibility degree. This makes no sense in practice. It was also suggested in [Prade 86] to define such a distribution thanks to two sets (possible and necessary elements). However, we do not account, in this way, for the most general possibility distribution on P(D).

In this context it is important to take into account a knowledge expressed in a cumulative way. For instance, if the "division" class features an attribute: set_of_regiment representing the regiments depending of this division, one could only have, at first, only a partial knowledge about an instance D from the division class, such as:"R1 certainly belongs to D", "R2 possibly belongs to D" and finally "a regiment subordinated to D has been identified and it can be R3 or R4 in an exclusive way".

In this case, we suggest **a certain number of operators**, provided with a precise semantic, allowing to **build, in an iterative way, fuzzy collections**. In order to achieve this aim, we introduce the following concepts :

- the **fuzzy union**, generalizing the operation of classic sets union on the set of fuzzy collections. This operation corresponds to the insertion of a new element in the set of the existing possible elements, but also the reinforcement of certitude on an already present element.
- the **alternative rule**, corresponding to the addition of a new hypothesis, properly weighted.
- the **quotient and difference operators** enabling, respectively, to remove a more or less possible element, and to remove an existing hypothesis from a fuzzy collection.
- the **necessary (a, c) constructor**, defining from an element "a" of the domain D, the fuzzy collection made up of a single singleton {a} assigned with a possibility degree equals to 1 and with the empty set assigned with 1-c (all the other elements of P(D) having a possibility degree equals to 0).
- the **Possible (a, c) constructor**, defining a fuzzy collection made up of the singleton {a} assigned with a possibility degree equals to c, with the empty set assigned with a possibility degree equals to 1;
- the **alternative constructor**, defining a fuzzy collection by direct enumeration of p weighted subsets.

In [Pollet 94] we present these various operators enabling to manipulate the fuzzy collections.

6. Principles of implementation

The use of the relational technology has been intentionally chosen because of its industrial maturity, and also because of the existence of a solid standard (SQL 2).This choice does not exclude the future implementations using Object Oriented DBMS, which would be easier thanks to the higher level concepts offered by these softwares.

The general principles of implementation of the model above SQL2 softwares are based on classic principles. The storage of fuzzy real and discrete attributes with finite domain is based on [Dubois 88]. As for the discrete infinite type "integer", the operators presented in the present model require a transposition of this technique. So, a unique list of values is stored, but this one can be interpreted in two ways according to a selector : list of possible values, or also list of the excluded values of the possible universe. It is possible to store pairs (possible value, possibility degree) in a specific "vertical" design implementations of these structures in large, non structurated attributes, provided with all the DBMS. A particular problematic is the need for translating the selection requests applied to the extended DBMS, into SQL 2 requests. The chosen principle consists in getting out tuples corresponding to objects *possibly* able to satisfy the request, followed by the processing of the selected tuples into RAM memory. In order to have an efficient extraction, the issue of defining indexes on fuzzy attributes has to be studied.

7. Conclusion

By using existing results from theoric works in the field of the advanced DBMS, from the object approach, from the fuzzy sets and from the possibility theory, the present study enabled to show that it was possible to build a **unified data model**, including in a common and homogeneous object approach, the concepts of time, space, semantic links and imperfection of knowledge, as part of the **possibility theory**. We showed then the practical possibility of implementing it above the existing DBMS. Then, the problematic is to know how building implementations that may be compared in terms of performance with the classic DBMS, a problematic for which different solutions have only begun to be studied.

Bibliographie

[Andres 89] Andres V., "Filtrage sémantique dans une base de données imprécises et incertaines: un syteme souple autorisant la formulation de requêtes composites pondérées", Thèse, Toulouse, 1989.

[Bouchon-Meunier 92] B. Bouchon-Meunier, "Représentation et traitement de l'incertitude". DGA, Sciences et Défense 92.

[Cayrol 82] Cayrol M. Farreny H., Prade H., "Fuzzy pattern matching", Kybernetes n°11, 103-116, 1982.

[Chen 93] Chen Z.Q., "Systeme de représentation et d'exploitation de connaissances temporelles incertaines", Thèse, Orsay 1993.

[Dubois 87] Dubois D., Prade H., "On incomplete conjunctive information". Proc. of the North American Fuzzy Information Processing Society Workshop, Purdue Univ. W. Lafayette, 1987.

[Dubois 88] Dubois D., Prade H., "Théorie des possibilités", Ed Masson, 1988.

[Galibourg 88] Galibourg M., "Mises en oeuvre d'opérateurs flous dans une base de données",Thèse Rennes 1988.

[Gardarin 92] Gardarin G., Valduriez P. "ESQL2 - Une extension de SQL 2 pour gérer les Bases de Données Objets et Déductives".

[Graham 87] Graham I., Jones P.L., "A theory of fuzzy frames", BUSEFAL (L.S.I., Univ. Paul Sabatier, Toulouse) N°31, 109-132, N°32, 120-135, 1987.

[Kelman 93] Kelman A. , "Optimisation de trajectoires - Préparation de missions aériennes". Troisièmes journées nationales". Applications des Ensembles Flous", Nïmes1993.

[Mouaddib 92] N. Mouaddib, O. Foucaut, P. Subtil, D. Vilmin, "Firms : un système orienté objet pour la représentation des information nuancées. Application aux sciences économiques". 2èmes journées nationales "Les applications des ensembles flous". Novembre 1992.

[Pivert 91] Pivert O., "Contribution à l'interrogation flexible de bases de données: expression et évaluation de requètes floues", Thèse, Rennes, 1991.

[Pollet 92] Pollet Y., "Les bases de données opérationnelles: une nouvelle voie pour la représentation des connaissances dans les SIC", Avignon 92, Conférence "Défense et Sécurité Civile". Juin 92.

[Pollet 93 a] Pollet Y.,"Une nouvelle approche pour la représentation des données dans les Systèmes d'Information et de Communication".1er congrès biennal AFCET. Coll "Bases de Données", 1993.

[Pollet 93 b] Pollet Y., Bregeault L., Bridon Ph., "Bases de Données pour les systèmes de fusion de données". Troisièmes journées nationales "Applications des Ensembles Flous", Nïmes1993.

[Pollet 94] Pollet Y., Robidou.S, "An approach for the representation of multivalued attributes in fuzzy datbases". FUZZ-IEEE/IFES'95 Workshop on Fuzzy Database Systems and Information Retrieval, Yokohama 1995.

[Prade 86] Prade H., Testemale C., " Representation of soft constraints and fuzzy attribute values by means of possibility distributions in databases", The Analysis of Fuzzy Information, (J. Bezdzk, Eds), CRC Press, 1986.

[Rossazza 90] Rossazza J.P., "Utilisation de hierarchie de classes floues pour la eprésenttaion de connaissances imprécises et sujettes à exeptions: le syteme SORCIER", Thèse, Toulouse, 1990.

[Torasso 89] Torasso P., Console L., "Approximate Reasonning and Protypical knowledge", International Journal of Approximate Reasonning, Vol 3, N°2, 155-177, mars 1989.

[Vignard 85] Vignard P., "Un mécanisme d'exploitation à base de filtrage flou pour une représentation centrées objets", Thèse, Grenoble, 1985.

FOUNDATIONS OF FUZZY LOGIC

FUZZY BETWEENNESS RELATIONS

Joan JACAS & Jordi RECASENS*

Sec. Matemàtiques i Informàtica. ETSAB
Univ. Politècnica de Catalunya
Diagonal 649, 08028 Barcelona. Spain

ABSTRACT

The aim of this paper is to introduce the idea of fuzzy betweenness relation on a set X. This is done by generalizing the definition of betweenness relation proposed by Menger. It is proved that a separating T-indistinguishability operator on X (with T a strict archimedean t-norm) generates a fuzzy betweenness relation on X and reciprocally, every fuzzy betweenness relation on X defines a separating T-indistinguishability operator on X. Moreover, it is proved that the crisp part of a fuzzy betweenness relation is a classical (metric) betweenness relation.

Keywords: Archimedean t-norm, betweenness, T-indistinguishability operator.

1. Introduction.

The idea of betweenness appears in many contexts like, among others, hierarchical decision processes [9], classification theory [8] or choice theory [6]. In [5], K. Menger defined the concept of (metric) betweenness in such a way that if (X, m) is a metric space, then given three different elements x, y, z of X, y is said to be between x and z if $m(x,y)+m(y,z) = m(x,z)$. Nevertheless, for practical pourposes this definition is too strict because it ignores any kind of uncertainty and/or lack of information within the data or the context. In this paper we introduce the concept of fuzzy betweenness in order to obtain a more useful tool when vagueness or uncertainty is involved in the process.

On the other hand, in [7] it is proved that every separating T-indistinguishability operator on a set X (with T a strict archimedean t-norm) generates a betweenness relation on X. In [12], Valverde proved a representation theorem for T-indistinguishability operators that roughly speaking states that a fuzzy binary relation E defined on a set X is a T-indistinguishability operator if and only if it can be generated in a natural way by a family of fuzzy subsets of X. In [1] it is also proved that E can be generated by a single fuzzy set (E is then called unidimensional) if and only if the betweenness relation generated by E is total. If E is "close" to a unidimensional operator, then it will "almost" generate a total betweenness relation. This is another context in which the concept of fuzzy betweenness can be useful.

Therefore, it seems interesting a sistematic study of this kind of fuzzy relations that since now, it has been done only in a very partial way [3,4].

After a section of preliminaries on t-norms and T-indistinguishability operators, fuzzy betweenness relations will be defined. It will be proved that such relations are generated by

*Research partially supported by DGICYT p.n. PB91-0334-CO3-02

separating T-indistinguishability operators and reciprocally, every fuzzy betweenness relation on X defines a separating T-indistinguishability operator on X.

We conclude with some comments relating the dimension of a T-indistinguishability operator and the cardinality of its associated betweenness relation.

2. Preliminaries.

For the sake of completeness, in this section some well known properties and definitions of t-norms and T-indistinguishability operators will be given. A more complete exposition on this topics can be found in [2,10,13].

Definition 2.1. A t-norm is a map $T : [0,1] \to [0,1]$ that for all x, y, z on [0,1] satisfies the following properties:

2.1.1. Associativity: $T(T(x,y),z) = T(x,T(y,z))$.

2.1.2. Commutativity: $T(x,y) = T(y,x)$.

2.1.3. $T(x,1) = x$.

2.1.4. T is an non-decreasing function in both variables.

Definition 2.2. A t-norm is archimedean if and only if $T(x,x) < x$ for all x in (0,1).

Given a t-norm T, for each $n \in \mathbb{N}$ we can define recursively the nth power T^n of T i.e. $T^0(x) = x$ and $T^n(x) = T(T^{n-1}(x), x)$.

Definition 2.3. A strict archimedean t-norm is an archimedean t-norm T such that $T^n(x) \neq 0$ for any $n \in \mathbb{N}$ and $x \in (0,1)$.

The most well known strict archimedean t-norm is the t-norm product Π defined by $\Pi(x,y) = xy$.

The following theorem characterizes strict archimedean t-norms:

Theorem 2.4. (Ling) A t-norm T is strict archimedean if and only if there exists a continuous and strictly decreasing function $t : [0,1] \to [0,+\infty]$ with $t(0) = +\infty$ and $t(1) = 0$ such that

$$T(x,y) = t^{-1}[t(x) + t(y)].$$

The function t is termed an additive generator of T. An additive generator of the product t-norm is $-\ln(x)$.

Definition 2.5. Given a continuous t-norm T, its quasi-inverse $\widehat{T}$ is defined by

$$\widehat{T}\ (x|y) = \sup\{\alpha \in [0,1] | T(\alpha, x) \leq y)\}.$$

The quasi-inverse of the product t-norm is $\widehat{\Pi}\ (x|y) = \text{Min}(1, y/x)$.

Proposition 2.6. Given a strict archimedean t-norm T and an additive generator t of T, then $\widehat{T}(x|y) = t^{-1}[t(y) - t(x)]$.

Definition 2.6. [11] Given a t-norm T, a T-indistinguishability operator in a set X is a fuzzy binary relation in X such that, for all x, y, z in X, it satisfies:

2.6.1. Reflexivity: $E(x, y) = 1$.

2.6.2. Simmetry: $E(x, y) = E(y, x)$.

2.6.3. T-transitivity: $T(E(x, y), E(y, z)) \leq E(x, z)$.

If for any x, y in [0,1], $E(x, y) = 1$ implies that $x = y$, then E is called a separating T-indistinguishability operator.

3. Fuzzy Betweenness relations.

Menger [5] defines a (metric) betweenness relation in a set X as a ternary relation B (i.e. a subset B in $X \times X \times X$) satisfying the following properties for all x, y, z in X:

3.0.1. If $(x, y, z) \in B$ then $x \neq y \neq z \neq x$.

3.0.2. If $(x, y, z) \in B$ then $(z, y, x) \in B$.

3.0.3. a) If $(x, y, z) \in B$ and $(x, z, t) \in B$ then $(x, y, t) \in B$.

b) If $(x, y, z) \in B$ and $(x, z, t) \in B$ then $(y, z, t) \in B$.

3.0.4. If $(x, y, z) \in B$ then $(y, z, x) \notin B$ and $(z, x, y) \notin B$.

If $(x, y, z) \in B$ it is said that y is between x and z. If for every given triplet of different elements x, y, z in X, one of them always lies between the other two, then the relation B is termed a total betweenness.

Following these ideas, we propose a generalization of the concept of betweenness in the setting of Fuzzy Sets Theory.

Note: In the sequel, unless it is esplicitely stated, T will represent a strict Archimedean t-norm.

Definition 3.1. A fuzzy betweenness relation (with respect to T) in a set X is a fuzzy ternary relation, i.e. a map

$$X \times X \times X \rightarrow [0,1]$$
$$(x, y, z) \rightarrow xyz$$

that, for all x, y, z in X, satisfies the following properties:

3.1.1. $xxy = 1$.

3.1.2. $xyz = zyx$.

3.1.3. a) $T(xyz, xzt) \leq xyt$.

b) $T(xyz, xzt) \leq yzt$.

3.1.4. If $x \neq y$, then $xyx < 1$.

Proposition 3.2. a) If $xyz = 1$ and $y \neq z$, then $xzy < 1$.

b) If $xyz = 1$ and $x \neq y$, then $zxy < 1$.

Proof: From 3.1.3.b) and 3.1.4., it follows $T(xyz, xzy) \leq yzy < 1$ and therefore $xzy < 1$. In a similar way we can prove b). ■

Corollary 3.3. If $xyz = 1$ and $x \neq y \neq z \neq x$, then $xzy < 1$ and $zxy < 1$.

Corollary 3.4. The crisp part of a fuzzy betweenness relation in the set of triples of different elements of X is a classical betweenness relation in X.

Proposition 3.5. If E is a separating T-indistinguishability operator on a set X with $E(x,y) \neq 0$ for all x, y in X, then the fuzzy ternary relation on X defined by

$$xyz = \widehat{T}\ (E(x,z)|T(E(x,y),E(y,z)))$$

is a fuzzy betweenness relation.

Proof:

a) 3.1.1. $xxy = \widehat{T}\ (E(x,y)|T(E(x,x),E(x,y))) = \widehat{T}\ (E(x,y)|E(x,y)) = 1$.

b) 3.1.2. is immediate.

c) 3.1.3.a) Applying the preceding definition to 3.1.3.a we have the following inequality

$$\begin{aligned}&T(\widehat{T}\ (E(x,z)|T(E(x,y),E(y,z))), \widehat{T}\ (E(x,t)|T(E(x,z),E(z,t))))\\ &\leq \widehat{T}\ (E(x,t)|T(E(x,y),E(y,t))).\end{aligned}$$

In order to prove this inequality, we express it in terms of an additive generator t of T. Then the mentioned inequality can be written as:

$$\begin{aligned}&t^{-1}[t[\widehat{T}\ [E(x,z)|T(E(x,y),E(y,z))]] + +t[\widehat{T}\ [E(x,t)|T(E(x,z),E(z,t))]]]\\ &\leq \widehat{T}\ [E(x,t)|T(E(x,y),E(y,t))].\end{aligned}$$

Using the representation of the quasi-inverse given in the proposition 2.6, we have

$$\begin{aligned}&t^{-1}\ [t[t^{-1}[t[T(E(x,y),E(y,z))] - tE(x,z)]] + t[t^{-1}[t[T(E(x,z),E(z,t))] - tE(x,t)]]]\\ &\leq t^{-1}[t[T(E(x,y),E(y,t))] - tE(x,t)],\end{aligned}$$

applying t to both sides of the inequality and simplifying, we obtain the equivalent expression

$$\begin{aligned}&t[T(E(x,y),E(y,z))] - tE(x,z) + t[T(E(x,z),E(z,t))]\\ &\geq t[T(E(x,y),E(y,t))].\end{aligned}$$

If we sustitute the t-norm T by its representation,

$$\begin{aligned}&t[t^{-1}[tE(x,y) + tE(y,z)]] - tE(x,z) + t[t^{-1}[tE(x,z) + tE(z,t)]]\\ &\geq t[t^{-1}[tE(x,y) + tE(y,t)]],\end{aligned}$$

and the above inequality takes the form

$$tE(x,y)+tE(y,z)-tE(x,z)+tE(x,z)+tE(z,t)$$
$$\geq tE(x,y)+tE(y,t).$$

If we simplify and apply t^{-1} to both sides, we finally obtain

$$t^{-1}[tE(y,z)+tE(z,t)] \leq E(y,t)$$

This last inequality expresses the T-transitivity of E.

d) 3.1.3.b) can be proved in a similar way as 3.3.1.a).

e) 3.1.4. Finally, since $E(x,y) \neq 0$, whenever $x \neq y$, we have $xyx = \widehat{T}\ [E(x,x)|T(E(x,y),E(x,y))] = T(E(x,y),E(x,y)) < 1$. ■

It is worth noticing that given a fuzzy betweeness relation defined by means of a T-indistinguishability operator E, since

$$xyx = \widehat{T}\ (E(x,x)|T(E(x,y),E(x,y))) = T(E(x,y),E(x,y)),$$

the T-indistinguishability E can be recovered from the fuzzy betweenness relation.

Moreover, the following theorem shows that every fuzzy betweenness relation in a set X determines a T-indistinguishability operator E in X.

Nevertheless, before proving the preceding asumption, we state the following lemma that will be used in the proof.

Lemma 3.6. For every three elements x, y, z in X the following inequality holds:

$$xyx \leq xyz.$$

Proof. From 3.1.3.a) it follows $xyx = T(xyx, xxz) \leq xyz$. ■

Theorem 3.7. Let $f : [0,1] \to [0,1]$ be the function defined by

$$f(x) = y \quad \text{iff} \quad x = T(y,y).$$

If xyz is a fuzzy betweenness relation in a set X, then the fuzzy relations E' and E in X defined by $E'(x,y) = xyx$ and $E(x,y) = f(xyx)$ are both T-indistinguishability operators in X.

Proof.

It is easy to prove that if E' is a separating T-indistinguishability operator then E is also such an operator. Therefore, it is only necessary to prove the theorem for E'.

Reflexivity: $E'(x,x) = xxx = 1$ follows from 3.1.1.

Simmetry: From 3.1.3.b) $E'(x,y) = xyx = T(xyx, xxy) \leq yxy = E'(y,x)$.

Transitivity: From lemma 3.4. and 3.1.3.b), $T(E'(x,y), E'(y,z)) = T(xyx, yzy) =$

$T(yxy, yzy) \leq T(yxz, yzy) \leq T(yxz, yzx) \leq xzx = E'(x, z)$.

Separability: If $x \neq y$, then $E(x, y) = xyx < 1$. ■

4. Dimension and cardinality of betweenness relations.

Given a continuous t-norm T, every fuzzy subset h of X generates a T-indistinguishability operator E_h on X in the following way:

$$E_h(x, y) = \hat{T}(\text{Max}(h(x), h(y))|\text{Min}(h(x), h(y))).$$

The infimum of a family $\{E_i\}_{i\in I}$ of T-indistinguishability opeators in X also is such an operator. Reciprocally, any T-indistinguishability operator E can be constructed in this way [12] i.e. given E there exists a family $\{h_i\}_{i\in I}$ of fuzzy subsets of X such that

$$E = \text{Inf}_{i\in I} E_{h_i}.$$

Different families can generate the same operator. Using the cardinal of the indices of these families, we can introduce the idea of the dimension of E:

Definition 4.1. [1] The dimension of a T-indistinguishability operator E is the minimum of the cardinals of the generating families of E.

In [1] it is proved that a T-indistinguishability operator E in a set X (with $E(x, y) \neq 0$ and T an archimedean t-norm) is one-dimensional if and only if the (crisp) betweenness relation B on X defined by $(x, y, z) \in B$ iff $T(E(x, y), E(y, z)) = E(x, z)$ is total.

If the dimension of E is "high", it means that E is generated by "many" fuzzy subsets, each one giving a one-dimensional T-indistinguishability operator, and therefore, a total betweenness relation. Some of these relations will be overlaped by the others and therefore we have the following principle: the "higher" the dimension of a T-indistinguishability operator, the "lower" is the cardinal of the betweenness relation that determines.

On the other hand, if X is a finite set (with card $X = n$) and the dimension of E is "low" it will be more convenient to store a minimal generating family of E rather then the matrix of E. In this sense, it is useful to work with "low" dimensional T-indistinguishability operators. If E is a "high" dimensional operator, defining a fuzzy betweenness relation with many values close to 1, then there will be a operator E' "close" to E defining a crisp betweenness relation with a "high" cardinal. In this case, instead of storing E, it will be preferable to store a minimal generating family of E'.

Example. Let us consider the probabilistic relation (T=prod) E defined on a set X (card $X = 5$) by the following matrix

$$\begin{pmatrix} 1.00 & 0.74 & 0.67 & 0.50 & 0.41 \\ 0.74 & 1.00 & 0.87 & 0.65 & 0.53 \\ 0.67 & 0.87 & 1.00 & 0.74 & 0.60 \\ 0.50 & 0.65 & 0.74 & 1.00 & 0.80 \\ 0.41 & 0.53 & 0.60 & 0.80 & 1.00 \end{pmatrix}$$

The associated fuzzy betweenness relation is given by the table

x	y	z	xyz	x	y	z	xyz	x	y	z	xyz	x	y	z	xyz
1	2	3	0.96	2	3	1	0.79	3	4	1	0.55	4	5	1	0.66
1	2	4	0.96	2	3	4	0.99	3	4	2	0.55	4	5	2	0.65
1	2	5	0.96	2	3	5	0.98	3	4	5	0.99	4	5	3	0.65
1	3	2	0.79	2	4	1	0.44	3	5	1	0.37	5	1	2	0.57
1	3	4	0.99	2	4	3	0.55	3	5	2	0.37	5	1	3	0.46
1	3	5	0.98	2	4	5	0.98	3	5	4	0.65	5	1	4	0.26
1	4	2	0.44	2	5	1	0.29	4	1	2	0.57	5	2	1	0.96
1	4	3	0.55	2	5	3	0.37	4	1	3	0.45	5	2	3	0.77
1	4	5	0.98	2	5	4	0.65	4	1	5	0.26	5	2	4	0.46
1	5	2	0.29	3	1	2	0.57	4	2	1	0.96	5	3	1	0.98
1	5	3	0.37	3	1	4	0.45	4	2	3	0.76	5	3	2	0.98
1	5	4	0.66	3	1	5	0.46	4	2	5	0.43	5	3	4	0.56
2	1	3	0.57	3	2	1	0.96	4	3	1	0.99	5	4	1	0.98
2	1	4	0.57	3	2	4	0.76	4	3	2	0.99	5	4	2	0.98
2	1	5	0.57	3	2	5	0.77	4	3	5	0.56	5	4	3	0.99

It can be shown that the dimension of the relation E is 3. Nevertheless, it is "close" to the one-dimensional E' given by the matrix

$$\begin{pmatrix} 1.00 & 0.76 & 0.67 & 0.50 & 0.40 \\ 0.76 & 1.00 & 0.88 & 0.66 & 0.53 \\ 0.67 & 0.88 & 1.00 & 0.75 & 0.60 \\ 0.50 & 0.66 & 0.75 & 1.00 & 0.80 \\ 0.40 & 0.53 & 0.60 & 0.80 & 1.00 \end{pmatrix}$$

whose associated fuzzy betweenness is represented in the following table

x	y	z	xyz	x	y	z	xyz	x	y	z	xyz	x	y	z	xyz
1	2	3	1	2	3	1	0.78	3	4	1	0.56	4	5	1	0.64
1	2	4	1	2	3	4	1	3	4	2	0.56	4	5	2	0.64
1	2	5	1	2	3	5	1	3	4	5	1	4	5	3	0.64
1	3	2	0.78	2	4	1	0.43	3	5	1	0.36	5	1	2	0.58
1	3	4	1	2	4	3	0.56	3	5	2	0.36	5	1	3	0.45
1	3	5	1	2	4	5	1	3	5	4	0.64	5	1	4	0.25
1	4	2	0.43	2	5	1	0.28	4	1	2	0.58	5	2	1	1
1	4	3	0.56	2	5	3	0.36	4	1	3	0.45	5	2	3	0.78
1	4	5	1	2	5	4	0.64	4	1	5	0.25	5	2	4	0.43
1	5	2	0.28	3	1	2	0.58	4	2	1	1	5	3	1	1
1	5	3	0.36	3	1	4	0.45	4	2	3	0.78	5	3	2	1
1	5	4	0.64	3	1	5	0.45	4	2	5	0.43	5	3	4	0.56
2	1	3	0.58	3	2	1	1	4	3	1	1	5	4	1	1
2	1	4	0.58	3	2	4	0.78	4	3	2	1	5	4	2	1
2	1	5	0.58	3	2	5	0.78	4	3	5	0.56	5	4	3	1

In this case, it is clear that the crisp part of the fuzzy betweenness is a total crisp one since its cardinal is $\binom{25}{3}$.

Concluding remarks

The procedure for the effective construction of the unidimensional operator E' "close" to a given T-indistinguishability operator E (T being a strict t-norm) can be found in [3]. If E is "far" from being unidimensional, it is an open question to find a feasible procedure to approach E by means of a "low-dimensional" operator. This is one of the applications where the study of the cardinality of the associated betweenness relation can be a useful tool.

Similar ideas, will be used in a future work in order to approximate fuzzy preorders by means of unidimensional ones.

REFERENCES

[1] Jacas, J., *On the generators of T-indistinguishability operator.* Stochastica XII-1, (1988), 49-63.

[2] Jacas, J., *Similarity Relations-The Calculation of Minimal Generating Families.* Fuzzy Sets and Systems (1990), 151-162.

[3] Jacas, J., Recasens, J., *On some kind of probabilistic relations*, In: B. Bouchon-Meunier et al. eds., Uncertainty in Intelligent Systems, North-Holland, 1993, 171-178.

[4] Katz, M., *Inexact Geometry*, Notre Dame J. of Journal Logica 21 (1980).

[5] Menger, K., *Untersuchungen über allgemeine Metrik.* Mat. Ann. 100 (1928), 75-113.

[6] Ovchinnikov, S., *Choice Theory for Cardinal Scale*, In: M.M. Gupta and E. Sanchez, Fuzzy Inf. and Decision Processes, North-Holland, Amsterdam (1982), 323-336.

[7] Recasens, J., Sobre la representació i generació de relacions d'indistingibilitat. Ph.D. Dissertation. Prog. Matemàtica Aplicada, UPC, Barcelona, 1992.

[8] Ruspini, E., *A theory of Mathematical Classification.* Ph. D. Dissertation. Dept. of System Science. School of Engineering. Univ. of California. Los Angeles (1977).

[9] Saaty, T.L., *The Analytic Hierarchy Process*, McGraw- Hill, (1980).

[10] Schweizer, B., Sklar, A., *Probabilistic Metric Spaces*, N. Holland, Amsterdam (1983).

[11] Trillas, E., *Assaig sobre les relacions d'indistingibilitat.* Primer Congrés Català de Lògica Matemàtica. Barcelona (1982), 51-59.

[12] Valverde, L., *On the structure of F-indistinguishability operators.* Fuzzy Sets and Systems 17 (1985), 313-328.

[13] Zadeh, L.A., Similarity relations and fuzzy orderings. Inform. Sci. 3, (1971), 177-200.

GENERATORS AND DUAL T-INDISTINGUISHABILITIES

Dionís BOIXADER & Joan JACAS*

Sec. Matemàtiques i Informàtica. ETSAB
Univ. Politècnica de Catalunya
Diagonal 649, 08028 Barcelona. Spain

ABSTRACT

The main idea of this paper is the introduction of a duality between the elements of a set X and its fuzzy subsets. This duality induces a useful interpretation of the elements as fuzzy subsets and viceversa. As a natural consequence of this interpretation, the set of the fuzzy sets over $[0,1]^X$ can be viewed as a classical extension of X. In the case that X is endowed with a T-indistinguishability operator E and considering only the set H of the generators of E, the operator E can be naturally defined over the mentioned extension. On the other hand, a dual interpretation of the representation theorem, leads to a similar theorem but with a set of points as generators. Finally, it is also shown that T-similarities over $[0,1]^H$ are a suitable tool in order to built a sound theoretical background for the study of the inference by similarity.

Keywords: T-indistinguishability operator, Representation Theorem, generator, Compositional Rule of Inference (CRI), Duality.

0. Introduction.

One of the difficult points of the foundation of the Fuzzy Sets Theory its the fact that the starting set is a classical one with crisp elements. In this paper, by means of the concept of duality between the elements of a set X and those of its fuzzy power set $[0,1]^X$, both, points and fuzzy sets can be considered as fuzzy sets, eventhough defined over different sets. Therefore, when defining fuzzy sets over a crisp set X, the construction transmits, in some way, the fuzziness to the elements of X. After a preliminary section, the introduction of this idea will be the main concern of section 2. In section 3, it is shown that set of fuzzy sets over $[0,1]^X$, can be considered as an extension of X and as a consequence, we study the compatibility of a T-indistinguishability operator over the extended set and obtain a dual Representation Theorem for similarity relations over $[0,1]^X$ having points as generators.

In section 4, we introduce the concept of proper similarity and the largest one, termed natural similarity, turns out to be a useful tool for approximate reasoning via the generalized CRI since it verifyes the property that the natural similarity between the hypothesis is always greater or equal than the similarity of the theses. We close our paper with a summary.

1. Preliminaries.

In this section we recall some concepts related T-indistinguishability operators widely

*Research partially supported by DGICYT p.n. PB91-0334-CO3-02

used in different contexts [2,6,7].

Let T be a continuous t-norm, $\hat{T}$ will represent its *quasi- inverse* [9] defined by

$$\hat{T}(x|y) = \sup\{\alpha \in [0,1] \mid T(x,\alpha) \leq y\}.$$

Definition 1.1. Given a t-norm T, [1,8] a *T-indistinguishability operator* E on a set X is a fuzzy relation on X that satisfies

1.1.1. $E(x,x) = 1$, $\forall x \in X$ (*Reflexivity*),

1.1.2. $E(x,y) = E(y,x)$, $\forall x,y \in X$ (*Symmetry*),

1.1.3. $T(E(x,y), E(y,z)) \leq E(x,z)$, $\forall x,y,z \in X$ (*T-transitivity*).

In Valverde [9] it is proved that any T-indistinguishability operator can be constructed starting from a family of fuzzy sets.

Lemma 1.2. Given a fuzzy subset h of a set X, the fuzzy relation E_h defined by

$$E_h(x,y) = \hat{T}\left(\text{Max}(h(x),h(y)) \mid \text{Min}(h(x),h(y))\right),$$

is a T-indistinguishability operators on X.

In particular

$E_T(x,y) = \hat{T}(\text{Max}(x,y) \mid \text{Min}(x,y))$,

$x,y \in [0,1]$, is a T-indistinguishability in the unit interval and with this in mind, we can use the following notation $E_h = E_T \circ (h \times h)$.

Representation Theorem 1.3. [3,6,9] A fuzzy relation on a set X is a T-indistinguishability operator if and only if there exists a family $\{h_i\}_{i\in I}$ of fuzzy subsets of X such that

$$E = \inf_{i\in I} E_{h_i} = \inf_{i\in I} E_T \circ (h_i \times h_i).$$

Theorem 1.3 suggests the following definition:

Definition 1.4. Given a T-indistinguishability operator E on X, a *generator* [3,5] of E is a fuzzy set of X that belongs to a generating family of E in the sense of the preceding theorem.

Denoting by H_E the set of generators of E, it is clear that $h \in H_E$ if and only if $E_h \geq E$.

Definition 1.5. The minimum of the cardinals of the generating families of a T-indistinguishability operator E is called its *dimension* and a family with this cardinality is termed a *basis* of E. [3,4]

The operator proposed in the next definition will be a useful tool in the study of the structure of the set H_E.

Definition 1.6. Given a t-norm T and a T-indistinguishability operator E on a set X, ϕ_E will denote the operator from $[0,1]^X$ to $[0,1]^X$ defined by

$$\phi_E(h)(x) = \text{Sup}_{y\in X}\{T(E(x,y),h(y))\}.$$

With this definition, a fuzzy subset h of X is a *T-eigenvector* [3] of E if and only if $\phi_E(h) = h$.

In Jacas & Recasens [6] the equivalence between generators and T-eigenvectors is proved. Namely,

Theorem 1.7. h is a generator of a given T-indistinguishability operator E, if and only if, h is a T-eigenvector of E.

Therefore H_E can be viewed either as the set of the generators of E or as the set of its T-eigenvectors.

2. Duality

In this section we introduce the concept of duality between the elements of a classical set X and its fuzzy subsets. The definition of this duality principle is independent from any possible T-indistinguishability defined over X. Further, it will be shown that it is always compatible with them.

We recall a classical definition

Definition 2.1. Given a non-empty set X and a family H of fuzzy subsets of X ($H \subseteq [0,1]^X$), we say the H a separating family of X if, for any $x, y \in X$, there exists $h \in H$ such that $h(x) \neq h(y)$.

For any $H \subseteq [0,1]^X$, let us consider the mapping $\psi_H : X \to [0,1]^H$, such that, for any $x \in X$, $\psi_H(x)$ is the fuzzy set over H given by:

$$\begin{aligned} \psi_H(x) : H &\longrightarrow [0,1] \\ h &\longrightarrow \psi_H(x)(h) = h(x) \end{aligned} \qquad (1)$$

Note. In the sequel, if there is no ambiguity for the set H, we shall use the notation $\psi_H(x) = x^*$ and, therefore $\psi_H(x)(h) = x^*(h) = h(x)$.

Proposition 2.2. ψ_H is an injective mapping if and only if, H is a separating family for X.

Proof. If ψ_H is injective, for any $x_1 \neq x_2$, $x_1^* \neq x_2^*$. Consequently, there exists $h \in H$ such that $x_1^*(h) \neq x_2^*(h)$. From the definition of ψ_H it trivially follows $h(x_1) \neq h(x_2)$ therefore, H is a separating family.

Conversely, if H is a separating family and $x_1^* = x_2^*$, for some $x_1 \neq x_2$ in X, then for any $h \in X$, $x_1^*(h) = x_2^*(h)$ and $h(x_1) = h(x_2)$ and $x_1 = x_2$. ∎

From this proposition, it follows that, given a separating family H of fuzzy subsets of X, the mapping ψ_H induces an identification between the elements $x \in X$ and the elements x^* of $[0,1]^H$. Therefore, we can state the folllowing principle.

Duality principle: By means of the mapping ψ_H, we can identify the set X with the set

$\psi_H(x)$. Under this assumption, the elements of $[0,1]^H$ can be considered as points of X and the elements of X as fuzzy subsets of H.

This duality shows that the fuzzy subsets have always an underlying meaning as points of a certain set and conversely any point of a set X can be interpreted as a fuzzy subset over a suitable set.

The next two schemas give a graphical interpretation of this principle.

Let $H = \{h_1, h_2\}$ be a separating family of the set $X = \{x_1, x_2, x_3, x_4\}$ represented in figure 1. By duality, the elements of X can be interpreted as fuzzy subsets of H as shown in figure 2.

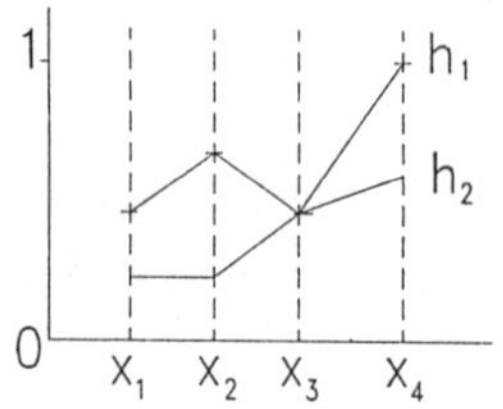

fig. 1.

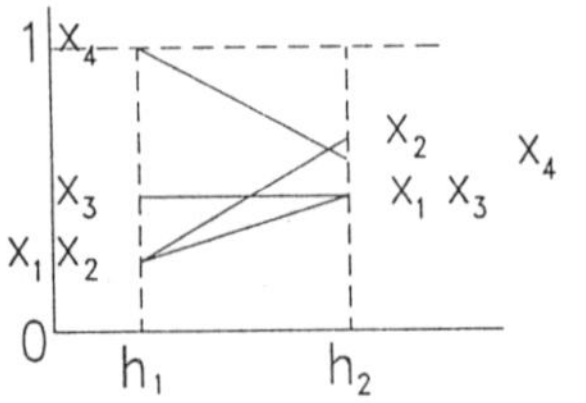

fig. 2.

Let us note that when $X = \{x_1, \ldots, x_n\}$ is a finite set, and the set $H = \{h_1, \ldots, h_k\}$ is represented by a $n \times k$ matrix $M = (m_{ij})$ where $m_{ij} = h_j(x_i)$, this matrix represents $\{x_1, \ldots, x_n\}$ as elements (points) and h_1, h_k as fuzzy sets, and the transposed one $M^T = (m_{ji})$ represents the set $\{x_1, \ldots, x_n\}$ as fuzzy sets and h_1, h_k as elements (points).

Given an arbitrary set H, ψ_H is not necessarily a surjective mapping. Nevertheless, it is possible to find a coherent interpretation of the elements of the set $\mathcal{X} = [0,1]^H - \psi_H(X)$ in the following way: given $\alpha \notin \psi_H(X)$, we add a new element a to X ($X' = X \cup \{a\}$) and we extend every $h \subseteq H$ to X' taking $h(a) = \alpha(h)$. The figure 3 ilustrates this extension.

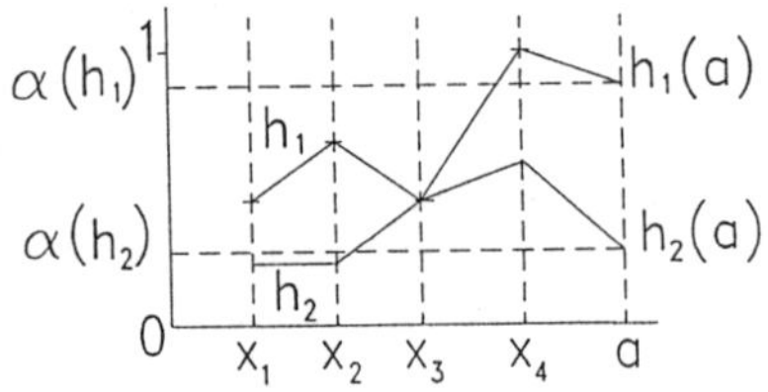

fig. 3.

In this way, we have $\psi_H(a) = a^* = \alpha$ and if ψ_H is injective, the set $[0,1]^H$ can be considered as an extension of X with respect to the classical set inclusion.

Note: In this procedure, we can evidently consider another element $a' \notin X$, $a' \neq a$ such that $\alpha(h) = h(a')$ $\forall h \in H$. In this case we have $h(a) = h(a') \forall h \in H$ and, if we impose the restriction that the family H has to separate points over the extended set $X' = X \cup \{a, a', ...\}$, we get into a contradiction. Therefore, for every $\alpha \in X$, we can select a unique "a", such that H is a separating family of extended set X'.

3. Duality and T-indistinguishability.

In this section, we show firstly that the T-indistinguishability relations in a set X, are compatible with the duality principle and as a result, it will follow a way to built such relations over the fuzzy power set of X.

Throughtout the rest of this section, X will be a non empty set, E a T-indistinguishability and T, a continuous t-norm. $H_E \subseteq [0,1]^X$ will represent the set of generators of E [3] and $\psi_{H_E} : X \to [0,1]^{H_E}$ defined in (1). With this notation, we have the following proposition.

Proposition 3.1 ψ_{H_E} is injective, if and only if, E is a separating-indistinguishability operator.

Proof. From proposition 2.1, it follows that ψ_{H_E} is injective iff H_E is a separating set of X, which is equivalent to the statement that E is separating relation. Efectively, from the representation theorem, we have $E = \text{Inf}_{h\in H_E} E_T \circ (h \times h)$ therefore if $E(x_1, x_2) = 1$ for some $x_1, x_2 \in X$, $E(x_1, x_2) = \text{Inf}_{h\in H_E} E_T(h(x_1), h(x_2)) = 1 \Leftrightarrow h(x_1) = h(x_2)\ \forall h \in H_E$. ■

In the sequel, only separating T-indistinguishabilities will be considered and consequently, we will be able to identify the inicial set X with the set $\psi_{H_E}(X)$.

The representation theorem shows that any T-indistinguishability E is generated by a family H of fuzzy subsets of X. On the other hand, the duality principle suggests that, in a similar way, a subset $X_0 \subseteq X$ of points generates a T-indistinguishability over $[0,1]^X$ that can be restricted over any subset $H \subseteq [0,1]^X$. This is proved in the following proposition.

Proposition 3.2. Given $X_0 \subset X$, and $H \subseteq [0,1]^X$, the fuzzy relation. $\tilde{E} : H \times H \to [0,1]$ defined by

$$\tilde{E}(h_1, h_2) = \text{Inf}_{x\in X_0} E_T(h_1(x), h_2(x)).$$
$$\forall h_1, h_2 \in H,$$

is a T-indistinguishability over $[0,1]^H$.

Proof. Let us consider the mapping ψ_H, with the previously defined notation ($\psi_H\,(x) = x^*$), and taking $X_0^* = \{x^* /\ x \in X_0\}$, we have

$$\tilde{E}(h_1, h_2) = \text{Inf}_{x\in X_0} E_T(h_1(x), h_2(x))$$
$$= \text{Inf}_{x^*\in X_0^*} E_T(x^*(h_1), x^*(h_2))$$

since x^* are fuzzy sets over $H(x^* : H \to [0,1])$, the representation theorem assures that $\tilde{E}$ is a T-indistinguishability. ■

Given E, let $H_E \subseteq [0,1]^X$ be the generators set, we can consider the fuzzy relation E^* defined over $[0,1]^{H_E}$ given by

$$E^*(\alpha, \beta) = \text{Inf}_{h\in H_E} E_T(\alpha(h), \beta(h)),$$
$$\alpha, \beta \in [0,1]^{H_E}$$

The preceding proposition assures that E^* is a T-indistinguishability. The next theorem expresses the links between E and E^*.

Theorem 3.3. Given a separating T-indistinguishability in X, and E^* generated by H_E, the mapping

$$X \xrightarrow{\psi_{H_E}} [0,1]^{H_E}$$
$$x \longrightarrow \psi_{H_E}(x) = x^*$$

is an injective morphism between (X, E, T) and $([0,1]^{H_E}, E^*, T)$.

Proof. It follows inmediately from proposition 2.1 and de definition of E^*. ■

This theorem shows that $[0,1]^{H_E}$ is not only an extension of X as a set but also as a similarity structure.

Definition 3.4. Given a separating T-indistinguishability E, in X, a "*pointwise representation*" of E is the pair $((H, E'), \varphi)$ where $H \subset [0,1]^Y$, for some set Y, E' is a T-indistinguishability defined over H by

$$E'(h_1, h_2) = \text{Inf}_{y \in Y} E_T(h_1(y), h_2(y)),$$

and $\varphi : X \to H$ is an isometry of T-indistinguishabilities i.e. $E(x_1, x_2) = E'(\varphi(x_1), \varphi(x_2))$ and φ is bijective.

As a consequence of this definition, it can be proved that any separating T-indistinguishability E has a pointwise representation. The concept of dimension [5] of E can also be redefined as the cardinal of a minimal set of points that generates E.

Let $\tilde{E}$ be a separating T-indistinguishability over $H \subseteq [0,1]^X$ and $H_{\tilde{E}} \subseteq [0,1]^H$ the set of its generators.

Definition 3.5. $\tilde{E}$ is termed a *proper* indistinguishability if $\psi_H(x) \subseteq H_{\tilde{E}}$.

If we interpret $[0,1]^H$ as an extension of X (identifiying X with $\psi_H(X)$) and we also define the elements of H over this extension, then "to be proper" can be understood as a principle of coherence that is: we impose the condition that the set of generating points of must contain al least, the set X.

Lemma 3.6. If $H \subset [0,1]^X$ is a separating family of X, the T-indistinguishability $\overline{E}$ over H defined by

$$\overline{E}(h_1, h_2) = \text{Inf}_{x \in X} E_T(h_1(x), h_2(x))$$

is proper.

Using the same notation, we have

Theorem 3.7. If F is any T-indistinguisha-

bility over $H \subseteq [0,1]^X$, the following statements are equivalent.

a) F is proper

b) $F \leq \overline{E}$

c) $\text{Sup}_{g \in H} T[F(h,g), g(x)] = h(x)$ for any $x \in X$, $h \in H$.

Proof.

(a) $\Rightarrow$ (b)

Follows from the fact that $\psi_H(x) \subseteq H_F$ (being F proper).

(b) $\Rightarrow$ (a)

Is immediate.

(a) $\Leftrightarrow$ (c)

To prove this equivalence we use the characterization of the generators [3]; ($\alpha \in [0,1]^H$ is a generator of $F \Leftrightarrow \phi_F(\alpha) = \alpha$),

$$\phi_F(\alpha)(h) = \mathrm{Sup}_{g \in H} T[\alpha(g), F(h,g)]$$

taking $\alpha = x^*$, we have

$$\phi_F(x^*)(h) = \mathrm{Sup}_{g \in H} T[x^*(g), F(h,g)] = x^*(h),$$

and finally

$$\mathrm{Sup}_{g \in H} T\left[g(x), F(h,g)\right] = h(x).$$

∎

This result entails that $\overline{E}$ is the greatest of the proper indistinguishabilities, and it will termed *the natural indistinguishability* over H.

4. Natural T-indistinguishabilities and CRI

Let us we recall the operator:

$$\begin{aligned} \phi_E(h) : X &\longrightarrow [0,1] \\ X &\longrightarrow \phi_E(h)(x) = \sup_{u \in X} T[E(x,u), h(u)] \end{aligned}$$

In particular, if $h = \{x\}$, then $\phi_E(\{x\}) = \{E(x,u)\}_{u \in X} = E_x$. With this notation, we have the following result.

Proposition 4.1. Given E, $H \subseteq [0,1]^X$ a separating family and $\overline{E}$ its associated natural indistinguishability, then

a) $E(x,y) = \overline{E}\left(\phi_E(\{x\}), \phi_E(\{y\})\right)$

b) $\overline{E}(\mu,\nu) \leq \overline{E}\left(\phi_E(\mu), \phi_E(\nu)\right)$

Proof.

a) It follows from the fact that the set $\{E_x\}_{x \in X}$ generates E [9].

b) From the properties of the quasi-inverse of T, and the definition of ϕ_E, we have

$$\begin{aligned} &\hat{T}[\phi_E(\mu)(x) \mid \phi_E(\nu)(x)] \geq \\ &\geq \mathrm{Inf}_{u \in U} E_T[\mu(u), \nu(u)] = \\ &= \overline{E}(\mu,\nu) \end{aligned}$$

and in a similar way

$$\hat{T}(\phi_E(\nu)(x)) \mid \phi_E(\mu)(x)) \geq \overline{E}(\nu,\mu),$$

so

$$E_T[\phi_E(\mu)(x), \phi_E(\nu)(x)] \geq \overline{E}(\nu,\mu)$$

and therefore,

$$\overline{E}(\phi_E(\mu), \phi_E(\nu)) \geq \overline{E}(\nu, \mu).$$

■

Next, we show that the natural indistinguishabilities are adequate tools when applying the CRI to the inference process.

Given A, $A' : U \to [0,1]$ and $B : V \to [0,1]$ applying the generalized Compositional Rule of Inference (CRI) [2] we obtain $B' : V \to [0,1]$ in the following way

$$B'(v) = \text{Sup}_{u \in U} T[A'(u), I(A(u), B(v))]$$

where I is the R-implication associated to the t-norm T [2].

Proposition 4.2. If $B', B'' : V \to [0,1]$ are the conclusion associated to the assumptions A and A' via the CRI respectively, then

$$\overline{E}_V(B', B'') \geq \overline{E}_U(A', A'')$$

where $\overline{E}_V$ and $\overline{E}_U$ are the natural indistinguishabilities over $[0,1]^V$ and $[0,1]^U$ respectively.

Proof. For any $v \in V$, we have

$$\begin{aligned}
&\hat{T}(B'(v) \mid B''(v)) = \\
&= \hat{T}(\text{Sup}_{u \in U} T[A'(u), I(A(u), B(v))] \mid \\
&\text{Sup}_{w \in U} T[A''(w), I(A(w), B(v))] \geq \\
&\text{Inf}_{u' \in U} E_T[A'(u), A''(u)] = \\
&= \overline{E}(A', A'')
\end{aligned}$$

In a similar way, we can prove that

$$\hat{T}(B''(v) \mid B'(v)) \geq \overline{E}(A', A'')$$

and, as a consequence

$$E_T(B'(v), B''(v)) \geq \overline{E}(A', A'')$$

for any $v \in V$, and the theorem follows immediately. ■

The application of natural indistinguishabilities combined with the CRI entails that the degree of similarity between the hipothesis leads to a degree of similarity between the thesis greater or equal to the former one which is a desirable property since it is a natural extension of the classical reasoning.

Summary.

We have shown that when defining a T-indistinguishability E over $H \subseteq [0,1]^X$ (H separating), the duality between points an fuzzy sets gives a clear picture of the situation

a) E is a proper similarity $E \leq \overline{E}$ (being $\overline{E}$ the natural similarity), or

b) The similarity E does not take into account the information given by some points of X (even all of them!) via the values that the elements $h \in H$ take over them.

It is also interesting to point out the natural similarity $\overline{E}$ is the greatest between the proper ones and therefore, it is the similarity that assumes the least amount of information.

On the other hand, the natural similarities are a suitable tool for the approximate reasoning since the CRI is contractive under their application.

REFERENCES

[1] Alsina, C., Trillas, E., Valverde, L. (1983) On some logical connectives for fuzzy set theory, *J. Math. Annal. Appl.* 93, 15-26.

[2] Godo, L., Jacas, J., Valverde, L. (1991) Fuzzy Values in Fuzzy Logic, *International Journal of Intelligent Systems*, 6, 199-212.

[3] Jacas, J. (1988) On the generators of a T-indistinguishability operator, *Stochastica*, 12, 49-63.

[4] Jacas, J. (1990) Similarity relations. The calculation of minimal generating families, *Fuzzy Sets and Systems*, 35, 151-162.

[5] Jacas, J., Recasens, J. (1992) Eigenvectors and generators of fuzzy relations, *Proc. of FUZZ-IEEE'92*, San Diego, 687-694.

[6] Jacas, J., Recasens, J. (1994) Fixed points and generators of fuzzy relations. Journal of Mathematical Analysis and Applications, vol. 186, n. 1, 21-29.

[7] Jacas, J., Valverde, L. (1990) On Fuzzy Relations, metrics and cluster Analysis in: J.L. Verdegay and M. Delgado Eds. Approximate Reasoning Tools for Artificial Intelligence, ISR 96 (Verlag TUV, Rheinland), 21-38.

[8] Klawonn, F., Kruse, R. (1992) Equality relations as a basis of Fuzzy Control, *Proc. of Fuzz-IEEE'92*, San Diego.

[9] Schweizer, B., Sklar, A. (1983) *Probabilistic Metric Spaces*, (North-Holland, Amsterdam).

[10] Valverde, L. (1985) On the structure of F-indistinguishability operators, *Fuzzy Sets and Systems*, 17, 313-328.

ON CONDITIONING IN SIMILARITY LOGIC

Francesc ESTEVA, Pere GARCIA, Lluís GODO

Institut d'Investigació en Intel·ligència Artificial, (IIIA)
Spanish Council for Scientific Research, (CSIC)
Campus de la Universitat Autonoma de Barcelona,
08193 Bellaterra, Barcelona, Spain
e-mails: {esteva,pere,godo}@iiia.csic.es

ABSTRACT

The similarity-based model of possibilistic and fuzzy reasoning mainly relies on the use of the so-called pairs of implication-consistency measures, that are shown to be possibility envelopes. This notion is the possibilistic counterpart of lower and upper probabilities and it is used in this paper to introduce within the similarity framework several conditioning methods. A preliminary and comparative study of them is also presented in a general setting, as well as the application of the different methods to a typical reasoning scenario with fuzzy variables.

1. Introduction

The similarity-based model for possibilistic and fuzzy reasoning [9, 10] combines logic with metric notions. To represent incomplete and vague knowledge, the model relies on the use of the so-called *evidential sets*, i.e. sets of possible worlds that are compatible with the evidence, together with a *similarity relation* describing the resemblance among possible worlds. The similarity relation permits to describe how far (dissimilar) or how close (similar) are arbitrary subsets of possible worlds from those of the evidential set. In what follows we consider a continuous t-norm T and a T-similarity relation S defined on the set of possible worlds W, i.e., a function

$$S: W \times W \to [0, 1]$$

which is reflexive, i.e. $S(w, w) = 1$, symmetric, i.e. $S(w, w') = S(w', w)$, and T-transitive, i.e. $T(S(w, w'), S(w', w'')) \leq S(w, w'')$.

Given a t-norm T, its corresponding R-implication function R_T is defined by residuation of T as:

$$R_T(x, y) = Sup\{ c \in [0, 1] \,/\, T(x, c) \leq y \}.$$

Throughout the paper we will also use infix notations for T and R_T, namely we will write $x \otimes y$ for $T(x, y)$, and $x \varnothing y$ for $R_T(y, x)$.

More precisely, the similarity-based formalism is based on the use of the so-called consistency and implication measures which bound by above and below respectively the distance of set of worlds (describing a proposition) from the set of evidential worlds. We start out from the original Ruspini's definitions of *unconditioned Consistency and Implication measures* [9, 5]. For any evidential subset $E \subset W$, we define respectively:

$$C_E(p) = Sup_{w \in E} Sup_{w' \vdash p} S(w, w')$$

$$I_E(p) = Inf_{w \in E} Sup_{w' \vdash p} S(w, w')$$

In [5], the formalization of the language and semantics of *similarity logic* is done as follows:

Language: given a Boolean algebra of Propositions L, the extended language SL is defined as $SL = \{(p, [\alpha, \beta]) \mid p \in L \text{ and } \alpha, \beta \in [0, 1], \alpha \leq \beta \}$

Semantics: A SL-model on W is a pair (S, E), where S is a T-similarity relation on the set W of possible worlds, and E is a subset of W. The satisfaction relation is defined as

$$(S, E) \vDash (p, [\alpha, \beta]) \text{ if, and only if, } \alpha \leq I_E(p) \text{ and } C_E(p) \leq \beta.$$

In this paper we are interested in investigating the problem of how to update the information conveyed by the consistency and implication measures at the light of new information, in other words how conditioning can be performed in similarity logic. In the original work of Ruspini [8], and later considered in [4], a single body of evidence was considered as the unique source of knowledge about the truth of single propositions (unconditioned knowledge) and that of information relating the truth of several propositions (conditional knowledge). Recently, in [6] the formalism has been extended to deal with inference in the more usual situation where the bodies of evidence that originate conditional and unconditional information can be different. In all these previous works, a first definition of conditional measures was proposed and proved to behave correctly w.r.t. two formulations of a generalized modus ponens rule. That definition is the following one:

$$C_E^0(p \mid q) = Sup_{w \in E}\{\pi_w(p) \varnothing \pi_w(q)\}$$

$$I_E^0(p \mid q) = Inf_{w \in E}\{\pi_w(p) \varnothing \pi_w(q)\}$$

being $\pi_w(p) = Sup_{w' \vDash p} S(w, w')$. This definition was specially interesting for providing a justification of the soundness of that inference rule on the basis of the similarity model of fuzzy logic. Moreover, this was the main purpose for that definition. On the other hand, although well suited for reasoning with fuzzy variables (see sections 4 and 5), the above definition may have some difficulties in a general setting since, for instance, it may happen that $I_E(p \mid q) > 0$ while being p and q mutually exclusive.

In this paper, the departure point is different. We let the behaviour of conditionals w.r.t. modus ponens as a secondary goal, and we concentrate on the problem of conditioning as an updating information procedure by itself and study several options somehow related to conditioning formulas proposed in the literature for other pairs of fuzzy measures, such as possibility and necessity measures, belief and plausibility measures, and mainly upper and lower probabilities. The main reason for that has been to realize that, on the one hand, consistency measures C_E are themselves possibility measures, and thus, plausibility measures as well, but on the other hand, a pair of consistency and implication measures are in fact a pair of *upper and lower possibilities*. That is, given an evidential set E, the consistency measure C_E is the supremum of a family of possibility measures, while the implication measure I_E is the infimum of the same family of possibilities. Namely, given a T-similarity relation S on the set of possible worlds W and a world w the function

$$\pi_w: W \rightarrow [0,1]$$

defined as

$$\pi_w(w') = S(w, w')$$

is a normalized possibility distribution on W. Then, it is immediate to see that the consistency and implication measures can be expressed as

$$C_E(p) = Sup\{\ \pi_w(p) \mid w \in E\}$$

$$I_E(p) = Inf\{\ \pi_w(p) \mid w \in E\}$$

where π_w denotes the possibility measure induced by the possibility distribution with the same name π_w, i.e.

$$\pi_w(p) = Sup\{\ \pi_w(w') \mid w' \vDash p\}$$

Therefore the pair $(I_E,\ C_E)$ is just a possibility envelope, or in other words, I_E and C_E are respectively the lower and upper possibilities of the family of possibility measures $\{\pi_w \mid w \in E\}$. It is worth noticing that the underlying set of possibility measures

$$Pos(E) = \{\Pi \text{ possibility measure} \mid I_E \leq \Pi \leq C_E\}$$

associated to the pair $(I_E,\ C_E)$, has also I_E and C_E as its lower and upper possibilities, that is,

$$C_E = Sup\{\ \Pi \mid \Pi \in Pos(E)\ \}$$

$$I_E = Inf\{\ \Pi \mid \Pi \in Pos(E)\ \}.$$

This is similar to the case of pairs of belief-plausibility measures with respect to their associated family of probabilities (see section 2).

The paper is organized as follows. After this introduction, we will follow in section 2 with a review of main conditioning forms for pairs of fuzzy measures, namely Dubois-Prade's conditioning definition for possibility and necessity measures, the two Dempster's definitions of conditioning belief and plausibility measures, and Moral-Campos' conditioning method for upper and lower probabilities. Next, in section 3, taking as reference the mentioned ways of conditioning, we propose and compare several conditioning formulas for the similarity-based measures in a general framework.

Finally, in section 4, results are particularized and discussed into the fuzzy variables reasoning framework. We conclude with some final remarks.

2. Background on conditioning fuzzy measures

As mentioned above, in this section we are going to make a brief review of main conditioning methods proposed for several fuzzy measures. In what follows, such measures are supposed to be defined on top of a boolean algebra L of propositions.

2.1. Conditioning possibility measures

Given a possibility measure Π and a proposition q, Dubois and Prade propose to define a new possibility measure $\Pi(./q)$ obtained by updating Π from q as follows

$$\Pi(p \mid q) = \Pi(p \wedge q) \oslash \Pi(q),$$

See [2] and [3] for a detailed justification. This definition satisfies a Bayes theorem-like expression, e.g.

$$\Pi(p \wedge q) = \Pi(q) \otimes \Pi(p \mid q)$$

Moreover if we denote the conditional measure $\Pi(./q)$ by Π_q, it also holds that

$$\Pi_{q \wedge r}(p) = \Pi_q(p \mid r) = \Pi_r(p \mid q)$$

i.e., the conditioning formula works properly with iterated conditionings. Finally, the corresponding conditioned necessity measure is defined in this case by duality as

$$N(p \mid q) = 1 - \Pi(\neg p \mid q)$$

2.2. Conditioning in the Theory of Evidence

Within the framework of the Theory of Evidence, Dempster introduced in [1] two different ways of conditioning. For a pair of belief and plausibility measures (Bel, Pl), the first form of conditioning is a generalization of the Bayes formula for the plausibility measure, i.e.

$$Pl_1(p \mid q) = \frac{Pl(p \wedge q)}{Pl(q)}, \text{ if } Pl(q) \neq 0$$

$$Bel_1(p \mid q) = 1 - Pl_1(\neg p \mid q), \text{ if } Pl(q) \neq 0$$

The other definition was given in terms of the underlying convex set of probability distributions $\mathcal{P} = \{P \text{ probability measure} \mid Bel \leq P \leq Pl\}$ assosiated to the pair (Bel, Pl). The conditioned belief and plausibility measures are then computed by conditioning all the probabilities, i.e.

$$Pl_2(p \mid q) = sup\{ P(p \mid q) \mid P \in \mathcal{P}\}$$

$$Bel_2(p \mid q) = inf\{ P(p \mid q) \mid P \in \mathcal{P}\}$$

Notice that $(Bel_1(./q), Pl_1(./q))$ and $(Bel_2(./q), Pl_2(./q))$ are both new belief-plausibility pairs, and that the inequalities

$$Bel_2(./q) \leq Bel_1(./q) \leq Pl_1(./q) \leq Pl_2(./q)$$

always hold.

2.3. Conditioning lower and upper probabilities

The Dempster's second way of conditioning can be directly applicable to upper and lower probabilities [7], that is, given a family of probability distributions $\mathcal{P}$, conditioning to q is just to consider the set of conditioned distributions $\mathcal{P}_q = \{P(. \mid q) \mid P \in \mathcal{P}\}$, and then to compute the upper and lower probabilities as

$$u(p \mid q) = sup\{ P(p) \mid P \in \mathcal{P}_q \} = sup\{ P(p \mid q) \mid P \in \mathcal{P}\}$$

$$l(p \mid q) = inf\{ P(p) \mid P \in \mathcal{P}_q \} = inf\{ P(p \mid q) \mid P \in \mathcal{P}\}$$

The first Dempster's definition of conditioning can be also generalized for convex sets of probabilities [8] taking into account that the convex set associated to the pair $(Bel_1(./q), Pl_1(./q))$ is the convex set generated by

$$\mathcal{P}'_q = \{ P(.\,/\,q) \,/\, P \in \mathcal{P}, P(q) = Sup_{P_i \in \mathcal{P}} P_i(q)\}$$

where $\{P_1, ..., P_n\}$ is the (finite) set of extreme probability distributions of $\mathcal{P}$.

In [7] and [8] it is shown that, from the upper and lower probabilities point of view, the resulting probability interval using the first Dempster formula is too informative, while the second formula produces intervals too uninformative. These reasons led to Moral and de Campos to introduce a new method of conditioning, intermediate between the above two. In their own words [8], the idea behind this new method is the following one. As we have already mentioned, if we want to update the upper and lower probability of a finite family of probability distributions $\mathcal{P} = \{P_i\}$, by the evidence of a proposition q, we have to consider the set $\mathcal{P}_q = \{P_i(./q) \,/\, P_i \in \mathcal{P}\}$ of conditioned probability measures. But knowing that q is true not all the conditioned probabilities have the same possibility. For instance, if $P_i(q)$ is very small, it is no easy that, given q, P_i be the true distribution. Then they associate to each P_i a possibility proportional to $P_i(q)$. That is, they define a normalized possibility distribution Pos in the set $\mathcal{P}_q$:

$$Pos(P_i(./\,q)) = \frac{P_i(./q)}{\max\{P_i(q) \,/\, P_i \in \mathcal{P}\}}$$

that induces a possibility measure POS in the power set of $\mathcal{P}_q$:

$$POS(T) = Sup\{Pos(t) \,/\, t \in T\}, \textit{for every } T \subseteq \mathcal{P}_q$$

and the dual necessity measure NEC is defined as usual as $NEC(T) = 1 - POS(\neg T)$. Then, within this approach, the result of conditioning is not a single value nor an interval but a possibility distribution on [0,1] defined as

$$\pi_{p/q}(x) = Sup_{P_i \in P} \{Pos(P_i(./\,q)) \,/\, P_i(p\,/\,q) = x\}$$

A conditional uncertainty interval is then defined by means of a fuzzy integral w.r.t. the above POS and NEC measures. Namely, the Choquet and Sugeno integrals of a function g w.r.t. to a measure h takes the form

$$In_C(g\,/\,h) = \int_0^1 h(\{x \,/\, g(x) \geq \alpha\})d\alpha.$$

$$In_S(g\,/\,h) = \underset{\alpha \in [0,1]}{Sup} (h(\{x \,/\, g(x) \geq \alpha,\}) \wedge \alpha)$$

respectively. Thus, the resulting interval using the Choquet integral has de following extreme points:

$$u_C(p\,/\,q) = In_C(g_p \,/\, POS)$$

$$l_C(p\,/\,q) = In_C(g_p \,/\, NEC)$$

and using the Sugeno integrals these extreme points are

$$u_S(p\,/\,q) = In_S(g_p \,/\, POS)$$

$$l_S(p\,/\,q) = In_S(g_p \,/\, NEC)$$

where, for each p, the function g_p is taken as

$$g_p(P_i(./q)) = P_i(p\,/\,q)$$

In [6] it is shown that these integrals satisfy the following duality conditions:

$$In_C(g\,/\,POS) + In_C(1 - g\,/\,NEC) = 1$$

$$In_S(g\,/\,POS) + In_S(1 - g\,/\,NEC) = 1$$

An interesting point is that when POS is a *crisp* possibility distribution of the form

$$Pos(t) = \begin{cases} 1, & \textit{if } t \in A \\ 0, & \textit{otherwise} \end{cases}$$

being $A \subset \mathcal{P}_q$, then it holds that

$$In_C(g\,/\,POS) = In_S(g\,/\,POS) = Sup\{g(t) \,/\, t \in A\}.$$

In particular, for the case of belief-plausibility pairs, [7] also shows that for

$$A = \{P_i(./q) \,/\, Pos(P_i(./q)) = 1\}$$

we recover the first Dempster definition while when we take

$$A = \{P_i(./q) \,/\, Pos(P_i(./q)) > 0\}$$

is the second Dempster definition which is recovered.

3. Conditional Similarity measures

Now we turn our attention to the similarity-based framework and more particularly to the consistency and implication measures. In this section we will first discuss the extension of the two Dempster ways of conditioning, and later the extension of Moral-Campos conditioning method. Finally, a last conditioning method, arising from considerations totally different from the previous ones, is considered.

3.1. Dempster-like conditional similarity measures

Noticing that C_E is a possibility measure, we can think of the following conditioning formula for Consistency measures:

$$C_E^1(p/q) = C_E(p \wedge q) \varnothing C_E(q)$$

which is the analog form of first Dempster's way of conditioning plausibility measures, although here it would be not necessary to impose the condition $C_E(q) > 0$. But on the other hand, I_E is neither a possibility nor a necessity measure, so at this point there is no hint to look for a similar way of conditioning for I_E measures. However, we will come back to this point later on.

Considering C_E and I_E as upper and lower possibility measures, conditioning by a proposition q can be therefore defined as the upper and lower conditional possibility measures respectively, in a similar way to the second Dempster's conditioning for belief and plausibility measures. That is, we define:

$$C_E^2(p/q) = Sup_{w\in E}\, \pi_w(p/q) = Sup_{w\in E}\{\pi_w(p \wedge q) \varnothing \pi_w(q)\}$$
$$I_E^2(p/q) = Inf_{w\in E}\, \pi_w(p/q) = Inf_{w\in E}\{\pi_w(p \wedge q) \varnothing \pi_w(q)\}$$

From this perspective, it makes no sense to consider those worlds of E that make impossible q. Therefore, and for the sake of simplicity, we will only consider, from now on, in the above conditioning formulas worlds w for which $\Pi_w(q) \neq 0$. Notice that, contrary to upper and lower probabilities, the upper envelope of a family of possibilities is itself a possibility measure, as it is the case of C_E. However, the upper measure $C_E^1(./q)$ of the set $\{\Pi_w(./q) / w\in E\}$ of conditional possibility measures does not coincide with the conditional possibility $C_E^2(./q)$ associated to the upper possibility C_E of $\{\Pi_w / w\in E\}$. Finally, and analogously to upper and lower probabilities, it is easy to check that the inequality

$$C_E^2 \leq C_E^1$$

also holds in this framework.

3.2. Conditioning upper and lower possibility measures: adapting the Moral-Campos' method

Analogous arguments of Moral and de Campos for the case of upper and lower probabilities can be now applied to the case of upper and lower possibilities. Given an evidential subset E, to update the upper and lower measure of the family of possibility measures $\Gamma = \{\pi_w / w \in E\}$ with the evidence of a proposition q, we have to consider the set of conditional possibility distributions $\Gamma_q = \{\pi_w(./q) / w \in E\}$. But again, knowing that q is true, not all the conditioned measures have the same possibility to be the true distribution. This is modelled by assigning to each π_i a possibility degree equal to $\pi_i(q)$. Normalizing by means of the residuated implication we get in the set Γ_q the following possibility normalized distribution :

$$Pos(\pi_i(./q)) = \pi_i(q) \varnothing sup\{\pi_w(q) / w \in E\} = \pi_i(q) \varnothing C_E(q)$$

The induced possibility measure in the power set of Γ_q will be denoted again by *POS* and its dual necessity measure by *NEC*.

Then, the result of conditioning is the following possibility distribution on the set of possible values:

$$\pi_{p/q}(x) = Sup_{w\in E}\{Pos(\pi_w(./q)) / \pi_w(p/q) = x\},$$

which has high resemblance with the notion of fuzzy truth-values (see for instance [4] for the use of such values in the same framework), except that worlds are restricted to those belonging to *E*.

Following the method of Moral-Campos, the above possibility distribution can be transformed into a conditional uncertainty interval by using Choquet integrals

$$C_E(p \mid q) = In_C(h_p \mid POS)$$
$$I_E(p \mid q) = In_C(h_p \mid NEC)$$

or Sugeno integrals

$$C_E(p \mid q) = In_S(h_p \mid POS)$$
$$I_E(p \mid q) = In_S(h_p \mid NEC)$$

where, for each p, now the function h_p is defined as

$$h_p(\pi_w(. \mid q)) = \pi_w(p \mid q)$$

To keep the analogy with the case of upper and lower probabilities, next we are interested in knowing the results of the above integrals when the above possibility distribution *Pos* is upper approximated by its support and when it is lower approximated by its core. Since the support and the core are crisp sets, we can calculate indistinctly using Choquet or Sugeno integrals (see previous section).

1) Upper approximation: consider, instead of *Pos*, the following distribution Pos^* corresponding to its support.

$$Pos^*(\pi_w(./q)) = \begin{cases} 1, & \text{if } Pos(\pi_w(.|q)) > 0 \\ 0, & \text{otherwise} \end{cases}$$

In this case we have the following result, which is completely analogous to the case of belief-plausibility functions.

Proposition 1. Let $C_E^*(p \mid q) = In_C(h_p \mid POS^*)$ and $I_E^*(p \mid q) = In_C(h_p \mid NEC^*)$ be. Then it holds:

$$C_E^*(p \mid q) = C_E^2(p \mid q)$$
$$I_E^*(p \mid q) = I_E^2(p \mid q).$$

2) Lower approximation: consider now, instead of *Pos*, the following distribution Pos_* corresponding to its core.

$$Pos_*(\pi_w(./q)) = \begin{cases} 1, & \text{if } Pos(\pi_w(.|q)) = 1 \\ 0, & \text{otherwise} \end{cases}$$

For this case the results are not so nice, since only inequalities can be obtained.

Proposition 2. Let $C_{*E}(p \mid q) = In_C(h_p \mid POS_*)$ and $I_{*E}(p \mid q) = In_C(h_p \mid NEC_*)$ be. Then it holds:

$$C_{*E}(p \mid q) \le C_E(p \wedge q) \oslash C_E(q) = C_E^1(p \mid q)$$
$$I_{*E}(p \mid q) \le I_E(p \wedge q) \oslash C_E(q).$$

As a summary, we list below the following relationships involving the conditional measures we have considered up to now:

a) $C_{*E}(p \mid q) \le C_E(p \mid q) \le C_E^*(p \mid q) = C_E^2(p \mid q)$

b) $C_{*E}(p \mid q) \le C_E^1(p \mid q) \le C_E^2(p \mid q)$

c) $I_E^2(p \mid q) = I_E^*(p \mid q) \le I_E(p \mid q) \le I_{*E}(p \mid q)$

3.3. Conditioning by constraining the evidential set

A third conditioning method can be envisaged in the similarity framework. As we have already mentioned, in this framework evidence is modeled as an evidential subset E, being $E \subset W$, together with a T-similarity relation S on WxW. If we know that certain proposition q is true, we can think of updating the consistency and implication measures by considering q as an additional evidence that restricts the set of possible worlds to those that satisfy q. This amounts to consider $E' = E \cap [q]$, being $[q]$ the set of worlds satisfying q, as the new evidential set and define:

$$C_E^3(p \,/\, q) = C_{E \cap [q]}(p)$$
$$I_E^3(p \,/\, q) = I_{E \cap [q]}(p),$$

whenever $E \cap q \neq \emptyset$. Notice that in this case, restricting the set of evidential worlds always has as consequence to constrain the uncertainty interval delimited by I_E and C_E, i.e. it holds

$$[I_E^3(p \,/\, q),\ C_E^3(p \,/\, q)] \subset [I_E(p),\ C_E(p)]$$

According to its definition, this conditioning method has as the advantage w.r.t. to previous ones that, if we observe the evidential set, i.e. if $[q] = E$, then there is no actual updating, that is, abusing the language, we have:

$$C_E^3(p \,/\, E) = C_E\,(p)$$
$$I_E^3(p \,/\, E) = I_E\,(p).$$

On the negative side, we have to mention that it can be the case that $C_E^3(p \,/\, q) \geq I_E^3(p \,/\, q) > 0$ while being $[p] \cap [q] = \emptyset$, or that the definability condition $E \cap q \neq \emptyset$ usually turns out to be too strong (see section 4). A comparison between I^3 and I^0 was given in [4]. Furthermore the following inequalities hold:

$$I_E^2(p \,/\, q) \leq I_E^0(p \,/\, q) \leq I_E^3(p \,/\, q) \text{ and } I_E^2(p \,/\, q) = I_E^0(p{\wedge}q \,/\, q) \leq I_E^3(p{\wedge}q \,/\, q)$$
$$C_E^3(p \,/\, q) \leq C_E^0(p \,/\, q) \text{ and } C_E^2(p \,/\, q) = C_E^0(p{\wedge}q \,/\, q) \geq C_E^3(p{\wedge}q \,/\, q).$$

4. Application to fuzzy variables

In this section we want to apply the previous methods of conditioning to the fuzzy reasoning framework. In [4] it is shown how the similarity relation S defined on the set of possible worlds allows us to define the fuzzy set $\tilde{p}$ for which the set $[p]$ of worlds satisfying the crisp proposition p is the set of prototypes, i.e. the set of worlds with maximum membership degree. Namely, the membership degree of a given world w to the fuzzy set $\tilde{p}$ is defined as $\mu_{\tilde{p}}(w) = Sup\{\ S(w,\ w') \,/\, w' \models p\}$. In terms of possibility measures, $\mu_{\tilde{p}}(w)$ is nothing but $\pi_w(p)$. Therefore, every crisp proposition has associated a fuzzy proposition as a result of a fuzzifying process as soon as similarity relation is introduced in the set of possible worlds.

In fuzzy logic, the unconditioned information typically constraints values of a (fuzzy) variable, while conditional information relates values of one variable with those of another variable. Consider two variables X and Y in the universes U and V respectively. To use the similarity based-model in such a scenario, first of all it is necessary to define similarity structures in the spaces U and V, as well as in the product space $U{\times}V$, as it is done in [5] and summarized next.

4.1. Similarities in Product Spaces

In the spaces U and V, we will be interested in atomic propositions of the form $[X \text{ is } A]$ or $[Y \text{ is } B]$, where A is a subset of U and B a subset of V. The elementary propositions corresponding to this set of atomic propositions are of the form $[X = x]$, i.e. statements declaring that the value of the variable X is x. Analogously, elementary propositions regarding the variable Y have the form $[Y = y]$. In the case of the product space $U{\times}V$, atomic propositions are of the form $[X \text{ is } G]$, where G is an arbitrary subset of the product space. In this case, elementary propositions are the members of $W = U{\times}V$, which correspond to propositions of the form $[X = x] \wedge [Y = y]$. Clearly, every proposition, in either U, V or $U{\times}V$, corresponds to a disjunction of elementary propositions and every elementary proposition corresponds to a possible world in that space.

The marginal propositional structures for one variable are the projections, into U or into V,of the propositional structure for $U{\times}V$. Conversely, propositional structures in either U or V may be extended by means of the *cylindrical extensions* $A^* = A{\times}V$ and $B^* = U{\times}B$, and *composition*, $A{\times}B = A^* \cap B^*$, into the propositional structure of the product space. These relations provide, in addition, the bases for the derivation of similarity measures in either U or V from knowledge of a similarity measure in $U{\times}V$ and viceversa. See [5] for a detailed description of the relations between such structures. Now we include only those properties needed for next section.

- If $M_U = (S_U, E_U)$ and $M_V = (S_V, E_V)$ are similarity models in U and V respectively, then $M = (S, E)$ defined by

$$E = E_U \times E_V,$$

$$S((a, b), (a', b')) = min(S_U(a, a'), S_V(b, b'))$$

is a similarity model in $U{\times}V$, called the *cartesian product* of M_U and M_V.

- Let $M = (S, E)$ a similarity model on $U{\times}V$, and let E_U and E_V be the projections of E on U and V, respectively. Then, the following equalities hold:

$$C_E(X = a) = C_{E_U}(X = a)$$
$$C_E(Y = b) = C_{E_V}(Y = b)$$

Furthermore, if M is a cartesian product model, where $E = G \times H$, being $G \subset U$ and $H \subset V$, then

$$C_E([X = a] \wedge [Y = b]) = min(C_G(X = a), C_H(Y = b))$$
$$I_E([X = a] \wedge [Y = b]) = min(I_G(X = a), I_H(Y = b))$$

We devote the rest of the section to present the results of the different conditioning methods we have considered so far for similarity measures when applied in a typical fuzzy environment with variables taking values in product spaces with product similarity models.

4.2. Conditioning fuzzy variables in product spaces

In what follows, we will consider variables X and Y taking values on U and V respectively, as well as a *cartesian product model* $M = (S, E)$ in $U{\times}V$.

We begin then with two propositions which provide some simplifications of the expressions of the Dempster-like conditioning forms. With the above assumptions and the results from the previous subsection, we must checked the following propositions .

Proposition 3. The following equalities hold:

a) $C_E^2(Y{=}b \,/\, X{=}a) = C_E(Y{=}b) \oslash I_E(X{=}a)$

b) $I_E^2(Y{=}b \,/\, X{=}a) = I_E(Y{=}b) \oslash C_E(X{=}a)$.

c) $C_E^1(Y{=}b \,/\, X{=}a) = C_E(Y{=}b) \oslash C_E(X{=}a)$.

Proposition 3 c) implicitly shows, in particular, the equivalence of the conditioning formulas $C_E^0(.| q)$, mentioned in the introduction section, and $C_E^1(.| q)$ for the present framework.

In a similar way, the expressions of the upper and lower approximations of $C_E(./ X{=}a)$ and $I_E(. / X{=}a)$ (see propositions 1 and 2) can be also simplified. In fact the expressions for the upper approximations are just a direct consequence of proposition 3, but the expressions we get for the lower approximations are worthwhile since they provide us with the missing definition of $I^1(./ X{=}a)$ (see previous section), although valid only in the present framework.

Proposition 4. With the usual assumptions,the following expressions for the upper approximations hold:

$$C_E^*(Y{=}b \,/\, X{=}a) \;= C_E^2(Y{=}b \,/\, X{=}a) = C_E(Y{=}b) \oslash I_E(X{=}a)$$
$$I_E^*(Y{=}b \,/\, X{=}a) \;= I_E^2(Y{=}b \,/\, X{=}a) = I_E(Y{=}b) \oslash C_E(X{=}a),$$

and for the lower approximations:

$$C_{*E}(Y{=}b \,/\, X{=}a) \;= C_E^1(Y{=}b \,/\, X{=}a) = C_E(Y{=}b) \oslash C_E(X{=}a)$$
$$I_{*E}(Y{=}b \,/\, X{=}a) \;= I_E(Y{=}b) \oslash C_E(X{=}a).$$

Observe that now we get an equality for the lower approximation measures, in spite of proposition 2 where only inequalities appeared. This leads us to propose the following expression

$$I_E^1(Y{=}b \,/\, X{=}a) = I_E(Y{=}b) \oslash C_E(X{=}a)$$

as definition for $I_E^1(. \,/\, X{=}a)$. Notice that the lower and upper approximations $I_{*E}(Y{=}b \,/\, X{=}a)$ and $I_E^*(Y{=}b \,/\, X{=}a)$ take the same value $I_E(Y{=}b) \oslash C_E(X{=}a)$. Therefore, $I_E(./X{=}a)$ must also coincide with them and, at the same time, $I^1(. \,/\, X{=}a)$ and $I^2(. \,/\, X{=}a)$ coincide as well. So, and as summary, in our present framework we have very analog interval inclusions as in the case of upper-lower probability pairs, as it can be observed in the following inclusionships:

$$[I_{*E}(b \,/\, a), C_{*E}(b \,/\, a)] = [I^1{}_E(b \,/\, a), C_E^1(b \,/\, a)] \subseteq [I_E(b \,/\, a), C_E(b \,/\, a)] \subseteq$$
$$\subseteq [I_E^*(b \,/\, a), C_E^*(b \,/\, a)] = [I_E^2(b \,/\, a), C_E^2(b \,/\, a)]$$

where a and b denotes $[X=a]$ and $[Y=b]$ respectively. As a matter of example we show below how the interval $[I_E(b/a), C_E(b/a)]$ looks like when using three different t-norms and both the Choquet and Sugeno integrals. For obvious reasons, only the value of $C_E(b/a)$ is worth.

- Using the Choquet integral, the expressions for $C_E(b/a)$ are:

  for $T(x, y) = min(x,y)$:

$$C_E^1(b/a)\ [1 + C_E^2(b/a) - C_E^1(b/a)],$$

  for $T(x, y) = xy$:

$$C_E^1(b/a)\ (1 + ln\ \frac{C_E^2(b/a)}{C_E^1(b/a)}),$$

  and finally, for $T(x, y) = min(1, 1-x+y)$:

$$C_E^1(b/a) + \frac{(M-N)\ (2+N-M)}{2} + Q\ (1-M)$$

  where $M = C_E^2(b/a)$, $N = C_E^1(b/a)$ and $Q = 1 - C_E(X=a)$.

- Using the Sugeno integral, although slightly modified by changing in the expression the minimum operator by the t-norm T for which the similarity S of the product model is indeed a T-similarity relation, then it turns out that in any case the following equality holds:

$$[I_E(b/a), C_E(b/a)] = [I_E^1(b/a), C_E^1(b/a)].$$

5. Concluding remarks

Within the similarity-based formalism to possibilistic and fuzzy reasoning, we have been interested in this paper in how conditioning of the implication and consistency measures can be performed. Several methods have been considered, most of them by analogy to conditioning methods already proposed for other kind of fuzzy measures, having been proved the case of upper and lower probabilities of special interest. However nothing has been said so far about their behaviour w.r.t. inference patterns. Of special interest are two formulations of the Generalized Modus Ponens that have been considered in previous papers. In what follows, the set of propositions $\{q_i / i \in I\}$ is considered to be a partition, i.e. $\cup_{i\in I}[q_i] = W$. The first formulation of the pattern,

$$I_E(p) \geq Sup_{i\in I}\ I_E(q_i) \otimes I_E(p/q_i)$$
$$C_E(p) \leq Sup_{i\in I}\ C_E(q_i) \otimes C_E(p/q_i),$$

which we call *GMP1*, already appears in [8] and in [4], while the following extension, which we call *GMP2*, has been recently considered in [5]:

$$I_{E\cap F}(p) \geq Sup_{i\in I}\ I_E(q_i) \otimes I_F(p/q_i)$$
$$C_{E\cap F}(p) \leq Sup_{i\in I}\ C_E(q_i) \otimes C_F(p/q_i).$$

In this last formulation, different bodies of evidence for unconditioned information (E) and conditional information (F) are allowed, and thus, *GMP2* clearly subsumes the previous formulation *GMP1*.

In a general setting, the *GMP2* pattern is satisfied by the conditioned pairs $(I^0(./.), C^0(./.))$ and $(I^2(./.), C^2(./.))$ -and thus by $(I^*(./.), C^*(./.))$ too-, while *GMP1* is partially satisfied by C^1. However, in the fuzzy reasoning framework described in last section, the conditioned pair $(I^1(./.), C^1(./.))$ also satisfies *GMP2*, and thus, also the conditioned pairs $(I(./.), C(./.))$ given by fuzzy integrals. So, at this point, the behaviour of the fuzzy integral conditioned pairs, as well as the conditioning method by constraining the evidential set deserves further investigation.

Acknowledgements

The authors have been supported by the ESPRIT III Basic Research Action nº 6156 DRUMS II, and by the Spanish DGICYT project nº PB91-0334. They also wish to thank Serafín Moral, Enrique H. Ruspini and Llorenç Valverde for their useful comments.

REFERENCES

[1] Demsper A.P. (1967) Upper and lower probabilities induced by a multivalued mapping. Annals of Mathematical Statistics 38, 325-339.

[2] Dubois D., Prade H. (1990) The logical view of conditioning and its application to possibility and evidence theories. Int. J. Approximate Reasoning 4, 23-46.
[3] Dubois D., Prade H. (1990) Belief revision and updates in numerical formalisms: an overview with new results for the possibilistic framework. In Proc. of IJCAI'93, 620-625.
[4] Dubois D., Prade H. (1994). Similarity versus preference in fuzzy set based-logics. Rapport IRIT/94-10-R.
[5] Esteva F., Garcia-Calvés, P., Godo L. (1994) Relating and Extending Semantical Approaches to Possibilistic Reasoning. International Journal of Approximate Reasoning. North Holland, pp. 311-344.
[6] Esteva F., Garcia-Calvés, P., Godo L., Valverde L., Ruspini E.H. (1994) On Similarity Logic and the Generalized Modus Ponens. Proc. of the FUZZ-IEEE'94. IEEE Press, pp. 1423-1427.
[7] Moral S. (1992) Calculating Uncertainty Intervals from Conditional Convex Sets of Probabilities. Proc. of the 8th. Conference on Uncertainty in Artificial Intelligence, Morgan Kaufmann, 199-206.
[8] Moral S., de Campos L.M. (1991) Updating uncertain information. In: Uncertainty in Knowledge Bases (B. Bouchon-Meunier, R.R. Yager, L.A. Zadeh eds.) Springer Verlag, LNCS nº 521, Berlin, 58-67.
[9] Ruspini E.H. (1991) On the Semantics of Fuzzy Logic. Int. J. Approximate Reasoning, 5, 45-88.
[10] Ruspini E.H. (1991) Truth as Utility: A Conceptual Synthesis. Proc. of the 7th. Conference on Uncertainty in Artificial Intelligence, Morgan Kaufmann, 316-322.

VAGUENESS AND CARDINALITY
A UNIFYING APPROACH

Maciej WYGRALAK

Adam Mickiewicz University
Faculty of Math. and Comp. Sci.
Matejki 48/49, 60-769 Poznań, Poland

ABSTRACT

In this paper, we like to present a general, complete and easily applicable nonclassical cardinality theory which makes use of the infinite-valued Łukasiewicz logic. The theory pertains to vaguely defined objects, understood as objects that are separated from a universe by means of arbitrary sharp or vague properties, including subdefinite sets (incompletely known sets, in other words). We will discuss questions of equipotency of vaguely defined objects, inequalities between and arithmetical operations on the resulting generalized cardinal numbers which express the powers (cardinalities) of vaguely defined objects. In particular, the presented theory offers a common generalization of different existing approaches to the problem of cardinality of fuzzy sets, twofold fuzzy sets, partial sets, etc.

1. Introduction

The starting notion in our discussion is that of a property. Each nonantinomial property, which can be formulated in a natural language and makes sense for the elements of an infinite universe **M**, separates from **M** an object, generally distinct in each case. It will be called *a vaguely defined object* (*VD-object*, in short). One can easily notice that some properties admit even infinitely many intermediate states between the states of their fulfilment and nonfulfilment, as it is in the trivial example 'to be a tall man'. Clearly, a VD-object collapses to a set if the number of those intermediate states reduces itself to zero, i.e. if the separating property is sharp rather than vague (e.g. 'to be an odd number'). VD-objects which are not sets will be called *proper VD-objects*, and can be understood as 'nebular' objects in **M**.

In practice, on the other hand, the classical notion of a set can be insufficient even if an object is separated from **M** by a sharp property **p**. This happens whenever the status of some elements from **M** with respect to **p** is unknown or uncertain. The object, i.e. an incompletely known set, is then called *a subdefinite set* in **M**.

As one knows, the most frequently used approach to VD-objects is that proposed by L.A. Zadeh in which VD-objects are represented by precisely determined, at least theoretically speaking, membership functions $\mathbf{M}\to\mathbf{I}$, where $\mathbf{I} := [0,1]$, and are then called *fuzzy sets*. The other are, for instance, various derivative concepts of, say, type 2 fuzzy sets, ultrafuzzy sets, L-fuzzy sets, fuzzy sets with triangular norms, etc., and semisets. Instead, subdefinite sets can be possibilistically represented by means of twofold fuzzy sets or flou (partial) sets. In the next section, we will introduce a unifying approximative representation of VD-objects including subdefinite sets.

In this paper, we like to construct a complete, general and easily applicable cardinality theory for VD-objects, including subdefinite sets, which is based on the infinite-valued Łukasiewicz logic $Ł_\infty$ and the unifying approximative representation of VD-objects. There are good motivations for undertaking the problem of cardinality of VD-objects. They are both theoretical (an essential generalization of the classical cardinality theory for sets) and applicational (cardinal analysis of objects separated by means of imperfect information, communication with data/knowledge bases, modelling the meaning of imprecise quantifiers in natural language statements, probabilities of vague events, fuzzy topological spaces, metrical analysis of vague regions of grey images, etc.; see e.g. [8], [9]).

Let us define the following symbols:

$$\mathrm{GP}(\mathbf{D}) := \mathbf{I}^{\mathbf{D}},\ \ \mathrm{PS}(\mathbf{D}) := \{0,1\}^{\mathbf{D}},\ \ \mathrm{GP} := \mathrm{GP}(\mathbf{M})\ \text{ and }\ \mathrm{PS} := \mathrm{PS}(\mathbf{M}).$$

Throughout the paper, bold capital letters **A**, **B**, **C**, ... , **X**, **Y**, **Z** will denote sets, whereas A, B, C, ... , X, Y, Z symbolize the (membership) functions from GP. As usual,

$$\mathrm{Y} \subset \mathrm{Z} \ \Leftrightarrow\ \forall x \in \mathbf{M}: \mathrm{Y}(x) \le \mathrm{Z}(x).$$

Moreover, let

$$\mathrm{Y}_t := \{x: \mathrm{Y}(x) \ge t\} \ \text{ for } t \in (0,1] \qquad (t\text{-}\textit{level set of}\ \mathrm{Y})$$

and

$$\mathrm{Y}^t := \{x: \mathrm{Y}(x) > t\} \ \text{ for } t \in [0,1). \qquad (\textit{sharp}\ t\text{-}\textit{level set of}\ \mathrm{Y})$$

The characteristic function of a set $\mathbf{D} \subset \mathbf{M}$ will be denoted by $1_{\mathbf{D}}$. In particular, let

$$\mathrm{T} := 1_{\varnothing} \ \text{ and } \ \mathrm{M} := 1_{\mathbf{M}}.$$

Finally, let

$$[\mathrm{Y}]_i := \bigvee\{t: |\mathrm{Y}_t| \ge i\}.$$

We emphasize that i, j, ... , p, q will denote both finite and transfinite cardinal numbers, whereas $\mathrm{CN} := \{i: i \le |\mathbf{M}|\}$. One proves that

$$[\mathrm{Y}]_i = \bigwedge\{[\mathrm{Y}]_j: j < i\}$$

if i is a limit cardinal number.

2. Approximative representation of vaguely defined objects

Let us introduce the following unifying approximative representation of VD-objects by means of pairs (F, G) with F, G $\in$ GP such that

$$\mathrm{F} \subset 1_{\mathrm{ker(G)}}, \tag{1}$$

where $\mathrm{ker(G)} := \mathrm{G}_1$. They will be called *free representing pairs* (cf. [5-7]). The family of all free representing pairs will be denoted by **K**. Obviously, for each $\mathbf{D} \subset \mathbf{M}$, the pair $(1_{\mathbf{D}}, 1_{\mathbf{D}})$ belongs to **K** and represents the set **D** in the universe of free representing pairs. One easily notices that (1) is equivalent to

$$\forall x \in \mathbf{M}: \mathrm{F}(x) > 0 \Rightarrow \mathrm{G}(x) = 1, \tag{2}$$

and implies

$$\mathrm{F} \subset \mathrm{G}.$$

So, each (F, G) $\in$ **K** is an ultrafuzzy set, and we have

$$[\mathrm{F}]_i \le [\mathrm{G}]_i$$

as well as

$$[F]_i > 0 \Rightarrow [G]_i = 1$$

for each cardinal $i \in CN$. The VD-object represented by $(F, G) \in \mathbf{K}$ will be denoted by obj(F, G), while (F, G) itself can be interpreted in the following two ways:

(a) (*Possibilistic interpetation*) Each free representing pair can be understood as a twofold fuzzy set with F, G and (1) being interpreted in the language of necessity and possibility degrees (see e.g. [2]). Clearly, obj(F, G) is then a subdefinite set in **M**.

(b) (*Approximative interpretation*) One can assume that (cf. [5-8])

$$F = f(A) \quad \text{and} \quad G = g(A)$$

for some $A \in GP$ and

$$f, g\colon GP \to GP.$$

In this case, if F = G, obj(F, G) is a set, else obj(F, G) is a proper VD-object in **M**, whereas F and G, respectively, are lower and upper approximations of an imprecisely/subjectively determined membership function A, respectively (indeed, $F \in PS$ whenever $(F, G) \in \mathbf{K}$ and F = G). We accept that the imprecision/subjectivity is more or less total: either no bounds for A(x) can be given or only one of them can be established. In other words, for each $x \in \mathbf{M}$, we accept that

$$0 \le A(x) \le 1 \quad \text{or} \quad 0 \le A(x) \le a \quad \text{or} \quad b \le A(x) \le 1$$

with some $a, b \in \mathbf{I}$ which depend on x. This means that

$$f(A)(x) = 0 \ \text{ or/and } \ g(A)(x) = 1 \quad \text{for each} \quad x \in \mathbf{M},$$

which is equivalent to (1). So, the condition (1), which is usually associated in the literature only with twofold fuzzy sets, appears to be more universal than one could expected it to be.

Both the interpretations (a) and (b) suggest that the nonclassical cardinality theory which we like to construct can be applied to arbitrary VD-objects, including subdefinite sets. On the one hand, it will be a generalization of the cardinality theory constructed in [5-8], and, on the other hand, it will form a proper expansion of the nonclassical idea of cardinality which is formulated, but not developed, for ('finite') subdefinite sets in [2]. The following natural definitions will be used for (F, G), $(H, S) \in \mathbf{K}$ (cf. [2] and flou sets):

$$(F, G) * (H, S) := (F * H, G * S),$$

where $*$ denotes $\cup$ (*sum*), $\cap$ (*intersection*) or $\times$ (*cartesian product*), and

$$(Y \cup Z)(x) := Y(x) \vee Z(x) \quad \text{with} \quad \vee := \max,$$

$$(Y \cap Z)(x) := Y(x) \wedge Z(x) \quad \text{with} \quad \wedge := \min,$$

$$(Y \times Z)(x, y) := Y(x) \wedge Z(y).$$

Moreover, let

$$(F, G)' := (G', F') \quad \text{with} \quad Y'(x) := 1 - Y(x), \qquad (\textit{complement})$$

$$(F, G) = (H, S) \quad \text{iff} \quad F = H \ \text{ and } \ G = S, \qquad (\textit{equality})$$

$$(F, G) \subset (H, S) \quad \text{iff} \quad F \subset H \ \text{ and } \ G \subset S. \qquad (\textit{inclusion})$$

Generally, if $\mathbf{J} \neq \emptyset$ denotes a set of indices and $(F_e, G_e) \in \mathbf{K}$ for each $\mathbf{e} \in \mathbf{J}$, then

$$\underset{e\in J}{\ast}(F_e, G_e) := (\underset{e\in J}{\ast} F_e, \underset{e\in J}{\ast} G_e),$$

where $\ast \in \{\bigcap, \bigcup, \times\}$. Let $T^{\&} := (T, T)$ and $M^{\&} := (M, M)$. $(\mathbf{K}, \cup, \cap, ', T^{\&}, M^{\&})$ forms an infinitely distributive de Morgan algebra. Finally, let

$$\mathrm{obj}(F, G) = \mathrm{obj}(H, S) \Leftrightarrow (F, G) = (H, S),$$

$$\mathrm{obj}(F, G) \subset \mathrm{obj}(H, S) \Leftrightarrow (F, G) \subset (H, S),$$

$$\mathrm{obj}(F, G)' := \mathrm{obj}(G', F'),$$

$$\mathrm{obj}(F, G) * \mathrm{obj}(H, S) := \mathrm{obj}(F * H, G * S)$$

for $* \in \{\cup, \cap, \times\}$, and, generally,

$$\underset{e\in J}{\ast}\mathrm{obj}(F_e, G_e) := \mathrm{obj}(\underset{e\in J}{\ast} F_e, \underset{e\in J}{\ast} G_e),$$

3. Equipotencies between vaguely defined objects

The notion of equipotency seems to be the most central notion of any cardinality theory. Therefore, we start with the following definition.

Definition 3.1. Let $(F, G), (H, S) \in \mathbf{K}$. We say that obj(F, G) and obj(H, S) are equipotent (are of the same power or cardinality, in other words) and we write $(F, G) \sim (H, S)$ or $|(F, G)| = |(H, S)|$ iff

$$|F^t| = |H^t| \text{ and } |G^t| = |S^t| \text{ for each } t \in [0,1).$$

Obviously, $\sim$ is an equivalence relation; if $F = G$ and $H = S$, it collapses to the usual equipotency of two sets. One easily notices that

$$(F, G) \sim (H, S) \text{ iff } (T, G) \sim (T, S) \text{ and } (F, M) \sim (H, M).$$

Let us formulate two more characterizations of $(F, G) \sim (H, S)$. Let $\mathrm{supp}(Y) := \{x\colon Y(x) > 0\}$ with $Y \in GP$ (*support of* Y), and $FGP := \{Y \in GP\colon \mathrm{supp}(Y) \text{ is finite}\}$. If $Y \in FGP$, then $[Y]_i$ is simply the i-*th* element in the nonincreasingly ordered sequence od all positive values Y(x), including their possible repetitions.

Theorem 3.2. *Let* $(F, G), (H, S) \in \mathbf{K}$. *The following equivalences are then satisfied*:

(a) $(F, G) \sim (H, S) \Leftrightarrow \forall i \in CN\colon [F]_i = [H]_i \;\&\; [G]_i = [S]_i$.

(b) $(F, G) \sim (H, S) \Leftrightarrow \forall t \in (0,1]\colon |F_t| = |H_t| \;\&\; |G_t| = |S_t|$ provided that $G, S \in FGP$.

Proof. (a) can be verified by routine transformations. (b) is an immediate consequence of the definition of $\sim$ for VD-objects. ■

The following implications are simple consequences of Definition 3.1, namely:

$$(F, G) \sim (H,S) \Rightarrow |\mathrm{supp}(F)| = |\mathrm{supp}(H)| \;\&\; |\mathrm{supp}(G)| = |\mathrm{supp}(S)| \tag{3}$$

and

$$(F, G) \subset (H, S) \subset (W, Z) \;\&\; (F, G) \sim (W, Z) \;\Rightarrow\; (F, G) \sim (H, S) \sim (W, Z). \tag{4}$$

Clearly, (4) is a counterpart of the well-known Cantor-Bernstein theorem. Moreover, the following equivalences are satisfied ($p := |\mathbf{M}|$):

$$(F, G) \sim T^{\&} \;\Leftrightarrow\; (F, G) = T^{\&}$$

and

$$(F, G) \sim M^{\&} \;\Leftrightarrow\; [F]_p = 1.$$

Theorem 3.3. *Let* $(F_e, G_e) \sim (H_e, S_e)$ *for each* $\mathbf{e} \in \mathbf{J} \neq \emptyset$. *Then*

$$\underset{e \in J}{\times} (F_e, G_e) \sim \underset{e \in J}{\times} (H_e, S_e).$$

Moreover, if $(F_e, G_e) \cap (F_{e*}, G_{e*}) = (H_e, S_e) \cap (H_{e*}, S_{e*}) = T^{\&}$ *for each* $\mathbf{e} \neq \mathbf{e}^*$, *then*

$$\bigcup_{e \in J} (F_e, G_e) \sim \bigcup_{e \in J} (H_e, S_e).$$

Proof. The thesis follows from Definition 3.1 by routine transformations. ■

It is clear that the properties formulated in Theorem 3.3 will be useful when later discussing basic arithmetical operations on the generalized cardinals introduced in Section 4, which express the powers of VD-objects.

4. Generalized cardinal numbers

In the presented nonclassical cardinality theory for VD-objects, the power or cardinality of a VD-object obj(F, G) will be expressed by means of a weighted family of cardinals from CN or, equivalently, by means of a function $CN \rightarrow \mathbf{I}$ such that its value at each $i \in CN$ is identical to the weight assigned to i. Those functions will be called *generalized cardinal numbers* (*gc-numbers*, in short). Obviously, the question is how to determine the weights associated with the cardinals from CN. To this end, we propose to use the truth values of the following many-valued sentence, where $\mathbf{P}_i := \{\mathbf{D} \subset \mathbf{M} \colon |\mathbf{D}| = i\}$ with $i \in CN$:

$$\text{sent}(F, G, i) \;:=\; \exists_m \mathbf{B} \in \mathbf{P}_i\colon F \subset_m 1_{\mathbf{B}} \;\&_m\; \exists_m \mathbf{C} \in \mathbf{P}_i\colon 1_{\mathbf{C}} \subset_m G,$$

where $\exists_m$, $\subset_m$ and $\&_m$, respectively, are symbols of many-valued existential quantification, inclusion and conjunction in $Ł_\infty$, respectively (see [3]). Let us define a function

$$GCN\colon GP \times GP \rightarrow GP(CN)$$

such that

$$GCN(F, G)(i) := [\![\text{sent}(F, G, i)]\!], \tag{5}$$

where $(F, G) \in \mathbf{K}$ and $i \in CN$. So, GCN(F, G)(i) is the result of a procedure in which we look for the best upper approximation of F and the best lower approximation of G by means of the characteristic functions of sets of the power i. Clearly, if $G \in FGP$, then we get

$$GCN(F, G)(i) = [\![\exists_m \mathbf{B} \in \mathbf{P}_i\colon F \subset_m 1_{\mathbf{B}} \subset_m G]\!].$$

Trying to express GCN(F, G)(i) in a convenient form, one proves that

$$\mathrm{GCN}(F, G)(i) = [G]_i \wedge 1-[F]_{i^+} \tag{6}$$

for each (F, G) ∈ **K** and i ∈ CN; as usual, i^+ symbolizes the successor of i (so, $i^+ = i+1$ if i is finite). More precisely, we have

$$\mathrm{GCN}(F, G)(i) = \begin{cases} 1-[F]_{i^+} < 1, & \text{if } i < i^{\#}, \\ 1, & \text{if } i = i^{\#}, \\ [G]_i, & \text{otherwise,} \end{cases} \tag{7}$$

where $i^{\#} := \bigwedge\{i \in CN: [G]_i + [F]_{i^+} \leq 1\}$. So, the following implications are satisfied:

$$\mathrm{GCN}(F, G)(i) > 0 \Rightarrow |F_1| \leq i \leq |\mathrm{supp}(G)|$$

and

$$H \subset F \subset G \subset S \Rightarrow \mathrm{GCN}(F, G) \subset \mathrm{GCN}(H, S).$$

Moreover,

$$\mathrm{GCN}(F, G) = \mathrm{GCN}(T, G) \cap \mathrm{GCN}(F, M). \tag{8}$$

Theorem 4.1. *For each* (F, G), (H, S) ∈ **K**, *the following equivalence is satisfied*:

$$\mathrm{GCN}(F, G) = \mathrm{GCN}(H, S) \quad \textit{iff} \quad (F, G) \sim (H, S).$$

Proof. The thesis is a consequence of (7). ■

In virtue of Theorem 4.1 and the Tarski's axiomatic definition of cardinals, the values of GCN are suitable candidates for the functions expressing the powers of vaguely defined objects and, therefore, they can be really called gc-numbers. The generalized cardinals will be denoted by small Greek letters. If

$$\mathrm{GCN}(F, G) = \alpha \in \mathrm{GP}(CN),$$

we say that *the power or cardinality of* obj(F, G) *is equal to* α, and we write $|(F, G)| = \alpha$. Let

$$\mathrm{GCN}_{\mathbf{K}} := \{\alpha \in \mathrm{GP}(CN): |(F, G)| = \alpha \ \& \ (F, G) \in \mathbf{K}\}$$

and

$$\mathrm{GCN}_{\mathbf{K}*} := \{\alpha \in \mathrm{GCN}_{\mathbf{K}}: |(F, F)| = \alpha \ \& \ F \in PS\}.$$

The following particular cases of gc-numbers from $\mathrm{GCN}_{\mathbf{K}}$ should be mentioned:

(a) If (F, G) is interpreted as a twofold fuzzy set and G ∈ FGP, then $\alpha = |(F, G)|$ is identical with the cardinality defined by Dubois and Prade for 'finite' subdefinite sets (see (a) in Section 2 and [2]; cf. [6-7]).

(b) If the approximative interpretation of (F, G) is used (see (b) in Section 2), we get the following special cases of $\alpha = |(F, G)|$ (cf. reviews in [1], [4-7]):

(b1) $f(A) = T$, $g(A) = A$: 'fuzzy' cardinals defined for finite fuzzy sets by Blanchard and Zadeh, and extended to arbitrary fuzzy sets by Šostak (see e.g. [1], [7], [9]).

(b2) $f(A) = 1_{\ker(A)}$, $g(A) = A$, A ∈ FGP: 'fuzzy' cardinals introduced for finite fuzzy sets by Dubois (see e.g. [1]).

(b3) $F, G \in PS$: partial cardinals introduced by Klaua for partial sets (see [3]).

Each gc-number α from GCN_K is convex and normal (see (7)). Moreover, if $F, G \in PS$, $|supp(F)| = m$, $|supp(G)| = n$, and $\alpha = |(F, G)|$, then (6) implies

$$\alpha = \begin{cases} 1_{\{i\in CN:\ i\le n\}}, & \text{if } F = T, \\ 1_{\{i\in CN:\ i\ge m\}}, & \text{if } G = M, \\ 1_{\{i\in CN:\ m\le i\le n\}}, & \text{otherwise.} \end{cases} \tag{9}$$

Hence $|(1_D, 1_D)| = 1_{\{|D|\}}$; we shall use the symbol $\langle k\rangle := 1_{\{k\}}$ with $k \in CN$. As we see, there exists a proper correspondence between cardinals from CN and gc-numbers from GCN_{K^*}. Finally, if $\alpha \in GCN_K$ and $|(F, G)| = \alpha$ for some $(F, G) \in \mathbf{K}$, we put

$$\alpha_- := |(T, G)| \quad \text{and} \quad \alpha_+ := |(F, M)|.$$

In virtue of (8), we have

$$\alpha = \alpha_- \cap \alpha_+,$$

whereas Theorem 4.1 implies that this decomposition of α is unique, i.e.

$$\alpha = \beta \quad \text{iff} \quad \alpha_- = \beta_- \ \text{ and } \ \alpha_+ = \beta_+.$$

5. Inequalities

The possibility of making comparisons between gc-numbers and between the powers of VD-objects seems to be essential in every respect. Therefore, we propose to introduce the following natural definition (cf. Definition 3.1).

Definition 5.1. Let $(F, G), (H, S) \in \mathbf{K}$.
(a) We say that the power of obj(F, G) is less than or equal to the power of obj(H, S) and we write $|(F, G)| \le |(H, S)|$ iff

$$|F^t| \le |H^t| \ \text{ and } \ |G^t| \le |S^t| \quad \text{for each } t \in [0,1).$$

(b) We say that the power of obj(F, G) is less than the power of obj(H, S) and we write $|(F, G)| < |(H, S)|$ iff $|(F, G)| \le |(H, S)|$ and $(F, G) \nsim (H, S)$.

We immediately notice that $\le$ is only a partial order relation. In reference to Theorem 3.2(a), the following equivalence is fulfilled:

$$|(F, G)| \le |(H, S)| \ \Leftrightarrow \ \forall i \in CN: [F]_i \le [H]_i \ \&\ [G]_i \le [S]_i.$$

Moreover, the following properties hold true:

$$(F, G) \sim (H, S) \ \Leftrightarrow \ |(F, G)| \le |(H, S)| \ \&\ |(F, G)| \ge |(H, S)|, \tag{10}$$

$$|(F, G)| \le |(H, S)| \ \&\ (F, G) \sim (F^*, G^*) \ \&\ (H, S) \sim (H^*, S^*) \ \Rightarrow \ |(F^*, G^*)| \le |(H^*, S^*)|, \tag{11}$$

$$obj(F, G) \subset obj(H, S) \ \Rightarrow \ |(F, G)| \le |(H, S)|, \tag{12}$$

$$\exists(H^*, S^*) \subset (H, S): (F, G) \sim (H^*, S^*) \quad \Rightarrow \quad |(F, G)| \leq |(H, S)|, \tag{13}$$

$$|(F, G)| \leq |(H, S)| \quad \Rightarrow \quad |\mathrm{supp}(F)| \leq |\mathrm{supp}(H)| \ \& \ |\mathrm{supp}(G)| \leq |\mathrm{supp}(S)|, \tag{14}$$

$$|(F, G)| \leq |(H, S)| \quad \Leftrightarrow \quad |(T, G)| \leq |(T, S)| \ \& \ |(F, M)| \leq |(H, M)|. \tag{15}$$

The inverse implication in (13) is not generally true, which forms an anomaly in comparison with the classical cardinality theory. With reference to the Generalized Continuum Hypothesis, one should point out that for each transfinite k there exist VD-objects of the powers lying between $\langle k\rangle$ and $\langle 2^k\rangle$ (cf. [5, 8]).

Definition 5.2. Let $\alpha, \beta \in GCN_{\mathbf{K}}$.
(a) $\alpha \leq \beta \Leftrightarrow \exists(F, G), (H, S) \in \mathbf{K}$: $|(F, G)| = \alpha \ \& \ |(H, S)| = \beta \ \& \ |(F, G)| \leq |(H, S)|$.
(b) $\alpha < \beta \Leftrightarrow \alpha \leq \beta \ \& \ \alpha \neq \beta$.

In virtue of (11), $\alpha \leq \beta$ and $\alpha < \beta$ are well-defined. Clearly, $(GCN_{\mathbf{K}}, \leq)$ forms a poset, and $\alpha = \beta$ iff $\alpha \leq \beta$ and $\beta \geq \alpha$. Moreover, we have

$$\alpha \leq \beta \Leftrightarrow \alpha_- \leq \beta_- \ \& \ \alpha_+ \leq \beta_+ \tag{16}$$

and

$$\alpha_- \leq \beta_- \Leftrightarrow \alpha_- \subset \beta_- \ , \quad \alpha_+ \leq \beta_+ \Leftrightarrow \alpha_+ \supset \beta_+. \tag{17}$$

This and (9) imply that $(CN, \leq)$ and $(GCN_{\mathbf{K}^*}, \leq)$ are isomorphic. Finally, let

$$\alpha \vartriangle \beta := (\alpha_- \cap \beta_-) \cap (\alpha_+ \cup \beta_+) \quad \text{and} \quad \alpha \triangledown \beta := (\alpha_- \cup \beta_-) \cap (\alpha_+ \cap \beta_+).$$

We conclude that $(GCN_{\mathbf{K}}, \vartriangle, \triangledown, \langle 0\rangle, \langle p\rangle)$ forms a complete distributive lattice with $p := |\mathbf{M}|$. Looking at Definition 3.1, Definition 5.1 and Definition 5.2, it is clear that many-valued equipotencies and inequalities could be easily introduced using $Ł_\infty$. Nevertheless, this idea will not be developed in this paper (cf. [7, 8]).

6. Arithmetic operations

The following modified extension principle will be used to define addition, multiplication, and exponentiation of gc-numbers α and β from $GCN_{\mathbf{K}}$ (cf. [7]):

$$\alpha * \beta := (\alpha_- * \beta_-) \cap (\alpha_+ * \beta_+),$$

where

$$\begin{aligned}(\alpha_- * \beta_-)(k) &:= \mathrm{ext}(\alpha_-, \beta_-, k; \geq, *),\\ (\alpha_+ * \beta_+)(k) &:= \mathrm{ext}(\alpha_+, \beta_+, k; \leq, *),\\ \mathrm{ext}(\lambda, \mu, k; \bowtie, *) &:= \bigvee\{\lambda(i) \wedge \mu(j) : i * j \bowtie k\}\end{aligned} \tag{18}$$

and $\bowtie$ denotes a binary relation on cardinals. If $* = \cdot$, we shall write $\alpha\beta$ instead of $\alpha \cdot \beta$. One immediately sees that $\mathrm{ext}(\lambda, \mu, k; =, *)$ is the usual Zadeh's extension principle. After quite obvious reformulations, (18) can be used to define sums $\Sigma_{e \in \mathbf{J}}\alpha_e$ and products $\Pi_{e \in \mathbf{J}}\alpha_e$ of arbitrarily many components and factors, where $\alpha_e \in GCN_{\mathbf{K}}$ for each $\mathbf{e} \in \mathbf{J} \neq \emptyset$; usually, we shall use the simplified notation $\Sigma\alpha_e$ and $\Pi\alpha_e$. Applying Theorem 3.3, one shows that both $\Sigma\alpha_e$ and $\Pi\alpha_e$ are then well-defined and have classical-like interpretations. More precisely,

$$\Sigma\alpha_e = |\bigcup(F_e, G_e)| \quad \text{and} \quad \Pi\alpha_e = |\times(F_e, G_e)|,$$

where $|(F_e, G_e)| = \alpha_e$ for each $e \in J$ and (in the case of Σ)

$$(F_e, G_e) \cap (F_{e*}, G_{e*}) = T^{\&}$$

for each $e \neq e*$. In particular, we have

$$|(F, G)| + |(H, S)| = |(F, G) \cup (H, S)| + |(F, G) \cap (H, S)|,$$

and

$$|(F, G)| + |(H, S)| = |(F, G) \cup (H, S)| \quad \text{whenever} \quad (F, G) \cap (H, S) = T^{\&}.$$

One easily notices that both $\Sigma\alpha_e$ and $\Pi\alpha_e$ are commutative and associative. Some further properties are collected in the following theorem (cf. [7]).

Theorem 6.1. *The following properties are satisfied by each α, β, $(\alpha_e)_{e \in J}$ and $(\beta_e)_{e \in J}$:*

(a) $\alpha + \langle 0 \rangle = \alpha, \ \alpha\langle 1 \rangle = \alpha.$ *(neutral elements)*

(b) $\Sigma\alpha_e = \langle 0 \rangle$ *iff* $\alpha_e = \langle 0 \rangle$ *for each* $e \in J$,
$\Pi\alpha_e = \langle 0 \rangle$ *iff* $\alpha_e = \langle 0 \rangle$ *for at least one* $e \in J$.

(c) $\beta\Sigma\alpha_e = \Sigma\beta\alpha_e$. *(distributivity)*

(d) $\Sigma\alpha_e \le \Sigma\beta_e$ *and* $\Pi\alpha_e \le \Pi\beta_e$ *whenever* $\alpha_e \le \beta_e$ *for each* $e \in J$. *(monotonicity)*

(e) $\Sigma\alpha = \langle k \rangle\alpha \ \ (k := |J|)$.

(f) $\Sigma\alpha_e \le \beta\langle k \rangle$ *provided that* $\alpha_e \le \beta$ *for each* $e \in J$ *and* $|J| = k$.

(g) $\Sigma\alpha_e \le \Pi\alpha_e$ *whenever* $\alpha_e \ge \langle 2 \rangle$ *for each* $e \in J$.

(h) $\alpha\beta = \alpha + \beta = \beta$ *if* $\langle 1 \rangle \le \alpha \le \beta$ *and* $\beta \ge \langle \aleph_0 \rangle$. *(absorption)*

(i) $(\alpha\beta)(k) = (\alpha + \beta)(k)$ *for each transfinite* k *and* $\alpha, \beta \ge \langle 1 \rangle$.

Proof. All the results are consequences of (18) and Definition 5.2. ■

Moreover, one proves that $(CN, \Sigma, 0)$ and $(GCN_{K*}, \Sigma, \langle 0 \rangle)$ as well as $(CN, \Pi, 1)$ and $(GCN_{K*}, \Pi, \langle 1 \rangle)$ are isomorphic. The reader can easily construct a counterexample showing that the compensation law $\alpha < \beta \Rightarrow \exists\gamma\colon \alpha + \gamma = \beta$ does not generally hold.

Putting $* := \uparrow$ with $i \uparrow j := i^j$, the formula (18) could be applied to extend the exponentiation of cardinals to gc-numbers. However, the problem is how to interpret $\alpha \uparrow \beta$. We realize that any classical-like interpretation does not seem to be possible. Nevertheless, it is convenient to use at least $\alpha \uparrow \langle q \rangle$ with $q \in CN$.

Theorem 6.2. *The following laws are satisfied:*

(a) $\alpha \uparrow \langle k \rangle = \Pi\alpha$ *for* $k := |J|$.

(b) $\alpha \uparrow \langle i \rangle \le \beta \uparrow \langle j \rangle$ *whenever* $\alpha \le \beta$, $i \le j$, *and* $\beta \ge \langle 1 \rangle$.

(c) $\alpha \uparrow \Sigma\langle i_e \rangle = \Pi\alpha \uparrow \langle i_e \rangle$.

(d) $(\alpha \uparrow \langle i \rangle) \uparrow \langle j \rangle = \alpha \uparrow \langle ij \rangle, \ \ \alpha \uparrow \langle 1 \rangle = \alpha, \ \ \langle 1 \rangle \uparrow \langle i \rangle = \langle 1 \rangle$.

Proof. (a) follows from (18), and implies (b)-(d). ■

Finally, the difference $\beta - \alpha$ of two gc-numbers can be defined in the classical-like way via addition., i.e. $\beta - \alpha = \gamma$ iff γ is a unique gc-number such that $\alpha + \gamma = \beta$. Then

$$\alpha + (\beta - \alpha) = \beta, \quad \beta - \alpha \le \beta, \quad \text{and} \quad \beta - \langle 0 \rangle = \beta.$$

Closing this presentation of the nonclassical cardinality theory, one sees that, in many respects, the arithmetic and inequalities of gc-numbers have classical-like properties. This is convenient from the viewpoint of applications. Nevertheless, there are some essential differences: for instance, the failure of the compensation law and the existence of intermediate gc-numbers lying between $\langle i \rangle$ and $\langle 2^i \rangle$ (i - transfinite). They reflect in a proper way the specificity of VD-objects in comparison with sets, and, therefore, are easy to accept.

REFERENCES

[1] Dubois, D., and Prade, H. (1985) Fuzzy cardinality and the modelling of imprecise quantification, Fuzzy Sets and Systems, 16, 199-230.

[2] Dubois, D., and Prade, H. (1987) Twofold fuzzy sets and rough sets - Some issues in knowledge representation, ibid., 23, 3-18.

[3] Gottwald, S. (1989) Mehrwertige Logik. Eine Einführung in Theorie und Anwendungen, Akademie-Verlag, Berlin.

[4] Wygralak, M. (1986) Fuzzy cardinals based on the generalized equality of fuzzy subsets, Fuzzy Sets and Systems, 18, 143-158.

[5] Wygralak, M. (1991) Generalized cardinal numbers and their ordering. In: Uncertainty in Knowledge Bases, (B. Bouchon-Meunier, R.R. Yager, and L.A. Zadeh, Eds.), Lecture Notes in Computer Science, vol. 521, Springer-Verlag, pp. 183-192.

[6] Wygralak, M. (1991) Powers and generalized cardinal numbers for HCH-objects - Basic notions, Math. Pannonica, 3, 91-115.

[7] Wygralak, M. (1993) Generalized cardinal numbers and operations on them, Fuzzy Sets and Systems, 53, 49-85 (Erratum: ibid., 62, 375).

[8] Wygralak, M. (1993) A cardinality theory for vaguely defined objects - Problems of inequalities and applications, Acta. Appl. Math., 30, 1-33.

[9] Zadeh, L.A. (1983) A computational approach to fuzzy quantifiers in natural languages, Comput. and Math. with Appl., 9, 149-184.

THE STRUCTURE AND LIAPUNOV STABILITY OF INVARIANT SETS OF FUZZY RELATIONS[1]

Phil DIAMOND, Aleksej POKROVSKII[2]

Mathematics Department
University of Queensland
4072 Australia

ABSTRACT

We give a characterization of $\mathrm{Inv}(R)$, *the class of invariant fuzzy sets of a fuzzy relation* R *on* $\mathbf{X}$. *It is shown that* $\mathrm{Inv}(R)$ *can be described in terms of a lattice–theoretic hull of no more than* N *extreme elements, where* $N = |\mathbf{X}|$ *is the cardinality of* $\mathbf{X}$. *These ideas are used to investigate the Lyapunov stability of a class of fuzzy control systems.*

1. Introduction

For any finite set $\mathbf{X}$ denote by $\mathfrak{F}(\mathbf{X})$ the totality of *fuzzy sets*, that is, of functions $X : \mathbf{X} \to [0,1]$.

Let $\mathbf{X}$ be a fixed finite set of cardinality $|\mathbf{X}| = N$ and let R be a given *fuzzy relation* on $\mathbf{X}$. That is, $R \in \mathfrak{F}(\mathbf{X} \times \mathbf{X})$. The set $R \circ X,\ X \in F(\mathbf{X})$ is defined by max–min composition to be

$$(R \circ X)(x) \;=\; \bigvee_{y \in \mathbf{X}} (R(x,y) \wedge X(y)) \;.$$

Here, $a \vee b = \max(a,b)$ and $a \wedge b = \min(a,b)$. There is a fairly complete theory of solution for fuzzy equations of the form

$$R \circ X \;=\; Y \tag{1}$$

where Y is a *fixed* element from $\mathfrak{F}(\mathbf{X})$ [6].

For applications to control theory, especially with regard to notions like stability and controllability, possibly even more important are equations of the form

$$R \circ X \;=\; X. \tag{2}$$

A fuzzy set $X_* \in \mathfrak{F}(\mathbf{X})$ satisfying (2) is said to be *invariant* for the relation R. Sanchez [7] has used the term *eigen fuzzy set* for this concept. Invariant sets of fuzzy relations have been studied in [7] and very recently in [4]. Their structure appears to be more complicated than solution sets to (1). Below, further results demonstrate this rich structure.

[1]This research has been supported by the Australian Research Council Grant A 8913 2609.

[2]Permanent address: Institute of Information Transmission Problems, Russian Academy of Science, Moscow.

2. Structure of Inv(R)

For $X \in \mathfrak{F}(\mathbf{X})$ and $\alpha \in [0,1]$ denote by $\alpha \wedge X$ the element from $\mathfrak{F}(\mathbf{X})$ defined by

$$(\alpha \wedge X)(x) \;=\; \alpha \wedge X(x)\;, \qquad x \in \mathbf{X}\;.$$

For any set $\mathfrak{S} \subseteq \mathfrak{F}(\mathbf{X})$ the *lattice hull* of $\mathfrak{S}$, written as hull$(\mathfrak{S})$, will be defined as the totality of fuzzy sets of the form

$$\bigvee_{k=1}^{K} (\alpha^k \wedge X^k) \tag{3}$$

where $K \leq N$, and

$$0 \leq \alpha^k \leq 1\;,\; X^k \in \mathfrak{S}\;, \qquad k = 1, \ldots, K\;.$$

Write Inv(R) for the totality of invariant fuzzy sets of the fuzzy relation R.

Lemma 1. $\qquad \mathrm{hull}\,(\mathrm{Inv}(R)) = \mathrm{Inv}(R)$

Proof: First, $\mathrm{Inv}(R) \subseteq \mathrm{hull}\,(\mathrm{Inv}(R))$, since we write $K = 1$, $\alpha^1 = 1$ and $X^1 = X$ for any $X \in \mathrm{Inv}(R)$. Suppose that

$$X = \bigvee_{k=1}^{K} \left(\alpha^k \wedge X^k\right) \;\in\; \mathrm{hull}\,(\mathrm{Inv}(R))\;,$$

so that $R \circ X^k = X^k$, $k = 1, 2, \ldots, K$. Then for each $x \in \mathbf{X}$

$$\begin{aligned}
(R \circ X)(x) \;&=\; R \circ \left(\bigvee_{k=1}^{K} (\alpha^k \wedge X^k)\right)(x) \\
&=\; \bigvee_{y \in \mathbf{X}} \left(R(x,y) \wedge \left(\bigvee_{k=1}^{K} \alpha^k \wedge X^k(y)\right)\right) \\
&=\; \bigvee_{y \in \mathbf{X}} \left(\bigvee_{k=1}^{K} \alpha^k \wedge R(x,y) \wedge X^k(y)\right) \\
&=\; \bigvee_{k=1}^{K} \alpha^k \wedge \left(\bigvee_{y \in \mathbf{X}} R(x,y) \wedge X^k(y)\right) \\
&=\; \bigvee_{k=1}^{K} \alpha^k \wedge X^k(x) \;=\; X(x)\;,
\end{aligned}$$

and thus $\mathrm{hull}\,(\mathrm{Inv}(R)) \subseteq \mathrm{Inv}(R)$. □

The principal result of this paper is

Theorem 1. *Let the cardinality of the set* $\mathbf{X}$ *be* N. *Then there exist* $K \leq N$ *fuzzy sets* $X^1, \ldots, X^K \in \mathfrak{F}(\mathbf{X})$ *such that*

$$\mathrm{Inv}(R) = \mathrm{hull}\left\{X^1, \ldots, X^K\right\} . \tag{4}$$

An algorithm to choose elements $X^1, \ldots, X^K \in \mathfrak{F}(\mathbf{X})$ which satisfy (4) will be described in the following sections.

Theorem 1 is not really analogous to the result that the eigenspace of an $N \times N$ matrix A is a linear space of dimension $K \leq N$. In fact, there exist sets $\mathfrak{S} \subset \mathfrak{F}(\mathbf{X})$ such that $\mathrm{hull}(\mathfrak{S}) = \mathrm{hull}(\mathfrak{S}_*)$ implies $|\mathfrak{S}_*| = \infty$. A closer analogy, because of the estimate on the number of generators of the hull, is to the classical Carathéodory theorem on convex combinations in convex hulls ([5], p. 17 *ff.*).

Also a close analogy is the following: A subset $\mathfrak{S} \subset \mathfrak{F}(\mathbf{X})$ satisfying $\mathrm{hull}(\mathfrak{S}) = \mathfrak{S}$ might be thought of as a *convex* collection of fuzzy sets on $\mathbf{X}$. An element $X_* \in \mathfrak{S}$ is said to be *extreme* if

1. $\alpha \wedge Y = X_*,\ Y \in \mathfrak{S}$ implies that $Y = X_*$; and

2. there exists $\beta \in (0, 1]$ such that $\beta \wedge X_* = Y \vee Z,\ Y, Z \in \mathfrak{S}$ implies that either $\beta \wedge X_* = Y$ or $\beta \wedge X_* = Z$.

Then the following is a fuzzy analogue of Choquet's Theorem [1].

Proposition 2. *Any closed convex set* $\mathfrak{S} \subseteq \mathfrak{F}(\mathbf{X})$ *is the hull of its extreme elements.*

Proof: Let X be any given element from $\mathfrak{S}$. It is sufficient to construct extreme elements $X_1, \ldots, X_N \in \mathfrak{S}$ and values $\gamma_1, \ldots, \gamma_N \in [0, 1]$ satisfying

$$X = \bigvee_{n=1}^{N} (\gamma_n \wedge X_n) . \tag{5}$$

A fuzzy set $U \in \mathfrak{S}$ will be called *critical* if from $U = V \vee W,\quad V, W \in \mathfrak{S}$ it follows that either $V = U$ or $W = U$. First of all we construct critical elements $Y_1, \ldots, Y_N \in \mathfrak{S}$ such that

$$X = \bigvee_{n=1}^{N} Y_n .$$

Let $x_1, \ldots, x_N$ be an enumeration of $\mathbf{X}$. For each $n = 1, 2, \ldots N$, define the subset of $\mathfrak{S}$

$$S_n = \{Y \in \mathfrak{S} : Y(x_k) \leq X(x_k),\ k \neq n;\ Y(x_n) = X(x_n)\} .$$

For each $n = 1, 2, \ldots N$, define the fuzzy set Y_n as any element from S_n which minimizes the functional

$$\Phi(Y) = \sum_{k=1}^{N} Y(x_k) .$$

It easy to check that $Y_1, \ldots, Y_N$ are critical, and that $X = \bigvee_{n=1}^{N} Y_n$. Now, for each $n = 1, \ldots N$ define

$$T_n = \{Z \in \mathfrak{S} : \exists\, \beta(Z) \in [0,1] \text{ with } \beta(Z) \wedge Z = Y_n\} .$$

For each $n = 1, 2, \ldots N$ let $\dot{X}_n$ be any element from T_n maximizing the functional Φ. Let γ_n be any $\beta \in [0,1]$ satisfying $\beta \wedge X_n = Y_n$. By construction, the elements $X_1, X_2, \ldots, X_N$ are extreme and satisfy (5). □

Theorem 1 can be reformulated as:

Theorem 3. *The closed and convex set* Inv(R), *of all invariant elements of the fuzzy relation R, is the hull of no more than N extreme elements.*

The algorithm to be described below gives an explicit description of extreme elements of Inv(R).

Theorem 1 has a further consequence worth noting. Consider a crisp mapping $f : \mathbf{X} \to \mathbf{X}$. This mapping induces a fuzzy relation $\mathfrak{F}(f)$ defined by

$$\mathfrak{F}(f)(x,y) = \begin{cases} 1 & \text{if } \; y = f(x) , \\ 0 & \text{otherwise} . \end{cases}$$

Let C_f be the set of all cycles of the mapping f. Then $|C_f| \le N$ and

$$\mathrm{Inv}(\mathfrak{F}(f)) = \mathrm{hull}\, \{\chi(C) : \; C \in C_f\} , \tag{6}$$

where $\chi(C)$ denotes the characteristic function of the set C. Denote by $\mathcal{I}(f)$ the set of all $\mathbf{S} \subset \mathbf{X}$ defined by $f(\mathbf{S}) = \mathbf{S}$. Then (6) means that the totality Inv($\mathfrak{F}(f)$) of *fuzzy invariant* sets of the fuzzy relation $\mathfrak{F}(f)$ coincides with the totality $\mathfrak{F}(\mathcal{I}(f))$ of fuzzy sets of the set $\mathcal{I}(f)$. This, in turn, is an analogue of the decomposition of any set $T \in \mathcal{I}(f)$ into the union of cycles of f.

The general procedure for constructing the set $\{X^1, \ldots, X^K\}$ in Theorem will give precisely the set $\chi(C)$, $C \in C_f$ in this case. See also Example 1 of Section .

3. Lyapunov stability

Some idea of the application of the ideas of the last section to fuzzy control systems can be seen as follows. To introduce stability, notions of distance are required. Suppose that the finite set $\mathbf{X} \subset \Re^k$, and that $\mathfrak{F}(\mathbf{X})$ is given a metric ρ defined by

$$\rho(X,Y) = d_H(\mathbf{Gr}(X), \mathbf{Gr}(Y)) ,$$

where $\mathbf{Gr}(X) = \{(x, X(x)) : x \in \mathbf{X}\}$ is the *graph* of X and d_H is the *Hausdorff distance* between closed subsets of $\Re^k$ [3]. Note that the maximum exists because $\mathbf{X}$ is a finite subset of $\Re^k$.

An invariant fuzzy set $X \in \mathrm{Inv}(R)$ is said to be *Lyapunov stable* if for any $\varepsilon > 0$ there exists $\delta > 0$ such that

$$\rho(X_0, X) \le \delta$$

implies that

$$\rho(X_i, X) \le \varepsilon\ , \qquad i = 1, 2, \dots \tag{7}$$

where X_i are defined by

$$X_i = R \circ X_{i-1} \qquad i = 1, 2, \dots\ .$$

Proposition 4. *All invariant fuzzy sets $X \in \mathrm{Inv}(R)$ are Lyapunov stable.*

Proof: For $X \in \mathfrak{F}(\mathbf{X})$, $\delta \ge 0$ denote by $X^{(-\delta)}$ and $X^{(\delta)}$ the fuzzy sets defined by

$$X^{(-\delta)})(x) = 0 \vee (X(x) - \delta)\ , \qquad x \in \mathbf{X};$$

$$X^{(\delta)})(x) = 1 \wedge (X(x) + \delta)\ , \qquad x \in \mathbf{X}.$$

It suffices to note that for any $\delta > 0$ the set

$$\mathfrak{S} \;=\; \left\{X \in \mathfrak{F}(\mathbf{X}) :\ X^{(-\delta)}(x) \le X(x) \le X^{(\delta)}(x),\ x \in \mathbf{X}\right\} \subseteq\ \mathfrak{F}(\mathbf{X})$$

satisfies $R \circ \mathfrak{S} \subseteq \mathfrak{S}$. □

Observe that nonzero invariant fuzzy sets never have the property of asymptotic stability. The zero invariant fuzzy set is asymptotically stable if and only if $\mathrm{Inv}(R)$ consists only of the zero element. This will occur if and only if the square $N \times N$ matrix $A = (R(x_i, x_j))$ has a zero spectral radius. Here, $x_1, \dots, x_N$ is an enumeration of $\mathbf{X}$.

For any $X_0 \in \mathfrak{F}(\mathbf{X})$ the sequence

$$\mathbf{Tr}(X_0, R) \;=\; (X_0, X_1, \dots, X_i, \dots)\ , \quad X_i = R \circ X_{i-1}\ ,\ i = 1, \dots, N\ ,$$

is called the *trajectory* of R with *initial state* X_0. Clearly any trajectory is eventually periodic with some period $p(X_0, R)$.For some R and X_0 the integer $p(X_0, R)$ can be rather large.

4. The algorithm

Consider a nonempty set $\mathbf{S} \subseteq \mathbf{X}$. We say that $\mathbf{S}$ is *α–cyclic* for the relation R if there exists a finite sequence $(x(1), x(2), \ldots, x(k))$ in $\mathbf{S}$, $k \geq |\mathbf{S}|$ and $\mathbf{S} = \bigcup_{i=1}^{k} x(i)$, with

$$R(x(1), x(2)),\ R(x(2), x(3)),\ \ldots,\ R(x(k-1), x(k)),\ R(x(k), x(1)) \geq \alpha \ .$$

In other words (see also Example 1 below), the elements of an α–cyclic set $\mathbf{S}$ can be considered as the node set of a cycle of possibility α with respect to R. Denote by $A(R,\alpha)$ the set of α–cyclic subsets of $\mathbf{X}$ which are maximal with respect to inclusion. Clearly, distinct elements from $A(R,\alpha)$ are disjoint.

The range of R is a finite subset of $[0,1]$, of cardinality ν say, consisting of elements $0 \leq \alpha_\nu < \alpha_{\nu-1} < \ldots < \alpha_1 \leq 1$. Let us set

$$\mathcal{B}(1) \quad = \quad A(R, \alpha_1)$$

and denote by $\mathcal{B}(i)$, $i = 2, \ldots, \nu$ the set of elements from $A(R,\alpha_i)$ which does not intersect with elements from $A(R,\alpha_{i-1})$. For any $B \in \mathcal{B}(i)$ let $Z_B \in \mathfrak{F}(\mathbf{X})$ be the fuzzy set

$$Z_B(x) = \begin{cases} \alpha_i & \text{if} \quad x \in B\ , \\ 0 & \text{otherwise}\ . \end{cases}$$

Let $U_B = \lim_{m\to\infty} Z^m$, where $Z^1 = Z_B$ and $Z^{m+1} = R \circ Z^m$. Write $\mathcal{B}(R) = \bigcup_{i=1}^{\nu} \mathcal{B}(i)$.

Theorem 4. *The limit* $\lim_{m\to\infty} Z^m$ *exists for every* $B \in \mathcal{B}(R)$. *The fuzzy sets* U_B, $B \in \mathcal{B}(R)$ *are invariant for* R *and*

$$\mathrm{Inv}(R) \quad = \quad \mathrm{hull}\ \{U_B :\ B \in \mathcal{B}(R)\}\ . \tag{8}$$

By construction, the set $\mathcal{B}(R)$ contains not more than N elements. Consequently, Theorem 3 will follow from Theorem 1.

5. Proof of Theorem

We first prove the existence and invariance of U_B.

Lemma 2. *For each* $B \in \mathcal{B}(R)$, $U_B = \lim_{m\to\infty} Z^m$ *exists and is invariant for the relation* R.

Proof: Choose a positive integer $\mu \leq \nu$, and a subset B of $\mathbf{X}$, $B \in \mathcal{B}(\mu)$. By the definition of α–cyclic set we have

$$Z^1 = Z_B \leq R \circ Z_B = Z^2\ . \tag{9}$$

On the other hand, the mapping $X \to R \circ X$ is monotone. That is, $X \leq Y$ implies $R \circ X \leq R \circ Y$. Therefore, from (9) it follows that

$$Z^1 \leq Z^2 \leq \ldots \leq Z^m \leq \ldots \,. \tag{10}$$

where $Z^{m+1} = R \circ Z^m$. But the sequence Z^m, $m = 1, 2, \ldots$ contains only a finite number of distinct elements, because for any $x \in \mathbf{X}$ and any positive integer m, number $Z^m(x)$ must be of the form $R(x, y)$, $y \in \mathbf{X}$. This proves that for sufficiently large M we have

$$Z^{m+1} = Z^m, \qquad m \geq M \,. \tag{11}$$

The existence of the limit U_B and invariantness with respect to R will now follow.
□

It remains to prove (8). By Lemma 1

$$\mathrm{Inv}(R) \supseteq \text{hull}\,\{U_B : B \in \mathcal{B}(R)\} \,.$$

Now, it suffices to proof the inclusion

$$\mathrm{Inv}(R) \subseteq \text{hull}\,\{U_B : B \in \mathcal{B}(R)\} \,. \tag{12}$$

This will follow from:

Lemma 3. *For any invariant fuzzy set $X_* \in \mathfrak{F}(\mathbf{X})$ and each $x \in \mathbf{X}$ there exist $B_x \in \mathcal{B}(R)$ and $\alpha_x \in [0,1]$ satisfying*

$$X_*(x) = \alpha_x \wedge U_{B_x}(x) \,. \tag{13}$$

and

$$\alpha_x \wedge U_{B_x} \leq X_* \,. \tag{14}$$

Proof: From Lemma , without loss of generality we can assume that

$$X_*(x) = \bigvee_{y \in \mathbf{X}} X(y) \,. \tag{15}$$

Let β denote the greatest of the α_i satisfying $\alpha_i \geq X_*(x)$. Consider the set $A(R, \beta)$. For any $B \in A(R, \beta)$ denote

$$Z_B(x) = \begin{cases} \beta & \text{if} \quad x \in B \,, \\ 0 & \text{otherwise} \,. \end{cases} \tag{16}$$

For each $B \in A(R, \beta)$, U_B exists by the first part of the theorem that we have proved above. For each $D \in A(R, \beta)$, denote by $D^{\#}$ the set

$$D^{\#} = \{y \in \mathbf{X} : U_D(y) = \beta\} \,.$$

We need the further result:

Lemma 4. *Let x satisfy* (15). *Then there exists $B \in A(R,\beta)$ such that $x \in B^{\#}$ and that*

$$X_*(x) = X_*(z), \quad \text{for all} \quad z \in B . \tag{17}$$

Proof: For any $u \in \mathbf{X}$ write

$$H(u) \;=\; \{y \in \mathbf{X} : \; R(u,y) \geq \beta, \; X_*(y) = X_*(x)\} \;,$$

and put $H^{n+1}(u) = H(H^n(x))$, $n = 1, 2, \ldots$. Denote by γ the greatest value of R on $\mathbf{X} \times \mathbf{X}$ which is strictly less than β. Then $X_*(x) > \gamma$. Because X_* is an invariant fuzzy set, the previous inequality implies that the sets $H^n(x)$ are nonempty for all positive integer n. In particular, there exist elements

$$z_i \in \mathbf{X}, \qquad i = 0, 1, 2, \ldots, N+1$$

such that $z_0 = x_*$ and

$$R(z_{i-1}, z_i) \;\geq\; \beta, \quad X_*(z_i) = X_*(x_*), \quad i = 1, \ldots, N . \tag{18}$$

The sequence (18) has repeated elements because $|\mathbf{X}| = N$. Now, it is sufficient to choose as B the element of $A(R,\beta)$ which contains the elements from a periodic fragment of this sequence. Lemma 4 is proven. □

Now let us complete the proof of Lemma 3 and therefore of the theorem. Consider a set B with the properties in Lemma 4. By construction, this set B contains at least one element $B_0 \in \mathcal{B}(i)$, for some i such that $\alpha_i \geq \beta$. For each $x \in \mathbf{X}$ set

$$B_x = B_0 \qquad \text{and} \qquad \alpha_x = X_*(x) .$$

By the monotonicity of the relation R (see the proof of Lemma)

$$(\alpha_x \wedge Z_x^m)(x) = X_*(x), \qquad \alpha_x \wedge Z_x^m \leq X_* \;,$$

where Z_x^m are defined by

$$Z_x^1 = Z_{B_x}, \quad Z_x^{m+1} = R \circ Z_x^m .$$

Therefore,

$$X_*(x) \;=\; \alpha_x \wedge U_{B_x}(x) . \tag{19}$$

and

$$\alpha_x \wedge U_{B_x} \leq X_* . \tag{20}$$

The relations (19) and (20) coincide respectively with (13) and (14). Hence the theorem is proved. □

6. Examples

Example 1. Let R be a crisp relation. Then R can be treated as a multi–valued mapping of the set $\mathbf{X}$. That is, R induces a mapping $\widetilde{R} : \mathbf{X} \to 2^{\mathbf{X}}$ defined by

$$\widetilde{R}(x) = \{y : R(x,y) = 1\} .$$

Denote also by $\mathcal{G}_R$ a graph with the node set $\mathbf{X}$ and with the arc set

$$A = \{(x,y) : \ R(y,x) = 1\} .$$

Call a subet $S \subseteq \mathbf{X}$ cyclic if there exists a cycle [2] of the graph G_R which gous exactly throw nodes from the set S. Then $\mathcal{B}$ is the collection of maximal with respect to inclusion cyclic sets; U_B, $B \in \mathcal{B}(R)$ is the minimal set $S \subseteq \mathbf{X}$ which contains B and satisfies $RS = S$.

Example 2. Suppose that $\mathbf{X}$ contains only two elements x_1, x_2 and the fuzzy relation R is defined in the usual way by the matrix

$$A_R = \begin{pmatrix} 1.0 & 0.3 \\ 0.1 & 0.5 \end{pmatrix} .$$

Then $\nu = 2$; $\alpha_1 = 1.0$, $\alpha_2 = 0.5$. The set $\mathcal{B}_1$ contains a single element $\{x_1\}$ and the set $\mathcal{B}_2$ contains a single element $\{x_2\}$. Further, the limits U_B are given by

$$U_{\{x_1\}}(x_1) = 1.0, \quad U_{\{x_1\}}(x_2) = 0.1$$

and

$$U_{\{x_2\}}(x_1) = 0.3, \quad U_{\{x_1\}}(x_2) = 0.5 .$$

Example 3. Let $\mathbf{X}$ contain two elements x_1, x_2 again and suppose that the fuzzy relation R is defined by the matrix

$$A_R = \begin{pmatrix} 1.0 & 0.1 \\ 0.5 & 0.3 \end{pmatrix} .$$

Then $\nu = 2$; $\alpha_1 = 1.0$, $\alpha_2 = 0.3$. The set $\mathcal{B}_1$ contains a single point $\{x_1\}$ and the set $\mathcal{B}_2$ contains a single point $\{x_2\}$. Then

$$U_{\{x_1\}}(x_1) = 1.0, \quad U_{\{x_1\}}(x_2) = 0.5$$

and

$$U_{\{x_2\}}(x_1) = 0.1, \quad U_{\{x_1\}}(x_2) = 0.3 .$$

Examples 2 and 3 demonstrate, in particular, that the totality $\mathrm{Inv}(R)$ of fuzzy relations with strictly positive components $R(x,y)$, $x,y \in \mathbf{X}$ often has a number of extreme elements. It contrasts with the fact that a Markov chain with positive transient probabilities always has a unique stationary probability distribution.

6. Concluding remarks

Inv(R), the class of invariant fuzzy sets of a fuzzy relation on a finite set $\mathbf{X}$, was characterized in terms of the lattice hull of $K \leq |\mathbf{X}|$ fuzzy sets. If Lyapunov stability of the recurrence $X_i = R \circ X_{i-1}$ is defined in terms of appropriate metrics on $\mathfrak{F}(\mathbf{X})$, then all invariant fuzzy sets are Lyapunov stable. While no individual fuzzy state in Inv(R) is asymptotically stable, the set Inv(R) is itself an asymptotically stable invariant of the recurrence.

References

[1] L. Asimow and A.J. Ellis, "Convexity Theory and its Applications in Functional Analysis", Academic Press, New York, 1980.

[2] B. Carre, "Graphs and Networks", Clarendon Press, Oxford, 1979.

[3] P. Diamond and P.E. Kloeden, "Metric Spaces of Fuzzy Sets: Theory and Applications", World Scientific, Singapore, 1994.

[4] A. Di Nola, W. Kolodziejczyk and S. Sessa, Transitive solutions of relational equations on finite sets and linear lattices, in B. Bouchon–Meunier, R.R. Yager and L.A. Zadeh (Eds.) "Uncertainty in Knowledge Bases", Lecture Notes in Computer Science **521**, 173–182, Springer–Verlag, Berlin, 1991.

[5] S.R. Lay, "Convex Sets and their Applications", John Wiley, New York, 1982.

[6] W. Pedrycz, Numerical and applicational aspects of fuzzy relational equations. *Fuzzy Sets and Systems*, **11**, pp. 1 – 18, 1983.

[7] E. Sanchez, Eigen fuzzy sets and fuzzy relations. *Mathematical Analysis Appl.*, **81**, pp. 399 – 421, 1981.

ON FUZZINESS MEASURES VIA SUGENO'S INTEGRAL

Pietro BENVENUTI, Doretta VIVONA, Maria DIVARI

Dipartimento di Metodi e Modelli Matematici
per le Scienze Applicate
Università "La Sapienza"
Via A. Scarpa, 10, I00161 Roma, Italy

ABSTRACT

We introduce a notion of partial order for fuzzy sets in connection with their greater or smaller fuzziness. This allows us to give a simple basis to the theory of fuzziness measures.

Sugeno's integral permits to build very large classes of fuzziness measures.

Keywords: Fuzzy sets, fuzzy measures, fuzzy integrals, fuzziness measures

1. Introduction

Given an abstract set Ω and a σ-algebra $\mathcal{C}$ of crisp sets of Ω. Let $\mathcal{F}$ the family of all $\mathcal{C}$-measurable fuzzy sets [6]. With every fuzzy set F a membership function is associated, $f : \Omega \longrightarrow [0,1]$, that by definition is $\mathcal{C}$-measurable. From now on, we shall denote with capital letters the fuzzy events, and with lower case letters the corresponding membership functions. In $\mathcal{F}$ there are also the crisp sets C, whose membership functions coincide with characteristic functions χ_C.

De Luca and Termini [2], gave a first definition of fuzziness measure, based upon a condition of monotonicity, with respect to a natural relation of less or greater fuzziness. Then they considered some measures of fuzziness, which, for their properties, are close to mesures of entropy related to uncertainty of membership to the fuzzy set of points in the space Ω.

In this paper, revisiting the definition of De Luca and Termini, and taking into account some considerations due to Wang [5], we introduce small variations, however leading to a more exact order relation among fuzzy sets, which represents the idea of less or greater fuzziness.

Then we construct an axiomatic theory, leaning on this order relation, and in particular we study some fuzziness measures built by means of Sugeno's integral [4], with respect to an assigned fuzzy measures.

2. Orderings

In the family $\mathcal{F}$, the inclusion relation, which generalizes the usual one for crisp sets, defines a partial ordering:

(1) $$F_1 \subset F_2 \Longleftrightarrow f_1(\omega) \leq f_2(\omega) \quad \forall\, \omega \in \Omega\,.$$

Moreover it is possible to introduce a different order relation, called *fuzziness relation*: F_1 is *less fuzzy* than F_2, $(F_1 \prec F_2)$ or F_2 is *more fuzzy* than F_1, $(F_2 \succ F_1)$ if the following condition holds:

(2) $$\begin{cases} f_1(\omega) \leq \frac{1}{2} & \Longrightarrow f_1(\omega) \leq f_2(\omega) \leq \frac{1}{2} \\ f_1(\omega) \geq \frac{1}{2} & \Longrightarrow \frac{1}{2} \leq f_2(\omega) \leq f_1(\omega)\,, \end{cases}$$

which is equivalent to

$$\begin{cases} f_2(\omega) < \frac{1}{2} & \Longrightarrow f_1(\omega) \leq f_2(\omega) \\ f_2(\omega) > \frac{1}{2} & \Longrightarrow f_1(\omega) \geq f_2(\omega) . \end{cases}$$

The relation $F_1 \prec F_2$, or $F_2 \succ F_1$ can be concisely expressed by requiring:

$$(f_1(\omega), f_2(\omega)) \in C \quad \forall\, \omega \in \Omega ,$$

where C is the shaded closed set in Fig. 1.

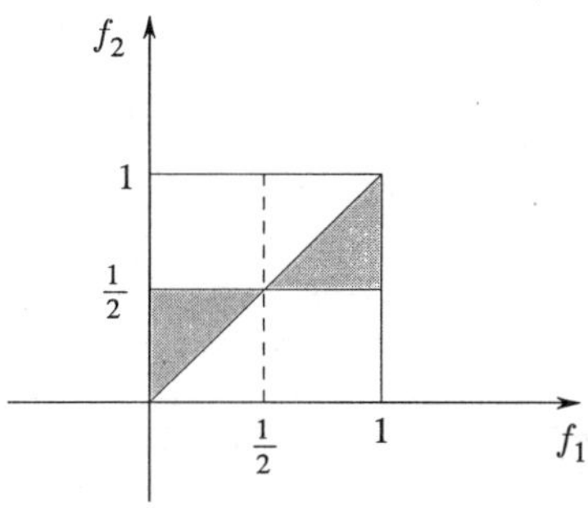

Fig. 1 – Order relation in fuzziness

This definition is similar to the one introduced by De Luca and Termini [2] who gave a first definition of fuzziness measure. We think that the present definition, giving rise to an order relation, improves both the one introduced by De Luca and Termini and the modified one given by Wang [5].

The fuzziness relation (2) is different and not comparable with the inclusion relation (1). In fact, it holds: $F_1 \cap F_1^c \subset F_1 \subset F_1 \cup F_1^c$, but these fuzzy sets are not comparable with respect to the fuzziness relation.

Moreover, if $f_1(\omega) \leq 1/2$ and $f_2(\omega) \leq 1/2$ for every $\omega \in \Omega$, the inclusion relation is equivalent to fuzziness relation, on the contrary, if $f_1(\omega) \geq 1/2$ and $f_2(\omega) \geq 1/2$ for every $\omega \in \Omega$, the fuzziness relation is opposite to the inclusion.

If the sets are crisp, the relation $C_1 \prec C_2$ implies $C_1 = C_2$. Two crisp sets, which are different, are not comparable with respect to the fuzziness relation.

With respect to this relation the crisp sets are minimal elements, in the sense that for every assigned crisp set C no fuzzy set F exists for which $C \succ F$: however, none of them is a minimum.

The fuzzy set $\Omega/2$, whose membership function is $f(\omega) = 1/2$ for every $\omega \in \Omega$, is the unique maximal element and it is the maximum.

PROPOSITION 2.1. *Given two fuzzy sets F_1 and F_2 there always exists the supremum with respect to the order $\prec$, and is the fuzzy set $F_1 \curlyvee F_2$, whose membership is the function*

$$(f_1 \curlyvee f_2)(\omega) = K[f_1(\omega), f_2(\omega)] \quad \forall\, \omega \in \Omega , \tag{3}$$

where $K(x, y)$ is the function defined in Fig. 2.

As to the infima we have the following

PROPOSITION 2.2. *Given two fuzzy sets F_1 and F_2, the infimum $F_1 \curlywedge F_2$ exists iff the following conditions hold:*

$$\begin{aligned} f_1(\omega) < \tfrac{1}{2} &\Longrightarrow f_2(\omega) \leq \tfrac{1}{2} \\ f_2(\omega) < \tfrac{1}{2} &\Longrightarrow f_1(\omega) \leq \tfrac{1}{2} . \end{aligned} \tag{4}$$

In this case the fuzzy set $F_1 \curlywedge F_2$ *is defined by*

$$(f_1 \curlywedge f_2)(\omega) = H[f_1(\omega), f_2(\omega)] \qquad \forall\, \omega \in \Omega\,,$$

where $H(x,y)$ *is defined in Fig. 3;*

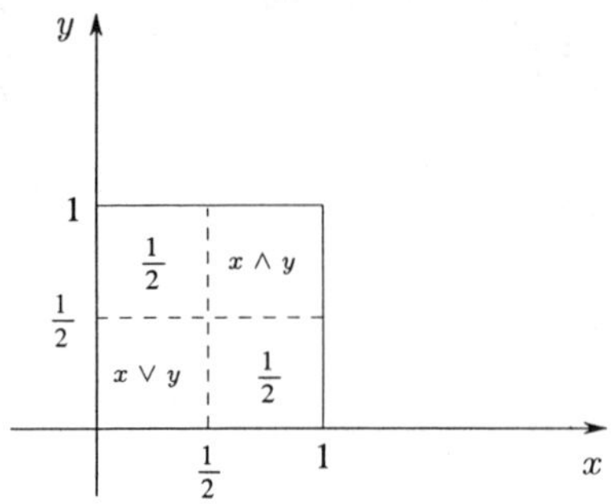

Fig. 2 – K function for supremum.

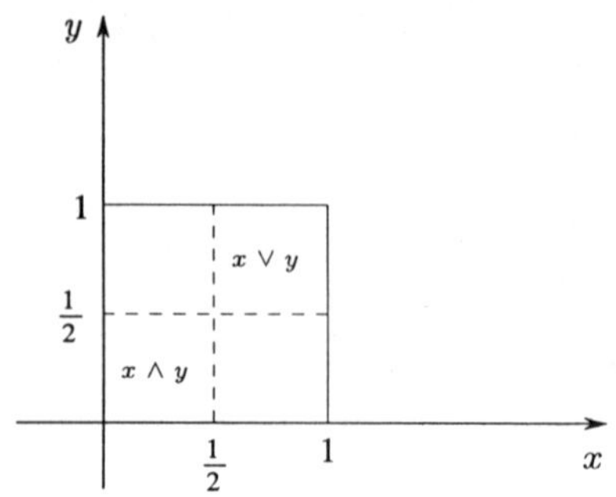

Fig. 3 – H function for infimum.

A sequence of fuzzy sets $\{F_n\}_{n\in\mathbb{N}}$ is *non-decreasing in fuzziness* (i.e. with respect to the fuzziness relation) if

$$F_1 \prec F_2 \prec \ldots \prec F_n \prec F_{n+1} \prec \ldots$$

This implies that the set Ω is decomposable into two subsets, C_1 and C_2, such that:

$$\begin{aligned} \omega \in C_1 &\Longrightarrow f_n(\omega) \le \tfrac{1}{2} \quad \text{and} \quad \{f_n(\omega)\} \text{ is non-decreasing} \\ \omega \in C_2 &\Longrightarrow f_n(\omega) \ge \tfrac{1}{2} \quad \text{and} \quad \{f_n(\omega)\} \text{ is non-increasing.} \end{aligned}$$

It follows that there always exists $\lim f_n(\omega) = f(\omega)$ for every $\omega \in \Omega$ and $f(\omega) = \lim f_n(\omega)$ is the membership function of the fuzzy set $F = \lim F_n = \curlyvee F_n$.

The sequence of fuzzy sets $\{F_n\}_{n\in\mathbb{N}}$ is *non-increasing in fuzziness* if

$$F_1 \succ F_2 \succ \ldots \succ F_n \succ F_{n+1} \succ \ldots$$

This implies that the set Ω is decomposable into two sets, C_1, C_2 as above, such that:

$$\begin{aligned} \omega \in C_1 &\Longrightarrow f_n(\omega) \le \tfrac{1}{2} \quad \text{and} \quad \{f_n(\omega)\} \text{ is non-increasing} \\ \omega \in C_2 &\Longrightarrow f_n(\omega) \ge \tfrac{1}{2} \quad \text{and} \quad \{f_n(\omega)\} \text{ is non-decreasing.} \end{aligned}$$

As above, then there exists $\lim f_n(\omega) = f(\omega)$ for every $\omega \in \Omega$ and $f(\omega) = \lim f_n(\omega)$ is the membership function of the fuzzy set $F = \lim F_n = \curlywedge F_n$.

3. Fuzziness measure and fuzzy measure

We define the *fuzziness measure* as a map $d : \mathcal{F} \to [0,1]$ with the following properties:

$[A_1]$ $d(C) = 0 \quad \forall\, C \in \mathcal{C}$,

$[A_2]$ $d(\Omega/2) = 1$

$[A_3]$ $F_1 \prec F_2 \Longrightarrow d(F_1) \le d(F_2)\,,\ \forall\, F_1, F_2 \in \mathcal{F}\,,$

$[A_4]$ $d(F) = d(F^c)\,,\quad \forall\, F \in \mathcal{F}\,,$ where F^c is the complement of F.

In particular, we recognize that the restriction of the fuzziness measure to sets of the kind $F = C/2$ (whose membership function is $(1/2)\chi_C$) is a *fuzzy measure* on $\mathcal{C}$. In fact, the function $\mu\colon C \to [0,1]$, defined by $\mu(C) = d(C/2)$, is monotone non decreasing with respect to the inclusion relation, and $\mu(\emptyset) = 0$.

So we call $\mu(C)$ the *fuzzy measure associated* with d and d is an *extension* of μ.

This measure is *regular* [6], because $\mu(\Omega) = 1$, but it doesn't satisfy those continuity properties required by many authors in the definition of fuzzy measure, unless the original fuzziness measure itself already satisfies them.

We say that the fuzziness measure d is *lower semicontinuous* if for every non decreasing (in fuzziness) sequence of fuzzy sets $\{F_n\}_{n\in\mathbb{N}}$, it holds: $\lim d(F_n) = d(\curlyvee F_n)$ (continuity from below).

Moreover, the function d is *upper semicontinuous* if for every non increasing (in fuzziness) sequence of fuzzy sets $\{F_n\}_{n\in\mathbb{N}}$, it holds: $\lim d(F_n) = d(\curlywedge F_n)$ (continuity from above).

These continuity properties of a fuzziness measure imply analogous properties of continuity for the fuzzy measure associated with d.

In fact, if d satisfies the continuity property for sequences non-decreasing (non-increasing) in fuzziness, then the associated fuzzy measure μ is continuous for sequences non-decreasing (resp. non-increasing) with respect to inclusion.

4. Fuzziness measure via Sugeno's integral

Let $(\Omega, \mathcal{C}, \mu)$ be a fuzzy measure space with μ continuous and $\varphi\colon \Omega \times [0,1] \to [0,1]$ a measurable function with the following properties:

$[B_1]$ $\quad \varphi(\omega, 0) = \varphi(\omega, 1) = 0$

$[B_2]$ $\quad \varphi(\omega, 1/2) = 1$

$[B_3]$ $\quad \varphi(\omega, \cdot) \begin{cases} \text{non decreasing in } [0, 1/2] \\ \text{non increasing in } [1/2, 1] \end{cases}$

$[B_4]$ $\quad \varphi(\omega, x) = \varphi(\omega, 1-x)\,, \quad \forall\, x \in [0,1]$

PROPOSITION 4.1. *Given a fuzzy measure μ and a function $\varphi(\omega, x)$ as above, the map $d\colon \mathcal{F} \to [0,1]$ defined by*

(6)
$$d(F) = ⨍_{\Omega} \varphi(\omega, f(\omega))d\mu$$

is a fuzziness measure, whose associated fuzzy measure is the given μ. The function φ determines the extensions from μ to d, it is called the generating function of d.

PROOF. Sugeno's integral, [4], is defined for every function $g\colon \Omega \longrightarrow [0,1]$ and for every $C \in \mathcal{C}$ by

$$⨍_C g d\mu = \bigvee_{x\in[0,1]} \{x \wedge \mu(C_x \cap C)\}\,,$$

where $C_x = g^{-1}([x,1]) = \{\omega \in \Omega \colon g(\omega) \in [x,1]\}$, and enjoys the following properties:

$[S_1]$ $$\fint_C g d\mu = \fint_\Omega (\chi_C \wedge g) d\mu\,,$$

$[S_2]$ $$\fint_C a d\mu = a \wedge \mu(C)\,, \quad \forall\, a \in [0,1]\,,$$

$[S_3]$ $$g_1 \leq g_2 \implies \fint_C g_1 d\mu \leq \fint_C g_2 d\mu\,,$$

$[S_4]$ $$g_n \uparrow g \implies \fint_C g_n d\mu \uparrow \fint_C g d\mu\,,$$

$[S_5]$ $$\fint_C \Big[\bigvee_{i\geq 1} (a_i \wedge g_i)\Big] d\mu = \bigvee_{i \geq 1} \Big[a_i \wedge \fint_C g_i d\mu\Big]\,,$$

provided that $\{g_i\}$ is a monotone non-increasing (non-decreasing) sequence of $\mathcal{C}$-measurable functions, and $\{a_i\}$ is a monotone non-decreasing (non-increasing) sequence of nonnegative real numbers.

From these properties it easily follows that the map (6) is a fuzziness measure i.e. enjoys the properties $[A_1]$-$[A_4]$.

PROPOSITION 4.2. *The fuzziness measure d defined by* (6) *enjoys the further properties:*

(a) $d(F) = d(F \cap F^c) = d(F \cup F^c) = d(F^c)\,,$

(b) $d(F \cup C) = d(F \cap C^c) = d(F) \qquad \forall\, C \in \mathcal{C} \qquad F \cap C = \emptyset,$

(c) $d(F \cap T(F)) = d(F) = d(F \cup [T(F)]^c)\,,$ *where $T(F)$ is the transom of F [3],*

(d) *if φ is continuous, i.e. the function $\varphi(\omega,\cdot)$ is continuous in $[0,1]$ for every ω, then d is continuous for monotone sequences: if $\{F_n\}$ is non-increasing (or non-decreasing) in fuzziness and F is the limit, then $d(F_n)$ monotonely converges to $d(F)$.*

PROPOSITION 4.3. *If the fuzzy measure μ is sup-additive [6], the following properties hold:*

(a) $d(F_1 \cup F_2) = d(F_1) \vee d(F_2)$,
if $F_1 \cap F_2 = \emptyset$, or, more generally, if $F_1 \cap F_2 \in \mathcal{C}$;

(b) $d(F_1 \curlyvee F_2) = d(F_1) \vee d(F_2)\,,$
if there exists $F_1 \curlywedge F_2$.

5. Homogeneous fuzziness measures

In this section we study those fuzziness measures which are homogeneous in Ω, i.e. they are built by functions which don't depend on ω: $\varphi(\omega, x) = \psi(x)$. The function ψ enjoys the following properties:

$[B_1']$ $\quad \psi(0) = \psi(1) = 0$

$[B_2']$ $\quad \psi(1/2) = 1$

$[B_3']$ $$\psi(x) \begin{cases} \text{non decreasing in } [0,1/2] \\ \text{non increasing in } [1/2,1] \end{cases}$$

$[B_4']$ $\quad \psi(x) = \psi(1-x)$

Then the fuzziness measure d is given by

(8) $$d(F) = \underset{\Omega}{\int\!\!\!\!-} \psi(f(\omega))d\mu \quad \forall\, F \in \mathcal{F}\,.$$

The fuzziness measure (8) enjoys further properties which are characteristic of the homogeneous choice of ψ.

PROPOSITION 5.1. *The fuzziness measure d defined by* (8) *satisfies the following additional properties:*

(a) $d(xC) = d(x\Omega) \wedge d(C/2) \quad \forall\, x \in [0,1], \quad \forall\, C \in \mathcal{C}$,

(b) $d\Big(\bigvee_{1\le i\le n} a_i C_i\Big) = \bigvee_{1\le i\le n} d(a_i C_i)$,
where $\{a_i\}$ *is any finite increasing sequence of real numbers in* $[0, 1/2]$ *and* $\{C_i\}$ *is any corresponding decreasing sequence of crisp sets.*

6. A representation theorem

Now we are ready for a representation theorem: we shall see that the conditions (a) of Prop. 4.2. and (a)-(b) of Prop. 5.1. are characteristic of those fuzziness measurés, that can be expressed in the form (8).

6.1. REPRESENTATION THEOREM. *Let d be a fuzziness measure enjoying the properties* $[A_1] - [A_4]$.
We assume, moreover, that the following additional properties are satisfied:

$[A_5]$ *d is continuous,*

$[A_6]$ $d(F) = d(F \cap F^c) = d(F \cup F^c) \qquad \forall\, F \in \mathcal{F}$,

$[A_7]$ $d(xC) = d(x\Omega) \wedge d(C/2) \qquad \forall\, x \in [0,1]$ *and* $\forall\, C \in \mathcal{C}$,

$[A_8]$ $d\Big(\bigvee_{1\le i\le n} a_i C_i\Big) = \bigvee_{1\le i\le n} d(a_i C_i)$,
where $\{a_i\}$ *is any finite increasing sequence of real numbers in* $[0, 1/2]$ *and* $\{C_i\}$ *is any corresponding decreasing sequence of crisp sets.*

Then there exists a continuous fuzzy measure on crisp sets, $\mu\colon \mathcal{C} \to [0,1]$ *and a continuous function* ψ *enjoying the conditions* $[B'_1] - [B'_4]$, *such that*

(9) $$d(F) = \underset{\Omega}{\int\!\!\!\!-} \psi(f(\omega))d\mu\,, \quad \forall\, F \in \mathcal{F}\,.$$

PROOF. Setting

$$\mu(C) = d(C/2) \quad \forall\, C \in \mathcal{C}\,,$$

by the properties $[A_1] - [A_3]$ of d, μ is a continuous fuzzy measure on $\mathcal{C}$.
Moreover, let

$$\psi(x) = d(x\Omega) \quad \forall\, x \in [0,1]\,;$$

it is immediate that ψ enjoys the properties $[B'_1] - [B'_4]$.
Let's now prove (9), with ψ and μ as just defined. By virtue of $[A_6]$, we can limit ourselves to fuzzy sets, whose membership is less than or equal to 1/2, and so we shall do from now on.

Now, let $x \in [0,1/2]$ and $C \in \mathcal{C}$ and consider the fuzzy set $F = xC$. Then by property $[S_2]$ we find

$$\int_\Omega \psi(x\,\chi_C(\omega))d\mu = \int_C \psi(x)d\mu = \psi(x) \wedge \mu(C) =$$
$$= d(x\Omega) \wedge d(C/2)\,.$$

Therefore, from $[A_7]$ we get that for the fuzzy set $F = xC$ the representation (9) is valid.

Now we suppose that the fuzzy set F is simple, i.e. F takes only a finite number of values: it is always possible to write it as:

$$F = \bigvee_{1\le i\le n} a_i C_i \text{ or } f(\omega) = \bigvee_{1\le i\le n} a_i \chi_{C_i}(\omega)\,,$$

with $\{a_i\}$ and $\{C_i\}$ satisfying the hypothesis $[A_8]$.

So we have by $[S_5]$:

$$\int_\Omega \psi(f(\omega))d\mu = \int_\Omega \bigvee_{1\le i\le n} \psi(a_i)\chi_{C_i} d\mu =$$
$$= \bigvee_{1\le i\le n} [\psi(a_i) \wedge \mu(C_i)] = \bigvee_{1\le i\le n} d(a_i C_i)\,.$$

By $[A_8]$ the representation (9) holds for any simple fuzzy set.

Now let's assume that F is any fuzzy set with $f(\omega) \le 1/2\ \forall\omega \in \Omega$, then, it is possible to find a sequence $\{F_n\}$ of simple fuzzy sets, increasing with respect to inclusion, and hence to the fuzziness order, whose limit is F. Then, by using continuity of d and of Sugeno's integral, we find that the representation (9) holds for F.

Finally by $[A_6]$ and $[B'_4]$ the representation (9) holds for any $F \in \mathcal{F}$.

REFERENCES

[1] N. Batle – E. Trillas: *Entropy and fuzzy integral*, J. Math. Anal. Appl., **69** (1979), 469-474.

[2] A. De Luca – S. Termini: *A definition of a non probabilistic entropy in the setting of fuzzy sets theory*, Inform. and Contr., **20** (1972), 301-312.

[3] E Roventa: *On the degree of fuzziness of a fuzzy set*, Fuzzy Sets and Systems, **36** (1990), 259-264.

[4] M. Sugeno: *Theory of Fuzzy Integrals and Applications*, Ph. D. Doctoral dissertation, Tokyo Inst. of Tech., 1974.

[5] Z. Wang – Zi-Xiao: *Fuzzy measures and measures of fuzziness*, J. Math. Anal. Appl., **104** (1984), 589-601.

[6] Z. Wang – G.J. Klir: *Fuzzy Measure Theory*, Plenum Press (1992).

[7] L.A. Zadeh: *Fuzzy sets*, Inform and Contr., **8** (1965), 338-353.

MONOTONOUS QUANTIFICATIONS AND SUGENO FUZZY INTEGRALS

Patrick BOSC, Ludovic LIETARD

IRISA/ENSSAT BP 447
22305 Lannion Cédex FRANCE

ABSTRACT

This paper is devoted to the interpretation of quantified statements of the type "Q X are A", where Q is a fuzzy quantifier, X is a set of elements and A a fuzzy predicate. The overall objective is to point out the connections which are existing between quantified statements and fuzzy integrals. In this paper, we show that Prade's interpretation of "Q X are A", when Q is a monotonous quantifier, is equivalent to a Sugeno fuzzy integral. One of the advantages of this view resides in the possibility to derive improved algorithms in order to compute quantified expressions.

1. Introduction

First order logic involves two quantifiers, the universal quantifier (∀) and the existential quantifier (∃), which are too limited to model any natural language quantified sentence [1]. For this reason, generalized quantifiers [1] and fuzzy quantifiers [14] have been introduced. Fuzzy quantifiers represent linguistic expressions such as "many", "at least 3" and are used in fuzzy quantified expressions. We can distinguish two main families of fuzzy quantified statements: "Q elements of set X are satisfying the fuzzy predicate A" ("Q X are A"), which will be considered in details in the following, and "Q elements of set X which satisfy the fuzzy predicate B also satisfy the fuzzy predicate A" ("Q X B are A").

Our prime interest concerns the computational aspects of fuzzy quantified expressions of the type "Q X are A". For instance, considering the statement "Q X are A" as representing a flexible query addressed to a database [8] (X refers to a set of items of the database), we would like to avoid the exhaustive scan of the set X at query evaluation time. In this paper, we investigate the possibility for defining the truth value of fuzzy quantified expressions (where Q is monotonic) in terms of Sugeno fuzzy integrals. In fact, one can find a strong similarity between the following two processes : 1) the specification of the truth value of "Q X are A" where A is a fuzzy predicate knowing the value of "Q X are A" when A is crisp, and 2) the specification of a measure for fuzzy sets (performed by a fuzzy integral) knowing the measure for regular sets. Moreover, this work is a first step to establish connections between fuzzy integrals and previous propositions [10,12,13,14] to define the truth value of fuzzy quantified statements. More precisely, in this paper, it is shown that Prade's proposition [10] is a Sugeno fuzzy integral (for monotonic quantifiers). We take advantage of some properties of Sugeno fuzzy integral to design algorithms involving heuristics, intended for improving the computation of monotonic fuzzy quantified statements. We illustrate the use of our proposition in the context of fuzzy queries addressed to ordinary relational databases [2,3].

In section 2, fuzzy quantified statements, their interpretation and their connection with Sugeno fuzzy integral [11] are presented. Section 3 is devoted to the introduction of Prade's proposition [10] related to the evaluation of fuzzy quantified statements. It is shown that his approach, when restricted to monotonous quantifiers, leads to compute a Sugeno fuzzy integral. We use this property in section 4 to incorporate some heuristics into an algorithm implementing the evaluation of monotonous quantified statements of the type "Q X are A". An example of use of this algorithm, in the field of database flexible querying [2,3] is given.

2. Fuzzy quantifications and integrals

2.1 Fuzzy quantified statements

Fuzzy quantifiers were first introduced by L.A. Zadeh [14] and it is possible to distinguish between the absolute quantifiers and the relative ones. Absolute quantifiers are expressed with a number (or a counter) while relative quantifiers are expressed by a proportion. Typical examples of respectively absolute and relative quantifiers could be "at least 3" or "at least half". Thus, an absolute quantifier is represented by a function Q from an integer range to [0,1] whereas a relative quantifier is represented by a function Q from [0,1] to [0,1]. In both cases, the value Q(j) is defined as the truth value of the statement "Q X are A" when exactly j elements from X fully satisfy A (whereas it is assumed that A is fully unsatisfied for the other elements). It is worth noticing that in the following, we deal with absolute quantifiers but our results are still valid for relative quantifiers (in each formula, it is necessary to turn every Q(i) into Q(i/n), n being the cardinality of the usual set X).

Monotonous fuzzy quantifiers are such that if Q is increasing, then: i) $Q(0) = 0$, ii) $\exists k$ such as $Q(k) = 1$ and iii) $\forall$ a,b if $a > b$ then $Q(a) \geq Q(b)$. When Q is decreasing: i) $Q(0) = 1$, ii) $\exists k$ such as $Q(k) = 0$ and iii) $\forall$ a,b if $a > b$ then $Q(a) \leq Q(b)$. Figure 1 gives an example of a monotonous increasing quantifier.

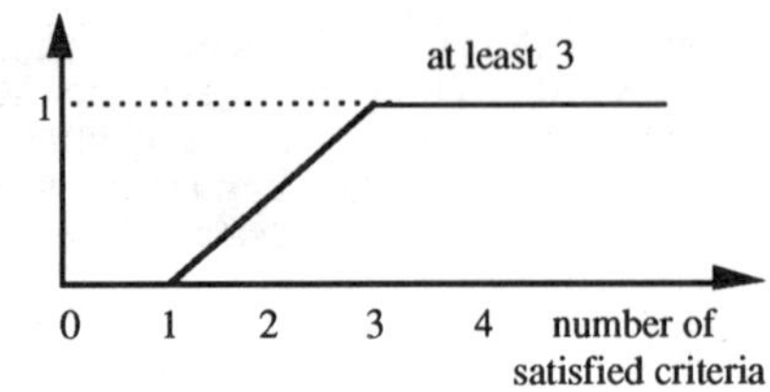

figure 1: the fuzzy quantifier "at least 3"

Example. Let Q be the fuzzy quantifier of figure 1. Let X be the crisp set made of 3 employees e1, e2, e3 working in the same department d and earning respectively \$48,000, \$45,000 and \$80,000. Let A be the fuzzy predicate "well-paid" described in figure 2. These employees satisfy "well-paid" according to the fuzzy set {0.9/e1 + 0.75/e2 + 1/e3}. The evaluation of the statement "at least 3 employees are well-paid" raises the problem of determining the value Q(N), N being the number of well-paid employees. This example illustrates the need for computing the value of Q when the number of satisfied criteria is not perfectly known, i.e. when the set of the elements from X which satisfy A is fuzzy (since A is a fuzzy predicate).

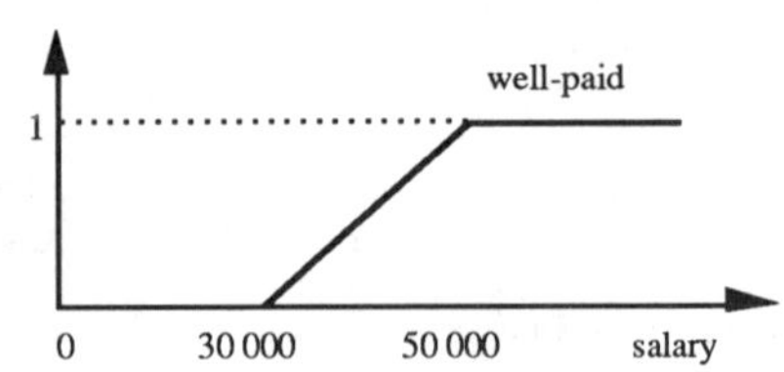

figure 2: the fuzzy predicate "well paid"

2.2 Sugeno integrals and increasing quantifiers

In this section, we point out that, given a measure defined on regular sets, the Sugeno fuzzy integral is a tool enabling the value of the measure on a fuzzy set to be calculated. Thus, if an increasing quantifier Q is defined as a measure which gives the truth value of "Q X are A" when A is crisp, we have a method for defining the truth value of the statement "Q X are A".

Let $\mathcal{P}(X)$ denote the powerset of $X=\{x_1, x_2, .., x_n\}$. Let us recall that a fuzzy measure [11] on a finite, discrete regular set X is a function g from $\mathcal{P}(X)$ to the unit interval which satisfies the following three axioms: i) $g(X) = 1$, ii) $g(\varnothing) = 0$ and iii) $\forall$ A, B such that $A \subseteq B$, $g(A) \leq g(B)$ (known as monotonicity).

A Sugeno fuzzy integral [11] denoted S_g defines a characteristic of a function h (from a universe X to the unit interval, thereby similar to a membership function) according to the measure g. Sugeno fuzzy integral is an averaging operator over the values taken by h. If we denote $h(x_i) = h_i$, and we order the h_i's ($h_1 \leq h_2 \leq ... \leq h_n$) we get:

$$S_g(h) = \max_i \min(h_i, g(A_i))$$

where $A_i = \{x_i, x_{i+1}, \dots, x_n\}$.

$S_g(h)$ is a median [9], i.e. it delivers the nth element of an ordered list of (2n-1) elements.

$$S_g(h) = \text{med}(h_1,\dots,h_n,g(A_2),g(A_3),\dots,g(A_n)).$$

Let h be the membership function of a fuzzy set H. It has been shown [7] that $S_g(h)$ can be considered the value of g for the fuzzy set H.

The definition of an increasing quantifier Q as a fuzzy measure g requires the definition of an underlying universe F such as:

$$g: \mathcal{P}(F) \rightarrow [0,1] \text{ such that: } g(E) = Q(\|E\|) \qquad (1)$$

where $\|E\|$ denotes the cardinality of the usual set E. As we need g(F) = 1, this definition requires:

$$Q(\|F\|) = 1. \qquad (2)$$

Since Q is increasing, the property of monotonicity is then satisfied.

When we want to evaluate "Q X are A", the satisfaction of property (2) requires to take F either as X (Q is increasing relative, or increasing absolute and $Q(\|X\|) = 1$), or as X plus dummy elements which do not satisfy A at all (when $Q(\|X\|) < 1$). The dummy elements are added so that $Q(\|X\|)$ reaches 1, but of course, they do not compete at all for the evaluation of "Q X are A" (for any dummy element z, $\mu_A(z) = 0$). In so doing, the meaning of "Q X are A" is not altered.

Let EA denote the set of elements from X satisfying the predicate A. The set EA is a (possibly fuzzy) subset of X and then of F. When EA is a regular set (the satisfaction degrees are either 1 or 0), the truth value of "Q X are A" is g(EA). In the general case, EA is a fuzzy set and the evaluation of "Q X are A" requires the computation of g for the fuzzy set EA, given by $S_g(EA)$. If the cardinality of F is k, these values are given by:

$$\begin{aligned} S_g(EA) &= \max_i \min(\mu_{EA}(x_i), g(F_i)) \\ &= \text{med}(\mu_{EA}(x_1),\dots,\mu_{EA}(x_k),Q(k-1),\dots,Q(1)) \end{aligned} \qquad (3)$$

where $F_i = \{x_i, x_{i+1}, \dots, x_k\}$ and $\mu_{EA}(x_1) \le \mu_{EA}(x_2) \le \dots \le \mu_{EA}(x_k)$.

It is important to notice that if $Q(n) \neq 1$, (k-n) dummy elements have been added to X in order to build the set F.

Example. The quantifier "at least 3" given in figure 1 may be represented using the measure g such that:

$g(E) = 0$ if $\|E\| \le 1$ $\quad$ $g(E) = 0.5$ if $\|E\| = 2$ $\quad$ $g(E) = 1$ if $\|E\| \ge 3$.

If we consider the previous example, X = {e1,e2,e3} and g(X) = Q(3) = 1. Therefore F is set to X and EA = {0.9/e1 + 0.75/e2 + 1/e3}. After an appropriate ordering (x_1 = e2, x_2 = e1, x_3 = e3), we get $\mu_{EA}(x_1) = 0.75$, $\mu_{EA}(x_2) = 0.9$, $\mu_{EA}(x_3) = 1$ and the truth value of "at least 3 employees are well-paid" is given by $S_g(EA)$:

$$\text{med}(\mu_{EA}(x_1), \mu_{EA}(x_2), \mu_{EA}(x_3), Q(2), Q(1)) = \text{med}(0.75, 0.9, 1, 0.5, 0) = 0.75.$$

3. Prade's approach

We first present the method advocated by Prade [10] for any fuzzy quantifier. Then, we distinguish increasing and decreasing quantifiers to show that both of them call on a Sugeno integral.

3.1 The general case

Prade's method involves two steps. The first one computes the fuzzy cardinality π of the set (denoted SA) of those elements of X which satisfy A. Then, the quantifier Q is considered a fuzzy constraint and the imprecise cardinality π is matched against it, thus giving the truth value of the fuzzy quantified statement "Q X are A".

Since SA is a fuzzy set, its cardinality can be represented as a possibility distribution π [6]. This possibility distribution is built from c the number of entirely satisfied criteria (possibly equal to 0) according to the following definitions : i) $\forall\, i < c \;\; \pi(i) = 0$, ii) $\pi(c) = 1$ and iii) $\forall\, j > c \;\; \pi(j)$ is the jth largest value $\mu_A(x)$. It should be noticed that the restriction of π to the values greater or equal to c is decreasing ($i \geq j \geq c \Rightarrow \pi(i) \leq \pi(j)$).

The second step of Prade's process performs a fuzzy pattern matching [5] between π, (the possibility distribution obtained beforehand) and the quantifier seen as a fuzzy constraint. Consequently, a couple of degrees is computed: Π the extent to which π possibly satisfies Q, and N the extent to which π necessarily (certainly) satisfies Q:

$$\Pi = \max_{u \geq c} \min\,(Q(u)\,,\, \pi(u)) \qquad\qquad N = \min_{u \geq c} \max\,(Q(u)\,,\, 1 - \pi(u)).$$

We can then conclude that the degree of truth of the quantified statement "Q X are A" is situated in the interval $[N\,,\, \Pi]$.

3.2 Monotonic quantifiers

In the following, we will show that if Q is increasing (resp. decreasing), the quantified statement "Q X are A" only requires the computation of Π (resp. N), which is a Sugeno integral.

If Q is increasing, we can write that N equals Q(c), c being the number of entirely satisfied criteria since:

$$\begin{aligned}
& i \geq j \geq c \Rightarrow Q(i) \geq Q(j) \text{ (Q is increasing)} \\
& \text{and } 1 - \pi(i) \geq 1 - \pi(j) \quad (\pi \text{ is decreasing}) \\
& \text{then } \max(Q(i), 1 - \pi(i)) \geq \max(Q(j), 1 - \pi(j)) \\
& \Rightarrow N = \min_{c \leq i} \max\,(Q(i)\,,\, 1 - \pi(i)) = Q(c).
\end{aligned}$$

Thus, N is the truth value of the quantified statement if we only consider the entirely satisfied criteria. When the other criteria are taken into account, this truth value can only increase since Q is an increasing quantifier. The useful information is brought by Π whose value takes into account all the values $\mu_A(x)$. Let us show that when Q is increasing, Π is given by $S_g(EA)$ (g is built from formulae (1) and (2)).

In fact, if we consider definition (3) for $S_g(EA)$:

$$S_g(EA) = \max_i \min(\mu_{EA}(x_i), Q(k-i+1))$$

because $g(F_i) = Q(k-i+1)$. If we remove the (k-n) dummy elements, we get:

$$S_g(EA) = \max_{i \geq k-n+1} \min(\mu_A(x_i), Q(k-i+1))$$

since the other elements satisfy $\mu_{EA}(x_i) = \mu_A(x_i)$. Now, we turn (k-i+1) into j, we get:

$$S_g(EA) = \max_{j\geq 1} \min(\mu_A(x_j), Q(j))$$

with $\mu_A(x_1) \geq \mu_A(x_2) \geq ... \geq \mu_A(x_n)$. If we consider the monotonicity of Q, we observe that:

$$\max_{c\geq j\geq 1} \min(\mu_A(x_j), Q(j)) = Q(c)$$

where c is the number of elements which fully satisfy A. Therefore:

$$S_g(EA) = \max_{j\geq c} \min(\mu_A(x_j), Q(j))$$

with $\mu_A(x_c) \geq \mu_A(x_{c+1}) \geq ... \geq \mu_A(x_n)$. The values $\mu_A(x_j)$ $(j \geq c)$ are nothing but $\pi(j)$ and finally $S_g(EA) = \Pi$.

Similarly, if Q is decreasing Π equals Q(c), where c is the number of entirely satisfied criteria. When all criteria are taken into account, the only changes affect N which still is calling on a Sugeno integral as shown below:

$$N = \min_{c\leq i} \max (Q(i) , 1 - \pi(i)) = 1 - \max_{c\leq i} \min (1-Q(i) , \pi(i)).$$

If we define the fuzzy measure g' such as:

$$\forall E, g'(E) = 1 - Q(\|E\|),$$

we get:

$$N = 1 - S_{g'}(EA).$$

4. Evaluation of quantified statements

In this section, we are interested in the effective computation of the truth value of quantified propositions of the type "Q X are A". The objective is to decide whether or not "Q X are A" is satisfied at a degree α (or more). The idea is to take advantage of properties established before: i) only Π or N is to be computed and ii) the Sugeno integral is a median, in order to derive heuristics avoiding the exhaustive scan of set X. We describe in details the algorithm for an increasing quantifier and we indicate how it should be modified to handle decreasing quantifiers. Finally, the use of these algorithms is illustrated in the scope of flexible queries against regular relational databases [2,3].

4.1 The algorithm and its heuristics

Let us recall that we want to know if the value of "Q X are A" is over α. This value is given by (3), i.e. $S_g(EA)$:

$$\text{med}(\mu_{EA}(x_1), .., \mu_{EA}(x_k), Q(k-1),...,Q(1)).$$

The values Q(1), ..., Q(k-1) are constants for a given quantifier Q and are computed once for all at the beginning of the algorithm. No ordering hypothesis is made about the satisfaction degrees of the elements of the set X (whose cardinality is n), which are the inputs of the algorithm which is basically iterative. Let us consider the situation at the end of the ith step of the loop. We are in a situation where (n-i) values are missing to compute $S_g(EA)$. If we assume the missing values to be zereos and $S_g(EA)$ is already over α, due to the monotonicity of Sugeno integrals, we can conclude that, whatever the actual values, the actual $S_g(EA)$ will be over α. Similarly, if we assume the missing values to be 1s and $S_g(EA)$ is less than α, then we can conclude that whatever the values of the missing elements, $S_g(EA)$ will be less than α. These two asumptions constitute the basis of the heuristics used in the algorithm.

The implementation requires two sorted arrays MI and MS of (2k-1) elements. At the ith step of the iteration, MI contains the previously treated elements $\mu_A(x)$'s, the values Q(1), ..., Q(k-1) and zeroes for the remaining elements (not yet accessed); MS contains the previously treated elements $\mu_A(x)$'s, the values Q(1), ..., Q(k-1), (n-i) values set to 1 (since not yet accessed elements are assumed to satisfy $\mu_A(x) = 1$) and zeroes for the remaining (dummy) elements. When moving from step i to step (i+1), a new satisfaction degree, $\mu_A(x)$ for the current element x, will replace a zero (resp. 1) in MI (resp. MS).

Let $MI[k]_i$ and $MS[k]_i$ be the values of the median element (MI[k] and MS[k]) at step i of the iteration. Then:

$$\forall i \;\; MI[k]_i \leq S_g(EA) \leq MS[k]_i \qquad (4)$$
$$MI[k]_n = MS[k]_n = S_g(EA) \qquad (5).$$

Property (4) relies on the fact that S_g is monotonic and then:

$$S_g(y_1, y_2, ..., y_n) \geq S_g(y_{i_1}, y_{i_2}, ..., y_{i_j}, 0, ..., 0)$$
$$S_g(y_1, y_2, ..., y_n) \leq S_g(y_{i_1}, y_{i_2}, ..., y_{i_j}, 1, ..., 1)$$

where $\{i_1, i_2, ..., i_j\}$ is any subset of $\{1,2,..., n\}$. Property (5) is based on the fact that at the end of the iteration (i = n), the contents of MI and MS are equal and that $S_g(EA)$ is the median element. Therefore, the algorithm is :

```
% MI contains k values 0, Q(1),..., Q(k-1) and MS contains (k-n) values 0, Q(1), ..,
  Q(k-1), n values 1.  MI and MS are in ascending order %
for i from 1 to n do:
   if MI[k] ≥ α then exit success endif;
   if MS[k] < α then exit failure endif;
   ordered insertion of μA(xi) in MI and MS;
   % replacing a 0 in MI and a 1 in MS%
endfor;
exit success: set X satisfies "Q X are A" ≥ α
exit failure: set X satisfies "Q X are A" < α.
```

We now consider a decreasing quantifier Q. Let us recall that the truth value of "Q X are A" is given by 1 - $S_{g'}(EA)$, where $\forall E$, $g'(E) = 1 - Q(\|E\|)$. We define the increasing quantifier Q' as: $\forall i$, $Q'(i) = 1 - Q(i)$. We compute the measure g' from Q' following formulae (1) and (2). The final result for "Q X are A" is 1 - $S_{g'}(EA)$. Here again, we incorporate heuristics to know wether or not the satisfaction degree for "Q X are A" reaches α. Since we have $\forall i \;\; MI_i \leq S_{g'}(EA) \leq MS_i$, the property $\forall i \;\; 1 - MI_i \geq 1 - S_{g'}(EA) \geq 1 - MS_i$ is valid, and used to build the heuristics. The previous algorithm is affected as follows:

if MI[k] ≥ α **then exit** success (resp. **if** MS[k] < α **then exit** failure)

is replaced by:

if (1 - MS[k]) ≥ α **then exit** success (resp. **if** (1 - MI[k]) < α **then exit** failure).

4.2 Use for database fuzzy querying

In relational database systems, the notion of imprecision can be introduced at two levels: for representing imprecise or uncertain data and to allow for flexible queries. We will consider the second aspect, that is to say that the data are assumed to take their values in ordinary universes, whereas queries may contain imprecise conditions.

In the context of an extended relational language supporting imprecise querying such as SQLf [2,3], queries are viewed as fuzzy predicates. A query is associated with a threshold α intended for the retrieval of the elements belonging to the α-cut. In such a language, it seems natural to have the ability for expressing queries involving fuzzy quantified statements of the type "Q X are A" as illustrated in the following example.

Example. Let us consider a relation which gathers information about a company and the query "find the departments satisfying at least 3 of their employees are well-paid, with an overall degree over 0.45". The fuzzy quantifier "at least 3" and the predicate "well-paid" are pictured in figure 1 and 2 respectively. For each department, we have to compute the truth value of the statement "at least 3 employees are well-paid" (using $S_g(EA)$ since the quantifier is increasing) to determine if it is over 0.45. We still consider the department d with the 3 employees e1, e2 and e3 described beforehand. The computation of $S_g(EA)$ by hand yields 0.75 and department d is qualified (≥ 0.45). We now show that our algorithm produces the same result without an exhaustive scan of the three employees.

We have k = n = 3, thus the size of the arrays MI and MS is 5. The initialization gives:

MI = 0 0 0 **0 .5** MS = **0 .5** 1 1 1

where the values corresponding to the quantifier are written in bold characters. We observe that MI[3] < 0.45 and MS[3] > 0.45, and no conclusion can be drawn to stop the process. We have to take into account the first employee.

First access: e1 is assumed to be read first. He satisfies the predicate "well-paid" at degree 0.9. This value is inserted into both MI and MS:

MI = 0 0 **0 .5** .9 MS = **0 .5** .9 1 1

Here MI[3] < 0.45 and MS[3] > 0.45, and we must continue.

Second access: e2 is assumed to be read. He satisfies the predicate "well-paid" at degree 0.75. This value is inserted into both MI and MS:

MI = 0 **0 .5** .75 .9 MS = **0 .5** .75 .9 1.

Since MI[3] > 0.45 it is possible to conclude that departement d satisfies the query with a degree greater than 0.45. Only 2 employees have been read and one access has been saved.

5. Conclusion

In this paper, we have considered the definition of the truth value of fuzzy quantified statements of the form "Q X are A" where A is a fuzzy predicate. The problem raised here concerns the determination of an imprecise cardinality representing the number of elements satisfying A. First, we have pointed out the connection between an increasing quantifier and the Sugeno fuzzy integral. In fact, there is a strong analogy between our problem (application of Q to an imprecise cardinality knowing its value for a precise one), and the ability of Sugeno fuzzy integral to extend the application of a measure to a fuzzy set. In this setting, an increasing quantifier Q plays the role of a measure used in the Sugeno integral. Then, we have shown that Prade's proposition to define the truth value of "Q X are A" (based on a two-layer fuzzy pattern matching) reduces to a Sugeno integral S when Q is increasing and is 1 - S, when Q is decreasing. The next point has been devoted to the effectiveness of the calculation (algorithms) of quantified statements interpreted as Sugeno integrals. We focus on a framework in which the elements of set X (whose cardinality is assumed to be known a priori) are accessed independently and we have to decide whether "Q X are A" is satisfied at a degree greater than a specified threshold α (α-cut of the quantified statements). Our objective is to stop the computation as soon as possible (thus obviating the entire access to X) and is the same as that pursued in [4] where an OWA-based interpretation was chosen [13]. To reach our goal, heuristics stemming from the monotonicity property of the Sugeno fuzzy integral are applied. The implementation of this algorithm only needs two arrays where the satisfaction degrees are stored step by step. Finally, an example has been given to illustrate a potential use of the algorithm in the area of relational databases fuzzy querying. In this context, the major interest of the heuristics involved in the algorithm is to save secondary memory accesses.

This paper proposes a somewhat general approach for interpreting quantified statements of the form "Q X are A" in terms of the value of a function (Q) for an imprecise argument. Here, Q is represented as a set function. Other interpretations of Q can be envisaged, especially to recover other propositions such as OWA-based [13] and ΣCount [14] interpretations.

REFERENCES

[1] J. Barwise and R.Cooper (1981) Generalized quantifiers and natural language, Linguistic and philosophy 4, 159-219
[2] P. Bosc, M. Galibourg, G. Hamon (1988) Fuzzy querying with SQL: Extension and implementation aspects, Fuzzy Sets and System 28, 3, 333-350.
[3] P. Bosc, O. Pivert (1995) SQLf: A relational database language for fuzzy querying, IEEE Transactions on fuzzy Systems to appear.
[4] P. Bosc, O. Pivert (1992) On the evaluation of fuzzy quantified queries in a database management system, Proc. North American Fuzzy Logic Processing Society, Puerto Vallarta, (Mexico), 478-487.
[5] M. Cayrol, H. Farreny and H. Prade (1980) Possibility and necessity in a pattern-matching process, Proc. IXth International Congress on Cybernatics, Namur, Belgium, 53-65.
[6] D. Dubois and H.Prade (1985) Fuzzy cardinality and the modeling of imprecise quantification, Fuzzy Sets and Systems 16, 3, 199-230.
[7] M. Grabisch T. Murofushi and M. Sugeno (1992) Fuzzy measure of fuzzy events defined by fuzzy integrals, Fuzzy Sets and Systems 50, 293 - 313.
[8] J. Kacprzyk and A. Ziolkowski (1986) Databases queries with fuzzy linguistic quantifiers, IEEE Transactions on Systems, Man and Cybernetics 16, 474-478.
[9] A. Kandel, W.J. Byatt (1978) Fuzzy sets, fuzzy algebra, and fuzzy statistics, proceeding of the IEEE 66, 1619-1639.
[10] H. Prade (1990) A Two-Layer Fuzzy Pattern Matching Procedure for the Evaluation of Conditions Involving Vague Quantifiers, Journal of Intelligent and Robotic Systems 3, 93-101.
[11] M. Sugeno (1974) Theory of fuzzy integrals and its applications, Thesis, Tokyo institute of technology.
[12] R.R. Yager (1983) Quantifiers in the formulation of multiple objective decision functions, Information sciences 31, 107-139.
[13] R.R. Yager (1988) On Ordered Weighted Averaging Aggregation Operators in Multicriteria Decisionmaking, IEEE Transactions on Systems, Man, and Cybernetics 18, 183-190.
[14] L.A. Zadeh (1983) A computational approach to fuzzy quantifiers in natural languages, Computer Mathematics with Applications 9, 149-183.

FUZZY SETS, DIFFERENCE POSETS AND MV-ALGEBRAS

Radko MESIAR
Slovak Technical University Bratislava
Radlinskeho 11, 813 68 Bratislava
SLOVAKIA

ABSTRACT

Difference posets of fuzzy subsets are studied. It is shown that the difference Θ of fuzzy subsets is generated by a normed generator g. Consequently the classes of full fuzzy g-difference posets and of semisimple MV-algebras coincide up to an isomorphism. Finally, states, observables and their expected values on fuzzy difference posets are studied.

Keywords: difference poset, fuzzy connectives, MV-algebra

1. Introduction

One of the latest generalizations for the abstract model of the quantum logic approach to the foundations of quantum mechanics are difference posets, see [12,13]. Several interesting results on difference posets can be found e.g. in [6,7,18,19,22,24]. Recall that difference posets includes MV-algebras, orthomodular posets, quantum logics, orthoalgebras etc.

Definition 1 A difference poset (D-poset) is a partially ordered set **L** with a partial ordering $\leq$, greatest element **1**, and with a partial binary operation $\ominus$: **LxL** $\to$ **L**, called difference, such that, for a, b $\in$**L**, $b \ominus a$ is defined if and only if $a \leq b$, for that the following axioms hold for a, b, c $\in$ **L**:

(DP1) $a \leq b \Rightarrow b \ominus a \leq b$;

(DP2) $a \leq b \Rightarrow b \ominus (b \ominus a) = a$;

(DP3) $a \leq b \leq c \Rightarrow c \ominus b \leq c \ominus a$ and $(c \ominus a) \ominus (c \ominus b) = b \ominus a$.

A D-poset **L** is called a complete D-poset (σ-D-poset) if, for any sequence $\{a_i\}$ in **L** with $a_1 \leq a_2 \leq \ldots$, the least upper bound $\vee_{i=1}^{\infty} a_i$ exists in **L**. □

In what follows, we write $(b \ominus a) = c$ if $(b \ominus a)$ is defined and $c = (b \ominus a)$. It is easy to see that if **L** is a D-poset then there exists also a minimal element $\mathbf{0} \in \mathbf{L}$, $\mathbf{0} = \mathbf{1} \ominus \mathbf{1}$. To exclude trivial case of a singleton $\mathbf{L} = \{a\}$, when $\mathbf{1} = \mathbf{0} = a$, we suppose throughout the rest of the paper that $\mathbf{0} \neq \mathbf{1}$. The following properties of the difference operation $\ominus$ have been proved in [13], see also [6,24] :

(α) $a \notin \{\mathbf{0},\mathbf{1}\} \Rightarrow \mathbf{1} \ominus a \notin \{\mathbf{0},\mathbf{1}\}$;

(β) $(b \ominus a) = b \Leftrightarrow a = \mathbf{0}$;

(γ) $(b \ominus a) = \mathbf{0} \Leftrightarrow a = b$;

(δ) $b \leq c \ominus a \Rightarrow (c \ominus a) \ominus b = (c \ominus b) \ominus a$;

(ε) $a < b \leq c \Rightarrow c \ominus b < c \ominus a$

Note that a D-poset can be equivalently defined through the axioms (DP1), (DP2) and (δ).

For an arbitrary but fixed element $a \in \mathbf{L}$ we define

$$a^{\perp} := \mathbf{1} \ominus a$$

We have:

(i) $(a^{\perp})^{\perp} = a$;

(ii) $a \leq b \Rightarrow b^{\perp} \leq a^{\perp}$.

The elements a and b from **L** are orthogonal, denoted by $a \perp b$, if and only if $a \leq b^{\perp}$ (or $b \leq a^{\perp}$). We define a partial binary operation $\oplus$ on **L** for orthogonal elements a and b ,

$$a \oplus b = (b^{\perp} \ominus a)^{\perp} \quad .$$

It is evident that if $a \perp b$ then $a \oplus b = b \oplus a$, i.e., $\oplus$ is a commutative operation and that $(a \oplus b) \ominus a = b$.

Note that an axiomatic approach to the structures based on a partial binary operation $\oplus$ leads to the so called effect algebras [8,9]. Each D-poset can be treated as an effect algebra with $\oplus$ given as above. Vice versa, each effect algebra can be treated as D-poset, where $a^{\perp}$ is the unique element for which $a \oplus a^{\perp} = \mathbf{1}$, and for $a \leq b$ the difference $\ominus$ is defined by $b \ominus a := (a \oplus b^{\perp})^{\perp}$. Hence all results for D-posets are valid for effect algebras, and vice versa.

Another general structure deeply investigated in the last years is the class of MV-algebras. MV-algebras (multi-valued algebras) were introduced by Chang [4], see also [1,5,9,20,21], in order to provide an algebraic proof of the completness theorem of infinite-valued logic (Lukasiewicz logic).

Definition 2 An MV-algebra is a structure $\mathbf{A} = (\mathbf{M}, \oplus, {}^{*}, \mathbf{1}, \mathbf{0})$, where $\mathbf{M}$ is a non-empty set, $\mathbf{0}$ and $\mathbf{1}$ are two distinct elements of $\mathbf{M}$, $\oplus$ is a binary operation on $\mathbf{M}$, and * is a unary operation on $\mathbf{M}$, satisfying the following axioms:

(MV1) $(a \oplus b) \oplus c = a \oplus (b \oplus c)$;
(MV2) $a \oplus \mathbf{0} = a$;
(MV3) $a \oplus b = b \oplus a$;
(MV4) $a \oplus \mathbf{1} = \mathbf{1}$;
(MV5) $(a^{*})^{*} = a$;
(MV6) $\mathbf{0}^{*} = \mathbf{1}$;
(MV7) $a \oplus a^{*} = \mathbf{1}$;
(MV8) $(a^{*} \oplus b)^{*} \oplus b = (a \oplus b^{*})^{*} \oplus a$ (Lukasiewicz axiom).□

Note that in the Chang's original definition three other binary operations were introduce, namely $\otimes$, $\vee$ and $\wedge$, which can be derived from axioms (MV1)-(MV8),

$$a \otimes b = (a^{*} \oplus b^{*})^{*}$$

$$a \vee b = (a^{*} \oplus b)^{*} \oplus b \quad \text{and}$$

$$a \wedge b = (a \oplus b^{*}) \otimes b.$$

We can introduce a partial order on $\mathbf{M}$ putting $a \leq b$ iff $b = a \vee b$ (or equivalently $a = a \wedge b$). For more details see [9].

Example 1
(a) Every finite chain $\mathbf{C} = \{x_0, \ldots, x_n\}$, $x_0 < \ldots < x_n$, is a uniquely defined D-poset [18]. Indeed, $\mathbf{1} = x_n$, $\mathbf{0} = x_1$, $x_i \ominus x_j = x_{i-j}$ whenever $j \leq i$. Further, $x_i^{\perp} = x_{n-i}$ and $x_i \oplus x_j = x_{i+j}$ whenever $i + j$ is less or equal n.

(b) Let $\mathbf{A}$ be an MV-algebra, $\mathbf{A} = (\mathbf{M}, \oplus, {}^{*}, \mathbf{1}, \mathbf{0})$. We define a partial binary operation Θ on $\mathbf{M}$ defined for $a \leq b$, $b \,\Theta\, a = (a \oplus b^{*})^{*}$. Then the set $\mathbf{M}$ equipped with the herited partial order and with the difference Θ is a D-poset.

2. Fuzzy sets and difference posets

For a given singleton $\mathbf{X} = \{x\}$, the system of all fuzzy subsets of $\mathbf{X}$ coincides with the unit interval $[0,1]$. For a difference poset defined on $[0,1]$ we have the following result.

Theorem 1 Let **L** = [0,1] be a D-poset with respect to a natural ordering on [0,1] with a given difference $\ominus$. Then there is unique normed generator g, g:[0,1]→[0,1], g(0) = 0, g(1) = 1, g is continuous strictly increasing, such that

$$b \ominus a = g^{-1}(g(b) - g(a)) \text{ for } a \le b .$$

Proof. It is evident that **1** = 1 and **0** = 0 . For non-orthogonal elements a and b we put $a \oplus b := 1$ and hence we extend the partial binary operation $\oplus$ for any elements a and b from **L**. Then $\oplus$ is a commutative, associative, non-decreasing binary operation on [0,1] and $a \oplus 0 = a$,$a \oplus 1 = 1$ for each $a \in [0,1]$, what means that $\oplus$ is a triangular conorm (t-conorm), see e.g. [19]. Further, $a \oplus a^{\perp} = (a \ominus a) = 1$ for each $a \in [0,1]$ and $a \oplus a > a$ for each $a \in]0, 1[$. Similarly, we can extend the partial binary operation $\ominus$ to whole **L** = [0,1] putting $b \ominus a := 0$ whenever a > b . Recall that then $\ominus$ corresponds to the difference operation suggested by Weber [27] induced by the t-conorm $\oplus$. In the rest of this proof, we denote by $\oplus$ and $\ominus$ the above mentioned binary operations defined on **L** = [0,1]. Note that the extended difference $\ominus$ fulfills the property (δ) for any a, b, c ∈**L**, but the axioms (DP2) and (DP3) holds true only in their original form as written above.

Let $\{a_n\} \subset [0,1]$ and let *lim* $a_n = a$. If b <a, then $b \ominus a_n = 0$ up to some finite number of indexes n and hence *lim* $(b \ominus a_n) = 0 = b \ominus a$. Denote by $i_n = inf(a_m ; m \ge n)$ and by $s_n = \sup(a_m ; m \ge n)$. Then $i_n \le a_n \le s_n$ for each n and consequently $b \ominus s_n \le b \ominus a_n \le b \ominus i_n$. Further $\{b \ominus s_n\}$ is a non-decreasing sequence and $\{b \ominus i_n\}$ is a non- increasing sequence.
For b = a , $b \ominus s_n = 0$ for each n . Put c:= *inf* $(b \ominus i_n)$. Then $i_n = b \ominus (b \ominus i_n) \le b \ominus c$, what implies $a = b \le b \ominus c$ and consequently c = 0 . But this means that *lim* $(b \ominus a_n) = 0 = b \ominus a$.
For b > a , put c as above and d = *sup* $(b \ominus s_n)$. It is evident that $i_n \le a$, i.e. $b \ominus i_n \ge b \ominus a$ for each n. Consequently $c \ge b \ominus a$. As above we can show that $a \le b \ominus c$. But then $a \le b \ominus c \le b \ominus (b \ominus a) = a$ and hence $a = b \ominus c$ and $c = b \ominus a$. Similarly we prove that $d = b \ominus a$. Now, it is evident that *lim* $(b \ominus a_n) = b \ominus a$.
Analogously, we can prove the continuity of the binary operation $\ominus$ and consequently the continuity of the t-conorm $\oplus$. But this means that $\oplus$ is a continuous nilpotent t-conorm [25] and by Ling [14] it has unique normed generator g so that

$$a \oplus b = g^{-1}(\min(1, g(a) + g(b))) , \quad a, b \in [0,1] .$$

Now, it is easy to see that $b \ominus a = g^{-1}(g(b) - g(a))$ whenever $a \le b$. Note that $a^{\perp} = g^{-1}(1 - g(a))$ and that a and b are orthogonal if and only if $g(a) + g(b) \le 1$. □

For other details see [18,19]. Immediately we get the following result.

Corollary 1 **L** = [0,1] is a D-poset (with respect to $\ominus$) if and only if it is an MV-algebra (with respect to $\oplus$ and ⊥).

Systems of fuzzy subsets of a given universe **X** equipped by fuzzy connectives induced by a complementation operator **c**, a t-norm **T** and a **c**-dual t-conorm **S** forms an MV-algebra (where $\oplus$ = **S**, $\otimes$ = **T** and * = **c**) if and only if **S** is generated by the same normed generator g as **c**, (and **T** is **c**-dual to **S**), i.e.,
S(a,b) = $g^{-1}(min(1,g(a)+g(b)))$, **c**(a) = $g^{-1}(1 - g(a))$ and **T**(a,b) = $g^{-1}(max(0,g(a)+g(b)-1))$.

Fuzzy MV- algebras coincide, up to an isomorphism, with the class of Archimedean MV-algebras, i.e. with the class of semisimple MV-algebras, see [1]. Recall that an Archimedean MV-algebra is an MV-algebra with the Archimedean property

"if $n.a = a \oplus \ldots \oplus a \le b$ for all $n \in \mathbf{N}$ then $a^* \oplus b^* = a^*$, i . e., $a = \mathbf{0}$ or $b = \mathbf{1}$"

Hence the fuzzy MV-algebras form a special well distinguished subclass of MV-algebras. Recall that any fuzzy MV-algebra is a difference poset, too.

The usual approach to introducing the fuzzy connectives (union, intersection, complementation) is based on the pointwise extension of some operations (t-conorm, t-norm, strong negation) on the unit interval [0,1] to the space $[0,1]^X$. We adopt the same approach to the fuzzy difference operation. Hence we may conclude that a difference $B \ominus A$ of two fuzzy subsets of **X**, $A \leq B$, is defined pointwisely by means of a difference $\ominus$ on the unit interval [0,1]. Analogously to the case of t-norms and corresponding fuzzy intersections (t-conorms and fuzzy unions, strong negations and fuzzy complementations), we will not distinguish (in notation) the difference on [0,1] and its extension to $[0,1]^X$.

Due to Theorem 1, for a given difference $\ominus$ on [0,1], there is unique normed generator g so that for $A, B \in [0,1]^X$, $A \leq B$, it is

$$(B \ominus A)(x) = g^{-1}(g(B(x)) - g(A(x))) \ , x \in X .$$

For a given t-norm **T**, Butnariu and Klement [3] introduced a **T**-clan and a **T**-tribe of fuzzy subsets of **X**. Analogously, for a given difference $\ominus$ on [0,1], i.e., for a given normed generator g, we will introduce a g-difference poset of fuzzy subsets of **X**.

Definition 3 Let g be a normed generator. A g-difference poset **F** of fuzzy subsets of **X** (briefly a fuzzy g-D-poset) is a subsystem of $[0, 1]^X$ containing the greatest element $\mathbf{1_X}$ ($\mathbf{1_X}(x) = 1$ for all $x \in \mathbf{X}$) such that for each $A, B \in \mathbf{F}$, $A \leq B$ (i.e. $A(x) \leq B(x)$ for all $x \in \mathbf{X}$), the fuzzy subset $B \ominus A$ is contained in **F**, where $(B \ominus A)(x) = g^{-1}(g(B(x)) - g(A(x)))$ for all $x \in \mathbf{X}$. □

Another characterization of a fuzzy g-D-poset is the following: a difference poset **F** of fuzzy subsets of **X** with the greatest element $\mathbf{1_X}$ is a fuzzy g - D- poset if and only if for each $A, B \in \mathbf{F}$, $A \leq B$, $x \in \mathbf{X}$, the value $(B \ominus A)(x)$ depends only on values A(x) and B(x), and a modified version of the above Archimedean property holds, namely

"if $b \ominus n.a = b \ominus a \ominus \ldots \ominus a$ is defined for all $n \in \mathbf{N}$ then $a = \mathbf{0}$"

If ,in a difference poset **L** of fuzzy subsets of **X** (containing $\mathbf{1_X}$), fulfilling the above modified Archimedean property, the value $(B \ominus A)(x)$ depends on values A(x), B(x) and on x, then **L** will be called a fuzzy difference poset. If a system **L** of fuzzy subsets of **X**, $\mathbf{1_X} \in \mathbf{X}$, is not a fuzzy difference poset (see the following example) then either the elements of **L** cannot be treated pointwisely or, for some $x \in \mathbf{X}$, the restriction of the difference $\ominus$ to the values $\{A(x), A \in \mathbf{L}\}$ cannot be extended to a difference on [0,1], see [19]. Then **L** will be called simply a difference poset of fuzzy subsets.

Example 2

a) Let **X** = {x,y} be a two-point universe. Let **L** = {(1,1), (0,0), (2/3,2/3), (2/3,1/3), (1/3,3/4), (1/2,1/2)}. The difference $\ominus$ on **L** is defined as follows: $A \ominus (0,0) = A$; (1,1) - (2/3,2/3) = (1/3,3/4); $(1,1) \ominus (1/3,3/4) = (2/3,2/3)$; $(1,1) \ominus (2/3,1/3) = (1/2,1/2)$; $(1,1) \ominus (1/2,1/2) = (2/3,1/3)$.

Then **L** is a D-poset of fuzzy subsets which is not a fuzzy D-poset and the difference $\ominus$ cannot be defined pointwisely. Note that **L** is also a quantum logic.

b) Let **X** = {x} be a singleton and let $0 = a_0 < a_1 < a_2 < \ldots < b_2 < b_1 < b_0 = 1$ be given real constants. Put $L = \{a_n, b_n; n = 0, 1, \ldots\}$ and define $a_n \ominus a_m = a_{n-m}$ whenever $n \geq m$, $b_n \ominus a_m = b_{n+m}$ for all n, m and $b_n \ominus b_m = a_{m-n}$ whenever $n \leq m$ (in other cases, the difference $\ominus$ is not defined). Then **L** equipped with the usual order of reals and the difference $\ominus$ is a difference poset (of fuzzy subsets of **X**) with pointwisely defined difference $\ominus$. However, if there is an extension of $\ominus$ to the whole interval [0,1], then by Theorem 1 there is some normed generator g so that $g(a_1) \rangle 0$ and $g(a_n) = n.g(a)$ exceeds 1 for some $n \in \mathbf{N}$, a contradiction. □

It is easy to see that a fuzzy g-D-poset **F** is a D-poset. Further, put $g\mathbf{F} = \{gA;\ A \in \mathbf{F}\}$, where $gA(x) = g(A(x))$, $x \in \mathbf{X}$. Then $g\mathbf{F}$ is a difference poset of fuzzy subsets of **X** with respect to the usual difference - introduced by Kôpka [12], (B - A)(x) = B(x) - A(x), i.e. $g\mathbf{F}$ is a fuzzy i-D-poset, where i is the identity on [0,1].

Recall that Kôpka and Chovanec [13] defined a D-poset **L** of fuzzy subsets as a difference poset of fuzzy subsets of **X** containing $\mathbf{1_X}$ as its greatest element. The difference $\ominus$ on **L** may be not induced by a difference on [0,1], in general, see Example 2. The same can be true even for fuzzy

D-posets. However, based on Belluce's results [1], we can show that each fuzzy D-poset **L** is isomorphic with some fuzzy i-D-poset **F**.

In [17], we have studied the systems of fuzzy subsets of a given universe **X** which are quantum logics in the traditional sense (see e.g. Varadarajan [26]). Due to Theorem 1, we have the following result.

Corollary 2 A system **L** of fuzzy subsets of **X** is a quantum logic in the traditional sense if and only if **L** is a fuzzy D-poset and if $a \le a^{\perp}$ for some $a \in \mathbf{L}$ then $a = \mathbf{0}$. □

Note that by [22] a system **L** is a D-poset such that $a \le a^{\perp}$ is true only if $a = \mathbf{0}$ if and only if **L** is an orthoalgebra of Randal and Foulis [23]. Hence a system **L** of fuzzy subsets of **X** is a quantum logic in the traditional sense if and only if **L** is an orthoalgebra.

The difference Θ defined on fuzzy subsets of a fuzzy g-D-poset **F** cannot be , in general, extended to a binary operation Θ defined for all elements $A, B \in \mathbf{F}$.

Example 3

Let **F** be the system of all even subsets of the space $\{1, 2,..., 2n\}$, $n > 1$. Then **F** is a D-poset (of crisp subsets, and hence also a fuzzy g-D-poset for all normed generators g) with respect to the inclusion and the set difference operation \ , $A \backslash B = \{ x \in \mathbf{X};\, x \in A \text{ and } x \notin B \}$ for each pair $A,\ B \in \mathbf{F}$ such that $B \subset A$. It is easy to see that difference on **F** cannot be extended to the case of an arbitrary A and $B \in \mathbf{F}$. Take e.g. $n = 2$, $A = \{1,2\}$ and $B = \{2,3\}$. What A\B should be? Applying pointwisely the extended difference on [0,1] one gets $A \backslash B = \{1\}$, what corresponds to the common difference of sets. However, $\{1\}$ is not an element of **F** and consequently A\B cannot be defined.

3. Full difference posets

We are interested in D-posets admitting an extension of the partial binary operation of the difference to a full binary operation. The latest will be called a full difference and we will use again the notation Θ.

Definition 4 A full difference poset (full D-poset) is a partially ordered set **L** with partial ordering $\le$, maximal element **1** and with binary operation Θ, called full difference, such that for all $a, b, c \in \mathbf{L}$ the following axioms hold:

(FDP1) $b \ominus a \le b$;

(FDP2) $a \le b \Rightarrow c \ominus b \le c \ominus a$;

(FDP3) $b \ominus (b \ominus a) = a \Leftrightarrow a \le b$;

(FDP4) $(c \ominus a) \ominus b = (c \ominus b) \ominus a$.

It is easy to see that each full D-poset is also a D-poset. Note that if **L** is a full D-poset then $b \ominus a = \mathbf{0}$ if and only if $a \ge b$ .Further, the class of full D-posets includes the class of MV-algebras,where the difference of two elements a and b is defined by $b \ominus a := (a \oplus b^*)^*$.

If L is a full D-poset which is a lattice, too, then by [5] it can be treated also as an MV-algebra (to be more precise, we are able to introduce the operations $\oplus$ and * so that L becomes an MV-algebra). A similar results follows if we strengthen the axiom (FDP3) to

(FDP3′) $b \ominus (b \ominus a) = a \ominus (a \ominus b)$.

Futher, each totally ordered D-poset corresponds to an MV-algebra, too. This is, e. g., the case of D-poset **L** from Example 2b).

In the case of a fuzzy full D-poset, we are able to prove directly that fuzzy full D-posets and fuzzy MV-algebras and hence also semisimple MV-algebras coincide. We will show this for fuzzy full g-D-posets only.

Theorem 2 Let **F** be a fuzzy full g-D-poset of fuzzy subsets of a given universe **X**. Then **F** is a semi-simple MV-algebra where

$$A \oplus B = (B^{\perp} \ominus A)^{\perp} = \mathbf{1} \ominus ((\mathbf{1} \ominus B) \ominus A)$$
$$A \otimes B = B \ominus A^{\perp} = B \ominus (\mathbf{1} \ominus A)$$
$$A^* = A^{\perp} = \mathbf{1} \ominus A \ . \qquad \square$$

Proof. Due to Belluce's results in [1], it is enough to show that **F** is an MV-algebra. For this, it is enough to show that , for any $x \in \mathbf{X}$, the restriction of **F** from **X** to $\{x\}$ is an MV-algebra. Fix an element $x \in \mathbf{X}$ and denote by $\mathbf{F}_x$ the restriction of **F** to $\{x\}$. For all a, b $\in$ **F** it is $b \ominus a = g^{-1}\ (g(b) - g(a))$ whenever $a \le b$ and $b \ominus a = \mathbf{0}$ whenever $a \ge b$, i.e. $b \ominus a = g^{-1}\ (max\ (0, g(b) - g(a)))$. Then

$$a \oplus b = g^{-1}\ (min\ (1, g(a) + g(b)))$$
$$a \otimes b = g^{-1}\ (max\ (0, g(a) + g(b) - 1))$$
$$a^* = g^{-1}\ (1 - g(a)) \ .$$

It is easy to see that $(a \otimes b^*) \oplus b = max\ (a, b)(= a \vee b)$ and that **F** is an MV-algebra isomorphic to the Lukasiewicz MV-algebra [1].

By the previous theorem, any fuzzy full g-D-poset **F** is closed under fuzzy complementation induced by * and by fuzzy intersection induced by $\otimes$. It is easy to see that then the system **gF** is closed under usual Zadeh fuzzy complementation and under bold Giles fuzzy intersection, i.e. **gF** is a $\mathbf{T}_\infty$-clan of Butnariu and Klement [3]. If, in addition, **F** is also complete, i.e., **F** is a fuzzy full σ-g-D-poset, then **gF** is a $\mathbf{T}_\infty$ -tribe [3] (and vice versa).

We have just obtain a simple definition of a fuzzy MV-algebra using only one binary operation $\ominus$. The problem whether any full D- poset is an MV-algebra remains open.Some partial results can be found in [5,9]. However, any notion and/or result from the D-poset theory can be immediately formulated in the framework of MV-algebras. In the fuzzy case, some specific results are true due to the special structure of the fuzzy D-posets (fuzzy MV-algebras).

4. Observables and states on fuzzy difference posets

Definition 5 [13] Let $\mathbf{L}_1$ and $\mathbf{L}_2$ be two D-posets. A mapping $\mathbf{m}: \mathbf{L}_1 \to \mathbf{L}_2$ is called a *homomorphism* of D-posets if and only if it preserves the partial ordering and the difference operation. A homomorphism **m** is called a *σ-homomorphism* of D-posets if it preserves the least upper bounds of countable sequences.
A *σ-homomorphism* $\mathbf{m}: \mathbf{L} \to [0,1]$, where **L** is a given D-poset and the D- poset [0,1] is equipped with the usual ordering and the usual difference on [0,1] (i.e., $b \ominus a = b - a$ for $0 \le a \le b \le 1$), is called a *state* on **L**.
A *σ-homomorphism* $\mathbf{x}: \mathcal{B}\ (\mathbf{R}) \to \mathbf{L}$, where **L** is a given D-poset and $\mathcal{B}\ (\mathbf{R})$ is a D-poset of all Borel subsets of the real line equipped with the inclusion ordering and set difference, is called an observable on **L**.

It is easy to see that, for any state **m** on **L** and any observable **x** on **L**, the composite mapping $\mathbf{m} \circ \mathbf{x}: \mathcal{B}\ (\mathbf{R}) \to [0,1]$, is a probability measure on $\mathcal{B}\ (\mathbf{R})$ and hence we are able to introduce an expected value of **x** with respect to **m**,

$$\mathbf{E_m}\ (\mathbf{x}) = \int t\ d(\mathbf{m} \circ \mathbf{x})$$

where the right hand side is the Lebesgue-Stieltjess integral (if it exists). In the case of fuzzy D-posets, we have the following results.

Theorem 3 Let **F** be a fuzzy full σ-g-D-poset. Then any state **m** on **F** is determined by its values on ⊕-idempotents of **F** (these are exactly the crisp sets contained in **F**),

$$\forall A \in \mathbf{L}: \mathbf{m}(A) = \int gA \, d\mathbf{M}$$

where **M** is a probability measure defined on the σ-algebra of all crisp subsets of **X** contained in **F**.

Proof. Recall that g**F** is a $\mathbf{T}_\infty$ -tribe of Butnariu and Klement [3]. It is easy to see that the system of all crisp subsets of **X** contained in **F** coincides with those from g**F** (crisp subsets are the only g-invariant elements of **F**). By [3], this system forms a σ-algebra, which will be denoted by $\mathcal{C}$. Further, all elements of g**F** are measurable with respect to $\mathcal{C}$ and consequently also all elements of **F** are $\mathcal{C}$-measurable fuzzy subsets of **X**.
Let **m** be a state on **F**. Then **p**, **p**(U) =**m** (g^{-1} U), U ∈ g**F**, is a state on g**L** and consequently it is a $\mathbf{T}_\infty$ -measure on g**F** [3]. By Butnariu [2], **p** is determined by its restriction **P** to $\mathcal{C}$, $\mathbf{P} = \mathbf{p}|\mathcal{C}$.

$$\forall U \in g\mathbf{F}: \mathbf{p}(U) = \int U \, d\mathbf{P}$$

Note that the restriction **M** of **m** to $\mathcal{C}$ coincides with **P**, **M** = **P**. Now, it is easy to see that for A ∈ **F** it is **m**(A) = **p**(gA) what implies the result. □

Note that by the previous theorem any σ-additive measure on a complete semi-simple MV-algebra **A** is determined by its values on the σ-idempotents of **A**. In [15], see also [11], we have shown that an observable **x** of a $\mathbf{T}_\infty$ -tribe **F** corresponds to a fuzzy-valued random variable **X** of Klement [10]. Further, we have introduced a fuzzy expected value of **x** in the latest case [16] as the expected value of **X**, see [10]. In the case of a general fuzzy full σ-g-D-poset **F**, any observable **x** on **L** corresponds to an observable g**x** on the $\mathbf{T}_\infty$ -tribe g**F**. For any state **m** on **F**, $\mathbf{m} \circ \mathbf{x} = \mathbf{m} \circ g^{-1} \circ g \circ \mathbf{x} = \mathbf{p} \circ g\mathbf{x}$, i.e. the expected value of **x** with respect to **m** is exactly the same as the expected value of g**x** with respect to $\mathbf{p} = \mathbf{m} \circ g^{-1}$. Similarly we introduce a fuzzy expected value of **x** with respect to **m** as the fuzzy expected value of the observable g**x** on $\mathbf{T}_\infty$ -tribe g**F** with respect to $\mathbf{T}_\infty$ -measure **p**,

$$\mathbf{FE_m}(\mathbf{x}) = \int g\mathbf{X} \, d\mathbf{M}$$

where $\mathbf{M} = \mathbf{m}|\mathcal{C}$ is a probability measure and g**X** is a fuzzy valued random variable on (**X**,$\mathcal{C}$) corresponding to the observable g**x**, i.e., for all x ∈ **X** it is $(g\mathbf{X}(x))(t) = g((\mathbf{x}(]-\infty,t[))(t))$, t ∈ **R**, and the right-hand side (if it exists) is a fuzzy real number resulting from the Klement's integral of g**X** with respect to **M** [10]. Note that $\mathbf{FE_m}(\mathbf{x})$ can be understood as a probability distribution function and then the corresponding expected value of this distribution coincides with the expected value $\mathbf{E_m}(\mathbf{x})$ [16]. It is obvious that all these results can be rewritten for the complete semisimple MV-algebras, too.

References

[1] Belluce, L.P., Semisimple algebras of infinite-valued logic and bold fuzzy set theory, Canadian J. Math. 38 (1986), 1356-1379.

[2] Butnariu, D., Values and cores for fuzzy games with infinitely many players, Intern. J. Game Theory 16 (1987), 43-68.

[3] Butnariu, D. and Klement, E.P., Triangular norm-based measures and their Markov kernel representation, J. Math. Anal. Appl. 162 (1991), 111-143.

[4] Chang, C.C., Algebraic analysis of many valued logics, Trans. Amer. Math. Soc. 88 (1957), 467-490.

[5] Chovanec, F. and Kôpka, F., D-lattices, submitted to Intern. J. Theor. Physics.

[6] De Lucia, P. and Pap, E., Nikodym convergence theorem for uniform space valued functions defined on D-poset, preprint.

[7] Dvurečenskij, A. and Riečan, B., Decomposition of measures on orthoalgebras and difference posets, Intern. J. Theor. Physics 33 (1994).

[8] Foulis, D.J. and Bennett, M.K., Effect algebras and unsharp quantum logics, Found. Physics 24 (1994)
[9] Giuntini, R., Quantum MV-algebras, preprint.
[10] Klement, E.P., Integration of fuzzy valued functions, Revue Roumaine Math. Pures Appl. 30 (1985), 375-384.
[11] Kolesárová, A. and Riečan, B., T-fuzzy observables, Tatramountains Math. Publ. 1 (1992), 73-82.
[12] Kôpka, F., D-posets of fuzzy sets, Tatramountains Math. Publ.1 (1992), 83-88.
[13] Kôpka, F. and Chovanec, F., D-posets, Math. Slovaca 44 (1994), 21 - 34.
[14] Ling, C.H., Representation of associative functions, Publ. Math. Debrecen 12 (1965), 189-212.
[15] Mesiar, R., Fuzzy measurable functions, Fuzzy Sets and Systems 59 (1993), 35-41.
[16] Mesiar, R., Fuzzy logics and observables, Intern. J. Theor. Physics 32 No7 (1993), 1143-1151.
[17] Mesiar, R., h-fuzzy quantum logics, Intern. J. Theor. Physics 33, No7 (1994), 1417 - 1426.
[18] Mesiar, R., Differences on [0,1], Tatramountains Math. Publ. 6 (1994).
[19] Mesiar, R. and Pap, E., Extensions of differences on [0,1], Internat. J. Theor. Physics 33 No12 (1994), 2317 - 2323.
[20] Mundici, D.:Interpretation of AFC *-algebras in Lukasiewicz sentential calculus, J. Funct. Anal. 65 (1986), 15 - 53.
[21] Mundici, D.: Uncertainty measures in MV-algebras, and states of AFC*-algebras, Abstracts of 13th Linz seminar on fuzzy set theory (1991), 42-43.
[22] Navara, M. and Pták, P., Difference posets and orthoalgebras, preprint.
[23] Randal, C. and Foulis, D., New definitions and theorems, University of Massachusetts Mimeographed Notes, Amherst, Massachusetts, 1979.
[24] Riečanová, Z. and Bršel, D., Contraexamples in difference posets and orthoalgebras, Internat. J. Theor. Physics, 33 (1994).
[25] Schweizer, B. and Sklar, A., Probabilistic metric spaces, North Holland, New York, 1983.
[26] Varadarajan, V.S., Geometry of Quantum Theory, Van Nostrand, New York, 1968.
[27] Weber, S., A general concept of fuzzy connectives, negations and implications based on t-norms and t-conorms, Fuzzy Sets and Systems 11 (1983), 115-134.

COMPOSITIVE INFORMATION MEASURE OF A FUZZY SET

Carlo Bertoluzza[a] - Teofilo Brezmes[b] - Gloria Naval[b]

[a]Dip.Informatica e Sistemistica - Universitá di Pavia
Corso Strada Nuova 65 - 27100 Pavia - ITALY

[b]Departamento de Matemáticas - Universidad de Oviedo
calle Calvo Sotelo s/n - 33071 Oviedo - Spain

Abstract In this paper the infomation associated to an event $\widetilde{A}$ (crisp or fuzzy) of a space $(\Omega, \mathcal{S})$ measures the degree of surprise which arises when this event occurs. J.Kampé de Fériet and B.Forte gave in 1967 ([5]) an axiomatic definition of this concept, in which the information measure J is associated directly to the crisp event A of a measurable space $(\Omega, \mathcal{S})$. We will propose here a possible way to measure the information in the fuzzy context. We proceed by supposing that an axiomatic information measure is defined on the crisp events (crisp subsets): the information of a fuzzy subset will be constructed by means of the informations of their α-cuts.

1. Preliminaries

Fuzzy Sets. Let $L = \{0 = \alpha_0 < \alpha_1 < \ldots < \alpha_n = 1\}$ a finite subset of the interval $[0,1]$ such that $0 \in L$ and $\alpha \in L \Longrightarrow 1-\alpha \in L$. An *L-fuzzy subset* of a given universe Ω is a map $\widetilde{A}$ from Ω to L. The family of the fuzzy subsets of Ω is equipped with union and intersection operations and a partial ordering (inclusion) $\sqsubseteq$ defined by

$$(\widetilde{A} \cup \widetilde{B})(\omega) = \sup\{\widetilde{A}(\omega), \widetilde{B}(\omega)\} \qquad (1.1.a)$$

$$(\widetilde{A} \cap \widetilde{B})(\omega) = \inf\{\widetilde{A}(\omega), \widetilde{B}(\omega)\} \qquad (1.1.b)$$

$$\widetilde{A} \sqsubseteq \widetilde{B} \iff \widetilde{A}(\omega) \leq \widetilde{B}(\omega) \qquad (1.1.d)$$

$$\forall \omega \in \Omega$$

For each value $\alpha_i \in L$ the $\alpha_i - cut$ of $\widetilde{A}$ is the crisp set A_i defined by

$$A_i = \{\omega \in \Omega | \widetilde{A}(\omega) \geq \alpha_i\} \qquad (1.2)$$

Information measure Let $(\mathcal{T}, \preceq)$ a lattice with minimum (m) and maximum (M). An *information measure* over $(\mathcal{T}, \preceq)$ is a map

$$J^* : \mathcal{T} \longrightarrow \mathbb{R}^+$$

with the following properties

$$J^*(M) = 0 \qquad (1.4)$$

$$J^*(m) = +\infty \qquad (1.5)$$

$$x \preceq y \Longrightarrow J^*(x) \geq J^*(y) \qquad (1.6)$$

$$\forall (x,y) \in \mathcal{T} \times \mathcal{T}$$

The lattice underlying the Forte-Kampé de Fériet measure (J) is an algebra of crisp subsets of an universe Ω, whereas the fuzzy information $\widetilde{J}$ which we try to introduce is defined over an algebra $\widetilde{\mathcal{S}}$ of fuzzy subsets. The maximum and tha minimum are Ω and $\emptyset$ in both the cases.

The family of the α-cutcompletely determines the fuzzy set $\widetilde{A}$: then any information measure associated to $\widetilde{A}$ may be viewed as depending on the family $\{A_1 \ldots A_n\}$. In particular we suppose that a set theoretic information J is defined over the crisp space $(\Omega, \mathcal{S})$ and that the information $\widetilde{J}(\widetilde{A})$ depends on the crisp informations $J_i = J(A_i)$ of their α-cut:

$$\widetilde{J}(\widetilde{A}) = G_n(J_1 \ldots J_n) \tag{1.9}$$

2. Basic Properties

We can easily recognize that the function G_n is defined on a suitable subset of $\Delta_n = \{(x_1 \ldots x_n) \mid 0 \leq x_1 \leq x_2 \leq \ldots \leq x_n\}$, and that it satisfies the following properties

$$G_n(0 \ldots 0) = 0 \tag{2.3}$$

$$G_n(+\infty \ldots + \infty) = +\infty \tag{2.4}$$

$$x_i \leq y_i \quad i = 1 \ldots n \implies \tag{2.5}$$

$$\implies G_n(x_1 \ldots x_n) \leq G_n(y_1 \ldots y_n)$$

We reserve a particular attention to the two classes of fuzzy subsets whose membership functionassumes only one value α_k or only two values α_r, α_S (with s¿r) different from zero.Let us consider the two following functions

$$g_k(u) = G_n(\underbrace{u \ldots u}_{k\ times}, +\infty \ldots + \infty) \tag{2.6}$$

$$g_{rs}(x, y) = G(\underbrace{x \ldots x}_{r\ times}, \overbrace{y \ldots y}^{t\ times} + \infty \ldots + \infty) \tag{2.7}$$

where $t = s - r$. If we pose $u = J(A)$, $x = J(A' \cup A'')$, $y = J(A'')$, then$g_k(u)$ and $g_{rs}(x, y)$ represent respectively the informations of the fuzzy subsets $\widetilde{B} = \alpha_k A$ and $\widetilde{C} = \alpha_r A' \cup \alpha_s A''$, whose membership functions assume respectively only one (α_k0 or only two (α_r, α_s) values different from zero. The functions g_k and g_{rs} have the following properties

$$g_n(0) = 0 \tag{2.8}$$

$$g_k(+\infty) = +\infty \tag{2.9}$$

$$x' < x'' \implies g_k(x') \leq g_k(x'') \tag{2.10}$$

$$k < m \implies g_k(x) \geq g_m(x) \tag{2.11}$$

$$g_{rn}(0, 0) = 0 \tag{2.12}$$

$$g_{rs}(+\infty, +\infty) = +\infty \tag{2.13}$$

$$x' \leq x'' , y' \leq y'' \implies \tag{2.14}$$

$$\implies g_{rs}(x', y') \leq g_{rs}(x'', y'')$$

$$g_{rs}(x, x) = g_s(x) \tag{2.15}$$

$$g_{rs}(x, +\infty) = g_r(x) \tag{2.16}$$

$$g_{rs}(x, x) \leq g_{rs}(x, y) \leq g_{rs}(y, y) \tag{2.17}$$

The study of these functions will result the key point in order to determine the function G_n, and consequently the fuzzy information $\widetilde{J}$.

3. Compositive Measures

Definition 3.1. A set theoretic information measure is said to be *compositive*, if there exists a function of two variables

$$F \ : \ \mathbb{R}^+ \times \mathbb{R}^+ \longrightarrow \mathbb{R}^+ \tag{3.1}$$

such that

$$A \cap B = \emptyset \implies J(A \cup B) = F[J(A), J(B)] \tag{3.2}$$

P.Benvenuti, B.Forte and J.Kampé de Fériet gave a complete characterization of the compositive measures. They proved that the general form of F is, roughly speaking, a suitable combination of the two extremal forms

$$F_i(x,y) = \inf(x,y) \tag{3.3}$$

$$F_S(x,y) = f^{-1}[f(x) + f(y)] \tag{3.4}$$

where $f(x)$ is any continuous and strictly decreasing function with $f(+\infty) = 0$. The compositive measure are the best known and the most utilized in the field of the set theoretic information theory; therefore it seems quite natural textend this notion to the fuzzy ones.

Definition 3.2. Formally we say that a fuzzy information is *compositive* if there exists a function

$$\widetilde{F} \ : \ \mathbb{R}^+ \times \mathbb{R}^+ \longrightarrow \mathbb{R}^+ \tag{3.5}$$

such that

$$\widetilde{A} \cap \widetilde{B} = \emptyset \implies \widetilde{J}(\widetilde{A} \cup \widetilde{B}) = \widetilde{F}[\widetilde{J}(\widetilde{A}), \widetilde{J}(\widetilde{B})] \tag{3.6}$$

It is easy to recognize that the function $\widetilde{F}$ presents the same characteristics of the function F of definition 4.1, and therefore its general form coincides with the one given in [7]. Moreover, if we express the fuzzy information $\widetilde{J}$ in terms of the information functions G , then it is easy to recognize that the following compatibility relations must hold

$$\widetilde{F}[G(x_1 \ldots x_n), G(y_1 \ldots y_n)] = G[F(x_1, y_1) \ldots F(x_n, y_n)] \tag{3.7}$$

$$\widetilde{F}[g_k(x_k), g_k(y_k)] = g_k[F(x_k, y_k)] \tag{3.8}$$

$$\widetilde{F}[g_{rs}(x_r, x_s), g_{rs}(y_r, y_s)] = g_{rs}[F(x_r, y_r), F(x_s, y_s)] \tag{3.9}$$

The next two sections are devoted to the study of the two classes of compositive fuzzy information which correspond to the two extremal cases specified by the expression (3.3) and (3.4). In particular we suppose that a composition law F (of the form (3.3) and (3.4) respectively) is defined over the crisp information space $(\Omega, \mathcal{S}, J)$ and also that the fuzzy information space is compositive. Starting from these hipotesis, we will determine the information measure $\widetilde{J}(\widetilde{A})$. We use the one level subsets in order to determine the form of the fuzzy composition law $\widetilde{F}$. The two levels subsets are used in the case of the type-inf composition law, in order to build up a recursive process which allows us to construct the function G_n. In the case of a type-M composition law the function G_n is determined directly by solving a suitable functional equation.

4. Type-inf composition law

Suppose that the crisp information space $(\Omega, \mathcal{S}, J)$ is compositive with composition law given by

$$F(x,y) = \inf(x,y) \tag{4.1}$$

It is easy to recognize, using (3.8) that the function $\widetilde{F}$ is of the same type, that is

$$\widetilde{F}(u,v) = \inf(u,v) \tag{4.2}$$

for all u, v in the domain $\Gamma = [g_k(0), +\infty)$ .Then the compatibility equations (3.7) and (3.9) assume the form

$$\inf[G(x_1 \dots x_n), G(y_1 \dots y_n)] = G[\inf(x_1, y_1) \dots \inf(x_n, y_n)] \tag{4.3}$$

$$\inf[g_{rn}(x_r, x_n), g_{rn}(y_r, y_n)] = g_{rn}[\inf(x_r, y_r), \inf(x_n, y_n)] \tag{4.4}$$

We start by determining the form of the function g_{rs} , that is the function which determines the information of the two-level subsets.

norTheorem 4.1 Let $\tau(y)$ a (i) not decreasing, (ii) lower semicontinuous function with (iii) $\tau(0) = 0$, (iiii) $\tau(t) \leq t$ and let $g(t)$ a (e) continuous (ee) not decreasing function , and let us pose $\overline{\tau}(x) = \sup\{y \mid \tau(y) < x\}$. Then any function $g_{rs}(x,y)$ defined by

$$g_{rs}(x,y) = \begin{cases} g[\overline{\tau}(x)] & \text{if } x \leq \tau(y) \\ g(y) & \text{if } x > \tau(y) \end{cases} \tag{4.5}$$

is a type-inf information function, and on the contrary any type-inf information function has the form (4.5), with a suitable choice of the functions τ, g .

It is easy to verify that a function of type (4.5) satisfy the compatibility equation (4.4), as well as the characteristic properties (2.12)—(2.17) of the information functions g_{rs} . So the "if" part of the theorem is proved.

To prove the "only if" part let us suppose that an information function g_{rs} is given satisfying the properties (2.12)—(2.17) and the compatibility equation (4.4). The functions g and τ defined by

$$g(y) = g_{rs}(y,y) \tag{4.6.1}$$

$$\tau(y) = \inf\{t \mid g_{rs}(t,y) = g(y)\} \tag{4.6.2}$$

fulfill the properties (e)—(eee) and (i)—(iiii) and the compatibility equation (4.4) forces g_{rn} to assume the form (6.1). The details of the proof may be found in [9], pp. 14-17.

The two extremal cases correspond to the two extremal choices of the function $\tau(t)$, namely $\tau(t) = t$ and $\tau(t) = 0$, which correspond to

$$g_{rs}(x,y) = g(x) = g[\inf(x,y)] \tag{4.7}$$
$$\widetilde{J}(\widetilde{A}) = \widetilde{J}(\{A_1,A_2\}) = g[\widetilde{J}(A_1)] \tag{4.8}$$

$$g_{rs}(x,y) = g(y) = g[\sup(x,y)] \tag{4.9}$$
$$\widetilde{J}(\widetilde{A}) = \widetilde{J}(\{A_1,A_2\}) = g[\widetilde{J}(A_2)] \tag{4.10}$$

In particular, if we choose $g(t) = t$, which corresponds to impose the condition $\widetilde{J}(A) = J(A)$, then we have

$$\widetilde{J}(\widetilde{A}) = \inf\{J(A'), J(A'')\} = \widetilde{J}(A') \tag{4.11}$$
$$\widetilde{J}(\widetilde{A}) = \sup\{J(A'), J(A'')\} = \widetilde{J}(A'') \tag{4.12}$$

We can start from the result just obtained to deduce the form of the information function G in the case where we have an arbitrary number n of different α−cuts. We describe in particular the case where $|L| = 3$, that is the case where the membership function may assume only four distinct values $\alpha_0 = 0 < \alpha_1 < \alpha_2 < \alpha_3 = 1$.

Let $G(x,y,z)$ $(x \le y \le z)$ the information function, and let us consider its restriction to the set of the points of type (y,y,z) . It is easy to prove that the function $\gamma(y,z) = G(y,y,z)$ have all the properties of the function g_{rn} which we analized previously. Then we can establish the following

Lemma 1. For each $z \in \mathbb{R}^+$, let $\tau(z) = \inf\{y | G(y,y,z) = G(z,z,z)\}$. Then, by posing $G(z,z,z) = g(z)$, we have

$$G(y,y,z) = \begin{cases} g[\overline{\tau}(y)] & \text{if } y \le \tau(z) \\ g(z) & \text{if } y > \tau(z) \end{cases} \tag{4.13}$$

We can also recognize that the function $\gamma_z(x,y) = G(x,y,z)$ shows all the properties of the functions g_{rs} , and therefore the following lemma holds:

Lemma 2. Let $\tau_z(y) = \inf\{x | G(x,y,z) = G(y,y,z)\}$. Then we have

$$G(x,y,z) = \begin{cases} g[\overline{\tau}_z(x)] & \text{if } x \le \tau_z(y) \\ G(y,y,z) & \text{if } x > \tau_z(y) \end{cases} \tag{4.14}$$

where $\overline{\tau}_z$ is defined like to $\overline{\tau}$ in theorem 4.1

We can observe that the function τ_z and τ are strongly related. In fact we have

$$\tau_z(y) = \inf\{x \mid G(x,y,z) = \begin{cases} g(y) & \text{if } y \le \tau(z) \\ g(z) & \text{if } y > \tau(z) \end{cases} \tag{4.15}$$

Combining (4.13),(4.14) and (4.15) we obtain the general form of the information function G , by means of the following procedure:

(a) fix three functions $\tau(z)$, $\tau_z(y)$ and $g(t)$,

(b) perform $G(y,y,z)$ by means of (4.13),

(c) verify the compatibility between τ_z , τ and g using (4.15),
(d) construct $G(x,y,z)$ according to (4.14).

The general case. The method used in the case $|L|=3$ is also allowable in the general case. The form of G may be obtained by means of a recurrent procedure of the type described in the previous case. We determine firstly $G(x_{n-1}\dots x_{n-1},x_n)$, then we determine $G(x_{n-2}\dots x_{n-2},x_{n-1},x_n)$, and so on, till to reach $G(x_1\dots x_n)$. At each step i we have to introduce a function τ_i , depending on the parameters $x_n\dots x_{n-i+1}$ which plays the role of the functions τ and τ_z in the case $n=3$. The family τ_i , together with the function $g(t)=G(t\dots t)$, determines completely the information function G .

5. Type-M composition law

Suppose that the crisp information space $(\Omega,\mathcal{S},J)$ has a type-M composition law with additive generator f , that is

$$F(x,y) \;=\; f^{-1}[f(x)+f(y)] \tag{5.1}$$

Then the fuzzy composition law $\widetilde{F}$ is also of type-M (possibly with a different generator $\widetilde{f}$), that is

$$\widetilde{F}(x,y) \;=\; \widetilde{f}^{-1}[\widetilde{f}(x)+\widetilde{f}(y)] \tag{5.2}$$

We proved this result (see [9], p.23) by recognizing that the set of the elements which are idempotent for the law $\widetilde{F}$, that is the set $\widetilde{\Delta} \;=\; \{\alpha\in \mathbb{R}^+ \mid \widetilde{F}(\alpha,\alpha)= \alpha\}$ contains only the values 0 and $+\infty$. The compatibility equation becomes in this case

$$\begin{aligned} \widetilde{f}^{-1}[\widetilde{f}(G(x_1\dots x_n))+\widetilde{f}(G(y_1\dots y_n)) &= \\ = G[f^{-1}(f(x_1+f(y_1))\dots f^{-1}(f(x_n)+f(y_n))] \end{aligned} \tag{5.3}$$

which may be written in the form

$$\theta(\xi_1\dots\xi_n) \;+\; \theta(\eta_1\dots\eta_n) \;=\; \theta(\xi_1+\eta_1\dots\xi_n+\eta_n) \tag{5.4}$$

where $\xi_i=f(x_i),\eta_i=f(y)i)$, $\theta(t_1\dots t_n)=\widetilde{f}\circ G[f(t_1)\dots f(t_n)]$.
Using a well known rwesult of functional equation, we recognize that the unique monotonic solution of this equation is ths function

$$G(x_1\dots x_n) \;=\; \widetilde{f}^{-1}[\sum_{i=1}^{n} a_i f(x_i)] \tag{5.5}$$

with a_i are non-negative fixed values.

We can use, in particular, (5.5) to deduce the form of the fuzzy information of the crisp subsets, thus obtaining

$$\widetilde{J}(A) \;=\; \widetilde{f}^{-1}[a\cdot f(J(A))] \tag{5.6}$$

where $a = \sum a_i$. We deduce immediatly, from the above relation, that the restriction of the fuzzy information to the class $\mathcal{S}$ of the crisp subsets reduces to the crisp infomation iff $\tilde{f}(t) = a \cdot f(t)$:

$$\tilde{J}(A) = J(A), \quad \forall A \in \mathcal{S} \iff \tilde{f}(t) = a \cdot f(t) \tag{5.7}$$

References

[1] J.Kampé de Fériet: Note di teoria dell'informazione. Quaderni dei Gruppi di ricerca del CNR, Ist.Mat.Appl. Roma 1972.

[2] N.Wiener: Cybernetics. Herrmann, Paris 1948.

[3] C.F.Picard: Aspects informatiques de l'information hyperbolique. Symposia Mathematica, vol 10, pp. 55-82 1975.

[4] B.Forte, J.Kampé de Fèriet: Information et Probabilité. C.R.A.S Paris, 265A, pp.110 and 142 1967.

[5] L.A.Zadeh: Fuzzy sets. Information and Control, 8 pp 338-353.

[6] G.Comyn and J.Losfeld: Définition d'une information composable sur un treillis. C.R.A.S. Paris, t.278 (25 février 1974), Série A, pp.633- 636.

[7] P.Benvenuti, B.Forte and J.Kampé de Fériet: Forme générale de l'operation de composition .... C.R.A.S Paris, 265 Serie A, 1967.

[8] J.Aczel:Lectures on functional equations and thei characterizations. A.P., New York, 1966, pg.348.

[9] C.Bertoluzza, T.Brezmes, M.L.Capodieci and G.Naval: Compositive information measures of a fuzzy set: two equivalent approaches. Manuscript.

UNCERTAINTY, FUZZY SETS, CONCEPTUAL SETS

Anio O. ARIGONI & Andrea ROSSI
Department of Mathematics, University of Bologna
P.zza P. S. Donato 5, 40127 Bologna, Italy
Fax: 051 354490

ABSTRACT

In a recent conference on fuzzy sets [4], Lofti Zadeh stressed the modest level of penetration of Fuzzy Set Theory in Artificial Intelligence. The possible causes of this are briefly investigated herein. Also possible extensions of some critical points of this theory are suggested, so that we can try to make it more flexible and suitable for different specific purposes, as for instance for inductive classification.

1. Introduction

In order to efficiently act man needs certainty about facts, situations and objects of the environment in which he operates. Therefore one of his main concerns is to reduce uncertainty, i. e., acquire specific information. Thus man devotes a great deal of resources to transform generic data he possesses, into information.

Formally, if a universe X formed by elements x, and a property P are given, rudiments of certainty may be drawn by deciding whether or not predicates p of the form "x is P" are true. Therefore, additional data that can improve such possibility are potential information that can reduce uncertainty on X.

Classically, most logicians and linguists share the convenient assumption according to which many (declarative) natural language predicates such as the one p, have either a binary truth value or no sense at all. But this is only a pretence; the main reason of this lays in the fact that many predicates may be only partially true. This is due to both the polysemy of natural linguistic means and the lack of sensible logical or mathematical tools suitable for evaluating the truth in topic. Thus, the boundary case in which classical Formal Logic is inadequate to evaluate most predicates, is accepted as realistic. Fuzzy sets partly makes up for it and, in the sixties, Zadeh introduced a specific theory regarding this [18], by extending Lukasievich's three valued logic [10].

It is well known that given a universal set X, a fuzzy set $\tilde{A}$ is a collection formed by elements x of X itself and singly associated to the relative membership function $\mu_{\tilde{A}}(x)$: real number ranging on the continuum [0, 1]. Since for whichever x of $\tilde{A}$ $\mu_{\tilde{A}}(x)$ is more informative than the binary truth value of the corresponding predicates p; every $\tilde{A}$ is more effective in dissipating uncertainty, compared to the mere list of elements that $\tilde{A}$ itself includes.

In Fuzzy Set Theory (FST), as in other theories derived from this one, for example in the one regarding (fuzzy) rough sets [3], the possible inclusion of each x in $\tilde{A}$ as well as the relative $\mu_{\tilde{A}}(x)$, are usually subjectively assessed [8, 9, 12, 18]. However, it can be presumed that for every x the relative membership evaluation is performed by taking into account all the other elements forming $\tilde{A}$ and the specific single attributes by which these are characterised.

2. Limits of FST

To develop the subject, we distinguish between *primitive fuzzy sets,* obtained directly by collecting elements of a universe, and others, that we call *composed fuzzy sets*, resulting from the combination of two or more sets of the former, throughout connectives.

The subjectivity followed in determining the values that are assumed by each $\mu_{\tilde{A}i}(x)$, may cause imprecision. Nevertheless, when $\tilde{A}_i$ is a primitive set, the constrains on the $\mu_{\tilde{A}i}(x)$, can be limited to the one according to which the bias of such values does not alter the expected ordering on the x forming $\tilde{A}_i$. Thus, although in this case part of the determined reduction of uncertainty may be only apparent, the introduced error may remain bearable.

Differently, when a fuzzy set $\tilde{S}$ is composite in that it derives from two or more primary fuzzy sets through MIN-MAX connectives, the rules applied to determine the values of $m_{\tilde{S}}(x)$ may imply unacceptable imprecision. In this case, in fact, the values of such $\mu_{\tilde{S}}(x)$ do not result as a semantic function of the contextual elements included in $\tilde{S}$ according to the implemented connective function; such values are instead only a numerical function of the membership values that every x of $\tilde{S}$ has with respect to the combined primitive sets $\tilde{A}_1, \tilde{A}_2$,.... Thus, if for instance the connection to be performed is $\tilde{A}_1 \cup \tilde{A}_2$,..., then $\mu_{\tilde{S}}(x)$ = MAX $(\mu_{\tilde{A}1}(x), \mu_{\tilde{A}2}(x), ...)$, b y definition. The $\mu_{\tilde{S}}(x)$ obtained for every x, depends so on the elements forming each primitive subsets $\tilde{A}_1$ and $\tilde{A}_2$, separately, rather than on those forming the context of the resulting $\tilde{S} = \tilde{A}_1 \cup \tilde{A}_2$,..., as differently it should be.

In synthesis, through MIN-MAX connectives the achieved membership values of the elements forming a composite set may be strongly biased. In this case, in fact, what is diffused throughout the chains of combinations to achieve this set, are merely numbers. Moreover, these do not regard neither the elements forming intermediate results at the ensuing stages of the implemented combinatorial chain, nor the ones forming $\tilde{S}$.

As an example of the possibility of falling into possible errors by applying directly connectives MIN-MAX, we consider the ambiguity arising in a case where is performed connection AND, as it is illustrated in Figure 1.

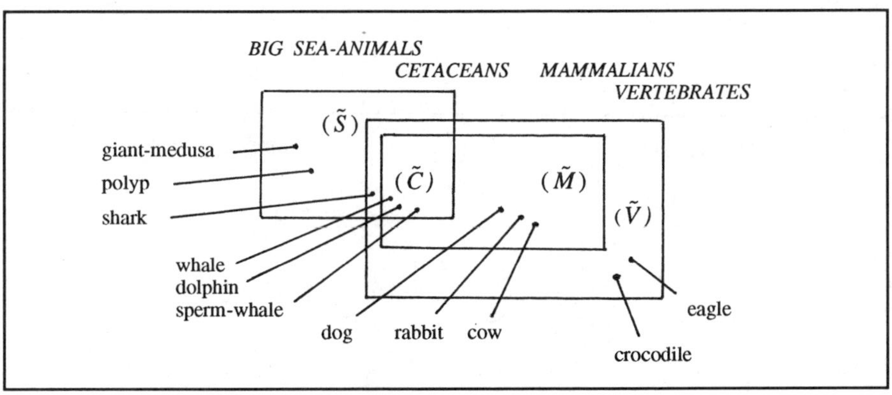

Fig. 1 - The reported diagram illustrates the possibility that two different values of membership, may result from considering a cetacean as either an element of the intersection of MAMMALIAN with BIG SEA-ANIMALS or of VERTEBRATES with BIG SEA-ANIMALS, respectively.

The following considerations show analytically the consequence of the case illustrated in the figure.

Since $\tilde{V} \supset \tilde{M}$, according to FST for every element x of $\tilde{M}$

$$\mu_{\tilde{M}}(x) < \mu_{\tilde{V}}(x).$$

In addition, when x is an element of $\tilde{C}$ which equal $\tilde{M} \cap \tilde{S}$ and is in $\tilde{V} \cap \tilde{S}$, then:

$$\mu_{\tilde{C}}(x) < \mu_{\tilde{M}}(x) < \mu_{\tilde{V}}(x) \text{ as well as } \mu_{\tilde{C}}(x) < \mu_{\tilde{S}}(x)$$

in addition,

$$\mu_{\tilde{C}}(x) = \text{MIN}\,(\mu_{\tilde{M}}(x)\,,\, \mu_{\tilde{S}}(x)\,) = \text{MIN}\,(\mu_{\tilde{V}}(x),\, \mu_{\tilde{S}}(x))$$

However, since it can be presumed that for every x of $\tilde{C}$, for example a whale,

$$\mu_{\tilde{S}}(x) > (\mu_{\tilde{M}}(x)\,,\, \mu_{\tilde{V}}(x))$$

then $\mu_{\tilde{C}}(x)$ depends on $\mu_{\tilde{M}}(x)$ and on $\mu_{\tilde{V}}(x)$, exclusively. Precisely:

$$\mu_{\tilde{C}}(x) = \left[\begin{array}{l} \mu_{\tilde{M}}(x) \\ \mu_{\tilde{V}}(x) \end{array} \right.$$

so that, being necessarily

$$\mu_{\tilde{M}}(x) \neq \mu_{\tilde{V}}(x)$$

the membership function $\mu_{\tilde{C}}(x)$ has ambiguous value.

The diagram reported in Figure 2 shows how the elements x of a composite fuzzy set $\tilde{L}$ and the relative membership function, are drawn from those of the primitive sets $\tilde{A}$ and $\tilde{B}$, and on the basis of $\mu_A(x)$ and $\mu_B(x)$.

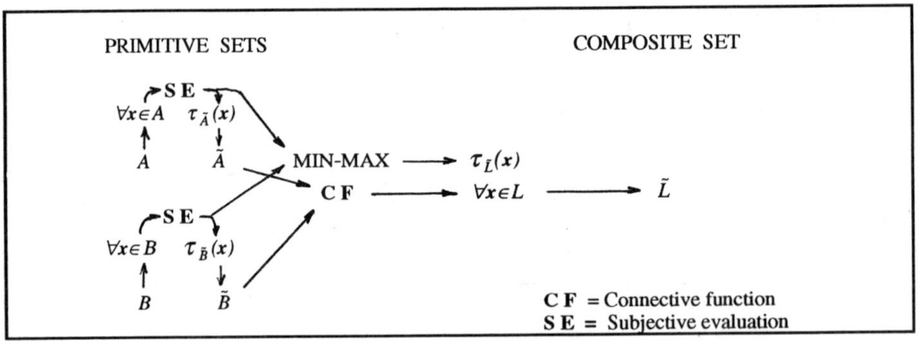

Fig. 2 - Derivation of the membership function of the elements of the composite fuzzy set $\tilde{L}$, obtained by combining those primitive $\tilde{A}$ and $\tilde{B}$.

3. Utilisation of FST in AI

In AI, specifically in both deductive and inductive classification, connective operators as well as the logical chains that derive from these and possibly involving a conspicuous number of logical steps, are often considered. Therefore, by direct application of Fuzzy Set Theory, and then of fuzzy logic, the membership function of the elements that form the resulting composite sets may be distort indeed. Thus, to keep with AI, such theory may need to be suited in some regards.

From the previous section, it is evident that the essential adaptation we propose for FST by the paper, consists in a particular way to evaluate the membership function of elements $\boldsymbol{x}$, to possible fuzzy sets. In Figures 2, which schematises a characteristic procedure of FS, the evaluation in issue is accomplished in two different ways; these depend on the type of considered fuzzy sets: by subjective evaluation (SE), for those primitive; by MIN-MAX for the composite ones.

On this line of reasoning, we propose a different manner to achieve a type of membership that may result in accordance with AI requirements. This is made up by calculations essentially based on semantic relations existing among the elements of each set. The operator implemented at this purpose is called conceptual evaluation (CE) and will be better analysed in Section 4. The fundamental module that derives by the utilisation of CE is illustrated in Figure 3. In this, it is assumed that two sets A and B constitute the domain of a connective function CF; the membership value for every $\boldsymbol{x} \in A$ (or $\boldsymbol{x} \in L$) and every subset of pairs singly formed by one $\boldsymbol{x}$ and the regarding membership, are respectively indicated by $\tau_A(\boldsymbol{x})$ and $\boldsymbol{A}$, or $\tau_L(\boldsymbol{x})$ and L.

By generalising the diagram shown in Figure 3 we obtain one consisting of a multiplicity of modules. The graphical representation of the considered functional dependencies is shown in Figure 4. This representation remarks the contextual dependence by $\tau_S(\boldsymbol{x})$ on the elements of S; the derivation of the $\boldsymbol{x}$ of S from $A_1, A_2, ..., A_n$ are attained through the implemented chain of connectives.

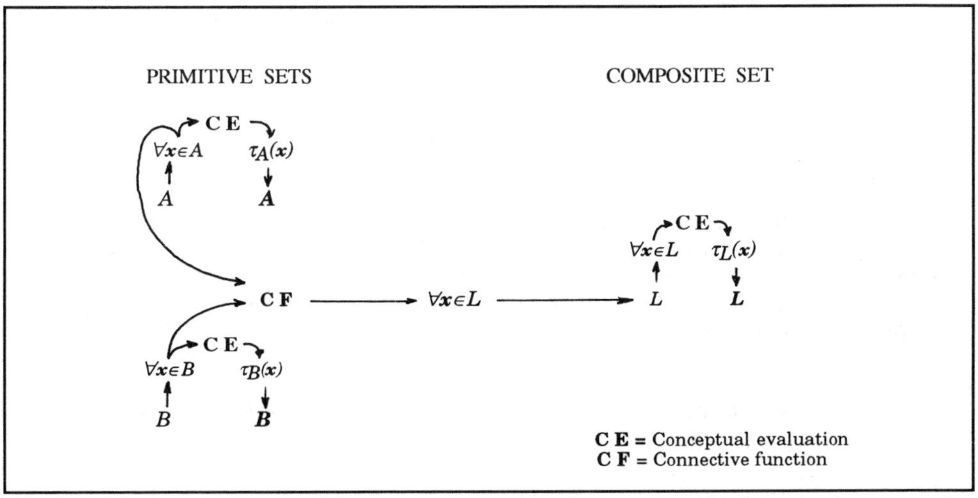

Fig. 3 - Connection of two sets A and B to derive the composite set L, through conceptual connection

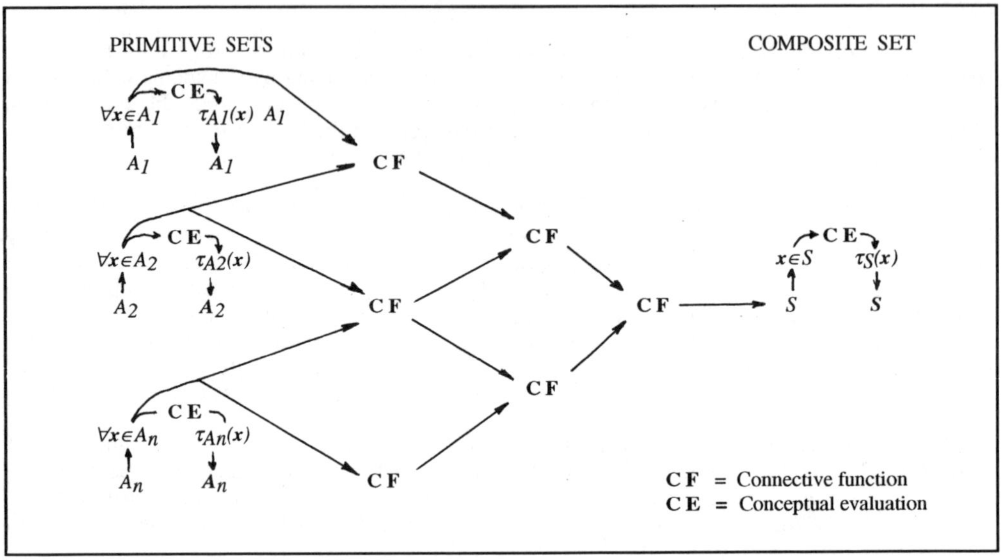

Fig. 4 - Derivation of the conceptual set S from the n primitive sets A_1, A_2, ..., A_n

4. Conceptual sets

The same evaluation performed for the sets appearing in Figures 3 and 4, is carried out by accounting the context of each considered set. It is remarked that in virtue of the illustrated procedure, what diffuses throughout the connection chains are elements that, collectively considered, reflect the global signification of the sets; that is, no numerical data are diffused. The latter can be drawn from such meaning, at whatever stage they may be requested, as necessarily it takes place at the composite one.

Every set whose elements are pairs $(x, \tau_{Ai}(x))$, is defined as *conceptual set*, and is indicated by $\boldsymbol{A}_i$, as is performed in the diagram of Figures 3 and 4.

The component **CE** —conceptual evaluation— that appears in the reported figures, regards a procedure consisting of two steps: *conceptual abstraction* and *numerical calculation*. These are outlined in the following sections.

4.1. Conceptual abstraction

It was assumed that the elements to be included in every set A_i of X are not casually determined, A_i is instead underlay by a *concept*; in other terms, the elements collected to form A_i are selected on the basis of specific criteria. The possible fuzziness of sets derives from the different degree that the elements forming such sets have in representing the mentioned concept. Such degree is function of the irrelevance of the single *attributes* characterising the distinct elements. Every element assumes so a specific *typicality* to the concepts A_1, A_2, ... concerning the sets $A_1, A_2, ...$; such typicality corresponds to the membership function before analysed in the early part paper, and it is indicated by $\tau_{Ai}(x)$.

For every attribute of an element x, the mentioned relevance depends on the context formed by x together with all other elements of the set to which this belongs. It depends, therefore, on the concept these elements can represent, given their positive typicality to it, just as it is postulated in Cognitive Sciences [11].

To the end of determining an element's typicality, the precise knowledge of the concept to which this is relative is required. The elements forming a set A_1 can only be instances of the concept underlying the set itself; the list of elements of this, which forms the *extensive definition* of the concept, does not constitute the relative knowledge, which remains so hidden. To achieve said knowledge, we propose the implementation of an abstraction procedure; this is sketched in the part *a)* of Figure 5 and it is outlined as follows.

Said procedure, reported in detail in [3], consists of applying to it a function $\mathcal{F}$, called *conceptual abstraction function* [1]. Such function is implemented by a process of *generalisation* of the elements of A_i. In the elements that result are evidenced the irrelevant attributes. The set so obtained is indicated by $A_i^{\mathcal{A}}$ (index $\mathcal{A}$ standing for *abstract*) and constitutes the *intensive definition* of A_i. This consists of a list of abstract elements: elements formed by joining the necessary conditions —relevant attributes— that make each of these elements representative of the concept.

The abstraction of A_i into $A_i^{\mathcal{A}}$ takes place on the basis of a process of *generalisation* of the elements forming the former, which hinges on determining the irrelevance of the relative attributes, as specified in [1, 2].

4 2. Calculation of typicality

If one specific attribute turns out to be irrelevant to an x that forms an A_i, it is not necessarily so in others elements as well. Thus, every attribute may result irrelevant in a variable number of elements, that which determines different degrees of irrelevance for the distinct attributes; this depends on which are the x forming A_i, then on the relative concept $\mathcal{A}_i$.

Therefore for every x of an A_i it can be implemented a procedure to calculate its typicality to $\mathcal{A}_i$, $\tau_{Ai}(x)$, based on the number of attributes that are relevant in it and on the relevance of these [4]. Such calculation of $\tau_{Ai}(x)$ follows from comparing each one of such x and a corresponding element of $A_i^{\mathcal{A}}$ in which are evidenced the irrelevant attributes that this has, as schematised in the part *b)* of Figure 5.

It could be shown that the typicality distribution on the elements concerning a concept, can be utilised to evaluate the *specificity* of the concept itself, as for possibility distribution is proposed in [19]

---

[1] Conceptual abstraction function derives from Abstraction Theory. This has been recently incepted by Tenengerg [16] and successively perfectioned by other scholars [5, 6, 7].

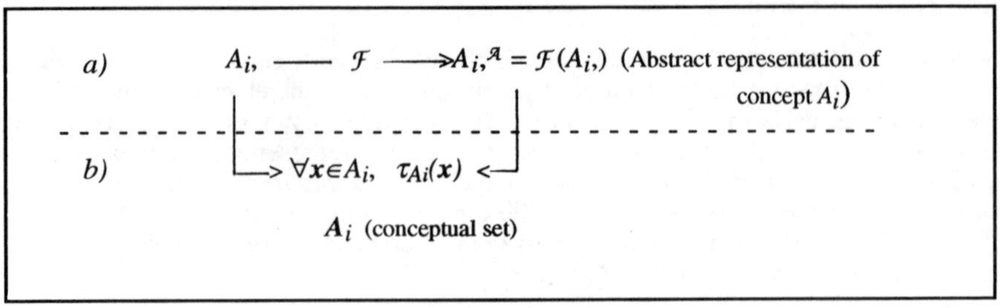

Fig. 5 - *a)* derivation of the abstract representation $A_i,^{\mathcal{A}}$ of the concept $\mathcal{A}_i$ underlying set A_i; *b)* evaluation of the typicality of the different elements forming A_i, so that it remains determined the conceptual set $\mathbf{A}_i$.

5. Conclusions

To avoid possible ambiguity and errors in evaluating the elements membership to sets, by the paper we propose an alternative procedure. Its validity for any type of sets, primitive and composed, ensures a uniform objectivity of the performed calculations.

It is made a fundamental assumption that regards the collecting elements of a general set into sets. This is carried out on the basis of specific criteria relative to the attributes that characterise such elements, standing that some of the former may be irrelevant to the mentioned collecting. In other words, the elements are collected into sets according to specific concepts.

The possibility by the elements of a so formed set, of representing the concerning concept, is conditioned by the number and the valence of the attributes that are relevant in each specific element. More formally, the conceptual formation of sets induces a contextual distribution of typicality depending on its elements. The different values this parameter may assume, are calculated by: *1st)* abstracting the intensive definition of the concept underlying the set, from the extensive definition of the concerned concept; *2nd)* comparing said intensive definition form to the extensive one, so that the irrelevant attributes of each element composing the set itself are brought into evidence. The numerical calculation can eventually be accomplished to determine the value of typicality for the single elements.

The so conceived sets, which keep the peculiarity of "fuzzy sets", have been defined as *conceptual sets.*

By making reference to the example illustrated in Figure 1 (Section 2), it is remarked that if the membership value of any element x to the set of cetaceans had been determined by considering the latter as a conceptual set, rather than a fuzzy one, its $\tau_C(x)$ would have not been ambiguous.

To confine the argument treated in the paper, the possibility of convoluting the considered semantic dimension of the elements —their typicality—, to that probabilistic, was avoided. Such convolution is however viable, as it results from further development of the topic.

REFERENCES

[1] A. O. Arigoni: "Mathematical development arising from 'semantic implication' and the evaluation of membership characteristic functions", *Fuzzy Sets and Systems*, No.4, 167-183,1980.
[2] A. O. Arigoni: "Transformational-generative grammar for describing formal properties", *Fuzzy Sets and Systems*, 8, 311-322, 1982.
[3] A. O. Arigoni & V. Maniezzo: "Conceptual abstraction: a context-depending simplification of concept representations", *Working Notes of the Approximation and Abstraction of Computational Theories '92"*, San Jose, Calif. 1992.
[4] A. O. Arigoni: "Typicality of concepts' instances: a semiotic way to its evaluation", Lect. Notes in Artificial Intelligence, E. P. Klement and W. Slany (Eds.), Springer-Verlag, 1993.
[5] A. Giordana, D. Roverso and L. Saitta: "Abstracting background knowledge for concept learning", *Prooc. of Machine learning-EWSL-91*, 1-13, Springer Verlag, 1991.
[6] F. Giunchiglia and T. Walsh:"A theory of Abstractation", *Artificial Intelligence* , 57, 323-389, 1992.
[7] J.R. Hobbs, : "Granularity", in Proceed. of 9th IJCAI Conference, 432-435, *Int. Joint Conf. on Artificial Intelligence*, 1985.
[8] L. I. Kunceva: "Fuzzy rough sets: application to feature selection", *Fuzzy Sets and Systems*, 51, 147-153 1992
[9] G. Lakoff: "Hedges: A study in meaning criteria and the logic of Fuzzy Concepts", *Jour. of Phil. Logic*, 2, 458-508, 1973.
[10] Michalski R.S., Carbonell J.G., Mitchell T.S. (eds.): "Machine learning: an artificial intelligence approach", Vol.2, Morgan Kaufmann, 1986.
[11] D. N. Osherson & E. E. Smith: "On the adequacy of prototype theory as a theory of concepts", *Cognition*, **9**, 52-72, (1981).
[12] Z. Pawlak: " Rough sets ", Int. Jour. *Comput. and Inf. Systems*, **11**, 341-356, 1982.
[13] E. Rosch: "Cognitive representation of semantic categories", *J. Exp. Psychology: general*, 104, 1975.
[14] E.M. Roth and C. B. Mervis: "Fuzzy set theory and class inclusion relations in semantic categories", *J. of Verbal Learning and Verbal Behaviour*, **22**, 509-525, 1983.
[15] P. Tabossi: "La rappresentazione mentale dei concetti", *Giornale Italiano di Psicologia*, **1**, 45-70, 1985.
[16] J.D. Tenenberg: "Preserving Consistency Across Abstraction Mappings", *IJCAI-10*, 1987.
[17] A. Vandierendonck: "Are category membership decision based on concept gradness ?", *Eur. Jour. of Cogni Phych.*, **3** (3), 343-362, 1991.
[18] L. A. Zadeh: "A note on prototype theory", *Cognition*, **12**, 291-297, 1982.
[19] R. R. Yager: "On the specificty of a possibility distribution", *Fuzzy Sets and Systems*, **50**, 279-292, 1992.

APPLICATIONS OF FUZZY SETS

FUZZY LOGIC APPLICATIONS IN CARDIOLOGY: STUDY OF SOME CASES

Senén BARRO*, Alberto BUGARÍN* and Paulo FÉLIX*
Ramón RUIZ** and Roque MARÍN**
Francisco PALACIOS***

*Dpt. Electrónica e Computación
Univ. de Santiago, SPAIN.
**Dpt. Informática
Univ. de Murcia, SPAIN.***UCI, Hospital
General de Elche, SPAIN.

ABSTRACT

In this work we comment on the necessity of having appropriate mechanisms for the representation and manipulation of uncertainty and vagueness in medicine. We present fuzzy theory as an adequate formal tool for this purpose, considering different examples in the domain of cardiology. The discussion of these examples follows an order of increasing complexity with respect to the knowledge representation and manipulation level in each case.

1. Introduction

The need for handling uncertainty in medicine has been profusely pointed out and addressed in the literature [11]. Despite their evident need, the introduction of mechanisms for the manipulation of uncertainty in the design of Expert Systems (ESs) in medicine has not always been considered. In many cases the problem has been solved by using *ad-hoc* techniques lacking solid mathematical formalism. However, one of the determinant factors for an ES to be successful is the inclusion of an adequate formalism for the representation and handling of uncertainty, that contemplates, among other requirements, the following: It must permit the explicit representation of the amount of evidence in order to support or reject a given hypothesis, facilitate the representation of the knowledge from the expert, evaluate to what extent certain pieces of knowledge are verified, facilitate the propagation of uncertainty and imprecision in the reasoning process, etc. If we add to this the need for methods with solid theoretical background, we find that the theory derived from fuzzy sets [13], from now on fuzzy theory, and, in particular, fuzzy logic, provide adequate tools for the manipulation of imprecision and uncertainty [4]. Fuzzy logic permits handling expressions that are close to those of natural language, and this, to a large extent, justifies its usefulness in the design of knowledge based systems. In the specific field of medicine, for example, it is evident that natural language is the main instrument for acquiring, storing, transferring and even manipulating knowledge.

In this work we will try to present and make evident the advantages that can be derived from the application of fuzzy theory and fuzzy logic to handling information in medicine through some examples in cardiology extracted from applications we have developed in previous work. We begin with event classification, in which events whose arrangement in time is irrelevant are handled. After this we consider cases in which an explicit account of time must be taken, whether in an absolute or in a relative manner.

2. Event detection and classification

We consider events as facts whose presence can be associated to a given instant of time and whose existence is indicated through the presence of specific feature values and/or characteristic morphological patterns.

In medicine the analysis of the evolution in time of certain phisiological variables is of great interest for the detection and later classification of specific events (for example beat detection on ECG signal). Nevertheless, the detection and classification of events is not an easy task, specially when considering its automation; electrocardiographic analysis and diagnosis by the cardiologist, for example, is usually based on visual impressions instead of detailed numerical values. Furthermore, the problems derived from the imprecision in the frontiers between the different categories to be differentiated, from the sometimes imprecise determination of the parameters on which the classification is based and the non deterministic relationship existing between these parameters and the different categories contemplated are frequent. To solve or mitigate these problems often leads to taking an approach that is based on trying to mimic the event classification process followed by a human expert. This approach, although it is not the only option, has been followed in many medical applications, even in the specific field of cardiology. Some examples are the generation of alarms in anaesthesia [12], arrhythmia detection [1] and diagnostics based on ECG analysis [5].

Briefly, this knowledge based approach is founded on the design of a set of Fuzzy Conditional Statements (FCSs), where the antecedent part defines prototypical conditions of the events to be classified and the consequent part represents the corresponding classification expressed by means of event-category membership degrees. This way, an FCS takes the following form:

$$R_s = \text{If } <F_1(e) \text{ is } P^s{}_1> \text{ and...and } <F_M(e) \text{ is } P^s{}_M> \text{ THEN } <\mu_{c_1}(e) = W^s{}_1>, \ldots , <\mu_{c_N}(e) = W^s{}_N>$$

where the symbol "< >" indicates that its content is optional, and $\mu_{c_n}(e)$ represents the membership degree of event e in category c_n. Each premise of the antecedent part of the s-th rule associates a condition $P^s{}_m$ to one of the features F_m. This condition constrains, usually in a fuzzy manner, the possible values of the universe U_m that can be taken by the feature, when it is evaluated over the event e to be classified, for this event to verify the premise. In many cases, the features take the form of linguistic variables. Thus, in a global manner, the antecedent part of each rule defines a fuzzy subset on $U = U_1 \times U_2 \times \ldots \times U_M$.

Based on this process, we have developed an application for the classification of beats according to their focus of origin [1]. The difficulties of this process is multiple: morphological dependence of the beats with the position of the electrodes, noise, variations in the position of the patient, the variations from one individual to another or the existence of cardiac pathologies. Nevertheless, the structure of FCSs goes well with the way in which medical knowledge is represented in this case; the definition of ventricular extrasystole is a good example of it [9]: "ventricular extrasystole are premature, wide QRS complexes (ventricular polarization) followed by a T wave (ventricular repolarization) with an axis opposite to that of the QRS. Characteristically, ventricular extrasystole are not preceded by P waves...". The need for the manipulation of linguistic concepts with clinical meaning, such as: premature beat or wide QRS complex, is obvious in order to retain the conceptual descriptions of the expert. For example, the value of "wide" for the attribute "width of the ventricular activity complex of the beat". In this case, the linguistic label "wide" can be associated with a fuzzy set to which the numerical values (in msec.) expressing the width of the QRS complex of the beat, belong to a greater or a lesser degree. This fuzzy set, defined over a numerical universe of discourse, attemps to encompass the concept of "Wide_QRS" the expert has and manipulates (figure 1).

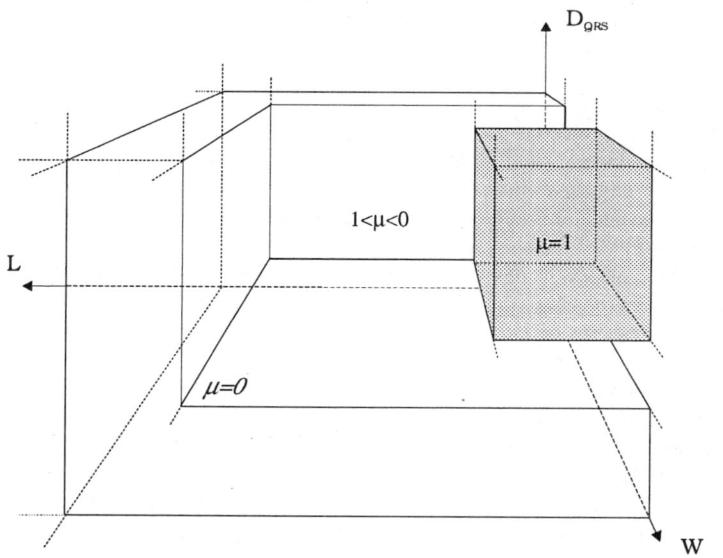

Figure 1. Fuzzy-cluster defined through the FCS IF (QRS_morphology very_aberrant) AND (QRS wide) AND (Beat premature) THEN $\mu_{Ventricular}$(beat)=1 (D_{QRS}, dissimilarity index normalized with respect normal QRS template; L, normalized distance between beats; W, width of the QRS complex in sec.)

3. Time as a fuzzy context

The chronology of facts has a crucial relevance for understanding the evolution of dynamic systems and is the key in monitoring tasks and in process control, planning, scheduling, etc. In the medical domain, we find multiple examples in which the diagnostic, prognostic and therapeutic evaluations not only depend on the known facts, but also on their temporal arrangement. If we have a real time clock, the manipulation of temporal structures defined over a discrete time axis can be achieved using an algebra of precise points. However, even in this case, a precise representation of time presents certain limitations; in many application domains of knowledge based systems, notably in medicine, there is not enough information available for determining with total precision and certainty the absolute time in which facts happen or the exact meaning of their temporal arrangement. Again, the approximate representation and reasoning model provided by fuzzy theory can achieve valid solutions for these problems [6,10]. In this sense, we have developed a model for the representation and manipulation of fuzzy temporal references to facts and knowledge in multiple application domains, mainly in medicine [3]. The basic concepts of this model are briefly presented, before going into the analysis of some examples where the link between the facts and their time of ocurrence is relevant.

3.1 Fuzzy time model

In our model we use the definition of date proposed by Dubois and Prade, to which we add the definition of time extent, as an additional primitive concept of the model. The definition of temporal entities is completed by introducing the concept of interval. Finally, we define fuzzy temporal constraint networks, for the representation of relative temporal relations based on the concept of distance between dates.

We consider *time* as projected onto a one-dimensional discrete axis $\tau = \{t_0, t_1, \ldots, t_i, \ldots\}$, where t_i represents a *precise instant* of time and $f_d = t_{i+1} - t_i$ is the *discretization factor*. Following Dubois and Prade [1989], we call *date* any fuzzy instant of time a represented by a possibility distribution π_a over τ, normalized and unimodal. This way, given a precise instant of time $t \in \tau$, $\pi_a(t) \in [0,1]$ represents the possibility of a being precisely t.

We introduce the concept of *time extent* in order to represent quantities of time, as, for example, the time elapsed between two dates. A fuzzy *time extent* b is represented by a possibility distribution, π_b, over the set of integer numbers $\mathbb{I}$, where the elements of $\mathbb{I}$ represent units of time f_d. This way, given $m \in \mathbb{I}$, $\pi_b(m) \in [0,1]$ represents the possibility of b being precisely m. For example, given an ordered pair of fuzzy dates (a, e), the time elapsed between them or the temporal distance between a and e, is given by a fuzzy time extent. This temporal distance is represented by means of a possibility distribution $\pi_{D(a,\ e)}$, over $\mathbb{I}$, where, given $m \in \mathbb{I}$, $\pi_{D(a,\ e)}(m)$ represents the possibility of the distance between the two dates being exactly m. As can be observed, introducing the time extent as one more primitive of the model allows talking not only about quantities about time but also about distances between temporal entities. This concept is used, as a mechanism for expressing relative temporal references and in the verification of relations among temporal entities. Finally, an *interval* is a period of time defined by its limits or extremes, which are dates, and its temporal duration, which is a time extent defined as the temporal distance between the extremes or limiting dates of the interval. We will denote as $I_{(a,e,d)}$ the interval limited by dates a and e, with a temporal duration d.

4. Episode detection

An episode is identified with the persistence in time of a fact or a set of facts. The definition of episode can have linked to it certain constraints on the time of persistence of the facts associated to the facts the episode refers to. That is the case of episodes of ichemia associated with a sustained deviation of the ST segment in the beats on the ECG. Nevertheless, as the minimum significant value and persistence of this deviation is not clear, the consideration of precise threshold levels of deviation and temporal duration, which is what is usually taken into account, does not seem adequate. Thus, and in a similar way as in the previous example for beat classification, a good way to establish the criteria for the description (detection) of ischemic episodes is through FCSs of the type:

If {(ST_Level high_depression) and (ST_Slope (positive or approx._null))}
DURING (more_than approx. 1 min)
Then Ischemia_Episode

There is, however, an important difference between this type of FCS and those used in the classification of beats: the need for the evaluation of the persistence in time of a series of facts, imposed by the use of the preposition "during". In an attempt to formalize the process of applying this FCS, we will introduce some definitions which have, notwithstanding, a more general application realm.

We consider that any numerical variable V, whose value in the instant $t \in \tau$ is $v(t) \in U$, has at least one representative in the domain of the linguistic variables. This linguistic variable L, takes values from a set of linguistic labels $\{l\}$, each of which is identified with a fuzzy subset of U, where $\mu_l(u) \in [0,1]$ represents the degree to which $u \in U$ belongs to this fuzzy subset. Or as an alternative, where $\mu_l(u)$ represents the degree to which u verifies label l. Given a temporal sequence of values for variable V, $S_V=\{v(t_i),..., v(t_j)\}$, whose temporal definition domain is represented as $\tau_{S_V}=\{t_i,..., t_j\}$, we define the application to this sequence of the "linguistic filter" associated to the linguistic value l, as the transformation of this sequence into the sequence: $F_l(S_V)=\{\mu_l(v(t_i)),...\mu_l(v(t_j))\}$. A linguistic filter is, consequently, an information abstraction mechanism.

Given a sequence S_V, a subsequence $e_l=\{v(t_m),..., v(t_n)\} \subseteq S_v$ constitutes an episode associated to linguistic value l, if and only if $\mu_l(v(t_k))>0$, $\forall\ t_k \,/\, t_m \le t_k \le t_n$ and there is no additional episode $e'_l \subseteq S_v \,/\, e_l \subset e'_l$.

Each element $v(t) \in e_l$ is embedded in a larger or smaller temporal context in which there is a certain degree of persistence of the property or linguistic value l. The degree to which an element $v(t) \in e_l$ is included within a context of duration $m \in \mathbb{I}$ where the linguistic value l persists is given by the expression below (which follows the minimum specificity principle):

$$\forall\ m \in \mathbb{I}\ \pi_{e_l,t}(m) = \max_{\substack{m=t_q-t_r \\ t_q,\, t_r \in \tau e_l}} \min_{t_r \le t \le t_q} \{\mu_l(v(t))\}$$

being τ_{e_l} the temporal domain where episode e_l is defined. This way, $\pi_{e_l,t}$ represents a temporal duration. This point is interesting because it permits the identification of situations defined through the persistence of a condition (or set of conditions) in time intervals of a given duration. Summarizing, we can make comparisons between a certain temporal extension π_d, taken as a preset persistence value, and $\pi_{e_l,t}$ according to the following expression:

$$\mu_{e_l,d}(t) = \max_{m \in \mathbb{I}} \min \{\pi_d(m), \pi_{e_l,t}(m)\}.$$

As an example, we will continue with the process of evaluating the previous FCS. For the sake of simplicity, we will consider that the existence of an ischemic episode, IE, is simply defined as the persistence of the condition: "(ST_Level high_depression)" during "more_than approx. 1 min". To determine the existence of IEs implies establishing what instants of the time axis are associated, and to what extent, to the situation described by the two conditions. Following the model we have introduced, the degree to which an IE is present in each instant $t \in \tau_{e_{high\_depression}}$, can be determined through the following expression:

$$\mu_{IE}(t)=\max_{m \in \mathbb{I}} \min\{\pi_d(m), \max_{\substack{m=t_q-t_r \\ t_q,\, t_r \in \tau e_l}} \min_{t_r \le t \le t_q}\{\mu_{high\_depression}(v(t))\}\}.$$

where d represents the time extent "more_than approx. 1 min" (50, 55, 60, ∞) and l represents the "high_depression" linguistic label associated with the variable "ST Level". Figure 2 displays the result from applying the previous expression.

This simple analysis can be extended to the evaluation of more complex situations, even to the consideration of sequences whose values are fuzzy or whose locations in time are fuzzy. Nevertheless, we are not going to go into this aspect any further, instead we are going to present some ideas about the representation and manipulation of fuzzy temporal relations as a patent necessity in many medical applications.

4.1 Fuzzy Temporal Constraint Networks

The temporal entities date and interval allow us to represent fuzzy temporal references expressed in a direct way: "21:15", "February, 12th", etc. However, for understanding the evolution of dynamic systems,it

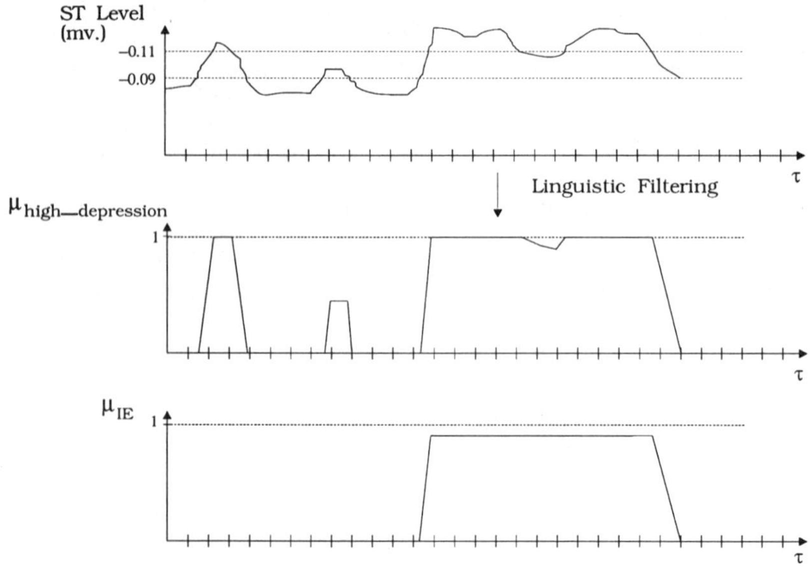

Figure 2. a) Sequence of ST values in time (f_d=10 sec.). b) Result from filtering using the "high_depression" linguistic label. c) Episodes associated to "high_depression" during "more_than approx. 1 min", where $\pi_{high\_depression} \triangleq (-0.09, -0.11, -\infty, -\infty)$.

is frequently necessary to specify the chronology of facts. This also frequently happensin monitoring tasks, process control, planning, scheduling, etc., where the structure of temporal relations between facts and knowledge must be preserved in the assumed computational representation. For this, we have introduced the concept of Fuzzy Temporal Constraint Networks (FTCN), that permits the representation and manipulation of fuzzy temporal relations among temporal entities. We define a Fuzzy Temporal Constraint Network (FTCN) L, as a finite set of n+1 nodes $\{N_0,...,N_n\}$, and a finite set of directed arcs P. Each node N_i is associated with an unknown date X_i and each arc $P_{(i,j)} \in P$, originating in node N_i and ending in node N_j, is associated with a time extent, $\pi_{P(i,j)}$, that reflects the existence of a (binary) constraint over the possible temporal distance between dates X_i and X_j. More specifically, given a precise duration $m \in \mathbb{I}$, $\pi_{P(i,j)}(m)$ represents the possibility of the temporal distance from date X_i to date X_j taking exactly a value of m. An interval $I_{(a,e,d)}$ is projected onto a FTCN through two nodes N_a and N_e, projecting the dates a and e, respectively, and through an arc $P_{(a,e)}$, whose associated time extent coincides with the duration d of the interval. That is $\forall m \in \mathbb{I}$, $\pi_{P(a,e)}(m) = \pi_d(m)$.

The objective of a FTCN is to maintain the temporal relations between temporal entities, and consequently between the facts associated with them, explicit. To this representation mechanism we have to add a "temporal expert" that constructs the network, detects inconsistencies, eliminates redundances (limiting the domains of possible values for each variable to the minimum set of values compatible with all the constraints) generates solutions, and responds to queries about the absolute and relative distribution in time of the temporal entities projected on the network [3,8]. As an example, figure 3 shows a FTCN that represents the facts of the following story (adapted from [3]): "5-6 months prior to current internment in Intensive Care Unit she started to have self limited palpitations. Some days later dizziness appeared and after that, but more than one month ago, dyspnea and asthenia started".

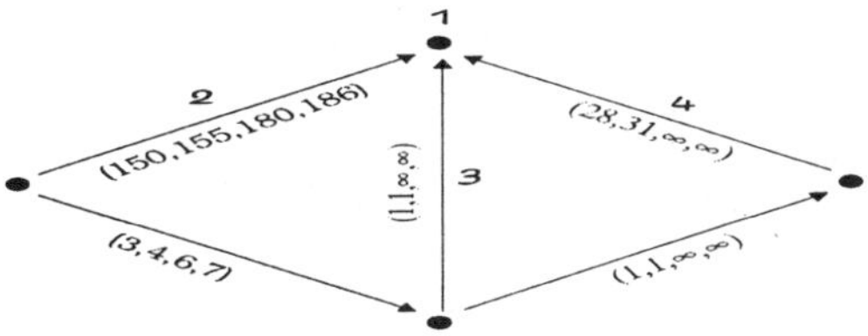

Figure 3. FTCN. Nodes represent dates and arcs represent distances between dates. Numbers represent facts (1: internment in ICU; 2: palpitations; 3: dizziness; 4: asthenia). Facts mapped to dates are displayed above nodes and facts mapped to intervals over the arcs. We assume trapezoidal distributions for representing time extents and f_d = 1 day.

As we said before, the temporal distribution of the facts can be absolutely relevant from the clinical viewpoint, even when there is not enough information for determining this distribution with total precision. In many cases, to reach a correct diagnosis implies evaluating the duration and arrangement of the facts in time. In the previous case, for example, the long duration of the dyspnea discards as a possible cause, an acute infection pathology of the respiratory tract. In a similar way, the temporal sequence of palpitations, dizziness and finally dyspnea and asthenia, suggests a bradycardia-tachycardia syndrome.

5. Conclusion and Discussion

The problems we have presented through different examples having to do with the management of uncertainty and vagueness, although included in the specific domain of cardiology, are frequent in other realms of medicine, and in information management in general. We have presented the cases according to increasing complexity, initially considering the presence of uncertainty associated to atemporal facts, then the need for combining the uncertainty in the facts, in their time of ocurrence and in their duration or persistence and finally considering the representation and manipulation of fuzzy temporal relationships between facts. In all of these cases, fuzzy theory has been an effective tool that has facilitated the adequate solution of different medical problems, specially in the field of cardiology, in which we have been working for a number of years.

We can finally make some considerations with respect to the impact of fuzzy theory on the design of KBSs in medicine. Medicine is a very adequate realm for the development of KBSs in general and ESs in particular, but, although a large number of implementations exist, very few have been completely accepted by the medical community. One reason of that is the lack of adequate manipulation of uncertainty, where fuzzy theory can be of great assistance. Nevertheless, even though fuzzy set theory and fuzzy logic have been widely and successfully employed in the design of controllers, the proliferation of fuzzy control solutions has been undoubtedly favored by the existence of a large number of environments and hardware & software tools developed for their design [TILShell, FuzzyControlDesk, FIDE, Aptronix]. This aspect, however, is not sufficiently covered in the case of ESs design, specially in those cases in which more complex strategies for the representation and handling of knowledge are required (temporal reasoning, for instance). In this sense we are working in the framework of the design of FRESH (Fuzzy and temporal Reasoning Expert system SHell) by three research groups belonging to the Universities of Murcia and Santiago de Compostela (Spain). Although FRESH is still in the stage of the initial specification of its functional and structural characteristics, it will try to be a shell for the development of knowledge based systems for applications in which the representation and management of time dependent information is required, providing solutions for those cases in which the events, their temporal labels and/or the applied knowledge necessary for reasoning with them are of an imprecise nature.

ACKNOWLEDGEMENTS

This work was supported by the Spanish CICyT under projects TIC-90-0496, TIC-91-0997, and TIC-92-0136 and by the Xunta de Galicia under project XUGA20601B92.

REFERENCES

[1] Barro, S., Ruiz, R. and Mira, J. "Fuzzy beats labelling for intelligent arrhythmia monitoring". Computers and Biomedical Research. N. 23, 240-258, 1990.

[2] Barro, S., Ruiz, R., Presedo, J., and Mira, J. "Grammatic representation of beat sequences for fuzzy arrhytmia diagnosis". Int. J. Bio. Computing, 27. 245-259, 1991.

[3] Barro, S., Marín, R., Mira, J. and Patón, A.R. "A Model and a Language for the Fuzzy Representation and Handling of Time". Fuzzy Sets and Systems, 61. 1994

[4] Bonissone, P. P. "Reasoning, Plausible". In: *Encyclopedia of Artificial Intelligence*, S.C. Shapiro (Ed.). John-Wiley and Sons. 854-863, 1990.

[5] Degani, R. and Bortolan, G. "Fuzzy decision-making in electrocardiography". Artificial Int. in Medicine, 1. 87-91, 1989.

[6] Dubois, D. and Prade, H. "Processing fuzzy temporal knowledge", IEEE Trans. on SMC, Vol.19, N. 4, 729-744, 1989.

[7] Godó, L., Jacas, J. and Valverde, L. "Fuzzy Values in Fuzzy Logic". Int. J. of Intelligent Systems, Vol. 6, 199-212, 1991.

[8] Marín, R., Barro, S. Bosch, A. and Mira, J. "Modeling the Representation of Time from a Fuzzy Perspective". Cybernetics and Systems, 25. 1994.

[9] Pucheu A., Lacroix H., Tonet J.L., Frank R., Fontaine G. and Grosgogeat Y. "Ventricular Arrhythmias". In: *Comprehensive Electrocardiology. Theory and Practice in Health and Disease*, MacFarlane and Veitch (Eds.). Vol. II, Cap. 25. Pergamon Press. 966, 1989.

[10] Qian, D. Q. "Representation and use of imprecise temporal knowledge in dynamic systems". Fuzzy Sets and Systems, Vol. 50, 59-77, 1992.

[11] Moskowitz, A. J., Kuipers, B. J. and Kassirer, J. P. "Dealing with uncertainty, risks, and tradeoffs in clinical decisions". Annals Internal Med., 108. 435-449, 1988.

[12] Schecke, Th., Rau, G., Popp, H.J., Kasmacher, H., Kalff, G., and Zimmermann, H.J. "A knowledge-Based Approach to Intelligent Alarms in Anesthesia". IEEE Eng. Med. Bio. Mag., December. 32-44, 1991.

[13] Zadeh, L.A. "Fuzzy Sets". Inform. and Control, Vol. 8, 338-353, 1965.

A FUZZY LOGIC BASED SYSTEM FOR THE QUANTIFICATION OF VISUAL INSPECTION IN CLINICAL ASSESSMENT

Elisabetta Binaghi, Maria Luisa Cirla, Anna Rampini

ITIM - CNR
Via Ampere 56,
20131 Milano, Italy

ABSTRACT

In this work we deal with medical diagnostic problems including visual inspection as a fundamental procedure with which to evaluate clinical signs and to formulate diagnostic conclusions. In this case discrepancies among physicians may arise in formulating diagnosis.

Our approach addresses this problem by combining the use of fuzzy sets as the representational framework with a machine-aided knowledge acquisition strategy which includes, neural networks.

In the paper, specifically, we illustrate the use of this approach in building a system that functions as an active support in the diagnosis of hormonal disorders.

For this application the solutions adopted have been integrated in a unified framework and implemented in an hybrid system.

1. Introduction

Fuzzy set techniques are widely used in the solution of various practical problems in medicine proceeding from the assumption that medical diagnosis is a cognitive process and the descriptions of clinical signs and subsequent decision activity are performed not on the basis of a mass of numbers but by "elicitation links"

between linguistic description of clinical signs and judgements [1]. The current literature provides various examples of medical expert systems with fuzzy inference models and of medical applications using fuzzy expert system shells [2].

In this work we deal with medical diagnostic problems including visual inspection as a fundamental procedure with which to evaluate clinical signs and to formulate diagnostic conclusions.

In this case the automation of the diagnostic procedure presents critical aspects. The assignment of medical significance to signs, that are evaluated by visual inspection, is based on complex and obscure criteria which may be difficult to assess directly and transfer to obtain a general consensus by medical community. Discrepancies among physicians may arise in formulating diagnosis even when the involved concepts are rigorously applied in a rationale scheme based on a shared background.

Our approach addresses this problem by combining the use of fuzzy sets as the representational framework with a machine-aided knowledge acquisition strategy which includes, neural networks and relaxed structured interviews for the generation of fuzzy knowledge. Fuzzy sets, as proposed by Zadeh [3], address the representation problem by providing adequate knowledge structures when the information is uncertain due to vagueness and imprecision [4]. Neural networks can facilitate the process of knowledge acquisition by transforming readily available knowledge, such as examples, into a useful form and by reducing ill defined interactions between expert and engineer.

We have defined several solutions for acquiring and structuring fuzzy knowledge and inference models in medical diagnosis, drawing on the strengths of the most suitable approach for each situation. A complete description of these tools and their application in several medical situations may be found in [5].

In this paper, specifically, we illustrate the use of these tools in building a system that functions as an active support in the diagnosis of hormonal disorders. For this application the solutions adopted have been integrated in a unified framework and implemented in an hybrid system based on neural networks and fuzzy set techniques, to combine the suitability and flexibility of fuzzy logic in representing cognitive processes with the effectiveness of neural networks in learning knowledge from examples [6].

2. Diagnosis of Hormonal Disorders in Women by Clinical Assessment of Hirsuties

In the complex clinical presentation of the pathology concerned, various indicators, such as hematochemical parameters can be easily determined, while the assessment of hirsuties which is essential for the formulation of diagnostic judgements, is a complex and sophisticated task. It includes, in fact, the assessment of body hair growth in women usually approached by visual inspection and examination of qualitative aspects in several body sites.

The overall diagnostic problem underlying this application is broken down into a hierarchy of meta-diagnostic tasks:

- first level: clinical assessment of hirsuties
- second level: diagnosis of hormonal disorders by aggregation of first level results and evaluation of hemato-chemical parameters

In this work we restrict our presentation to the first diagnostic level.

The clinical assessment of hirsuties in women have usually been approached as if there were a clear dividing line between hirsuties and the normal state. Body hair growth appears, however, to be a graded characteristic. The problem of assessment resolves itself into a selection of suitable sites for study, and the choice of a method for grading hair growth at any given site.

The methods commonly used for grading hair growth are based on a crisp evaluation of visual signs such as *area* and *density,* providing numerical gradings for each sites. Hair growth is scored according to the sum of the gradings obtained [7].

The resulting discrepancies among experts and unpredictable subjectivity in the diagnosis have suggested that a fuzzy medical expert system could better guarantee an objective, explicit and deterministic assessment of the diagnostic process.

3. Acquisition and Representation of Knowledge in the Domain of Hirsuties

The clinical assessment of body hair growth in women used in the study of hormonal disorders, proceeds by evaluating visual clinical signs, in 9 different body sites.

A metadiagnosis is made for each body site and partial diagnostic judgements are then aggregated to perform the final diagnosis.

3.1 Representation of clinical signs

The clinical signs involved in the evaluation of hair growth are *density*, *consistency* and *area.*

We proceed from the idea that the visual inspection of the above clinical signs is subjective in nature and that the cognitive representation of their evaluation has a qualitative linguistic character.

The difficulties encountered in creating fuzzy sets for these signs are due principally to the global visual perception underlying the clinical evaluation. Visual data, in the form of scenes and pictures, are often processed in visual terms alone, without any corresponding translation or recording as verbal labels or representations [8]. We addressed the problem by implementing a graphical interface displaying all the elements of the body as parts of an integrated whole, and not as separate and independent items and by adopting a relaxed elicitation strategy. Figure 1 shows the interface window for the acquisition of membership functions related to the sign *area* evaluated in the chin site.

The global region of interest was divided into 5 items. Progressive combinations of these items were displayed: they constituted the range on which the membership functions for the linguistic terms *significant, borderline, insignificant* are defined. Because it was impossible to procure explicit descriptions of the sign *area,* these terms were semantically related to a *meta-diagnosis* in which, together with the *area, consistency* and *density* parameters are evaluated by the experts. Following the membership function exemplification method [9], the experts were asked to express the compatibility of each term with each combination of items by answering *yes* or *no* and assigning a number from 0 to 5 to indicate their degree of confidence in the answer. The collected data were then processed to produce the representative value $\mu_t(x)$ of the membership function for the term t and the combination of area items x. The following formula were used:

$$\mu_t(x) = \frac{1}{2} + d \times (\frac{r}{10})$$

where r is the degree of confidence in the answer, d=1 if the answer was yes and d=0 if the answer was no. The degrees $\mu_t(x)$, quantifying the compatibility between the linguistic term t and the crisp value assumed by the clinical sign concerned for each expert, were aggregated and a fitting operation were applied on the rough distribution. Figure 2 shows membership functions for the sign *area* in the chin site obtained by applying a spline-based fitting operation on the representative values computed.

A fuzzy matching procedure was applied to compare membership functions values and expert judgements [5]. Global agreement, of 75%, evaluated on a training set of 50 sample values, was found. A partial agreement of 90% was found for clinical signs for which the fuzzy consistency [5] among experts was fully respected.

4. The neural stage

The metadiagnostic reasoning designed aimed to produce a diagnostic judgement for a given body site has been implemented with a connectionist model.

Knowledge underlying the diagnostic stage is complex, difficult to assess, and difficult to represent in symbolic form, due partially to the global visual perception underlying the clinical evaluation of each body site.

Neural network (NN) techniques can successfully address these problems by providing accurate learning mechanisms capable of generating and/or modifying knowledge structures autonomously. In particular NNs offer the means to overcome some of the major drawbacks of conventional knowledge acquisition techniques, such as the lengthy consultation with experts required in structured interview techniques, and the design of complex conceptual models when symbolic machine learning algorithms are adopted.

We propose for building NNs into the hybrid framework for medical diagnostic reasoning the use of a feed forward one hidden layer neural network with an adaptive, problem-dependent back-propagation learning algorithm. A detailed description of the NN model and of the learning mechanism adopted can be found in the references Binaghi et al. [10].

The reasoning activity involved in the evaluation of body hair growth has been modelled in any single given body site as a classification process which can assign an event, i.e. a set of linguistically qualified clinical signs (for example *area is significant, density is sparse, consistency is terminal*), to one or more of several predefined categories.

The implemented NN classifier must compute the degree of similarity (membership of possibility) of the event being analyzed to each one of the prototypal hypotheses defined in the model. Thus for each sub-diagnosis a link between the symbolic knowledge base and the connectionist stage is created. The knowledge base provides membership functions characterizing the clinical signs involved.

The sub-diagnosis (of a body part) is therefore characterized by a set of input events, a set of output hypotheses, and a NN implementing a fuzzy relationship between input and output.

The NN assumes a feed-forward topology with three layers. Each input node represents the value of an attribute of the object (or event) to be classified; in our approach each term linguistically qualifying a clinical sign is represented by a different neuron .

The degree of confidence in each of these terms (obtained by evaluating its membership) is represented by the state of activation of the neuron. Hidden neurons represent intermediate abstractions.

The output layer represents the hypothesis. In particular the state of activation of each output neuron represents the degree of possibility of the event under analysis to the hypothesis represented by the neuron.

In our application there is one diagnostic class, corresponding to the clinical assessment of body hair growth in women; the different hypotheses are linguistically qualified degrees of severity, ranging from *very low* to *very high.*

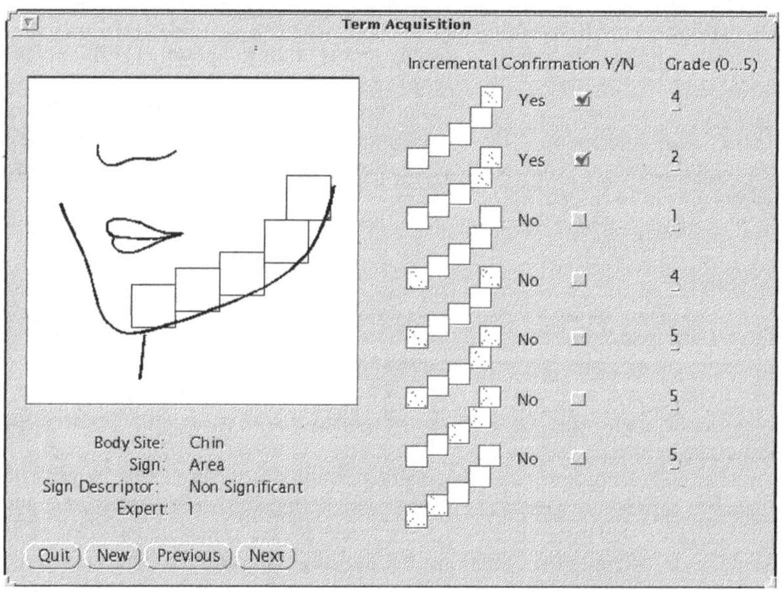

Figure 1 - Elicitation of the membership functions of the sign *area*

4.1 Problem dependent learning rule in feed-forward neural networks

The main drawback of the NN method lies in the fact that the generalized delta rule is computationally complex: when the training data are numerous the learning time can be very long. Moreover, the generalized delta rule can not guarantee that the global error minimum will be found, and the solution strategy depends on the initial network configuration (often assigned at random).

Several attempts have been made to improve the performance of back propagation algorithms; fast learning algorithms have been proposed to accelerate the training phase [11]

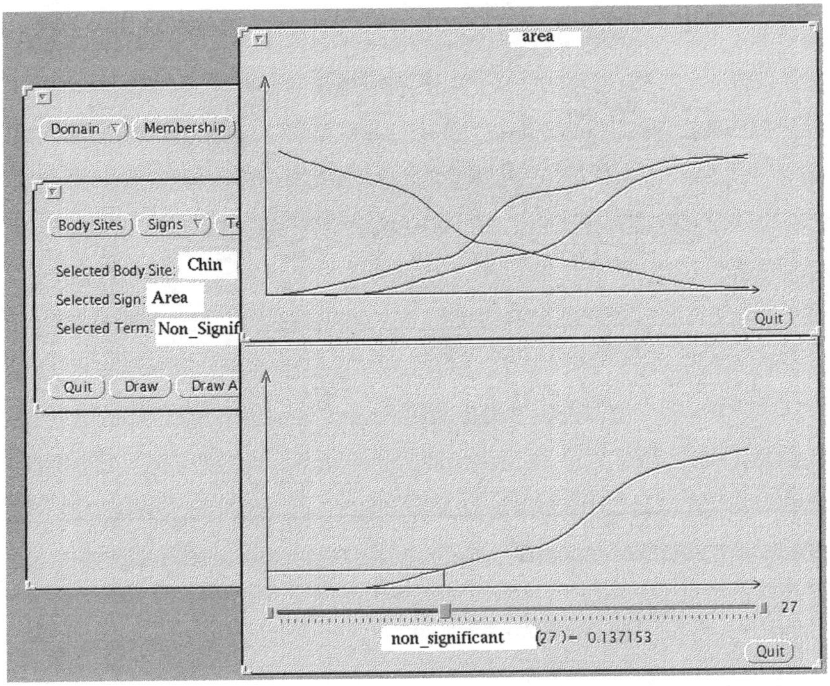

Figure 2 - Membership functions of the sign *area* in the chin site

Our approach integrates problem-dependent knowledge in the analytical formulation of the back propagation algorithm in order to control and subsequently improve the weight update procedure.

The central concept is that of subordinating the activation level of one or more hidden neurons to the presentation of certain input pattern configurations to input layer.

The fuzzy logic framework is used to support the representation of this knowledge, which may be uncertain and/or linguistically qualified. The flexibility of the fuzzy reasoning approach makes it easy to define and update rules until a reasonable performance is achieved, and to introduce new intuitive knowledge at different levels of abstraction in a natural and controlled way.

Fuzzy production rules, structured as evaluation-decision pairs, are used as knowledge structures to represent all the ingredients of the fuzzy reasoning process. In the antecedent of the rules, conditions related to input patterns are expressed in terms of compound fuzzy declarative propositions. The decision part of the rules contains fuzzy declarative propositions which describe linguistically the values that variables related to hidden neuron activation assume when the corresponding evaluation part of the rules is satisfied.

We have defined a forward chaining fuzzy inference mechanism to interpret these rules: it assigns a degree of confidence to each rule in the knowledge base, using a fuzzy matching procedure, and deduces the best hidden neuron activation through a defuzzification process applied to the variables in the consequent part of rules.

We use the fuzzy logic based learning algorithm to train the networks devoted to each diagnostic subproblem involved in the medical application concerned.

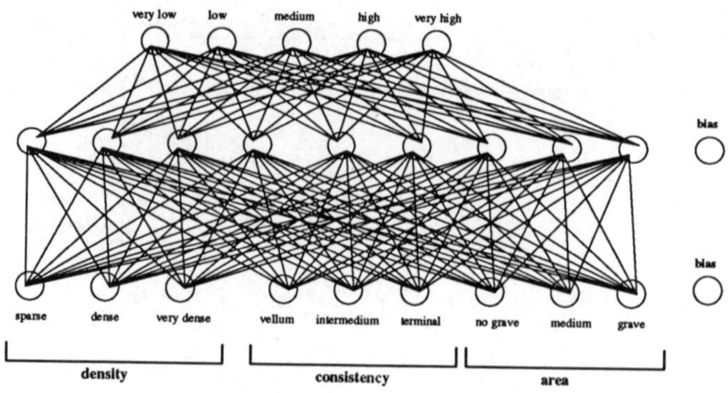

Figure 3- Neural network topology

If density.sparse is excited ***And*** consistency.vellum is excited ***Then*** hidden neuron-1 is excited
If density.sparse is relaxed ***And*** consistency.vellum is excited***Then*** hidden neuron-1 is relatively relaxed
If density.sparse is relaxed ***And*** consistency.vellum is relaxed ***Then*** hidden neuron-1 is relaxed

Table 1- Rules in the knowledge base

4.2. Neural network implementation

At the present stage of our experiments, preliminary versions of the NN classifiers have been implemented for each of the 9 body sites examined.

Figure 3 shows the topology of the neural network used. The number of input nodes is determined by the dimension of the input patterns, which are degrees of evidence specified for each term of the clinical signs.

Input neuron activation lies to the range [0,1]. Each output neuron represents a degree of possibility of assigning the current event to a certain degree of severity and also has activations in the range [0,1].

Each of the NNs implemented have been trained using training sets constituted by diagnosed clinical cases.

A total number of 400 examples have been considered.

The problem dependent learning rule has been implemented using different knowledge bases in function of the site.

For the chin for example, 1 hidden neuron has been constrained to the presentation to the input layer of evidence confirming a *low* degree of severity.

Table 1 lists a subset of rules in the knowledge base for the chin site.

Suitable membership functions have been elicited to characterize linguistic terms in the rules.

5. Preliminary results and conclusions

To assess the diagnostic accuracy of the implemented neural networks and their classification capability, a test set of examples, consisting of already diagnosed clinical cases referred to specific body sites, has been used. The testing data considered were 200. The diagnosis was confirmed in 82% of the cases.

Further experimentations necessary for a complete evaluation of the adopted solutions.

However results today are encouraging and confirm the validity of the approach. Flexibility and relaxed structure of the fuzzy logic framework ensure a representation of knowledge close to the mental representation of experts. The ability of NNs to learn from examples even in uncertain and unknown environment garantees good performances of the overall system.

We are currently engaged in experimental work to conclude the application by modelling the overall diagnostic process.

REFERENCES

[1] Szolovits P., Patil R.S., Schwartz W.B. (1988) Artificial Intelligence in medical diagnosis, Annals of internal medicine, 108, pp.80-87.

[2] Adlassnig K.P., Kolarz G., SheithauerW. (1985) Present State of the Medical Expert System Cadiag-2, Meth. Inform. Med., 24, 1, pp.13-20.

[3] Zadeh L.A. (1965) Fuzzy Sets, Information and Control, 8, pp.338-353.

[4] Zadeh L.A. (1987) The concept of Linguistic variable and its Application to Approximate Reasoning. In "Fuzzy Sets and Applications", (Yager R.R., Ovchinnikov S., Tong R.M., Nguyen H.T. Eds.), pp. 193-329. John Wiley & Sons.

[5] E. Binaghi (1993) Fuzzy Logic-Based Tools for the Acquisition and Representation of Knowledge in Biomedical Applications, in A. N. Venetsanopulos and S. Tzafestas Eds., Fuzzy Reasoning in Information, Decision and Control Systems, MBISE series, Kluwer Academic Publishers, pp. 238-253.

[6] Kandel A., Langholz G. (Eds.) (1992) "Hybrid Architectures for Intelligent Systems", CRC Press, Boca Raton (Florida) .

[7] Ferriman D.M., Gallwey J.D. (1961) Clinical Assessment of Body hair Growth in Women, J. Clin. Endocr. Metab., 21, pp. 1440-1447.

[8] Freedman J., Haber R.N. (1974) One Reason Why We Rarely Forget a Face, Bull. Psychon. Soc., 3, pp.107-109.

[9] Chameau J., Santamarina J. (1987) Membership Functions I: Comparing Methods of measurements, Int. J. of Approaximate Reasoning, pp.288-317.

[10] Binaghi E., Mazzetti A., Orlando R., Rampini A. (1993) Integration of Fuzzy Reasoning Techniques in the Error Back-Propagation Learning Algorithm, in Proceedings of 6th Italian Workshop on Neural Nets, Vietri sul Mare (Salerno).

[11] Rumelhart D.E., Hinton G.E., Williams R.J (1986) Learning Internal Representation by Error Propagation. In "Parallel Distributed Processing", (Rumelhart D.E., Mc Clelland J.L. Eds.), MIT Press, Cambridge MA.

MENTA / MD, FUZZY DECISIONAL ENGINE

Benoît LE BLANC, Gilles PALMER

Département d'Informatique Médicale
Université de Bordeaux 2
F 33076 Bordeaux cedex, France

ABSTRACT

This paper describes a decision making process in a fuzzy context. Decisions are considered as a vote result. Votes are done by specific production rules : decision rules. We build an expert system shell called MENTA, with a particular decisional engine for this vote process. This paper presents the engine, its rules and a medical application.

1. Introduction

Software realization for antibiotic prescription leads us to the development of a Decision Engine. It is based on fuzzy extension of the Social Choice Theory [1]. This Engine is using an independent set of Decision Rules, as like as usual Inference Engine with a knowledge base of Inference Rules. We associate this Decision Engine with a more conventional Inference Engine. Both are working in the framework of Possibilistic Logic [2]. Then, we have implement a cooperation between the two different engines, to solve two different kinds of decision making problems.

In a first step, this article presents the design of our Decision Engine. Then we describe our hierarchical organization of decisions, supported by a Directed Acycling Graph (DAG). We present at last the cooperation between the two engines. An overview of the medical application is given.

2. Decision Engine and Fuzzy Decision Making

2.1. Decision Engine

Let us consider a set of potential decisions, and a list of basic criteria for decision acceptation or decision refuse. The Arrow's Social Choice Theory is a very well known framework to study this situation. The research of the best solution is compared to a vote situation. Each Decision Criterion (DC) « votes » for the possible decisions (it gives a preferential order on the decisions). But the research of the final vote, taking into account all the votes, leads to the Impossibility Theorem : this research is incompatible with the respect of common sense axioms on vote equity.

A fuzzy extension of this principle is able to jump over this problem. In this case, the voters do not give an ordered list of choices, but a fuzzy set of preference.

Let us consider $\Delta = \{\delta_1, \delta_2, \cdots, \delta_n\}$ the set of all elementary decisions (the « candidates »), and $DC_1, DC_2, \dots, DC_m$ the set of all considerate Decision Criteria (the « voters »). For each DC_k the expert gives a fuzzy set of preference : $\Delta_k = \{\mu_1 / \delta_1, \mu_2 / \delta_2, \cdots, \mu_n / \delta_n\}$.

The membership values (μ_i) correspond to the pertinence of elementary decisions δ_i when DC_k is completely true (see below for modification of this degree during fuzzy evaluation of DC_k). So $\mu_i = 1$ when δ_i is quite appropriate and $\mu_i = 0$ when δ_i is strongly inappropriate.

To express the couples (DC_k, Δ_k), we have created a frame : the Decision Rule (based on ordinary Inference Rule syntax) : « if DC_k then Δ_k ». It is detailed as :

Vote Aggregation_Principle (i.e. « Dictatorship », « Veto » or « Democracy »)
If
Logical Fact $_1$
and Logical Fact $_2$
... *(Hypothesis part is DC_k : a conjunction of Logical Facts)*
and Logical Fact $_m$
Then
μ_1 δ_1
... *(Conclusion part is the extension description of the set Δ_k)*
μ_n δ_n

In this definition, the VOTE section is used to announce the rule participation [3] in the Final Vote (FV). FV is the fuzzy set obtained after activation of all pertinent rules.

In a dictatorial Vote, the rule imposes its conclusions to the final set : $FV \leftarrow (FV \cup \Delta_k)$,
In a veto Vote, the rule restricts the final set to its conclusions : $FV \leftarrow (FV \cap \Delta_k)$,
In a democratically Vote, the rule influences equitably the final set : $FV \leftarrow (\textit{average of } \Delta_k)$.

For basic coherence reasons, it is forbidden to use uncoordinated application of rules. A two times aggregation of rules is necessary [4]. In a first part, each set of same vote rules are aggregated ; then a final aggregation with the three result sets is made.

Hypothesis part of rule is exactly the same as hypothesis part of inference rules. DC_k is a criterion with logical determination. It is composed of a conjunction of elementary logical facts. Validity of these facts is obtained in a fuzzy way, by application of Possibility Theory.

2.2. Fuzzy Inference

When the expert writes a decisional rule, he expresses hypothesis with logical facts as « X is A » where « X » is the variable (*age* for instance) and « A » a fuzzy set (for example the fuzzy set which characterizes the *elderly person* notion).

The real and observed value of variable X is « X is A' ». Fuzzy relation between A' and A is measured with two numerical values : Poss and Nec (Possibility and Necessity values from Possibility Theory). With this two numbers it is possible to obtain automatically appreciation of well or not well matching between A (a theoretical value, given by the expert) and A' (observed value). According to this numbers, and considering a fixed threshold, the rule is or is not activated.

We have emphasized that the expression of a decisional rule postulates that the hypothesis part is considered as « completely true ». But fuzzy pattern matching between A and A' may appreciate differently this validity. Consequently, every rule application needs an adaptation of membership values given by the expert in the rules conclusion. For automatically modification of μ_i value, we apply a methodology derived from medical prescription.

Contra-indication situation : in front of a particular context we want to exclude a specific decision from the final result. The reasoning process tries to minimize the risk ; as soon as the contra-indication is *possible*, we wish to exclude the decision. So Possibility Measure is a natural estimator to appreciate this context.

Indication situation : some specific decision have to be required in front of special situations. In this case, the decision is only envisaged if the situation is *certain*. In this case, the exigency to be in the correct context of indication leads us to use the Necessity Measure.

We implement this two situations in our decisional process. Let us consider an example with a *veto* rule ($\cap$ operator), and let μ_i be the appreciation of pertinence for a member of this rule conclusion. Modification process permits to obtain μ'_i a corrected value of μ_i.

If we use *a contra-indication* context, the corrected value is :

$$(\text{Poss} = 1) \Rightarrow (\mu'_i \leftarrow \mu_i)$$
$$(\text{Poss} = 0) \Rightarrow (\mu'_i \leftarrow 1)$$

This formula should be interpreted as two extreme points of view. On the first hand, if it is strongly possible (without any indication about the necessity) that the real situation (A') matches with the described situation (A) then we take the exact value given by the expert (μ_i). On the second hand, if the described situation is quite impossible then we don't exclude any more the decision δ_i.

Indication context has a symmetrical aspect for corrected value :

$$(\text{Nec} = 1) \Rightarrow (\mu'_i \leftarrow \mu_i)$$
$$(\text{Nec} = 0) \Rightarrow (\mu'_i \leftarrow 0)$$

with a symmetrical interpretation.

For intermediary values of Poss (respectively Nec), many interpolation may be envisaged. In respect with our medical expert, we use linear modulation of corrected values. Some extensions of this problem are exposed in [4].

3. Decision Graph

Our application in medical field is a problem of antibiotics prescription. Consequently, our decision criteria are composed of biological or clinical data (age, renal or hepatitis insufficiencies, pregnancy for women, ...), and associated decisions correspond to drugs (in France, more than 200 different antibiotics are prescribed in ambulatory medicine). But it was useless to construct rules with every potential drugs in each conclusion parts. It was a strict and needless work : most part of drugs have some similar properties.

For knowledge expression (given by the expert) and rules manipulation (done by the engine), we first organized the sets of drugs in many trees based on pharmacological classification. For instance, Penicillin family is represented through the tree structure shown in Fig. 1.

This expression is difficult to obtain for all drugs families. Some of them are composed of several molecules, and therefore belong to several different families. Then, our antibiotic classification looks like a graph. Branches are directed (Family → SubFamily), and no cycle is accepted (great families are disposed upper, specified drugs are lower).

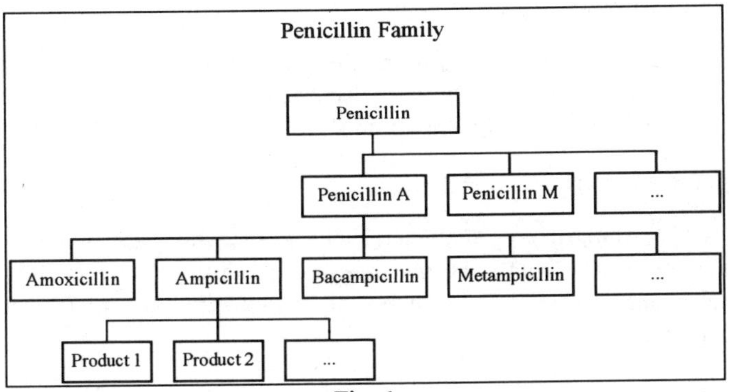

Fig. 1

Using this directed acycling graph (DAG) organization, the expert is able to specify in conclusion part of decision rules (any node of our graph) instead of elementary ones (the leaves).

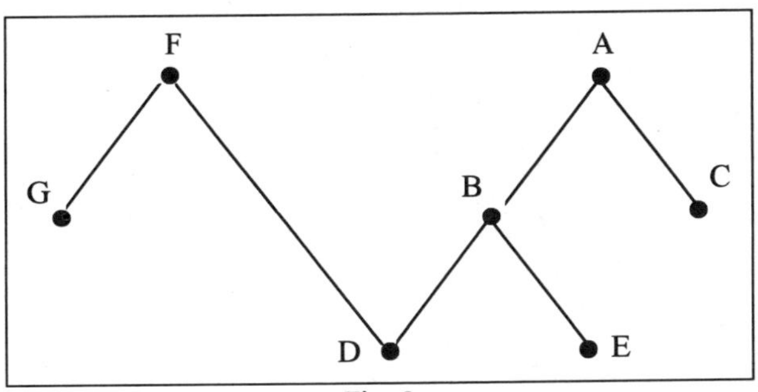

Fig. 2

The main problem consists in propagation of conclusion values through the graph. Given the graph in fig.2, and the two following rules :

Vote XX	**Vote XX**
If	**If**
DC_1	DC_2
Then	**Then**
0.3 F	0.8 B
	0.5 C

conclusion parts of rules are interpreted as :

Then	**Then**
0.3 G	0.8 D
0.3 D	0.8 E
	0.5 C

In respect with graph structure, information circulation is possible. Each node receives a membership value in a rule (locally), but it also inherits global properties (depending on its graph position). In an antibiotic context, such properties are drug contra-indications, side effects, interactions with other drugs, ... Properties are directly inherited, extended, or overwrote from upper nodes.

Because of this properties circulation, and because of our generic frame for rules (inference or decision), we used an object oriented implementation.

4. Decision Process extension

Two main extensions of our decision process are now considered. The first one is the possibility to interpret decision result in an usual forward chaining context. This is done by activation of our Inference Engine. The second one is the opportunity to manipulate association of decisions.

4.1. Inferences about decisions

When decision engine finishes its work into a decisional rule base, we obtain a fuzzy set of pertinent decisions : the Final Vote FV. At this time, it is interesting to investigate elected decision to obtain more conclusions. So we call our inference engine in a forward chaining session with a specific inference rule base.

These rules may be :

If
δ_i is an indicated decision
Then
...

or

If
δ_i is an eliminated decision
Then
...

Hypothesis parts of these rules can be respectively exposed with following standard fuzzy expressions :

« $\mu_{FV}(\delta)$ is A », where A is the fuzzy set

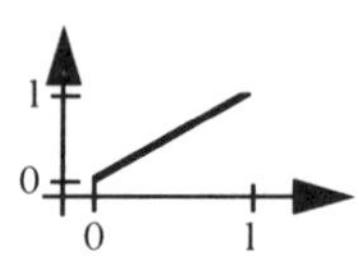

and

« $\mu_{FV}(\delta)$ is B », where B is the fuzzy set

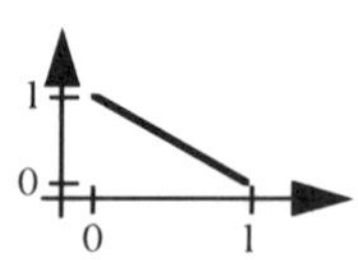

($\mu_{FV}(\delta)$ is the membership value of a decision δ in the Final Vote, elaborated in the decisional process).

If δ is an elementary decision form decision graph, then it is easy to calculus numbers Poss/Nec, measuring hypothesis pertinence. But if δ is an intermediary node in the decisional DAG, then hypothesis pertinence is more difficult to be estimated.

Given Fig.2 graph, and supposed that FV was calculated as : { 0.9/G, 0.8/D, 0.6/E, 0.3/C }. What is the way to estimate hypothesis pertinence of the following rule :

If
A is an indicated decision
Then
...

(A is not a direct component of FV, but all sub-decision of A are in FV).

The difficulty is into the correct interpretation of the sentence « *A is an indicated decision* ». Should we interpret it as « ***one*** *of the elementary decisions derived from A is in FV* », or as « ***all*** *of the elementary decisions derived from A are in FV* » ? This choice only depends on the context of the rule. So we propose two expressions for the expert knowledge : δ-is-totally-in-FV, or δ-is-partially-in-FV.

To estimate the fact « $\mu_{FV}(\delta)$ is A », we calculate $\mu_{FV}(\delta)$ by :

$$\min_{\delta_i \text{ final decision derived from } \delta} \{\mu_{FV}(\delta_i)\} \qquad \text{(in a δ-is-totally-in-FV context)}$$

or

$$\max_{\delta_i \text{ final decision derived from } \delta} \{\mu_{FV}(\delta_i)\} \qquad \text{(in a δ-is-partially-in-FV context)}$$

We have shown the possibility to use a decisional conclusion in a deductive process. Opposite cooperation is easier because all decisional hypothesis are composed of logical facts (which could be results of a deductive process).

Our system MENTA is able to alternately use these two engines. Exchange of conclusions is done dynamically. This cooperation between engines requires separated rule bases, and previous detection to prevent cycles (cycles in a rule base, and cycles from rule bases mutual call).

Particularities of our inference engine are presented in [5].

4.2. Associations of decisions

In ambulatory prescription, a General Practitioner must avoid the association of antibiotics. These critical therapies are only done in hospitals. So, the first version of our system worked with sample decisions. An hospital orientation leads us to tconsiderate multiple decisions.

In the case of the following rules (where conclusions are illustrated with antibiotics) :

Vote veto
If
  DC_1
Then
  1.0 $Antibiotic_1$
  0.0 $Antibiotic_2$

Vote veto
If
  DC_2
Then
  0.0 $Antibiotic_1$
  1.0 $Antibiotic_2$

These rules are interpreted as : « in front of DC_1 (a given bacterium), $Antibiotic_1$ is completely effective and $Antibiotic_2$ isn't » and « in front of DC_2 (another bacterium), situation is exactly the opposite ». If DC_1 and DC_2 are simultaneously true (both germs have been isolated), the decisional process concludes to a Final Vote : FV = { 0.0/$Antibiotic_1$, 0.0/$Antibiotic_2$ }. The two decisions are eliminated, whereas a prescription of both antibiotics may be a good therapy.

In front of this multi-decisional problem two solutions have been envisaged :
- the system tries to build dynamically a composition of decisions,
- or some associations may be directly anticipated in the conclusion of the decisional rules.

The problem of dynamic association of antibiotics is a very complex topic : some of these associations are very dangerous for human beings, others are prescribed considering their improved efficiency, ... Dynamic research of associations is a very costly operation. Moreover, it can only be processed at the end of the construction of FV set (if no decision is indicated).

So, in a first step we have decided to implement only « static associations » expressed by the expert in the rules.

Static associations are explored at the same time than other decisions. They correspond to specific situations in which the expert recognizes that no ordinary decision is sufficient.

We use an extended syntax for the explanation of these associations :

Vote XX
If
DC
Then
...

$$\mu_{ass(1,2)}\left(\delta_1,\delta_2\right)$$

where the membership value μ operates on the association $\delta_1 + \delta_2$.

The three kinds of vote are extended for the management of these situations. One example illustrates this process :

- given the above rule, if a second veto rule associates a membership value μ_2 to the same association, then result will naturally be equal to $\min(\mu,\mu_2)$;
- given the same rule, if a second veto rule associates a membership value μ_3 to the alone decision δ_1, meaning of veto sets a global result for association to $\min(\mu,\mu_3)$, even if the second rule operates just on one decision from the association.

5. AntibES : an Antibiotic Expert System

AntibES is our medical application of MENTA. It is mainly an utilization of the decision engine. Inference engine works on data to prepare the decisional consultation. A special inference rule base drives the complete software. Actual decision rule base is composed of about 400 rules.

A typical script of consultation is shown in Fig.3.

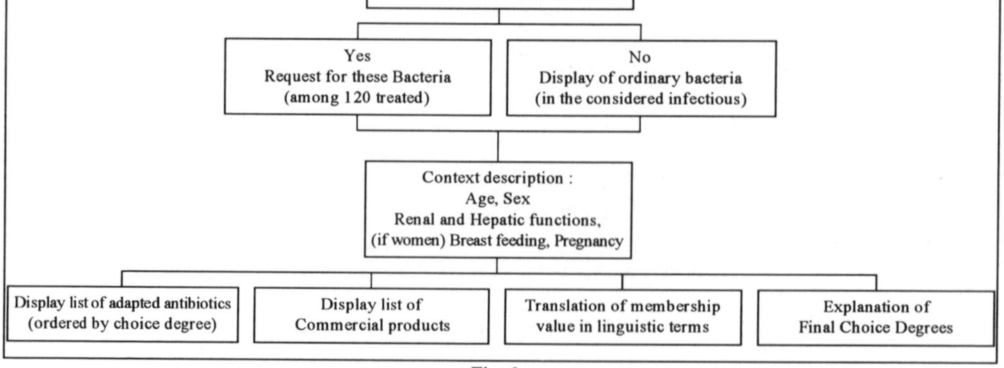

Fig. 3

The expert has first determinate the different diffusion sites of the antibiotics. For each of them, he wrote a specific decisional rule which associates the site and the potential antibiotics. The initial choice of infectious localization allows to select this first Decisional Criterion. Corresponding drugs are then imposed by dictatorial rules.

For each bacterium, the expert wrote fuzzy sets of adapted antibiotics (one rule for one infectious agent). Some specific rules have been added in front of particular germs or diseases. All of these rules are veto rule.

For each clinical or biological particularity, other veto rules are written for Final Vote limitation to useable drugs.

This expert system was presented to specialists during the Third Symposium about Informatic and Infectious Diseases (Nice - May 1992). More medical details are exposed in [6].

6. Conclusion

Fuzzy extension of Social Choice Theory has already be used in decisional system for industrial process management [7]. We have applied this approach to a medical problem, using a new rule syntax and a specific decisional engine.

We now investigate two main extensions : the multi-decisional approach (as discussed above), and the multi-criteria approach. The second one offers possibility to compute a more complex judgment for each decision. Our actual solution is based on calculus of only one value to appreciate a decision. This number have to summarize appreciations on different characteristics of a decision (antibiotics are judged on several features : efficiency of the action, poor side effects, cost, ...).
After the calculus of a degree of preference for each attribute of the decisions, we need to aggregate these results. By this way we join a more classical multi-criteria point of view for decision making. Some works about fuzzy approach for these aggregations ([8] for instance) give us theoretical directions for our problem.

REFERENCES

[1] K.J. ARROW, Social choice and individual values, (Wiley, 1963).
[2] D. DUBOIS and H. PRADE, Théorie des possibilités : Application à la représentation des connaissances en informatique, (Masson, Paris, 1988).
[3] D. DUBOIS and J.L. KONING, Social choice axioms for fuzzy set aggregation, Fuzzy Sets and Systems, 43(3), (1991) 257-274.
[4] G. PALMER and B. LE BLANC, Menta : un moteur décisionnel dans le cadre de la logique possibiliste, Le Flou 93, 3ème Journées Nationales sur les Applications des Ensembles Flous, Nîmes, 26-27 Octobre 1993 (EC2, Nanterre, 1993) 61-68.
[5] B. LE BLANC and G. PALMER, AntibES : an Expert System for antibiotic prescription based on fuzzy logic, Thirteen International Conference on Artificial Intelligence, Expert Systems and Natural Language, Avignon, 24-28 Mai 1993 (EC2, Nanterre, 1993) 123-132.
[6] P. MORLAT, Antibiotic Expert System (AntibES) : logiciel d'aide à la prescription d'antibiotiques établi en logique floue, Université de Bordeaux 2, Rapport de D.E.A. Biologie Santé, 1992.
[7] J.L. KONING, Un mécanisme de gestion des règles de décision antagonistes pour les systèmes à base de connaissances, Université de Toulouse III, Thèse de doctorat en Informatique, 1990.
[8] M. GRABISCH, Application des mesures et intégrales floues, Les applications des ensembles flous, Nîmes (EC2, Nanterre, 1992) 139-147.

FLEXIBLE RETRIEVAL OF SEMANTIC INFORMATION IN A MEDICAL IMAGES DATABASE

Marie-Christine JAULENT*, An YANG*, Patrice DEGOULET*

* Medical Informatics Department,
Broussais Hospital,
96 rue Didot,
75014 Paris, France

ABSTRACT

This paper addresses the problem of image content retrieval from imprecise queries in the context of a digital angiographies Database Management System. We present a flexible retrieval system that evaluates to what extent an image in the database corresponds to a query containing imprecise radiological criteria. The three main modules of the system are described: the image content representation module provides high level symbolic representations of the renal arteries depicted in the image, the query representation module is based on the fuzzy set theory and the retrieval process is described as a matching process between a query including imprecise criteria and symbolic image descriptions.

1 Introduction

An hospital of average size can produce several thousand images of different types per year. It can be conventional radiography, Scanner, IRM, etc. Evolving technology like PACS and MIDBMS (Medical Images Data Bases Management Systems) allows the storage and management of the produced images objects [1][2]. In order to be used as an efficient tool for consultation, training or medical research, a MIDBMS must support the retrieval of images based on radiological or clinical criteria. The main difficulty of this task is to access the content of the images, in particular the semantic information. This problem has generated a research area on image retrieval by content that deals both with the problem of the representation of semantics in images and the problem of the retrieval itself. In conventional medical image database, image retrieval by content is accomplished through indexing mechanisms using individual fields in the medical record of a patient or using information contained in the image header [3][4][5]. One main drawback of this approach is that the linguistic labels used as indexes can be subjective or not well-known.

In this paper, we are concerned with one main issue of content-based retrieval which is to yield cases of patients with similar diagnostic imaging examinations. The source of the target examination is an operator querying directly the system in natural language. An important goal of our work is to allow imprecise queries on image content. We present a flexible retrieval system that has been developed in the context of a digital angiographies database. Digital subtraction angiography can characterize both functional and morphological features of arterial or venous networks and are a valuable support for the physician to evaluate the severity of visible lesions. In the process of diagnostic imaging, the knowledge of similar cases may improve significantly the examination and support the best therapy prescription. In this context, our objective is to develop a system that evaluates to what extent an image in the database corresponds to a query containing imprecise radiological criteria.

The system is based on a fuzzy pattern matching procedure which takes image description and query description as input and maps them into an output space that evaluates the compatibility between them. The image representation problem and the query description are presented in the first part of the article. The second part of the article focuses on the matching procedure itself and our current results are presented.

2 Semantic retrieval in medical images databases

Image Content Retrieval can be defined as a matching problem between an image content representation and a representation of what is to be retrieved. This problem is related to two different research fields, the image and query description methods and the retrieval strategies. In the following, we describe first the image content representation that is attached to digital angiographies and second the queries that are considered in our application and their representation. In the next section, we will describe the retrieval procedure itself.

2.1.Image content representation

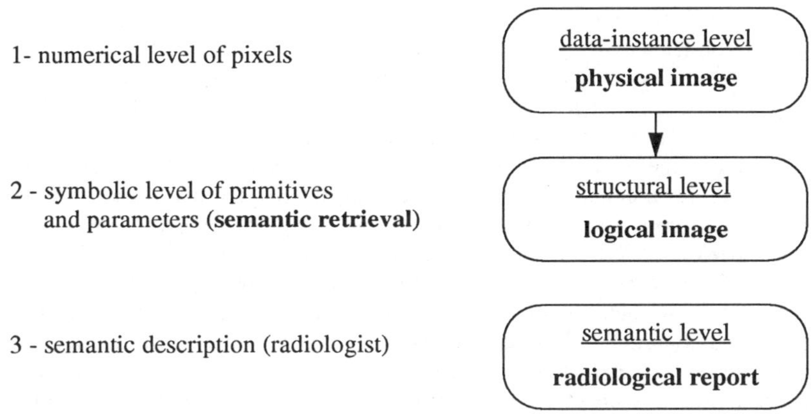

Figure 1 : Medical image object representation

A medical imaging department produces "objects". Each object includes at least the raw image and the text report attached to it. Some work aims at proposing generic models of medical image objects, which can be used to implement medical image databases [3]. They usually distinguish three main layers of an image object, the data-instance level, the structural level and the semantic level. These layers are organized into the same abstraction levels as within the classical image-understanding framework: the numerical level of pixels (the raw image), the symbolic level of primitives with their parameters and the highest level of semantic descriptions (the text report). The term "logical image" is usually used as a standard term for the symbolic information embedded at the structural level. Logical databases are then developed to manage logical images. The most elaborate methods of image content description in logical databases (that is at the structural level) incorporate mostly information from textual descriptions. For instance, the SENOBASE [6] proposes linguistics labels to describe logical images. Many recent researches focuses on the automatic and objective extraction of the logical image from the raw image. The main issue is to obtain at this level an information as meaningful as possible. For instance, in [7], a semi-automatic system to go from an input image to a semantic description of the image content is available. Other systems mix the structural and semantic level. For instance in [4] the images are classified into

S ***for stenosis***	degree = 45% length = 30% aspect = non-centered character= regular
D ***for aneurism***	aspect = non-centered size = 5%
N ***for normal segment***	length = 40%

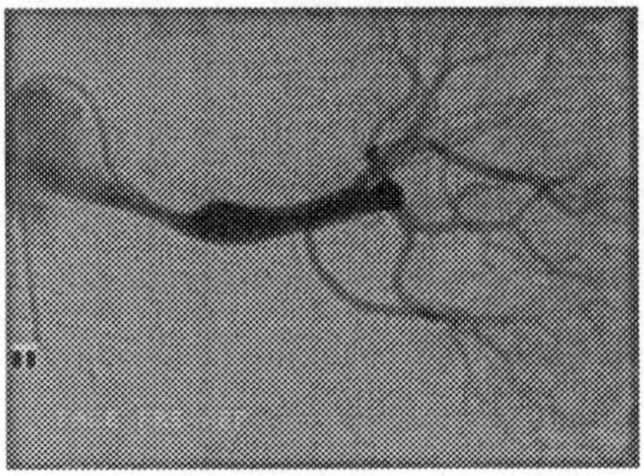

Figure 2 : The structural description **SDN** of a left renal artery

image classes based on information supplied by the user or obtained from the image itself. In the I^2C system [8], medical images are assigned to images classes (the logical representation) based on a set of user-defined attributes. To obtain automatically a meaningful logical representation of images is an important issue of image content representation in logical databases in particular to enable the system to insert images in bach mode without any user intervention.

In the particular domain of angiographic images, we adopt a multiscale representation scheme that convey the same idea of logical images (figure 1). The data-instance level contains the image itself, the structural level contains a symbolic (logical) description that is automatically extracted from the segmented image and the semantic level contains the radiological report. The image content retrieval is performed at the structural level. The database is composed of digital angiographic images that have been scannerised, numerised and segmented by automatic techniques.

In a previous work, we have developed a system that provides automatically a symbolic description of normal and abnormal aspects of the vascular structures present in the image [9]. The outline of the different vascular structures present in an image are syntactically analysed to extract a structural description in terms of the concepts of the domain [9]. The artery structural description is a list of segments (figure 2) representing the different normal and abnormal aspects of the artery. Each segment is quantified by the set of parameters that characterizes it. For instance, the degree, the length, the aspect and the character are the four parameters that characterize a segment **S** representing a stenosis [9]. Our contention is that this logical representation is meaningful enough to perform a good retrieval of similar cases.

2.2. Querying an image database

Intelligent image databases are perceived to provide high search capability. The indexing mechanism is based on a remarkable data structure that it is usually linked to the logical image. The indexes are defined at the level of the attributes [4] and content-based indexing is often equivalent to attribute-based indexing [8].

Our approach consists in using more flexible pattern recognition techniques than indexing mechanisms. Indeed, queries can be considered as lists of atomic requirements and used as templates for the data. The retrieval problem is then a recognition problem that can be solved by a pattern matching procedure where the query is the pattern or model. In classical recognition problems from visual data, regions or structures in the image are characterized by feature vectors that are related to the descriptive attributes in the model. The prerequisite is that the data and the model are organized into two identical structures. The matching is usually performed in three steps: low features (attributes), high features (concepts) and root features (the object).

2.3.Imprecise querying

Queries often refer to classes of image objects rather than a specific image of a given patient. For instance, a query can be of the type: "*Find all cases with centered stenosis on at least one renal artery non followed by a dilatation*". The classes are generally described by linguistic variables [7]. The motivation for considering imprecise querying, among others, arises from the fact that not all requirements in a query can be described linguistically with precision and from the fact that linguistic variables tend to be subjective. In [7], the query model is described in terms of propositions in the context of a magnetic resonance images application. The representation of the semantics of such propositions is in accordance with the calculus of "fuzzy restrictions" [7]. The most often, the linguistic variables can be translated and evaluated by means of some numerical descriptions based on the fuzzy sets approach.

Some retrieval systems uses fuzzy approaches to take into account the imprecision in queries. For instance, a fuzzy approach has been used in the domain of retrieval from a face photographs database [5]. In this system, the retrieval is based on mixed queries, linguistic and graphic. In the world of relational databases, some work has been done to develop a SQL-like language in order to allow a variety of imprecise queries [10]. In the context of medical images databases, fuzzy classification techniques have also been used to perform symbolic matching [7] [11]. The difference of our approach with these work is that our retrieval procedure is based on the well known fuzzy pattern matching procedure developed by Cayrol et al. [12]. The matching process uses membership degrees to evaluate the comparison at each level of the hierarchical representation of the query.

2.4.Query representation

In the domain of digital angiographies, we restrict our study in the hypertensive management context which means that we restrict the set of queries that can be written out. The radiological reports have been studied in order to identify the set of the relevant concepts that may appear in queries [13]. The main concepts corresponds to specific vascular lesions in the domain like *stenosis, aneurism, infiltration*, etc. Each concept can be characterized by a set of dimensions like for instance the *degree of a stenosis* or the *aspect of an aneurism.*

Two types of information intervene in a query: a topographical information that precises which part or structure of the image is under interest and some logical propositions that describes the different criteria to be satisfied by this structure. The query is encoded in a graph structure where the leaves are the criteria. A criterion is expressed by a couple (***Attribute Value***) where the ***Attribute*** refers to a specific dimension of a concept and the ***Value*** is the required value of the attribute. The criteria do not correspond only to precise numerical quantities like for instance "**a 80% degree stenosis**". Most often they are represented by linguistic and imprecise expressions like for instance, "**narrow stenosis**". The representation of a criterion is based on the fuzzy sets theory [14]. The figure 3 shows a representation in terms of trapezoid fuzzy sets for the criterion (*degree narrow*).

The universe of discourse associated to the attribute "degree" of the concept "stenosis" is the interval [0,100]. A stenosis with a degree 0 means that there is no lesion and a stenosis with a degree 100 means that there is an obstruction of the blood flow. The criterion is interpreted as a fuzzy set of [0,100] entirely defined by its membership function μ_{narrow}.

Besides the criteria used to described the required lesions, the query may include quantifiers (more, less, all, etc.) and logical operators (and, or, non) like for instance in the query: "*Find all centered and elongated stenosis on renal arteries non followed by a dilatation*". The graph representation integrates these quantifiers and logical operators (figure 4) and allow to formulate vari-

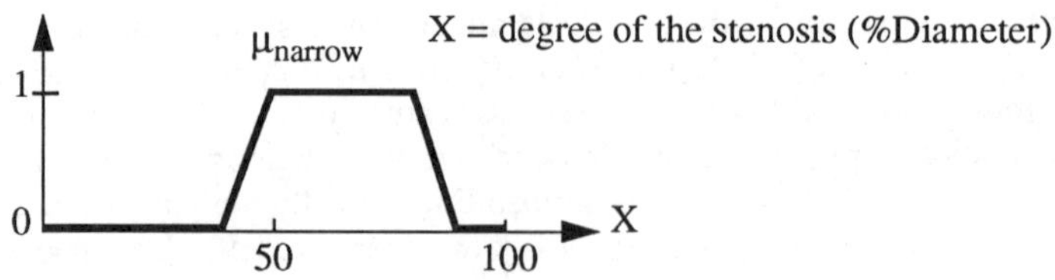

Figure 3 : Example of a fuzzy set for the criterion "narrow stenosis"

ous and sophisticated queries.

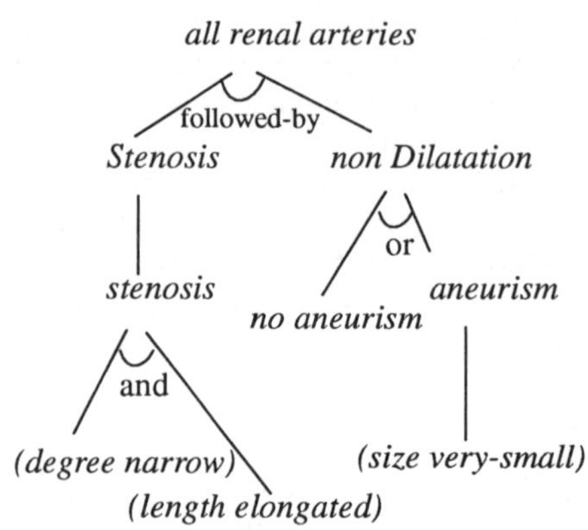

Figure 4 : Graph representation of a query

3 Description of the retrieval procedure

The fuzzy pattern matching has been introduced to take into account the imprecision attached to the pattern in the framework of the possibility theory [12]. Retrieval is then based on similarity rather than on an exact match. Our retrieval procedure is based on a fuzzy pattern matching procedure [2] which takes the image structural description (the datum) and the query description (the pattern) as inputs and provides as output an evaluation of the compatibility between them which is graded on the interval [0, 1]. This matching process involves quantifiable similarity that evaluates the compatibility between the two inputs. Both the structural description and the query description have an and/or tree-like representations whose leaves correspond to elementary components (respectively segment's parameters and criteria) called atoms. To each atom of the pattern is attached the membership function of a fuzzy set restricting the values which are more or less compatible with the meaning of the atom. The matching procedure is performed in three steps: first, an elementary compatibility is computed at the atomic level for each criteria, second an intermediate compatibility is computed at the concept level and third a global compatibility is established between the query and the structural description.

In the following, we will illustrate the different steps of the procedure by the example given in figure 2 and figure 4.

3.1.Elementary compatibility

The elementary compatibility is defined between a criteria in the query and the parameter's value in a labelled segment. To be a matching candidate, the parameter must correspond to the attribute of the criteria which means that the value of the parameter (D) and the value of the criteria (P) are both defined on the same universe of discourse.

On the one hand the criteria P is represented by a fuzzy set. On the other hand, the datum D may refers to a precise value (degree 45%) or to an imprecise value (character regular). When D is a precise value d, the elementary matching is successful as soon as $d \in P$. The degree of matching is then the grade of membership $\mu_P(d)$. More generally, when D is imprecise, this grade of membership becomes imprecise too. This notion of compatibility was first introduced by Zadeh and further used to develop compatibility models for semantic pattern matching procedures. In such models, the compatibility between P and D is expressed by two scalar measures: the **possibility** that the information provided by the datum is compatible with the information required by the pattern and the **necessity** that the information provided by the datum is compatible with the information required by the pattern.

By hypothesis, the pattern and the datum are represented by fuzzy sets those membership functions are μ_P and μ_D. The possibility measure $\Pi(P, D)$ and the necessity measure $N(P,D)$ are given by [12]:

$$\Pi(P, D) = \sup_{x \in X}(\min(\mu_P(x), \mu_D(x)))$$

$$N(P, D) = \inf_{x \in X}(\max(\mu_P(x), 1 - \mu_D(x)))$$

where X is the universe of discourse for the pattern and the datum. The measure $\Pi(P, D)$ estimates to what extent it is possible that P and D refer to the same value and $N(P,D)$ estimates to what extent it is certain that the value to which D refers is among the ones compatible with P. The duality possibility/necessity (i.e. the necessity of an event corresponds to the impossibility of the opposite event) is expressed by the relation: $N(P,D) = 1 - \Pi(\neg P, D)$ where $\mu_{\neg P} = 1 - \mu_P$ is the membership function of the complement of the fuzzy of values compatible with P.

Note that when D is precise (d) then $\Pi(P, d) = N(P,d) = \mu_P(d)$ which is then a regular number in [0, 1]. In this case, the trapezoid representation of the fuzzy set attached to the pattern allows the practical and simple calculus of the quantity $\mu_P(d)$. Indeed, the value of this quantity is given by the ordinate of the point I (figure 6) and expresses to what extent d is a possible value of P.

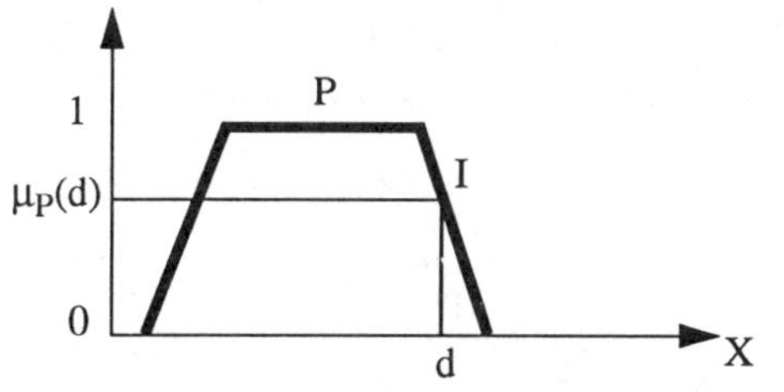

Figure 5 : Possibility calculus

In the example shown in figure 5, we assume that the corresponding concepts between the query and the structural description have been identified by a search procedure. For instance, for the concept "stenosis" in the query, the procedure searches for a corresponding segment **S** in the description. Two criteria are attached to the required concept stenosis: (degree narrow) and (length elongated). The values of the corresponding parameters in the segment **S** are precise values so that the two elementary compatibilities computed by the system are: $\mu_{narrow}(45)$ and $\mu_{elongated}(30)$.

3.2. Intermediate and global compatibility

The atomic measures of possibility and necessity are aggregated at the intermediate and global levels in order to obtain two global measures between the whole query and the whole structural description. There are various ways to perform this aggregation operation. For instance, a dempster-shafer like rule of combination is used in [5] and a fuzzy logic approach is used in [15]. In the fuzzy pattern matching procedure described in [12], the aggregation is performed using the "min" operation when the pattern expresses a conjunction of elementary requirements. In this case, the global possibility and necessity measures are given by:

$$\Pi(P_1 \times \ldots \times P_n, D_1 \times \ldots \times D_n) = \min_{i=1,n} \Pi(P_i, D_i)$$
$$N(P_1 \times \ldots \times P_n, D_1 \times \ldots \times D_n) = \min_{i=1,n} N(P_i, D_i)$$

where P_i and D_i are supposed to be defined on the same universe of discourse and where **x** denotes the Cartesian product defined for two fuzzy sets A and B by

$$\forall\, x \in X_A \text{ and } \forall\, y \in X_B,\ \mu_{AxB}(x,y) = \min(\mu_A(x), \mu_B(y))$$

The attributes in the pattern and the datum are supposed to be non-interactive. Moreover the aggregation depends on the salience of the criteria in the pattern. Many approaches have been studied to weight the criteria for importance [16].

In the current version of our system, we use the same "min" aggregation operation at the intermediate and global level and our contention is that the criteria are of equal importance. We assume that the structural constraints such as *followed by, before, after* have been validated in the sense that only the structural descriptions that verify (yes or no verification) these constraints are matched with the query. Finally, we use the 'max' operation for the aggregation at the level of an 'or' node in the query structure.

At the concept level, the compatibility between the segment **S** and the concept "**narrow and elongated stenosis**" is given by

$$\mu_{narrow\ and\ elongated\ stenosis}(S) = \min(\mu_{narrow}(45), \mu_{elongated}(30))$$

For the concept "**non aneurism**", the procedure will return 1 if there is no **D** segment in the structural description and 0 else. The compatibility between the segment **D** (if any) and the concept "**very-small aneurism**" is given by $\mu_{very\text{-}small\ aneurism}(\mathbf{D})$

At the root level, the compatibility between the query and the entire structural description is given by

$$\begin{aligned}\mu_{query}(\mathbf{SDN}) &= \min(\mu_{narrow\ and\ elongated\ stenosis}(\mathbf{S}), \mu_{\neg\ dilatation}(\mathbf{D}))\\ &= \min(\mu_{narrow\ and\ elongated\ stenosis}(\mathbf{S}), \max(\mu_{\neg\ aneurism}(\mathbf{SDN}), \mu_{very\text{-}small\ aneurism}(\mathbf{D})))\\ &= \min(0.50, \max(0, 0.95))\\ &= 0.50\end{aligned}$$

4 Results

Nowadays, the image database is composed of thirty patients images and we tested the system with fifteen different queries. The user is guided to formulate the query to the system using a textual interface to enter a set of criteria linked by structural constraints (followed by, before, after), logical operators (and, or, not) and quantifiers (all, at least,...). Given a query, the system searches the whole database. The result is then a ranking of the compatible images.The fuzzy sets approach allows the user to formulate imprecise queries that can be further refined according to the set of similar cases obtained.

Below, a very simple example of the session of the system is shown. Because of fitting problems, we restrict the database to three images.

CaseN° 3

```
ard
N 2
S ostial 0,5 centered regular 0,05
D distial centered 0,2
N 4
```

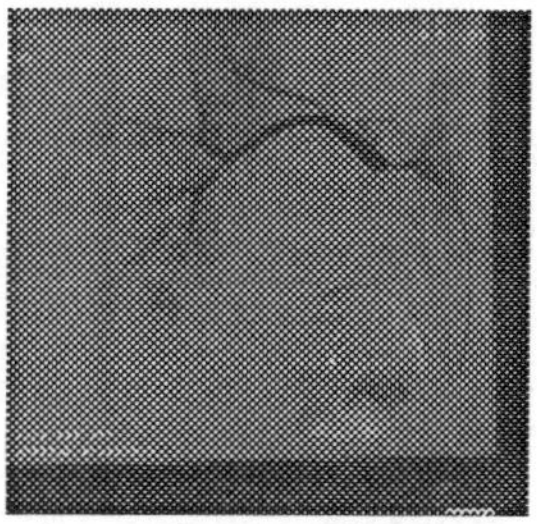

CaseN° 4

```
arg
N 2
S ostial 0,45 centered regular 0,1
D distial centered 0,2
N 4
```

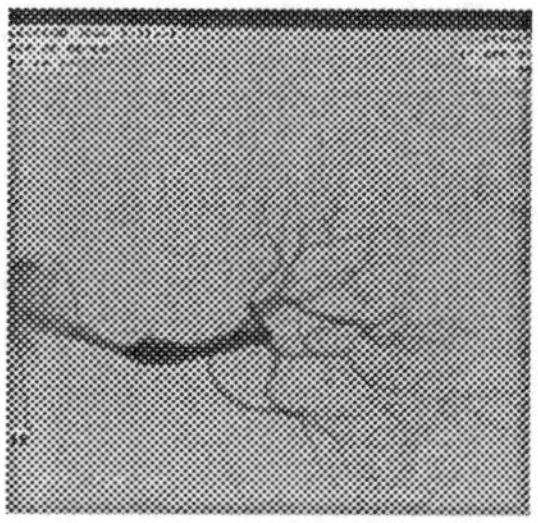

Case N° 7

```
ard
N 2
S ostial 0,7 not-centered regular 0,1
N 6
```

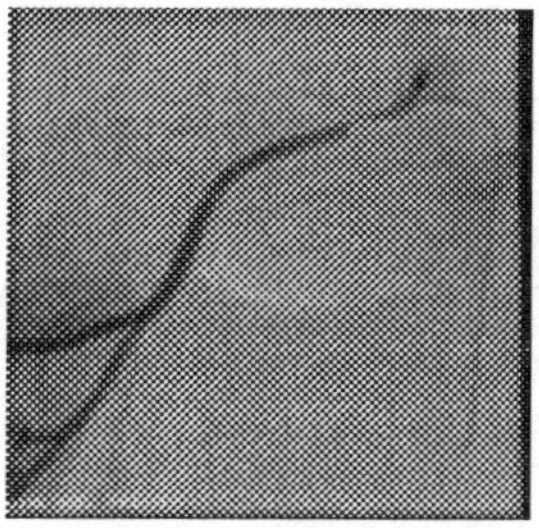

The considered query is the one provided in (figure 4).

The following sequence shows the results of the retrieval procedure on the restricted base. In couple (a b), a refers to the possibility measure and b refers to the necessity measure.

```
**************************RESULTS**************************************

compatibility between the query and case 3:
- compatibility with (centered and narrow stenosis): (1 1)
- compatibility with (non dilatation): (0.5 0.5)
- global compatibility (followed_by): (0.5 0.5)

compatibility between the query and case 4:
- compatibility with (centered and narrow stenosis): (0.8 0.8)
- compatibility with (non dilatation): (0 0)
- global compatibility (followed_by): (0 0)

compatibility between the query and case 7:
- compatibility with (centered and narrow stenosis): (0 0)
- compatibility with (non dilatation): (1 1)
- global compatibility (followed_by): (0 0)

*****************************************************************************
```

5 Conclusion

This paper presents an application of the fuzzy pattern matching procedure developed in [12] to the flexible retrieval of semantic information in a digital angiographies database. The purpose of the system is to retrieve from a natural description of a case, the ones in the images base that are similar. We adopt a three levels representation of the image, the physical image, the logical image and the radiological report. The logical image provides a structural and quantified description of the image. It is automatically extracted from the image. The interrogation is performed on the logical image. Queries are considered as lists of imprecise requirements and embedded in a tree structure. The result of the matching procedure provides a measure of the adequation of a given structural description to the query. The measure is expressed in terms of the possibility and necessity that the image object is a good answer to the query.

The primary results show that the fuzzy pattern matching procedure is well-adapted to the direct querying of the system in natural language. Moreover, the structural description what we extracted from the image seems to be meaningful enough to yield the right similar cases from the base. However, this last point has still to be validated on Large Image Database.

Many directions are foreseen to improve the presented work. The graph representation of the query should be generalized to conceptual graph representation. Indeed, one of our objective is to use the same query representation to retrieve similar cases from logical images or from radiological reports. Moreover, the same approach (the fuzzy pattern matching procedure) was used for the classification of renal artery lesions [13] and showed that the same retrieval procedure can be used to get similar cases from linguistic queries but also from other logical images.

REFERENCES

[1] Chang SK, Hsu A. Image Information Systems: where do we go from here ? *IEEE Transactions on Knowledge and Data Engineering*, 1992; 4(5):431-42.

[2] Garfagni H, Gibaud B, Stut W, Aubry F, Bizais Y. MIMOSA: A functional model of the management of medical images. *Medical Informatics*, 1994; 19(2):95-108.

[3] Aubry F., Chameroy V. Lavaire F. Ramond J.P. Saidane I.E., Giron A., Bizais Y. Todd-Pokropek A., Di Paola R. Medical Image Management Using a Semantic Approach: Image Description. In : *Proceedings of the International Symposium CAR'93. Computer Assisted Radiology.* H.U. Lemke, K.Inamura, C.C. Jaffe and R Felix (eds). Berlin: Springer-Verlag 1993; pp265- 271.

[4] Kostomanolakis S. Lourakis M. Chronaki C. Kavaklis Y. Orphanoudakis S.C. The Architecture of a System for the Indexing of Images by Content. In : *Proceedings of the International Symposium CAR'93. Computer Assisted Radiology.* H.U. Lemke, K.Inamura, C.C. Jaffe and R Felix (eds). Berlin : Springer-Verlag. 1993; pp279- 282.

[5] Iwamoto H. and Ralescu A. Towards a multimedia model-based image retrieval system using fuzzy logic. In : *SPIE Vol. 1827 Model-Based Vision.* H.N. Nasr & R.M. Larson (eds). Washington : SPIE. 1992; pp177-185.

[6] Magnin I.E., Vray D., Brémond. Early detection of breast cancer using computer assisted diagnosis. In: *Proc. 14th Annual Int Conf of the IEEE EMBS*, Piscataway, NJ: IEEE 1992; pp 849-850

[7] Dellepiane S., Venturi G. and Vernazza G. Model generation and model matching of real images by a fuzzy approach. *Pattern Recognition*, 1992; 25(2):115-137.

[8] Orphanoudakis SC, Chronaki C and Kostomanolakis S. I^2C: A system for the indexing, storage and retrieval of medical images by content. *Medical Informatics* 1994; 19(2):109-122.

[9] Jaulent MC, Paul JF, Boittin P, Degoulet P. Description and representation of segmented renal arteries from angiograms. In: *Proceedings of the 17th SCAMC 93.* New York: McGraw-Hill. 1993; 818–23.

[10] Pivert O. Expression et evaluation de requêtes floues dans les bases de données. In: *Les applications des ensembles flous.* Nanterre : EC2 Editions. 1992; pp285-292.

[11] Eisele T.P., Klein H-M, Ameling W., Günther R.W. : Classification of Medical Image Objects based on Fuzzy Sets. In : *Proceedings of the International Symposium CAR'93. Computer Assisted Radiology.* H.U. Lemke, K.Inamura, C.C. Jaffe and R Felix (eds). Berlin : Springer-Verlag. 1993; pp635-640.

[12] Cayrol M., Farreny H., Prade H. Fuzzy pattern matching. Kybernetes 1982; 11: 137-155.

[13] Jaulent MC., Degoulet P., A fuzzy model for diagnosing renal artery lesions. *IEEE Engineering in Medecine and Biololy Magazine*, 1994; 13(5):699-704.

[14] Zadeh L.A. Fuzzy Sets. *Information and control*, 1965; 8:338-353.

[15] Miyajima K., Ralescu A. Fuzzy logic approch to model-based image analysis. In : *SPIE Vol. 1827 Model-Based Vision.* H.N. Nasr & R.M. Larson (eds). Washington : SPIE. 1992; pp145- 156.

[16] Dubois D., Prade H. *Théorie des possibilités: applications à la représentation des connaissances en informatique.* Paris : Masson. 1985.

THE APPLICATION OF FUZZY METHODS TO IMAGES SEGMENTATION

Franco FONTANA, Paul C. SMITS and Silvana DELLEPIANE

Department of Biophysical and Electronic Engineering
Via all'Opera Pia 11a - I-16145 Genova ITALY

ABSTRACT

This paper describes a method that exploits fuzzy-set concepts to model successfully ambiguities and uncertainties inherent in assigning an image pixel to an object of interest for the purpose of image segmentation. The method relies on the choice of a set of seed points (one for each object of interest) and exhibits the particular novelty of being entirely independent of input parameters. Some results on medical images are reported, which illustrate the method's validity.

1. Introduction

The processing of real images can be regarded as an information transformation (and reduction) process to improve the accuracies of manual and automatic interpretations. Objects or features representing physical phenomena in the real world have to be extracted from an image under test.

Image segmentation is an important tool for reducing the information in image processing. The objective of image segmentation is to separate significant from insignificant information in order to achieve the data reduction needed to improve the interpretation process.

It is well known that in image segmentation the task of outlining *objectively* the boundary of an object is very difficult. The application of different techniques to the same image very seldom gives the same result, as there is no single way of accounting for inaccuracies or noise during processing. The use of fuzzy-set theory seems very appropriate for coping with ambiguities and uncertainties in the image model. Particularly in medical applications, objects are very often ill-defined, due to the physical properties of imaging systems and to small differences in energy absorption by different types of tissues. Therefore, the exact locations of contour points are almost impossible to detect.

The earliest studies of fuzzy sets are due to Zadeh (1965) [1], who first developed a suitable framework for handling uncertainties. His work was extended by Rosenfeld to image processing and pattern-recognition systems [2]. Basically, an image can be viewed as a fuzzy subset, where each value is normalized to the range [0,1]. The value of each image pixel in the set is regarded as a grade for deciding whether or not the pixel belongs to the set. Accordingly, various geometric properties have been generalized for fuzzy subsets (e.g., compactness, surroundness and connectedness, etc.) [3]. In particular, it has been shown that image ambiguities may be quantitatively measured in terms of indexes of fuzziness and non-fuzziness [4] and entropy [5]. These concepts served also as useful tools to detect thresholds for various bimodal and multimodal distributions [6,7]. In medical-image processing, fuzzy sets for contour detection have also been used in [8, 9].

However, in the literature, very few work deals with complex image-segmentation problems. Despite the usefulness of the proposed approaches for evaluation of region properties or for description of image ambiguities, they are not suitable for a straightforward application to the segmentation process.

In this paper, we present a method that makes use of fuzzy tools to obtain a fuzzy segmentation. In particular, each object of interest in an image may be assigned a set of contours, each marked with a reliability value. In an interactive session, the user is asked for selecting from the set the contour that best meets his requirements. This method is being evaluated within the framework of the COVIRA [12] project. Suitable applications are, for example, in the areas of radiotherapy planning and neuroradiological diagnosis, as an accurate delineation of targets in an image is basic to clinical intervention or to diagnosis. This task is usually accomplished via manual interaction.

Fuzzy tools deal mainly with topological and multivalued aspects related to the two basic processes of the proposed method: image scanning and pixel labelling.

Starting from a seed point located inside an object of interest, an image-scanning mechanism applied to the area surrounding around the seed area, can locate potential candidates for aggregation. Then, a novel method for pixel labelling allows us to obtain a multivalued result which is particularly appropriate for outlining complex structures in medical imaging.

The paper is organized as follows. The next section addresses the problem of multivalued segmentation, as compared with traditional region segmentation. In Section 3 the proposed method is detailed. A number of tests on medical images, both CT and MR modalities, are described in Section 4. In Section 5, some final remarks are made and conclusions are drawn.

2. The idea of multivalued segmentation

Let $\mathbf{Z}=\{\zeta(x,y)\}$ (defined on the integer lattice $L=\{(x,y):1<x,y<D\}$) represent the random field of the grey levels of an original image.

Traditional region segmentation algorithms generate a label field $\Lambda=\{\lambda(x,y)\}=\{\lambda(p)\}$, where a scalar label is associated with each pixel p with coordinates (x,y). The scalar label λ may assume values ranging from 1 to the number of regions NR. The field Λ is therefore a new image with NR levels, and is referred to as the segmented or labelled version of the original image.

This result allows one to detect a region, Reg_a, as a set of connected pixels characterized by the same label a:

$$Reg_a=\{p|\lambda(p)=a\}. \tag{1}$$

Then a region consists of a crisp set of points.

On the contrary, the proposed method performs a multivalued (or fuzzy) labelling, i.e., it associates a value ranging from 0 to 1 with each pixel. Such a value denotes the degree of membership in the region Γ^a identified by a reference point a. As a result, a fuzzy field $\boldsymbol{M}^a=\{\mu^a(x,y\}=\{\mu^a(p)\}$ is obtained, which represents the labelled image.

Therefore, the region Γ^a associated with the seed a is a fuzzy set defined on the set of $\forall p\in L$ for which:

$$\Gamma^a(p)=\{\mu^a(p)\}. \tag{2}$$

This set is made up of the seed point (characterized by a label value equal to 1) and of surrounding pixels with constant or decreasing membership values.

On the basis of this result, a segmented image can be interpreted as a topological map from which one can derive a set of object areas or contours. The highest membership values correspond to points that are most likely to belong to the searched object.

A multivalued segmentation result can be achieved for each seed (hence for each object).

Every point in Γ^a characterized by a membership value (in the field $\boldsymbol{M}^a$) higher than a fixed value α is defined as belonging to the "isoregion at level α of Γ^a". The related boundary is named "isocontour at level α of Γ^a".

In accordance with the theory of fuzzy sets [1], the isoregion at level α is then the α-level set (or α-cut) $IsoR_\alpha$:

$$IsoR_\alpha=\{p|\mu^a\geq\alpha\}. \tag{3}$$

Isoregions and isocontours will be detected after the labelling process. Through such definitions, it is possible to obtain various contours of the same object, not by changing the parameters used during processing but by deriving a graded result from only one processing session.

3. Method description

The segmentation problem is translated to a hypothesis-testing problem that includes the computation of the likelihood values and the test process. The former step corresponds to the actual labelling process, while the latter corresponds to isoregion selection, whose threshold can be adaptively selected. Applied to a single seed point, the likelihood value may be also interpreted as a membership value.

The detection is based on the observation of a first signal sample, namely, a seed point a, which will be used as a reference for the next processing steps. The process implies the assumption that the signal is characterized by some degrees of homogeneity and connectedness.

As in the classical hypothesis-testing problem, we define the null hypothesis H_0 and the alternative hypothesis H_1 as:

H_0 *: the current sample does not belong to the object identified by the seed point;*
H_1 *: the current sample belongs to the object identified by the seed point.*

The problem of labelling is thus associated to the computation of the conditional density function, $f(\sigma|H_1)$, where σ expresses the *state* of the system, that is, the collection of already taken decisions and observations.

Basically, our method relies on two main processes: the image scanning and the actual pixel labelling. With the former process, at each step, a set of candidate pixels is selected on the basis of some properties. The pixel labelling process lets the candidate to take on a label value by expressing explicitly $f(\sigma|H_1)$.

3. 1. Image scanning

The proposed image scanning is a non-iterative mechanism which can be compared to a tree search for best paths within the image, starting from a seed point (the root of the tree, in fact). The strategy of tree expansion (or growing), that is, the decision of which nodes are expanded first, is regulated by the label progressively computed at each leaf of the tree. At each step, only nodes characterized by highest label values are expanded, whereas nodes with lower label values are temporarily suspended. As processing goes on, all nodes are considered for expansions, even though the instant in which they become *active* (i.e., expanded) depends on the label values.

The advantages of such a mechanism are manifold: first, it is very fast since it is not iterative and does not suffer of any convergence or computation time problems. Secondly, it allows the propagation of both global and local contexts, which is usually a feature of iterative methods. Finally, a selective mechanism allows one to privilege certain directions among others, where most likely the object itself expands. For a formal and more complete description of the method, the reader is referred to [10, 11].

The scanning mechanism allows us to select at each step of the process a set of candidate pixels, which have to be labelled.

3.2. Pixel labelling

The value of the label is computed by making explicit $f(\sigma|H_1)$. As demonstrated in [11], if we consider the knowledge of the seed point as a conditioning event for $f(\sigma|H_1)$, we may express the label of each node in the tree as a nonlinear function of the label inherited by the father. That is, each site i takes on a label computed as:

$$f(\sigma / H_1) = \lambda_i = \lambda_{i-1} \cdot g(\Delta c). \tag{4}$$

where the function $g(\Delta c)$ is expressed by the parameter of *intensity connectedness*, introduced in [11], that properly takes into account intensity levels and region topology.

To understand the behaviour of $g(\Delta c)$, we shall impose some constraints. For example, if a region is perfectly homogeneous, then initial seed label should not be degraded.

The behaviour of this function is such that small differences in connectedness between two neighbouring pixels indicate a good probability for two pixels to belong to the same object. On the contrary, large differences indicate that two pixels do not belong to the same object.

The labelling function has therefore a decreasing behaviour with respect to growing differences in the degree of connectedness. However, no threshold is pre-set, but continuous variations from 1 to 0 are used to grade properly the various degrees of difference in connectedness.

In such a framework, the function g plays the role of a fuzzy restriction function, or, similarly, it may also be seen as a probability density function, as it describes how differences in connectedness are distributed with respect to two neighbouring points, under hypothesis H_1.

If no noise is assumed and the searched object is perfectly homogeneous and connected, it is reasonable to assume $g(\Delta c)$ as a delta function, centered in the origin. More realistically, we may assume that noise and uncertainties in the hypotheses cause the delta function to become a Gaussian function. Alternatively, based on the same assumption, we may model $g(\Delta c)$ as a sigmoid curve or as an exponential curve.

In the following experimentation, we adopted an exponential model, as $g(\Delta c) = exp(-\Delta c / \varepsilon)$. The parameter ε was estimated on the basis of some quantitative analyses and set to 75 for all the experiments.

4. Application to medical images

In this paper we will focus on the application of our method to medical images, both CT and MR. The result of the labelling is a map which should be interpreted according to a given α-level which, in turn, corresponds to a crisp region in the fuzzy set. The computation of the output map does not require any input parameter, nor it is sensitive to the particular position of the seed point inside the object of interest. It should be noticed that the labelling process is completely automatic, except for the final phase of contour selection.

When implemented with a user interface, as done within the COVIRA [12] project, the user has the possibility to scroll very quickly throughout the various level and to decide the best result according to a specific application.

Figure 1 shows a MR image where interesting structures are the grey and the white matters and the ventricles. A seed point was located inside the grey matter, as indicated in Figure 1 a. Three different α-levels (0.55, 0.60 and 0.70) are displayed in Figs. 1b-d, which denote the region-growth when the threshold becomes lower. It is interesting to note that different levels of resolution for the same structure may be obtained with just one processing session.

A CT image of the pelvic region is shown in Figure 2. In this image, various seed points were located to identify the different objects of interest in the image (the bladder, the bones, and the rectum). The α-level for each object were individually selected to extract the best contour for each object.

Finally, Figure 3 shows a section of an MR image; two seed points were located as indicated in the figure. The result demonstrates the accuracy in object delineation.

5. Conclusion

We presented a fuzzy method to the segmentation of complex structures applied to medical images. The method obtains a set of fuzzy contours as output as the consequence of the proposed image labelling. The results may be interpreted as alternative contours of the same object and, if the method is used interactively, the user has the possibility to scroll very quickly through the different levels to select the most appropriate for a specific application according to his own knowledge. Results have been presented on a variety of images, that demonstrate the validity of the approach.

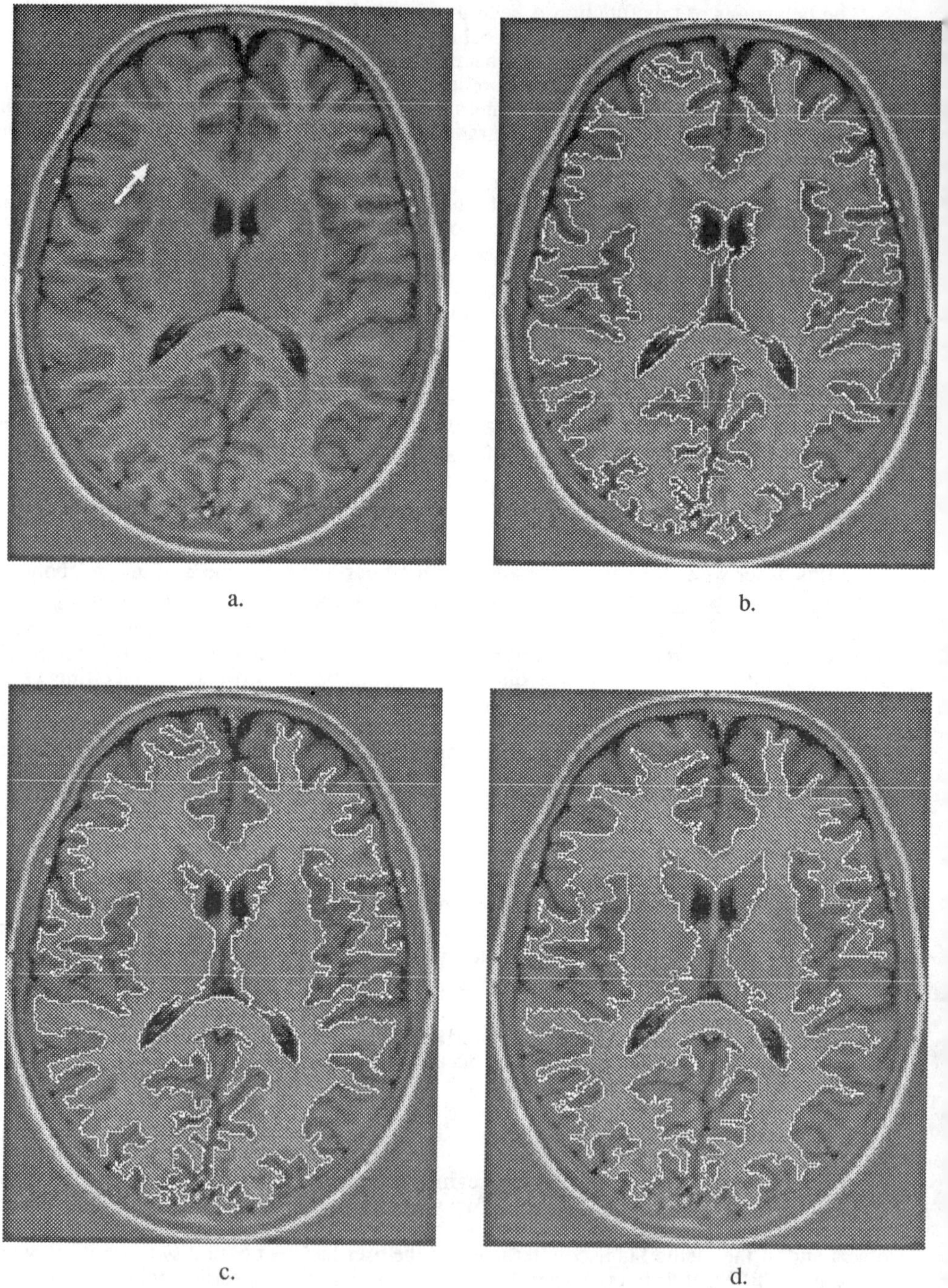

Figure 1 - Multilevel segmentation on a MR image. a) Original image; the seed point is indicated by the arrow. b-d) Isocontour at a-level 0.55, 0.60 and 0.70, respectively.

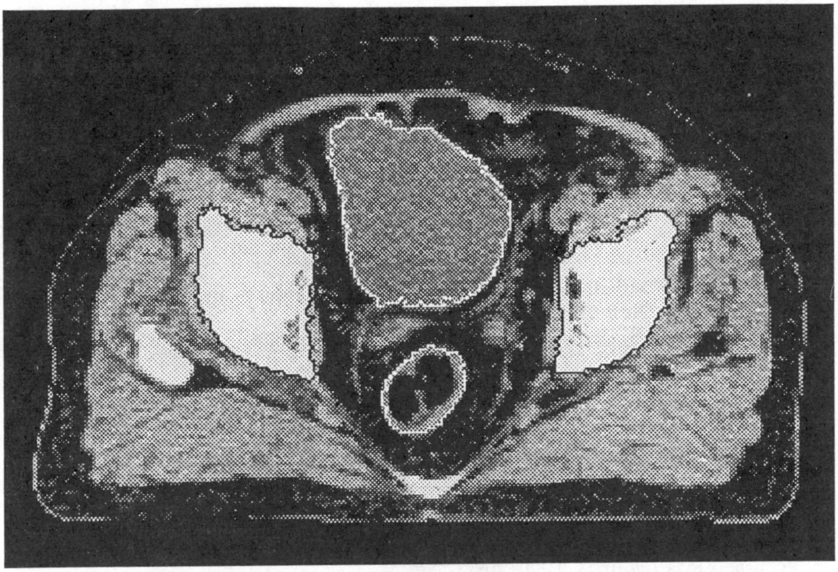

Figure 2 - Outlined bones, bladder and rectum of a CT image.

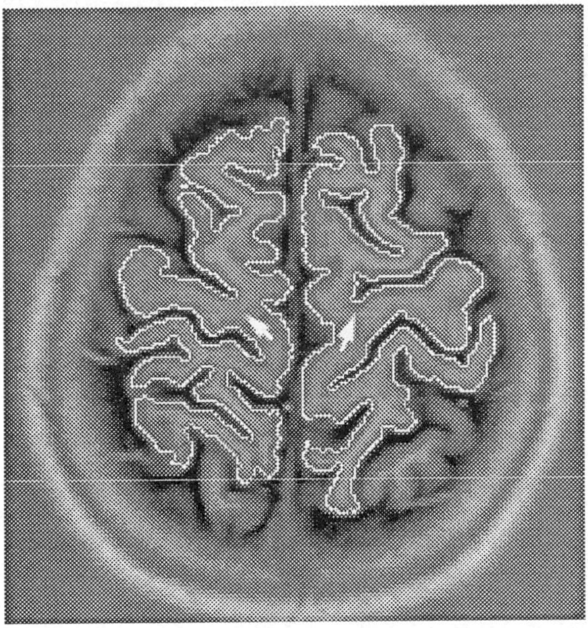

Figure 3 - Isocontours extracted at a-level 0.65 (left structure) and 0.55 (right structure). The arrows point to the posistion of the seedpoints.

ACKNOWLEDGEMENTS

This work was developed within the framework of the AIM (Advanced Informatics in Medicine) project COVIRA [12] (AIM2003). The images shown are part of the COVIRA reference data-set.

REFERENCES

[1] L.A. Zadeh, "Fuzzy sets", Information Control, Vol. 8, 1965, pp.338-353.
[2] A. Rosenfeld, "The fuzzy geometry of image subset", Pattern Recognition Letters, Vol. 2, 1984, pp.311-317.
[3] A. Rosenfeld, "On connectivity properties of grayscale pictures", Pattern Recognition, vol. 16, No.1, 1983, pp.47-50.
[4] A. Kauffman, "Introduction to the theory of fuzzy subsets - Fundamental theoretical elements", Vol. 1, Academic Press, New York, 1975.
[5] A. De Luca and S. Termini, "A definition of non probabilistic entropy in the setting of fuzzy set theory", Information Control 20, 1972, pp. 301-312.
[6] S.K. Pal, R.A. King and A.A. Hashim, "Automatic grey level thresholding through index of fuzzyness and entropy", Pattern Recognition Letters 1, 1983, pp.141-146.
[7] C.A. Murthy and S.K. Pal, "Fuzzy thresholding: mathematical framework, bound functions and weighted moving average technique", Pattern Recognition Letters 11, 1990, pp.197-206.
[8] S.K. Pal, R.A. King, "On edge detection of X-Ray Images using fuzzy sets", IEEE Trans. on Pattern Analysis and Machine Intelligence, vol. PAMI-5, No. 1, 1983.
[9] W. Menhard and M. Imme, "Contour detection using fuzzy sets", in Proceedings of Signal Progessing IV: Theory and Applications, J.L. Laucome, A.Chehikian, N. Martin and J. Malbos (Eds.), Elsevier Science Publishers B.V. (North Holland), 1988.
[10] F. Fontana and S. Dellepiane: "Context-Dependent Region Growing by a Label-Driven Method", Proceedings of 8th SCIA, May 25-28, 1993, Tromso, Norway, pp.1253-1260.
[11] Dellepiane, S., Fantana, F., Extraction of intensity connectedness for image processing, *Pattern Recognition Letters*, February 1995.
[12] COVIRA - Project A2003 of the AIM programme of the European Community, Consortium Partners: - Philips Medical Systems, Best (NL) and Madrid (E); Corporate Research, Hamburg (D) (prime contractor) - Siemens AG, Erlangen (D) and Munich (D) - IBM UK Scientific Centre, Winchester (UK) - Gregorio Maranon General Hospital, Madrid (E) - University of Tuebingen, Neuroradiology and Theoretical Astrophysics (D) - German Cancer Research Centre, Heidelberg (D) - University of Leuven, Neurosurgery, Radiology and Electrical Engineering (B) - University of Utrecht, Neurosurgery and Computer Vision (NL) - Royal Marsden Hospital/Institute of Cancer Research, Sutton (UK) - National Hospital for Neurology and Neurosurgery, London (UK) - Foundation of Research and Technology, Crete (GR) - University of Sheffield (UK) - University of Genoa (I) - University of Aachen (D) - University of Hamburg (D) - Federal Institute of Technology, Zurich (CH)

PREDICTIVE FUZZY MODEL OF GLYCAEMIC VARIATIONS

Pierre Yves Glorennec, Hubert Pircher
Département d'Informatique
INSA de Rennes
(glorenne,pircher)@irisa.fr

Jean Pierre Hespel
Unité Marcel Simon
CHR de Rennes

Abstract

Human glycaemia regulation is a complex phenomenon in which factors of different nature occur, for instance: glucose level and its variation, physical activity, morphology, time and composition of the meals, etc. In case of pancreas deficiency, diabetics have to control their glucose level thanks to injections of insulin, using empirical rules to determine the infusion rate. We propose a predictive fuzzy model of glycaemic variations, as a part of an automatic control system for diabetics. An associated learning procedure makes it possible to deal with interpersonal variations. This work about the treatment of diabetes is achieved in cooperation with the University Hospital Center of Rennes.

keywords: Fuzzy logic, predictive model, learning, insulin regulation, glycemia.

1 Introduction

In the human body, the glucose level depends on the insulin secretion rate, physical activity, composition and time of meals, mental stress, and so on. Among the diabetics, the insulin deficiency of the pancreas is compensated with daily injections, by syringe or pump. Doctors have built empirical rules to determine the insulin injection rate: there are qualitative rules expressing only tendencies (increasing or decreasing the previous rate). Moreover, these rules are not easy to use in actual conditions, because the human brain cannot manage the simultaneous variations of more than three significant variables.

Up to now, our study is made up of two parts:

- firstly, the expression of the knowledge of diabetes specialists thanks to fuzzy rules,
- secondly, the way of dealing with interpersonal variations and adapting the rule base to each patient.

For this purpose, we have at our disposal several diabetic-designed notebooks, from different patients, with incomplete and imprecise data. From these data, a fuzzy model is tuned for each patient, using a constrained learning algorithm, combining gradient descent and random walk. The constraints are used in order to preserve the semantics of fuzzy rules during the learning process.

Our paper is organized as follows. In section 2, we present the problem of glucose regulation in case of pancreas deficiency. In section 3, we show how fuzzy logic is a natural framework. In section 4, we present our model, whereas the learning procedure is developed in section 5. Lastly, we conclude with the work in progress.

2 Glucose regulation

In the human body, the pancreas controls the blood glucose rate through several hormones, the main one being insulin. Biological deficiency of insulin is liable for chronic hyperglycaemia which identifies diabetes mellitus. The diabetic's treatment consists in daily injections of insulin, in order to fill in the pancreas. But it is very difficult to mimic this organ and to determine the right insulin rate. Indeed, in human body, the secretion rate is very changing and complex, under the dependance of several factors. Insulin acts with delay and its efficiency decreases if the glucose concentration is high or is rising fastly. For an optimal regulation, an automatic control system should act in anticipation, before the glucose level rises, as for endogenetic insulin secretion.

For a long time, diabetes specialists have tried to educate their patients to the self adjustment of the injection rate. At the University Hospital of Rennes, a methodology is proposed to the diabetics. This is the so-called P.R.I.X. methodology[1] which takes into account the following elements so as to set the injection rate:

1. **P**rospective: what are the future events (about one or two hours) with foreseeable consequences on the glucose level ?
   - physical activity, which can be low (watching TV), medium (walking) or high (biking),
   - meal time: long before, near, just after or long after,
   - composition of the meal: glucidic, balanced, rich in fibers.
2. **R**etrospective: an analysis of the effects of insulin injected the day before makes it possible to modulate the present rate.
3. **I**mmediate. What is the present level of glycaemia ?
4. **X**, the mathematical unknown, to deal with personal data like mental stress (catching the next plane, talking to the boss...).

Diabetics learn to adapt their insulin rate themselves. They have a supervision notebook where they write down the times of blood samples and the level of glucose, and all the elements for PRIX methodology. Therefore they can adjust their treatment themselves and assess their evolution with their doctor.

Diabetics with an insulin pump have to set two parameters: the basic rate, $Ba(t)$ (about 50% of the daily insulin need) and a flash injection before meals, $Bo(t)$, the bolus rate, whose role is to compensate the glucidic supplies of meals. To set these parameters[2], the diabetics must evaluate too many variables and, despite PRIX methodology, regulate poorly their glycaemic level, see figure 1. A right level would be one gr/l, but, as they are afraid of hypoglycaemia, they prefer to underestimate the injection rate. Therefore, the mean glucose level is generally too high.

[1] "Prix" is the French word for "price": a good glucose regulation has quite a price...

[2] The basic rate is usually set once a day, whereas the bolus rate is set before each meal.

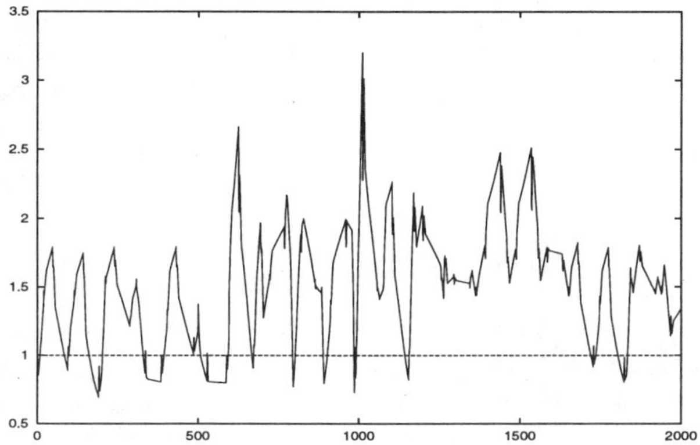

Figure 1: Actual samples of glycaemia

This problem of glucose control, taking into account exogeneous variables, has been tackled already by other searchers. In [14], a fuzzy system gives the insulin infusion rate from three inputs: glucose concentration, its rate of change and distance from meals. The authors observed better results than with a conventional control system. In [3], an expert system is proposed, combining diet manipulation and insulin administration.

Our purpose differs because we want to give more autonomy to diabetics in their every day life, not only in hospital conditions. In order to make the fixing of injection rate easier, $Ba(t)$ and $Bo(t)$, we have built from doctors' experience a model of glycaemic variations.

3 Why a fuzzy model ?

The relation between the planned activity, the composition of the meal, the present glucose level etc, at time t, and, on the other hand, the glucose level at time $t + \Delta t$ can be thought of as a nonlinear mapping from some input space to $\mathcal{R}$. To modelize this mapping, one can use either neural networks or fuzzy systems. But doctors imposed the following constraints on us.

No black box. All the decisions must be clearly understood and easily compared with the diabetic's or the doctor's intuition or knowledge.

Universality. The knowledge base must be unique. The rules are the same for all the diabetics: **quantitative** modifications, dealing with interpersonal differences, are allowed, but not **qualitative** ones.

Evolutivity. Let us give an example. The rule "when it is cold I put on my coat" is the same for all human beings, but not the notion of coldness. Therefore, the role of a learning process is 1) to respect the rules and 2) to adapt it for each individual. In other words, we need a learning algorithm respecting the semantics of the knowledge base.

Other constraints come from the very nature of some variables: composition of meal (fiber/glucide) or planned activity are essentially fuzzy. It is more efficient to characterize such situations by linguistic labels rather than by the crisp data from hypothetical sensor... In their notebooks, the diabetics wrote down only discrete values for these two items, but, in the experiments in progress, we asked them to refine their valuations using either linguistic terms (very, moderately, not very, etc) or continuous numerical values.

For all these considerations, we have built a fuzzy model of glycaemic variations.

4 Fuzzy modelization

Following the PRIX methodology, we have identified six significant input variables to build a one-step-ahead predictor of glycaemia. Other inputs, like mental stress or tension are too difficult to estimate in the present state of our work. The selected inputs are:

1. Glucose level at time t, $Gl(t)$.
2. Basic insulin rate, $Ba(t)$.
3. Bolus injection, $Bo(t)$.
4. Distance from the meal, $Dm(t)$.
5. Composition of the meal, $Cm(t)$.
6. Planned physical activity, $Act(t)$

The output is the predicted variation of glycaemia at time $t + \Delta t$, $\Delta\hat{Gl}(t + \Delta t)$, where Δt can be set from one hour to two hours.

For each input variable, we have chosen a strong fuzzy partition with triangles or trapezoids crossing at the grade 0.5, such as:

$$(\forall x)(\sum_i \mu_{A_i}(x) = 1) \tag{1}$$

where A_i is a fuzzy set and $x \to \mu_{A_i}(x)$ the associated membership function. This choice has two advantages:

- firstly, it reduces the number of parameters to be tuned in the learning process, see section 5,
- secondly, it makes the interpretation of fuzzy rules easier.

In the initial set of rules, the fuzzy labels associated to conclusions belong to the set $\{NVB, NB, NM, NS, Z, PS, PM, PB, PVB\}$, where P stands for "positive", N for "negative", V for "very", etc. However, this constraint of dealing with a reduced set of labels is relaxed during the learning process, for greater flexibility and adaptation to each diabetic's case. Following doctors' experience, we have chosen the following fuzzy partitions.

1. Five triangles for glycaemia, from HYPOglycaemia to HYPERglycaemia.
2. Three triangles for both Base and Bolus (Low, Normal, High).
3. Four trapezoids for meal nearness (Long Before, Near, Just After or Long After).

4. Three triangles for meal composition (Glucidic,Balanced, Rich in Fibers).

5. Three triangles for planned physical activity (Low, Normal, High).

These partitions would lead to 1620 rules, but a lot of occurrences between input labels are either irrelevant or without importance. From medical considerations, the rule base has only 64 rules, covering all the meaningful situations. Examples of such rules are:

$$\begin{array}{lll} \textbf{if } Gl \text{ is Low} & \textbf{and } Bo \text{ if High} & \textbf{then } \Delta\hat{Gl} \text{ is } NVB \\ \textbf{if } Gl \text{ is High} & \textbf{and } Bo \text{ if High} & \textbf{then } \Delta\hat{Gl} \text{ is } NS \end{array}$$

The first rule shows the risk of hypoglycaemia if too much insulin is injected when the glucose level is low; the second shows the relative inefficiency of insulin when the glucose level is high.

The Takagi-Sugeno inference [20] has been chosen because it is easy to implement and it has the universal approximation property [7]. The rule base was tested thanks to diabetic's notebook data, showing a good accordance with the reality. Figure 2 shows the actual glycaemia (solid line) and the predicted one (dotted line) for one patient, with $\Delta t = 2$ hours.

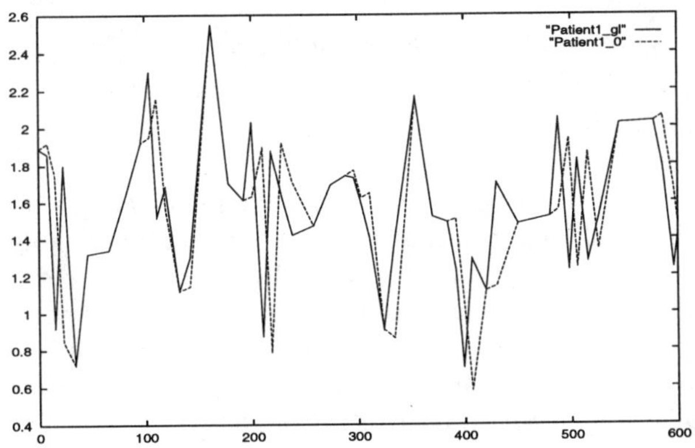

Figure 2: Predicted glycaemia **before** learning

In order to increase the robustness of our model and to match the interpersonal differences, we have performed a supervised learning.

5 Constrained supervised learning

We have collected data from diabetic's notebooks and divided the experimental data into two sets: one for learning, the other to test the generalization ability of the tuned fuzzy system.

Our model has the following form:

$$y(x) = \sum_{i=1}^{64} b_i \frac{\alpha_i(\mathbf{m}, \mathbf{x})}{\sum_j \alpha_j(\mathbf{m}, \mathbf{x})} \tag{2}$$

Variable	membership function	nb param.	
		with	without
Glycaemia	5 triangles	5	13
Base	3 triangles	3	7
Bolus	3 triangles	3	7
Meal nearness	4 trapezoids	6	12
Composition	3 triangles	3	7
Activity	3 triangles	3	7

Table 1: Number of parameters needed to optimize membership functions, with or without a strong fuzzy partition on input spaces.

where b_i is the crisp conclusion of rule i, $\mathbf{m}$ a vector specifying the exact location of membership functions and $\alpha_i(\mathbf{m}, \mathbf{x})$ the truth value of rule i for input vector $\mathbf{x}$.

As we can see from equation (2), *if membership functions are given*, i.e. if vector $\mathbf{m}$ is given, $\mathbf{b} = (b_1, \ldots, b_{64})^t$ is the solution of a linear set of equations. We have chosen to update $\mathbf{b}$ by stochastic gradient descent. Let $d(\mathbf{x})$ be the desired output for input vector $\mathbf{x}$. Then, b_i, $i = 1$ to 64, is updated by:

$$\triangle\, b_i = -\epsilon(y(\mathbf{x}) - d(\mathbf{x}))\frac{\alpha_i(\mathbf{m}, \mathbf{x})}{\sum_j \alpha_j(\mathbf{m}, \mathbf{x})} \tag{3}$$

where ϵ is a learning rate.

But, as we have chosen triangular or trapezoidal membership functions and minimum for conjunction, the global input-output mapping is not differentiable. Therefore, we can not apply gradient descent techniques for the whole set of parameters.

To optimize the location of membership functions, we used a mono-agent evolutionary strategy, proposed by Solis and Wets [Solis 81]. As already stated, fuzzy partition (see equation (1)) allows one to reduce the search space. Table 1 shows that the number of parameters to update in vector $\mathbf{m}$ decreases from 53 to 23.

Figure 3 shows the five membership functions for glycaemia, which are completely defined by the five parameters $(g_i)_{i=1}^{5}$. Adapting the membership functions to each individual case consists in finding an optimal value for these parameters.

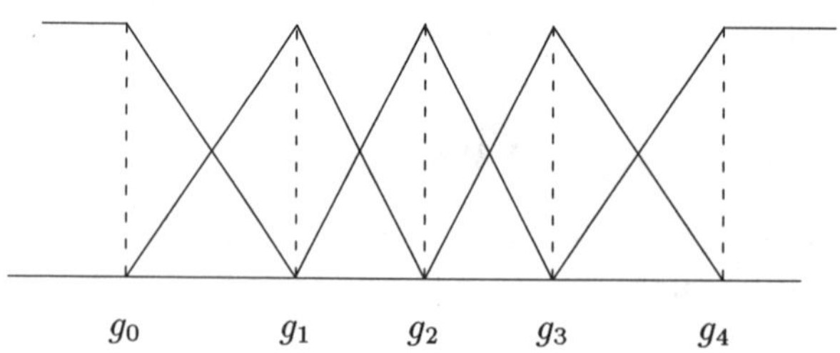

Figure 3: Fuzzy partition for glycaemia

Vector **b** is built by concatenation of parameters of the six fuzzy partitions. The basis of the algorithm is [10]:

1. Build an initial **m** vector by an a priori choice of membership functions.

2. Compute **b** by gradient descent.

3. Compute *error*

4. While $error \geq threshold$ do

   (a) Try another **m**

   (b) Deduce the corresponding **b**.

   (c) Compute *error*

   (d) Accept or not the new **m**.

The Solis and Wets algorithm is a random walk with memorization of the previous successful ways. As we start from a point in the search space presumably close to the optimal one, the algorithm quickly gives a solution. Moreover, we can prove convergence with probability one to the global minimum of error function.

Figure 4 shows the predictions of our model: actual glycaemia (solid line) Vs. predicted one (dotted line). Figure 5 shows the modified membership functions for one diabetic (dotted line); the standard case is shown with the solid line. We can see that the notions of "normal glycaemia" or "hyperglycaemia" are not the same in the standard case and in a particular individual.

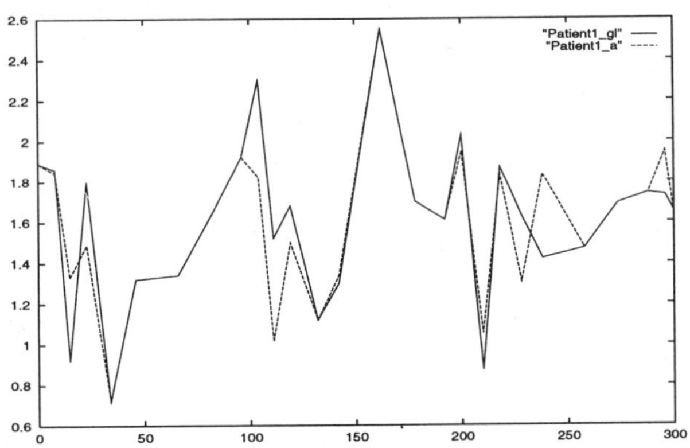

Figure 4: Predicted glycaemia **after** learning

6 Work in progress

In the present state of our work, the fuzzy model of glycaemic variations is thought of as an assisted-decision system. We have implemented the model into a programmable pocket calculator. At this date, one diabetic is using the model to test the consequences of guessed bolus rate, $Bo(t)$, and modulate this rate consequently. Before long, we will

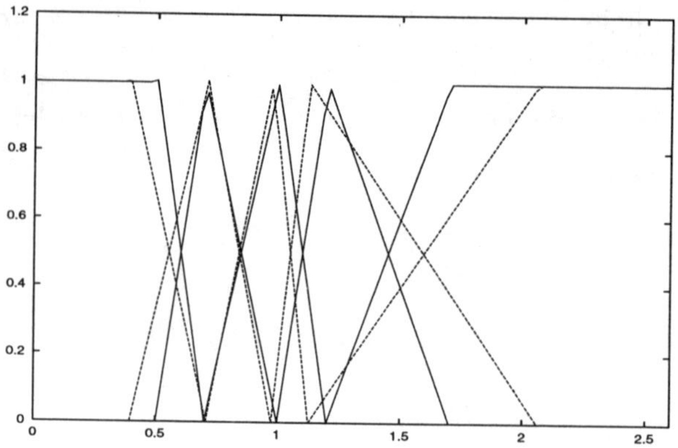

Figure 5: Standard fuzzy partition of glycaemia (solid line) and tuned one for a patient (dotted line).

be able to assess this method and compare the results from several diabetics using it.

We can derive several main lines for future work:

- Need for more accurate data for planned activity, *Act*, and meal composition, Cm, with continuous or semi-continuous values.
- Design of a fuzzy controller, in conjunction with the present fuzzy model.
- Design of a complete pocket control system, associated with insulin pump.

Another way to bring some autonomy to the diabetics is related with the impending availability of electronic glycaemia sensors, allowing more frequent samples and better control.

The first results are very promising: fuzzy rules capture the complex nature of glucose evolution. They can be tuned for a better fit of interpersonal differences. This encourages us to design the complete pocket control system.

References

[1] Akaiwa E., Takahashi M., Sanchez E., Bartolin R., Aurrand-Lions J.P., Yamakawa T. : "Hardware and software of fuzzy logic controlled cardiac pacemaker", *Proc. of the Int. Conf. on Fuzzy Logic and Neural Networks*, Iizuka, Japan, 1990, pp. 549-552.

[2] Albisser A.-M., Leibel B.-S., Ewart T.-G., Davidovac Z., Botz K., Zingg W., Schipper H., Gander R. : "Clinical control of diabetes by the artificial pancreas", *Diabetes*, 1974, 23, pp. 397-404.

[3] Ambrosiadou B.V., Alevizos M., Ziakas G. : "Decision support in diabetes management for optimal glycaemic control by insulin administration", *Proc. of SMC'93*, Le Touquet, oct. 1993, pp. 391-396.

[4] Arakawa K., Arakawa Y., Harashima H. : "Digital signal processing using fuzzy logic for biomedical signals", *Proc. of the Int. Conf. on Fuzzy Logic and Neural Networks*, Iizuka, Japan, 1990, pp. 95-98.

[5] Arita S., Tsutsui T. : "Fuzzy-logic control of blood pressure through inhalational anaesthesia", *Proc. of the Int. Conf. on Fuzzy Logic and Neural Networks*, Iizuka, Japan, 1990, pp. 545-547.

[6] Becker K., Käsmacher H., Juffernbruch K., Rau G., Kalff G., Zimmermann H.-J. : "An intelligent alarm system for decision-support in cardioanaesthesia : Knowledge base and user interface", *Proc. of First European Cong. on Fuzzy and Intelligent Tech.*, Aachen, Germany, 1993, pp. 1023-1026.

[7] Buckley J.J. "Sugeno type controllers are universal controllers", *Fuzzy Sets and Systems*, Vol. 53 no. 3, feb. 93.

[8] Cios K.J., Shin I. : "Using fuzzy sets to diagnose coronary artery stenosis", *IEEE*, 1991, pp. 57-63.

[9] Claude D. : "Automatique et régulation Biologique", in "New trends in nonlinear control theory", *Proc. of an Int. Conf. on Nonlinear Systems*, Descusse J., Fliess M., Isidori A., Leborgne D. (Eds.), Springer-Verlag (Pu.), Nantes, France, 1988, pp. 496-507.

[10] Glorennec P.Y., "Learning algorithms for Neuro-Fuzzy Networks", in *Fuzzy Control Systems*, A. Kandel et G. Langholtz (Eds), CRC Press, 1993.

[11] Hayashi Y. : "Neural expert system using fuzzy teaching input and its application to medical diagnosis", *Proc. of the 2nd Int. Conf. on Fuzzy Logic and Neural Networks*, Iizuka, Japan, 1992, pp. 989-993.

[12] Hsin-Lin L., Adachi Y., Suzuki S., Umeno M.: "Application of fuzzy theory to traditional Chinese medecine by using the pulse phenomena", *Proc. of the 2nd Int. Conf. on Fuzzy Logic and Neural Networks*, Iizuka, Japan, 1992, pp. 1021-1024.

[13] Jensch P. : "Multi-modal medical imaging by mathematical morphology and fuzzy approaches", *Proc. of First European Cong. on Fuzzy and Intelligent Tech.*, Aachen, Germany, 1993, pp. 826-830.

[14] Kageyama S., Mimura A., Ito K., Ikeda Y., Tsuki K., "Blood glucose control by a fuzzy control system", *Proc. of the Int. Conf. on Fuzzy Logic and Neural Networks*, Iizuka, Japan, 1990, pp. 557-560.

[15] Meier R., Nieuwland J., Hacisalihzade S., Steck D., Zbinden A. : "Fuzzy control of blood pressure during anaesthesia with Isoflurane", *IEEE*, 1992, pp. 981-987.

[16] Möller D.P.F. : "Fuzzy logic and its impact for medical applications", *Proc. of First European Cong. on Fuzzy and Intelligent Tech.*, Aachen, Germany, 1993, pp. 263-269.

[17] Shichiri M., Kawamori R. : "Optimized algorithms for closed-loop glycemic control systems", in *First International Symposium on Computer Systems for Insulin Adjustment in Diabetes mellitus*, Beyer J., Albisser M., Schrezenmeir J., Lehmann L. (Ed.), Novo Industrie GMBH (Pu.), 1985, pp. 171-183, ISBN 3-907507-01-0.

[18] Slama G., Klein J.-C., Tardieu M.-C., Tchobroutsky G. : "Normalisation de la glycémie par pancréas artificiel non miniaturisé", *La Nouvelle Presse médicale*, 1977, 6, no 26, pp. 2309-2313.

[Solis 81] Solis F., Wets J., "Minimization by random search techniques", *Mathematics of Operation Research*, Vol. 6, 1981.

[19] Soula G., Vialettes B., San Marco J.L., "PROTIS, a fuzzy deduction-rule system : Application to the treatment of diabetes", *Proc. MEDINFO 83*, Amsterdam.

[20] Takagi T., Sugeno M., "Fuzzy identification of systems and its applications to modeling and control", *IEEE Trans. on S.M.C.*, SMC-15 (1), jan. 1985.

[21] Weierich P., Wetzel D., Niemann H. : "AUDIGON - A medical expert system with new judgement functions for the fuzzy rulebase", *Proc. of First European Cong. on Fuzzy and Intelligent Tech.*, Aachen, Germany, 1993, pp. 472-477.

[22] Yoshida K., Hayashi Y., Imura A., Shimada N. : "Fuzzy neural expert system for diagnosing hepatobiliary disorders", *Proc. of the Int. Conf. on Fuzzy Logic and Neural Networks*, Iizuka, Japan, 1990, pp. 539-543.

A HYBRID NEURAL NETWORK WITH SYMBOLIC ACTION LAYER FOR MEDICAL DECISION SUPPORT

Donna L. HUDSON*, Maurice E. COHEN*,**, Malcolm F. ANDERSON*,***

*Univ. of California, San Francisco
2615 E. Clinton Avenue
Fresno, CA 93703

**California State University
Fresno, California 93740

***VA Medical Center
Fresno, California 93703

ABSTRACT

Neural network modeling has found many applications in the last decade, especially in medicine. It offers an efficient method for the development of decision tools directly from database information. This methodology has replaced in many cases the knowledge-based approach that gained popularity in the mid-1970's. Neural networks in most instances are not capable of symbolic processing. In this paper, a neural network structure is combined with a symbolic layer to produce a hybrid system for medical decision making which is illustrated in diagnosis and treatment of carcinoma of the lung.

1. Introduction

Neural network research has abounded in the last decade [1,2]. The new generation of neural networks have overcome the objections which had been leveled at this approach, which was first used in the late 1950's and early 1960's [3,4]. They offer an effective method for development of decision models directly from data, without expert intervention.

These new approaches have met with many successes, with a wide range of applications [5]. This is especially true in medicine, where new techniques are constantly sought which can deal with formerly intractable problems in medical decision making [6,7].

In the work described here, a neural network design based on a learning algorithm developed by the authors over the last decade is used as the basis for a hybrid system which incorporates an action layer [8,9]. This action layer using symbolic processing based on a grammar structure, and can be interpreted in many ways, depending on the intent of the network [10]. It can also be used to incorporate expert input into the data-driven knowledge base.

In this paper, the use of this hybrid network to combine diagnosis and treatment strategies is discussed, along with issues of handling of uncertainty in each stage of the network, both in the model development through supervised learning and in the symbolic layer. A specific medical example for carcinoma of the lung is given [11].

2. Neural Network Structure

The approach taken by the authors does not use the popular back propagation learning algorithm. Instead, a learning algorithm developed by the authors is used. The overall network design is discussed, followed by a summary of the advantages of the learning algorithm. Representation of uncertainties in the network are then discussed.

2.1. Network Design

The network structure basically comprises four layers, which can themselves consist of sub-layers. The basic layers are the input, hidden, output, and symbolic layers, as illustrated in Figure 1. The hidden layer itself can consist of more than one layer, depending on choices made in the selection of the learning algorithm. Figure 1 illustrates a network with two parts to the hidden layer, with each accounting for fractional contributions of input nodes. The actual structure of the hidden layer is discussed in the next section in conjunction with the learning algorithm. The third layer consists of the output nodes generated by the model developed through the learning algorithm. These output nodes then feed values to the symbolic layer, which is discussed in Section 3.

2.2. Learning Algorithm

The learning algorithm has been discussed in detail elsewhere [12]. It is based on a potential function approach to supervised learning. The mathematical foundation used vector spaces based on the class of Cohen orthogonal functions:

$$F_n(m,a_t;s) = \sum_{k=0}^{n} \frac{\prod_{i=0}^{n-1} (m+a_i+a_k)x^{a_k}}{\prod_{j=0}^{k-1} (a_j-a_k) \prod_{s=1}^{n-k} (a_{k+s}-a_k)} \qquad (1)$$

where

$$\prod_{j=0}^{k-1} (a_j-a_k) = (a_0-a_k)(a_1-a_k)\dots(a_{k-1}-a_k) \quad k \geq 1$$

$$\prod_{s=1}^{n-k} (a_{k+s}-a_k) = (a_{k+1}-a_k)(a_{k+2}-a_k)\dots(a_n-a_k) \quad n \geq k$$

A number of special cases can be derived from this orthogonal series. In order to obtain orthogonal polynomials, the series a_k must be chosen to assume integral values. However, it should be noted that this function is capable of generating non-integral series, which can be used to develop networks in which nodal values contribute to fractional powers. This function allows the user to design networks ranging from linear to higher-order, including the use of fractional contributions of nodes, as illustrated in Figure 1. The most common design is to use a function of the type

$$D_i(\mathbf{x}) = \sum_{i=1}^{n} w_i x_i + \sum_{i=1}^{n} \sum_{\substack{j=1 \\ i \neq j}}^{n} w_{ij} x_i x_j \qquad (2)$$

which includes all linear terms and all interactive terms.

2.3. Fuzzy Aspects of Network

As the network is described, fuzzy variables are not directly considered. In order to handle the management of uncertainty in a network such as this, a number of factors must be considered.

The network allows input variables to take any form, as long as the values are ordered. Thus they may be continuous, integer, binary, or categoric. Most input variables can be redefined to accommodate the degree of uncertainty which may be present. The network can accept a fuzzy number as an input instead of a continuous variable, in which case an interval algorithm is invoked which produces a decision model which encompasses the range of the fuzzy number [13]. Triangular fuzzy numbers are assumed in this algorithm. Categoric variables can be replaced with separate variables for each category, with each variable represented by a membership function for that category. Instead of binary input, a fuzzy value between 0 and 1 can be entered instead. Specific examples will be given later for lung cancer. Incorporation of approximate reasoning in the symbolic layer will be discussed in the following section.

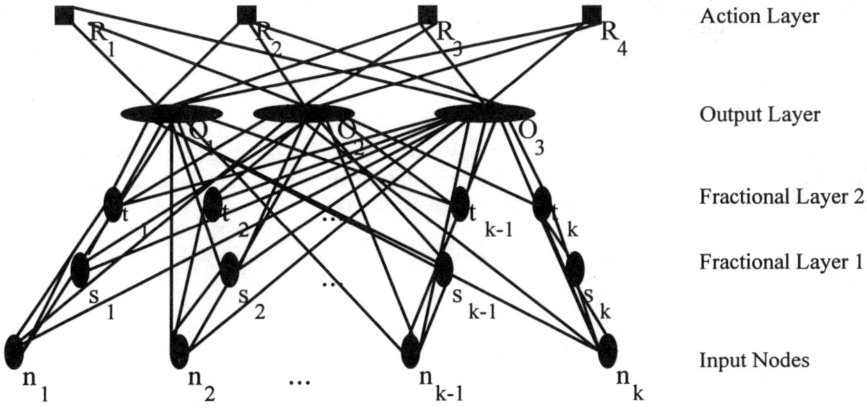

Figure 1: Neural Network Diagram

3. Symbolic Reasoning

3.1. Symbolic Layer

The nonlinear decision function produces a normalized numerical value between 0 and 1, inclusive. A separate decision function $D_i(\mathbf{x})$ is attached to each of the output nodes. A threshold value $T_i(\mathbf{x})$ (or values) is also attached to each level three node. If the value of the decision function exceeds the corresponding threshold value, a certain symbol is produced; if it does not, a different symbol is produced. This process can be extended by having any number of threshold values produce as many symbols as desired. An additional layer is added to the network which combines the symbols generated by adjacent nodes according to a well-structured grammar.

3.2. Definition of a Grammar

A grammar provides the rules by which a symbol can be combined. The following notation is used. A grammar G is defined by (V_n,V_t,P,S), where V_n are the variables, V_t are the terminals, P are productions, and S is the start symbol. A simple grammar would be

$$V_n=\{S\}$$

$$V_t=\{0,1\} \tag{3}$$

$$P=\{S\rightarrow 0S1, S\rightarrow 01\}.$$

The result of this grammar is to produce the sequence 0^n1^n [14].

3.3 Interpretation of Symbolic Layer

Consider an application in which it is necessary to choose among three conditions. It is possible that none, one, or a combination of these conditions are present. A neural network is set up to learn using data of known classification, determining which variables are pertinent to making the decision and the appropriate weighting factors for each of these variables. The decision algorithm is represented by a multi-layered network as shown in Figure 1. In the combined system described here, the values from the output level is used to produce an additional level, denoted the action level. In the case of medical decision making, level n-1 would determine which condition or conditions are present, and level n would determine what treatment or follow-up testing should be done based on these results.

Table I shows a sample grammar. A possible interpretation is the following. Each node has an potential excitatory action (represented by a) or inhibitory action (represented by b). This is determined by threshold values:

$$\text{If } D_i(\mathbf{x}) > T_i \quad \text{then "a" else "b".} \tag{4}$$

In the grammar, each node is numbered N_i. A response of "a" at node i indicates that condition C_i is present. To simplify the grammar, a sequential evaluation of nodes is assumed. If random evaluation is done, additional entries in the grammar are required.

For the example, four possible actions are assumed. One possible interpretation for a medical application could be:

r_1: No action
r_2: Re-do test
r_3: Do additional tests
r_4: Begin treatment

Table I: Sample Grammar

$V_n = \{S,A,B,C,D,T\}$

$V_t = \{a,b,r_1,r_2,r_3,r_4\}$

P:

$S \to N_1$	$N_1 \to aN_2$	$aaaT \to r_1$
	$N_1 \to bN_2$	$bbbT \to r_4$
	$N_2 \to aN_3$	$bbaT \to r_3$
	$N_2 \to bN_3$	$baaT \to r_3$
	$N_3 \to aT$	$abbT \to r_2$
	$N_3 \to bT$	$aabT \to r_2$
		$abaT \to r_2$
		$babT \to r_2$

4. Use in Medical Decision Making

4.1. Medical Application: Carcinoma of the Lung

The above structure will be illustrated with medical application involving carcinoma of the lung. In a retrospective and a prospective study, the authors investigated parameters which were useful in decision making in the diagnosis and treatment of carcinoma of the lung. A subset of the analysis is given here for illustrative purposes pertaining to the decision of whether or not surgery should be performed.

4.2. Crisp Implementation

The neural network model identified 12 parameters which were important in the surgery decision. Each variable and its type is listed below:

x_1:	Forced vital capacity	*continuous*
x_2:	Contraindications to surgery	*binary*
x_3:	Bronchoscopy results	*binary*
x_4:	Bronchoscopy type	*categoric*
x_5:	Mode of presentation	*categoric*
x_6:	Presence of local symptoms	*binary*
x_7:	Metastasis evident from bone scan	*binary*
x_8:	Largest tumor diameter, chest x-ray	*continuous*
x_9:	Contralateral involvement, chest x-ray	*binary*
y_1:	Needle biopsy cell type	*binary*
y_2:	Bronchoscopy cell type	*binary*
y_3:	Sputum cell type	*binary*

where the x_i's indicate the corresponding variable numbers in equation (5) and node numbers in Figure 2. The presence of oat cell carcinoma (confirmed by any of needle biopsy (y_1), bronchoscopy (y_2), or sputum analysis (y_3)) results in a no surgery decision. These are binary nodes indicating the presence or absence of this cell type. This decision is handled by output node O_1. A weight of -1 is attached to each branch. If any one of the input nodes for bronchoscopy, needle biopsy, or sputum analysis has a value of 1, the result, using equation 1, will be less than or equal to -1. If all are zero, the result will be zero.

Note that the data type is indicated for each variable. In the next section, the areas of uncertainty are analyzed for each variable, along with their possible representation as fuzzy numbers.

For node O_2, the nine parameters listed above must be taken into account, with appropriate weighting factors. The weighting factors are determined using the learning algorithm described above, and result is an equation of the form:

$$D(\mathbf{x}) = 124 + 40x_1 - 25x_2 + 9x_3 + ... - 28x_9 + 3x_1x_2 + 6x_1x_3 + ... - 2x_8x_9. \quad (5)$$

The interpretation is as follows: the multiplying coefficient for x_1 is w_1=40, which corresponds to n_1; the coefficient of x_1x_2 is $w_{1,2}$=3, which corresponds to $n_{1,2}$. These weights are in relation to output node O_2, and again are labeled only by the originating node, for simplicity of notation. The interactive nodes are formed at an intermediate level with equal weights of 1 from input node to intermediate node.

4.3. Symbolic Layer

Each output node, O_1, O_2, and O_3, has a threshold value associated with it, which can be interpreted as follows. If the threshold is exceeded for O_1, that indicates that oat cell carcinoma is present, and surgery is precluded. Note that for the crisp implementation the value of O_1 is either 0 or 1, but this is not necessarily true for the fuzzy implementation, as explained in the following section. If O_2 is greater than the threshold, then surgery should be performed. If O_3 exceeds the threshold, the patient should have no treatment due to the severity of the disease. The weights for this node are not explained in detail here. In general, the learning algorithm is run separately to produce models for each of the output nodes. In this case, the same database is used with different decisions as the objective, however, the output nodes could come from different databases.

In this example, the treatment plan is clear if the output nodes exceed their respective thresholds. The symbolic layer is used to combine outputs when the thresholds are not exceeded and the plan of action is consequently unclear. The four possible actions are:

R_1:	No treatment
R_2:	Surgery
R_3:	Chemotherapy
R_4:	More testing

4.4. Fuzzy Implementation

For variable x_1, a continuous measure of lung capacity is recorded. This measurement has certain accuracy restrictions which are dependent on the accuracy of the machine and also may vary if the test is repeated on the same patient. This measurement could more be more accurately represented as a fuzzy number. The same is true to a lesser degree for the other continuous variable, x_8, tumor diameter.

For all the binary variables, the yes/no response could be replaced with a degree of presence. If the three y_i variables are considered, it appears that they are inherently binary, as either oat cell carcinoma is present or absent. However, determination of cell type is not an absolute science and is subject to a degree of certainty, which could be used instead of the yes/no result. The same is true for several of the other binary variables.

The two categoric variables could be separated into separate variables for each category, with each variable representing the degree of presence of that category.

Deciding between the crisp and fuzzy implementations may be done experimentally by establishing both models to determine their relative accuracy. In some instances, the added information which may be obtained from fuzzifying the variables may make a significant difference in the outcome. In other cases, the crisp implementation may yield almost identical results.

4.5. Other Uses of the Symbolic Layer

Although the symbolic layer is used here for prescribing treatment, it may have many other interpretations. It can be used to connect expert-supplied rules with the neural network output, and also to incorporate fuzzy interpretations with an otherwise crisp neural network implementation. It can also be used to facilitate incorporation of an explanation capability similar to knowledge-based systems.

5. Conclusion

Neural networks offer an efficient approach to the development of decision support systems directly through accumulated data. In the approach described here, data of all types can be included in the network, including fuzzy data. The learning algorithm employed ensures that in the learning process the system will remain stable and will not degenerate into chaotic solutions. The method has been used in numerous medical applications with excellent results.

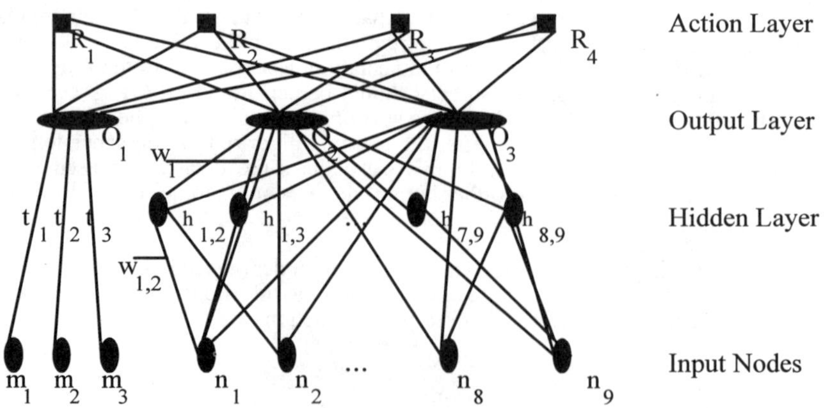

Figure 2: Neural Network for Carcinoma of the Lung

The addition of the symbolic layer to the basic neural network has a number of advantages. First, it facilitates the combination of separate models produced by learning processes which focus on different decisions. For instance, in the example given here, the learning algorithm is invoked three times to arrive at models for three different conclusions, which are then combined in the symbolic, or action layer, to produce a comprehensive decision support system. The resulting system is easy to use and provides fast and accurate results in practical applications.

REFERENCES

[1] S. Grossberg, Ed., *Neural Networks and Natural Intelligence*, MIT Press, Cambridge, MA 1988.

[2] D.E. Rummelhart, J.L. McClelland, and the PDP Research Group, *Parallel Distributed Processing*, vols. 1 and 2, MIT Press, Cambridge, 1986.

[3] N.J. Nilsson, *Learning Machines*, McGraw Hill, New York, 1965.

[4] F. Rosenblatt, *Principles of Neurodynamics, Perceptrons, and the Theory of Brain Mechanisms*, Spartan, Washington, 1961.

[5] S.I. Gallant, Connectionist Expert Systems, *Communications ACM*, **31 (2)**, 1988, 152-169.

[6] R.M.E Sabbatini, Applications of connectionist systems in biomedicine, in *MEDINFO 92*, K.C. Lun et al. (Eds.), Elsevier Science Publishers, 1992, 418-425.

[7] A.F. Rocha, A neural network for extracting knowledge from natural language data bases, *IEEE Trans. on Neural Networks*, **3(5)**, 1992, 819-828.

[8] D. L. Hudson, M. E. Cohen, M.F. Anderson, Use of neural network techniques in a medical expert system, *Int. J. of Intelligent Systems*, **6(2)**, 213-223, 1991.

[9] D.L. Hudson, M.E. Cohen, P.C. Deedwania, A neural network for symbolic processing, *IEEE Engineering in Medicine and Biology*, A.Y.J Szeto, R.M. Rangayyan, Eds., IEEE, 1993, 248-249.

[10] D.L. Hudson, M.E. Cohen, A nonlinear neural network combined with symbolic processing, *Nonlinear Theory and Its Applications*, 1993, 937-940.

[11] D.L. Hudson, M.E. Cohen, M.F. Anderson, Use of a neural network model to develop cost-effective testing strategies, *Proceedings, Int. Soc. for Computers and Their Applications*, D.C. Pheanis, Ed., 1993, pp. 84-88.

[12] D.L. Hudson, M.E. Cohen, Use of fractional powers to moderate neuronal contributions, *Proc. IEEE Conference on Neural Networks*, 1993, 517-522.

[13] M.E. Cohen, D.L. Hudson, Approaches to the handing of fuzzy input data in neural networks, *Proc. IEEE Conference on Fuzzy Systems*, 1992, 93-100.

[14] J.E. Hopcroft, J.D. Ullman, *Formal Languages and Their Relation to Automata*, Addison-Wesley, Reading, MA, 1969.

This work was supported in part by a grant from University of California Valley Medical Education Foundation.

APPLICATIONS OF FUZZY MATRICES IN AGRICULTURE

Peter W. EKLUND*, Ximing SUN**, and D.A. THOMAS**

*Department of Computer Science
The University of Adelaide
Adelaide 5005, Australia
email peter@cs.adelaide.edu.au

**Department of Environmental Sciences
The University of Adelaide
Adelaide 5005, Australia

ABSTRACT

The paper proposes a multi-level fuzzy evaluation function[1,2] which imposes a total order on crop alternatives. The model can be used to consider various input conditions and allows for flexible treatment of such issues as soil degradation, erosion and pollution. These issues and others are graded subjectively in terms of a membership function and partitioned into a matrix which is crossed with an aggregated weighted index for crop yield.

The resulting multi-level comprehensive fuzzy evaluation function considers environmental issues, management skills and economic benefits with biophysical suitability. It allows us to make recommendations for land usage.

1. Introduction

The technique described in this paper provides a mechanism for integrating multi-dimensional variables, determined from both empirical and expert knowledge, to give a overall numeric assessment of interactions between variables. The technique has a wide range of applications in many fields. As an example of how the technique can be applied, we demonstrate how to perform comprehensive land evaluation for selecting suitable cropping options in agriculture. The domain is interesting because decision makers base their choices on a wide range of considerations, e.g. biophysical land capability, environmental impacts, socioeconomic and political considerations, local infrastructure etc. Land evaluation needs a synthesis of this information to guide objectives which are easily constructed and viewed in a structure called an *interaction matrix*.

2. Comprehensive evaluation

Assume the object set $C=\{C_1,C_2,\dots,C_n\}$ and the object's attribute set $X=\{X_1,X_2,\dots,X_n\}$ where X_i is a fuzzy subset of C. Given the grade membership of each object to each attribute,

$$\mu_{C_i}(X_j)=\gamma_{ij} \qquad \gamma_{ij}\in[0,1],\ i\in[1,m],\ j\in[1,n]$$

the fuzzy relationship matrix of C to X is;

$$\underline{R}=\begin{array}{c} \\ X_1 \\ X_2 \\ \vdots \\ X_n \end{array}\begin{array}{c} \begin{array}{cccc} C_1 & C_2 & \cdots & C_n \end{array} \\ \left[\begin{array}{cccc} \gamma_{1,1} & \gamma_{1,2} & \cdots & \gamma_{1,n} \\ \gamma_{2,1} & \gamma_{2,2} & \cdots & \gamma_{2,n} \\ \cdots & \cdots & \cdots & \cdots \\ \gamma_{m,1} & \gamma_{m,2} & \cdots & \gamma_{m,n} \end{array}\right] \end{array} \qquad (1)$$

If weightings a_i are given to each attribute X_i, a normalised fuzzy set or a fuzzy vector on X is given by $\underline{A}=(a_1,a_2,\dots,a_n)$ where $\sum_{i=1}^{m} a_i=1$ and $0\le a_i<1$, i.e. there is more than a single weighting.

The comprehensive evaluation result $\underline{B}$ is:

$$\underline{B} = \underline{A} \circ \underline{R}, \qquad (2)$$

where $\underline{B}$ is a fuzzy set on the set of objects C, written $\underline{B} = (b_1, b_2, \ldots, b_n)$. The preferential object in C (B^*) in the fuzzy set $\underline{B}$ verifies $\mu_{\underline{B}}(B^*) = \max(b_1, b_2, \ldots, b_n)$.

There are a number of ways we can consider combining the matrices $\underline{A}$ and $\underline{R}$. They can be crossed using a number of algebraic variants[2] providing they are associative;

$$b_j = \bigvee_{i=1}^{m} (a_i \wedge \gamma_{ij}); \qquad (3)$$

$$b_j = \bigvee_{i=1}^{m} (a_i \cdot \gamma_{ij}); \qquad (4)$$

$$b_j = \sum_{i=1}^{m} (a_i \wedge \gamma_{ij}); \qquad (5)$$

$$b_j = \sum_{i=1}^{m} (a_i \cdot \gamma_{ij}); \qquad (6)$$

(3) and (4) we call primary factor determination types: the primary factor determines the result. (5) and (6) highlight the primary factors but also consider secondary factors. (4) and (6) accord weightings by considering all factors with equal influence. Note that these naturally are not the only composition operators. Generally we can use a $\perp$-T composition where $\perp$ and T are T-conorm and T-norm respectively but only that subclass of T-conorms and T-norms that are associative. Note that (3) and (4) always give a $b_j \in [0,1]$, i.e. $\underline{B}$ is a fuzzy subset of C even if $\sum_{i=1}^{m} a_i \neq 1$ with $a_i \in [0,1]$. (5) and (6) can only be used since $\sum_{i=1}^{m} a_i = 1$ otherwise $b_j \notin [0,1]$ and $\underline{B}$ would not be a fuzzy subset of C.

The four operators can be used to compute an evaluation under varying domain circumstances. (6) will be suitable in most domains but (4) should be used for Markovian processes for instance. (6) will resume to a convex weighting of the m fuzzy subsets of X_i [4]. The more the agreement between the models, the greater the confidence in the result.

3. Fuzzy multi-level comprehensive evaluation

In any complex system many factors may influence a final outcome and each factor needs to be considered. Such factors may also exist at several levels of influence. The difficulty with a first level comprehensive evaluation model, is that weightings, given each factor, are often difficult to assign. In addition, the weighting set $\underline{A}$ must satisfy the unity sum condition $\sum_{i=1}^{m} a_i = 1$ (although this is not true of a $\perp$-T composition operator, it is true of (5) and (6)).

The compound operation in the fuzzy matrix may be given by taking a *max* or *min* value. When these are used and the number of factors is large, weightings tend to have extremal influence and dominate grade membership in the matrix.

To counter this, a multi-level comprehensive evaluation model can be formulated[1]. The principle is to divide properties into factor sets. The first level of evaluation in each set is to obtain the fuzzy vectors (A_i). These vectors are then combined into a relationship matrix $\underline{R}$ for each group of factors. These vectors and matrices are then used to operate over the weighting set to obtain a high-level comprehensive evaluation $\underline{B}$.

The process of formulating a multi-level evaluation model is:

- factors set $X = \{X_1, X_2, ..., X_m\}, card(X) = m$
- groups of factors set,

$$X / C = \{\mathrm{X}_1, \mathrm{X}_2, ..., \mathrm{X}_p\} \text{ with } p \in [1, m]$$

- for every $i \in [1, p]$, every X_i is defined by $\mathrm{X}_i = \{X_l \in X\}$, with $card(\mathrm{X}_i) = k_i$ s.t. $\sum_{i=1}^{p} k_i = m$

In the example presented we have; $X = \{X_1, X_2, ..., X_8\}$, $m = 8$ with X_1 as biophysical factors; X_2 as economic factors; X_3 as management skills; X_4 as crop preference; X_5 as social status; X_6 as soil degradation; X_7 as soil erosion and X_8 as pollution. The following four (4) groups of factors ($p = 4$) are subsequently constructed;

- group biophysical factors;

$$\mathrm{X}_1 = \{X_1\},\ k_1 = 1$$

- group economic factors:

$$\mathrm{X}_2 = \{X_2\},\ k_2 = 1$$

- group social factors:

$$\mathrm{X}_3 = \{X_3, X_4, X_5\},\ k_3 = 3$$

- group environmental factors:

$$\mathrm{X}_4 = \{X_6, X_7, X_8\},\ k_4 = 3$$

We can verify that $\sum_{i=1}^{4} k_i = 8$.

A first level comprehensive evaluation on each X_i with k_i factors proceeds as follows. Suppose the weighing assigned to X_i factors is a vector $\underline{A}_i$ the evaluation relationship matrix is $\underline{R}_i$, i.e.,

$$\underline{A}_i \circ \underline{R}_i = \underline{B}_i = (b_{i1}, b_{i2}, ..., b_{in}).$$

The multi-level comprehensive evaluation model $\underline{B}*$ is given as;

$$\underline{B}* = \underline{A} \circ \underline{R} = A \circ \begin{bmatrix} \underline{A}_1 \circ \underline{R}_1 \\ \underline{A}_2 \circ \underline{R}_2 \\ \cdots \\ \underline{A}_i \circ \underline{R}_i \\ \underline{A}_p \circ \underline{R}_p \end{bmatrix} = (b_1, b_2, ..., b_n) \tag{7}$$

where $\underline{A}$: overall weighting set and $\underline{R}$: relationship matrix.

For example, in an agricultural domain, the most suitable crop for any one parcel of land should be selected by considering, the biophysical, economic (cost-benefit), social and environmental impacts of a given crop. Suppose we have five crops to choose from,

$$C = \{C_1, C_2, C_3, C_4, C_5\},$$

we can now evaluate the suitability of these crops in terms of the above considerations.

3.1 Biophysical evaluation

The biophysical evaluation vector can be obtained through our A^* (aggregated weighting index) from the interaction matrix[4]. We then ask the user for the desired yield for each crop. This information can be taken as the optimal value for the membership function for each crop and the membership function of biophysical suitability for each crop can be expressed,

$$\mu_{C_i}(A^*) = \begin{cases} 1 \\ \frac{1}{3}(2c_1 / 10^{c_2 \cdot A^*} + 1) & A/3 < A^* < A \\ 0 \end{cases} \quad (8)$$

where A is the aggregated weighted index for the users desired yield for each crop C_i. Where A^* in this case is the biophysical aggregated weighted index for a specific parcel of land. c_1 and c_2 are constants for a given production region.

The membership function can be obtained by linear conversion of the curve of yield against A^* which is generated from the interaction matrix for each crop[4]. From these considerations the biophysical evaluation final vector $\underline{B}_B = \underline{A}_B \circ \underline{R}_B$ (7).

3.2 Economic (cost-benefit) evaluation

The economic evaluation (or cost benefit) vector can be obtained from membership function μ_{CB}. As the estimated yield of each crop can be obtained using A^*.

The price return of the yield minus the fixed and variable costs can be used to estimate the gross margin for each crop. A satisfactory gross margin for each crop can be obtained by the user. Suppose GM' is the satisfactory gross margin for the user then the membership function (μ_{CB}) of economic evaluation is:

$$\mu_{CB}(GM) = \begin{cases} 1 & \text{if } GM \geq GM' \\ GM / GM' & \text{if } GM < GM' \end{cases} \quad (9)$$

where GM is the gross margin for each crop and GM' is the user's satisfactory gross margin.

The first-level economic evaluation vector $\underline{B}_c = \underline{A}_c \circ \underline{R}_c$ can be directly calculated from (7).

3.3. Social evaluation

If the social suitability of five crops is determined by three factors: management skills; individual preference; social status. Suppose the weightings assigned in these three factors are,

$$\underline{A}_s = (a_{s1} \quad a_{s2} \quad a_{s3}).$$

The fuzzy relationship matrix is (from (7)):

$$\underline{R}_s = \begin{matrix} & C_1 & C_2 & \cdots & C_5 \\ X_1 & \gamma_{1,1} & \gamma_{1,2} & \cdots & \gamma_{1,5} \\ X_2 & \gamma_{2,1} & \gamma_{2,2} & \cdots & \gamma_{2,5} \\ X_3 & \gamma_{3,1} & \gamma_{3,2} & \cdots & \gamma_{3,5} \end{matrix} \quad (10)$$

Then the first-level *social* evaluation vector B_s is;

$$\underline{B}_s = \underline{A}_s \circ \underline{R}_s = (a_{s1} \quad a_{s2} \quad a_{s3}) \begin{bmatrix} \gamma_{1,1} & \cdots & \gamma_{1,5} \\ \gamma_{2,1} & \cdots & \gamma_{2,5} \\ \gamma_{3,1} & \cdots & \gamma_{3,5} \end{bmatrix} \quad (11)$$

3.4 Environmental impact evaluation

Suppose there are three aspects which determine the environmental impacts: soil degradation (E_1); soil erosion (E_2); pollution (E_3). The weighting set for these three aspects is $\underline{A}_E$,

$$\underline{A}_E = (a_{E1}, a_{E2}, a_{E3}).$$

$\underline{A}_E$ can be obtained from the intensity index (figure 1) which is based on factor sets.

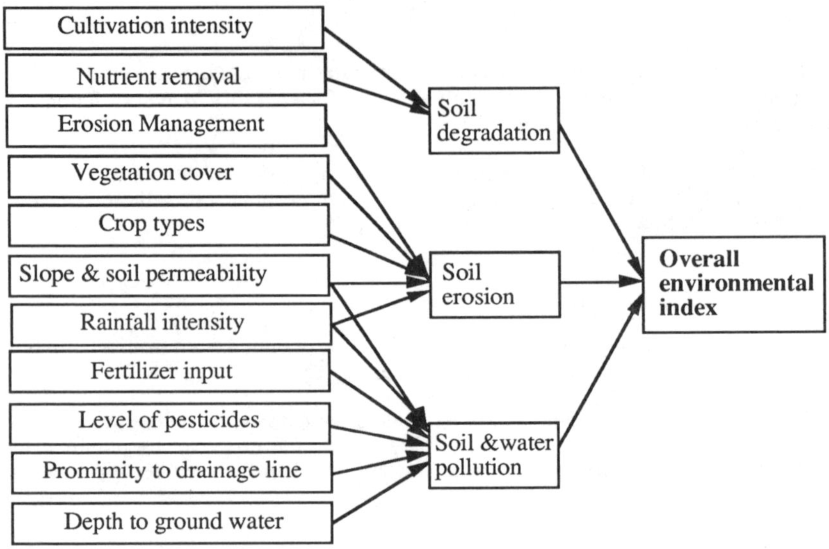

Figure 1: Intensity index based in factor sets.

δ_{ij} in the fuzzy relationship matrix can be determined: (i) for E_1 by cultivation intensity and nutrient removal; (ii) for E_2 by empirical data; (iii) for E_3 by fertiliser input and level of pesticide use (Figure 1).

Then the first-level environmental impact vector $\underline{B}_E$ is,

$$\underline{B}_E^* = \underline{A}_E \circ \underline{R}_E = (a_{E1}, a_{E2}, a_{E3}) \cdot \begin{bmatrix} \gamma_{1,1} & \cdots & \gamma_{1,5} \\ \gamma_{2,1} & \cdots & \gamma_{2,5} \\ \gamma_{3,1} & \cdots & \gamma_{3,5} \end{bmatrix} \tag{12}$$

Finally, the multi-level fuzzy comprehensive evaluation is $\underline{B}^*$, the fuzzy subset of C, where b_i is the degree of membership of C_i;

$$\underline{B}^* = \underline{A} \circ \underline{R} = A \circ \begin{bmatrix} \underline{A}_B \circ \underline{R}_B \\ \underline{A}_c \circ \underline{R}_c \\ \underline{A}_s \circ \underline{R}_s \\ \underline{A}_E \circ \underline{R}_E \end{bmatrix} = (b_1, b_2, b_3, b_4, b_5) \tag{13}$$

The most suitable crop is the one which has a degree of membership equal to $\max(b_1, b_2, \ldots, b_5)$.

4. Discussion

The methodology described gives a simple and effective means of incorporating interdisciplinary knowledge, both empirical and human, in one comprehensive model. The advantage of the methodology is that the user, by way of weightings, can estimate how the interplay of various practices will affect the end result. The users' active participation in discussion and a comparison of the results obtained is of great help in learning and understanding the decision process. Continual dynamic adjustment of weightings are used in the training phase and are set by a domain expert. Extensions to the work include constructing a knowledge acquisition tool defined by the methodology and automating the weighting assignment through symbolic and subsymbolic machine learning[5].

The use of aggregated knowledge with the incorporation of fuzzy membership constructions in models has been used successfully. It significantly increases the predictive value of crop yield estimates over conventional methods[3].

5. Conclusion

This paper describes a method based on fuzzy sets, that allows us to determine a preferential grain crop for a given set of circumstances. This technique is based on a multi-level fuzzy evaluation function that produces a total order crop alternatives under consideration.

The resulting multi-level comprehensive fuzzy evaluation function considers environmental issues, management skills and economic benefits with biophysical suitability and allows us to make recommendations for land usage. The paper demonstrates, by way of a simple example, how degradation, erosion and pollution, management skills and economic benefits can be reconciled with crop yields for optimal land use.

Acknowledgment

This research was supported by the Australian Research Council. Thanks to Dr. Zyed Zalila for detailed comments and suggestions on the paper. Any remaining errors in the paper are of course our own.

REFERENCES

[1] Chen, Y. Y., "A comprehensive evaluation model", Journal of fuzzy mathematics, 1, 1993.
[2] Wang, P.Z., *Fuzzy sets and their application*, Shanghai Technical Press, 1983.
[3] Sun, X., Thomas, D.A., Eklund, P.W., "Fuzzy sets in aggregated knowledge construction", IEEE Conf. on Intelligent systems and fuzzy logic, pp. 437-442, 1993.
[4] Zalila, Z., *Contribution a une theorie des relations floues d'ordre* N, PhD Thesis, Universite de Technologie de Compiegne, 1993.
[5] Eklund, P.W., Salim, A., Kirkby, S., "A Framework for incremental knowledge base update from additional data coverages", Proc. 7th Australasian Remote Sensing Conf. pp. 364-377, 1994.

FUZZY/CLASSIC HYBRID CONTROLLER FOR THE LONGITUDINAL CONTROL OF AN AUTONOMOUS VEHICLE

Zyed ZALILA
Université de Technologie de Compiègne
Unité de Recherches COSTECH - IDTH
BP 649
60206 Compiègne Cédex, FRANCE
E-mail: zalila@mx.univ-compiegne.fr

Patrick LEZY
RENAULT
Direction de la Recherche
Département Systémes Véhicules
9-11, avenue du 18 juin 1940
92500 Rueil Malmaison, FRANCE

ABSTRACT

This research on the longitudinal control of a motor vehicle was carried out as part of the Autonomous Intelligent Cruise Control (AICC) project, which is itself part of the PROMETHEUS European Community program. It has two objectives : a speed regulation : the driver sets a cruising speed, the vehicle reaches that speed and keeps it ; a distance control : when the AICC vehicle detecs a target vehicle, it decides whether it is necessary to modify the actuators instructions so as to place itself at a safety distance behind the followed vehicle.

These two operating modes are permanently active and must satisfy subjective criteria of comfort while ensuring a high robustness in control. These constraints are one reason why we had a hybrid approach in the conception of the regulator. Indeed, a classic type of control (by internal model) ensures a deceleration/acceleration regulation of the vehicle ; the deceleration/acceleration instruction being generated by a controller based on Fuzzy Logic methods. This original hybrid control guarantees the robustness criterion even if there is a variation of the structural parameters such as mass, and if there are model errors as well. It also enables the driver to personalize the running of his vehicle by selecting a driving style ("confortable", "sportslike",...).With the controller thus conceived we obtain quite satisfactory results within the desired limits of use.

***Keywords** : longitudinal control, motor vehicle, distance/speed regulation, fuzzy control, fuzzy supervision, hybrid regulator.*

1. INTRODUCTION

The study presented here deals with the longitudinal control of a motor vehicle ; it was carried out as part of the AICC project which is itself part of the Prometheus European community program. This research, a result of the collaboration between Renault and the Université de Technologie de Compiègne (UTC), was done during spring 1991 and was succintly explained in March 1992 at the Fuzzy Logic Meeting organized by Volkswagen Research Center in Wolfsburg (Germany).

It has two main objectives : a speed regulation which consists in reaching and keeping the vehicle equipped with AICC system at a cruising speed selected by the driver (cf. Figure 1) ; a regulation of the distance which separates the AICC vehicle from the one preceding it in order to place AICC at a security distance behind the followed vehicle (also called target vehicle) with a zero relative speed (cf. Figure 2). The basic informations which are relative distance and relative speed are given by a sensor placed in front of the AICC vehicle.

The chief idea behind the AICC project is mainly traffic regulation on highways in order to increase the steady flow. This concept "smart vehicles on smart highways" is also illustrated by the Californian project of "electronic trains of cars" consisting in virtually and automatically hooking the cars together ; the objective in view is to increase the hourly flow from 2500 to 8000 cars [VAR92].

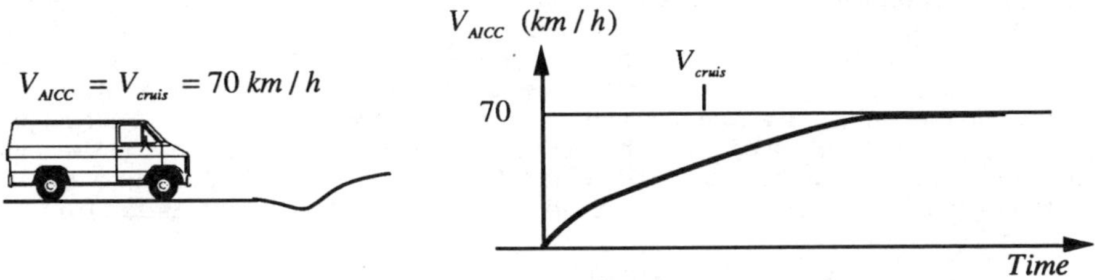

Figure 1 : Speed regulation

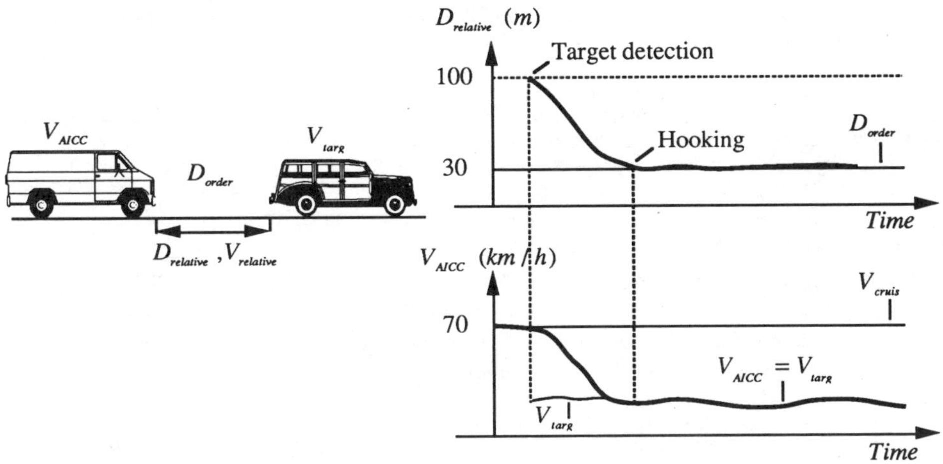

Figure 2 : Distance regulation

2. CONSTRAINTS

To reach these objectives it is necessary to take into acount four types of constraints :

• Comfort aspect : the two functionning modes (speed and distance regulations) being continuously active must make accelerations, braking and automatic changing of gears very easy and smooth in order to meet the subjective comfort criterion felt by the driver and the passengers as well, while ensuring a strong robustness to the regulation.

• Security aspect : at any time, the driver must be able to take again control of the car, and consequentely to easily disconnect himself from the AICC system.

• Equipment aspect : the developped algorithms must function in real time and must also be re-used in other projects. Furthermore, the longitudinal control system must be on board and portable on other vehicle of the make.

• Environment aspect : the system must take into account climatic (rain, snow, fog, sun, night) and traffic (steady flow, density) variations. Moreover it must adapt itself to different kinds of roads (secondary, main, highways) and to the profiles of the road (even, bend, downhill or climb).

3. DESIGN

The original solution adopted by RENAULT was a classic control theory approach ; its main advantages was to turn to usual and mastered techniques. Yet, this solution presented several drawbacks : on the one hand it inferred a rather abrupt driving with strong jerks, generally uncomfortable in the transition phases between the two functionning modes (speed regulation, distance regulation) ; on the other hand, this approach lacked robustness in that it did not adapt well to the great variety of scenarii suceptible to be met in real driving ; that it was very sensitive to any variation in the structural parameters of the vehicle (loaded car, worn-out brakes,...) and that it finally was not portable since the adaptation to another type of vehicle was far from being immediate.

The second solution, the one retained as part of this research, consisted in a hybrid approach coupling the fuzzy logic [MAM74] [SUG85a] [LEE90] and the classic control theory techniques. Thanks to this original approach, we thus turned to good account the whole modelization work initially developed by the control system engineers at RENAULT [REN90a] [REN90b] [LEZ91], by having it gain flexibility, robustness, even vis-à-vis model errors, and adaptability brought about by fuzzy logic for non linear dynamic systems [ZAD88] ; however it was done at the expense of an inescapable increase in the parameters to set.

In concrete terms, the regulation system is composed of a fuzzy rule-based supervisor controlling by order points a classic controller with internal model (cf. Figure 3). From the informations given by the LIDAR sensor (relative speed and relative distance with reference to the target vehicle) and the data fixed by the driver (cruising speed, order distance and driving style), the fuzzy supervisor determines the strategy to be applied (speed regulation, distance regulation or a combination of both) and gives the classic controller acceleration or deceleration Γ_o instructions. The controller in turn orders actuators acting on the throttle as well as on a system of additional brakes. A positive value of Γ_o is equivalent to a request to accelerate. In case it is negative, a small value of $|\Gamma_o|$ means that only an engine brake is needed ; so the classic controller only acts on the throttle by narrowing its opening. On the other hand a high value of $|\Gamma_o|$ means the necessity to act on the additional braking system in top of the engine braking.

Let us note that by default, the instruction distance is calculated from the speed of the Fuzzy AICC vehicle. Still, the driver can change this value any time, for example if he wishes to "stick" very close to the target vehicle even at a sustained speed.

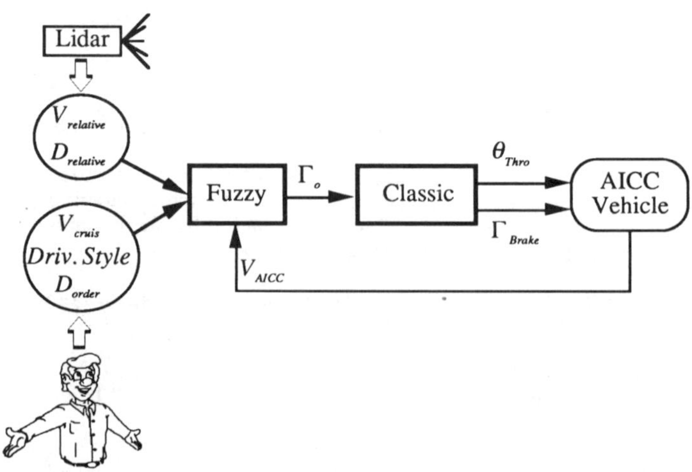

Figure 3 : Control law structure

4. Fuzzy strategy

The completion of a fuzzy controller goes through five steps [SAN89] [MAL91] :

- the definition of the input and output variables of the controller ;
- the cutout of those variable's ranges into fuzzy partitions consisting of linguistic labels (fuzzification) ;
- the elaboration of fuzzy rules of control, generally based on expert knowledge, which process those variables ;
- the choice of fuzzy logic operators for the evaluation and inference of those rules ;
- the choice of the method to obtain the real order to apply to the process from partial fuzzy conclusions (defuzzification).

Our fuzzy control box has three inputs : the difference between the relative distance and the order distance ($D_r - D_o$), the relative speed ($V_r = V_{AICC} - V_{targ}$) and the difference between the speed of the Fuzzy AICC vehicle and the cruising speed ($V_{AICC} - V_{cruis}$). The first two variables are the explicit inputs of the control table ; the last variable being used implicitly in the cells of the table labelled by Γ_{SR} (cf. Figure 5). The only output is the decelaration/acceleration order (Γ_o).

The fuzzy partition of $D_r - D_o$ is symmetrical and composed of five fuzzy intervals defined on the universe $[-100\,m, +100\,m]$: "Order very much overstepped", "Order overstepped", "Order reached", "Order not reached" and "Security not reached" (cf. Figure 4a). The V_r partition is also composed of five fuzzy intervals : "+++", "++", "+", "=" and "-" (cf. Figure 4b) ; but it is not symmetrical : it is indeed much more interesting to detail the situations in which the speed of the Fuzzy AICC vehicle is superior to the speed of the target vehicle, that is to say, situations in which the target vehicle is a potential obstacle to the following vehicle. In the opposite case, we are inevitably in speed regulation (no obstacles).

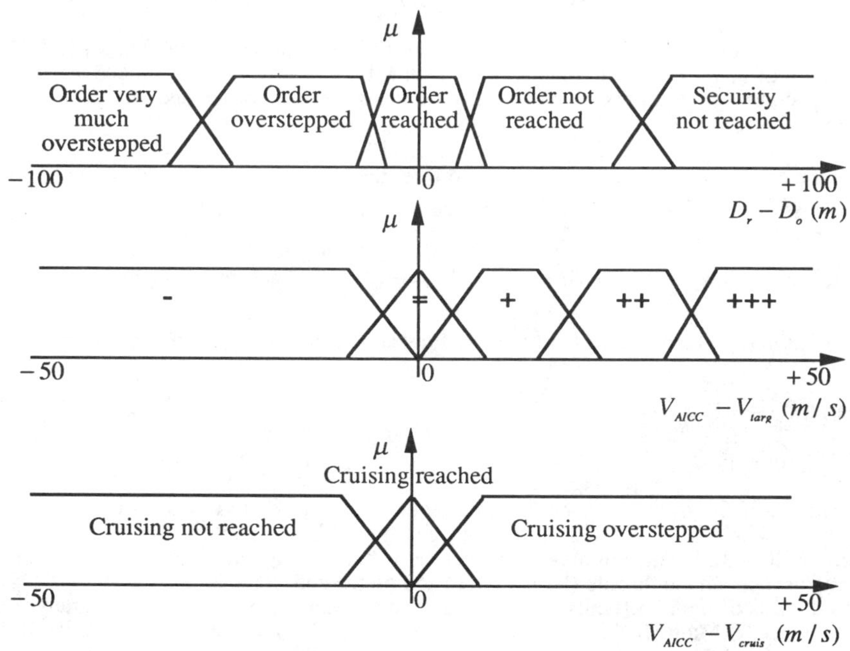

Figures 4a, 4b et 4c : Fuzzy partitions of the three input variables of the supervisor

The conclusions of the supervisor's fuzzy rules are crisp (real values) and can have seven different acceleration/deceleration values : "Strong" (Γ_S), "Medium-Strong" (Γ_{MS}), "Medium" (Γ_M), "Weak" (Γ_W), "Very Weak" (Γ_{VW}), "Nil" (Γ_{nil}) and "Speed Regulation" (Γ_{SR}) (cf. Figure 5a). The last value indicates that we are exclusively in speed regulation mode ; in this particular case Γ_o will be generated by a second fuzzy controller with one input variable ($V_{AICC} - V_{cruis}$) which is defined on a symmetrical fuzzy partition composed of three fuzzy intervals. This sub-controller, which induces a chain of fuzzy rules, produces a comfort acceleration/deceleration aiming at joining the cruising speed with the greatest flexibility and the minimum disturbance.

$D_r - D_o$ / $V_{AICC} - V_{targ}$	Order very much overstepped	Order overstepped	Order reached	Order not reached	Security not reached
+++	Γ_S	Γ_S	Γ_S	Γ_{MS}	Γ_M
++	Γ_S	Γ_S	Γ_M	Γ_W	Γ_{VW}
+	Γ_{MS}	Γ_W	Γ_{VW}	Γ_{VW}	Γ_{SR}
=	Γ_W	Γ_{VW}	Γ_{nil}	Γ_{SR}	Γ_{SR}
–	Γ_{nil}	Γ_{SR}	Γ_{SR}	Γ_{SR}	Γ_{SR}

$V_{AICC} - V_{cruis}$	Cruising not reached	Cruising reached	Cruising overstepped
Γ_{SR}	$+\Gamma_{comfort}$	Γ_{nil}	$-\Gamma_{comfort}$

Figures 5a et 5b : Fuzzy supervision tables

After several comparative tests, we finally kept the Zadeh primary operators [ZAD65] for the modelization of conjunction and disjunction (resp. MIN and MAX) ; they indeed were offering an interesting compromise between the quality of the generated order and the time necessary to calculate this order.

As all the conclusions are crisp, we chose to infer the command Γ_o by calculating the barycentre of the i possible conclusions C_i of the fuzzy rules, respectively weighted by the firing degrees α_i of these fuzzy rules [MAM77][BRA79] : $\Gamma_o = \frac{\sum_i \alpha_i . C_i}{\sum_i \alpha_i}$. Thereby, it is the so called Sugeno method [SUG85b] which is used to obtain the final conclusion of the fuzzy supervisor.

5. Results

5.1 The prototype

The Fuzzy AICC system was first adjusted by numerous series of tests on simulator. Then it was loaded on a RENAULT 25 with automatic gearbox. The sensor used is a Lidar with double scanning (RENAULT radar) in order to detect the vehicles which are more or less low on the road. The actuators acts on the throttle (RENAULT) and on the additional braking system (BOSCH). The embarked PC is of the 386 family to which is added a signal processing card to analyse the Lidar signal. Finally, a Man-Machine Interface enables the driver to set in a user-friendly way the cruising speed and the order distance, as well as the style of driving desired : quiet and comfortable driving, sportslike driving implying free accelerations/decelerations and brakings...

5.2 Real tests

The successful convincing results in simulation and in test centres enabled the testing of the Fuzzy AICC system on open road. Figues 7, 8 and 9 show recordings of speed tracking and speed regulation with different road profiles. The first graphs of each figure represent the cruising speed V_{cruis} and the Fuzzy AICC speed V_{AICC}. The second diagrams represent Γ_o, Γ_{AICC} and the deceleration order Γ_{Brake} for the BOSCH additional braking system . The third curves represent θ_{Thro} and the gear (multiplied by 10). Let us note that although the accelerometer detected real accelerations (Γ_{AICC}) appear relatively noisy they are not a source of discomfort ; indeed, the fluctuation frequency is such that the human body does not feel anything at all.

In the figure 7 scenario, the cruising speed is initialy set at $90\ km/h$ on a slope. The Fuzzy AICC vehicle meets the speed and even exceeds it given the downward configuration of the road. Following the change in the cruising speed desired by the driver ($V_{cruis} = 61\ km/h$), the system then finds itself quite beyond the speed order. Therefore it asks for an important deceleration which cannot be achieved by the engine braking only (θ_{Thro} at its minimal value of 6°) ; The additional BOSCH braking system is then activated ($\Gamma_{Brake} = -0.7\ m/s^2$). Both braking types are maintained until the Fuzzy AICC vehicle reaches its cruising speed. Even if the road profile becomes a climb, the system perfectly keeps the vehicle at $61\ km/h$ by disconnecting the additional braking ($\Gamma_{Brake} = 0\ m/s^2$). We can also notice that this automatic gear change is progressive, without jerks and that ensures a definite comfort for the driver and the passengers.

In the figure 8 scenario, the cruising speed is set to $61\ km/h$. Even on a climb configuration with bends, the Fuzzy AICC system reaches V_{cruis} in 20 seconds without overtaking it. Then it perfectly maintains itself at the speed by going up into third gear. The BOSCH braking system is never activated ($\Gamma_{Brake} = 0\ m/s^2$).

In the figure 9 scenario, the vehicle is on a highway ; the driver sets the cruising speed to the maximum speed authorized in France ($V_{cruis} = 131\ km/h$). The Fuzzy AICC vehicle reaches the speed in 55 seconds while being into fourth gear and with a throttle maximum opening ($\theta_{Thro} = 90°$). Reaching this speed is done smoothly before a correct stabilization to the desired value. At no time is the additional braking requested.

Figures 10, 11 and 12 show recordings done in target approach and distance regulation situations. The first graphs of each figure represent the order distance D_o and the relative one D_r. The second graphs show the cruising speed V_{cruis}, the Fuzzy AICC speed V_{AICC} and the target speed as well ($V_{targ} = V_{AICC} - V_r$). The third curves represent Γ_o and Γ_{AICC} . The fourth graphs show the brake deceleration order Γ_{Brake} for the BOSCH system . Finally, the last curves represent θ_{Thro} and the gear (multiplied by 10).

In the figure 10 scenario, a target car moving at $94\ km/h$ is detected by the Lidar after 6 seconds. The Fuzzy AICC vehicle then finds itself with a relative distance greater than $100\ m$ and a speed of $129\ km/h$. From this moment on, the value of the order distance previously fixed by the driver to $45\ m$ is calculated again by the system and stabilizes itself around $35\ m$. The system immediately asks for a maximum engine breaking (θ_{Thro} at its minimum value of 6°), supported by some exceptional actions on the additionnal breaking : during the first three seconds Γ_{Brake} equals twice $-1.4\ m/s^2$. About 25 seconds after the detection, the Fuzzy AICC vehicle smoothly hooks itself to the target at the desired order distance with a zero relative speed : an "electronic train of cars" is starting out. The approch as well as the distance regulation take place into fourth gear.

The figure 11 scenario corresponds to a downhill road section. A target is moving with a speed varying between $94\ km/h$ and $117\ km/h$ and at a relative distance moving between $40\ m$ and

50 m. We notice that the Fuzzy AICC vehicle does maintain itself in a thin margin around the order distance, while smoothing the target speed ; this in order not to slavishly reproduce the untimely speed changes of the followed vehicle. The BOSCH braking system was activated twice in order not to dangerously come close to the followed vehicle. Indeed, after the target speed change, the engine braking maximum ($\theta_{Thro} = 6°$) was not deemed sufficient to bring the relative distance back to the security distance.

In the figure 12 scenario, the Fuzzy AICC vehicle is moving at 140 km / h and detects a target moving at 86 km / h. It obviously comes close to the obstacle with a too high initial speed (initial $V_r = 54\ km / h$). The system immediately asks for a maximal engine breaking ($\theta_{Thro} = 6°$), supported by free decelarations thanks to the additional breaking (up to $-2.7\ m / s^2$). Inspite of that, the Fuzzy AICC vehicle enters into the security zone rather dangerously : it goes beyond the order distance by nearly 10 m and stays at this undesirable zone for almost 10 s. Once this time elapsed, the hooking takes place smoothly into fourth gear.

6. CONCLUSIONS AND PERSPECTIVES

In the present state of the research, the Fuzzy AICC system gives every reason for satisfaction in a great number of real configurations (cf. Figure 6). Yet, it cannot be used in cases when the initial relative speed is important wheareas the initial relative distance is close to the order distance ; since indeed for comfort reasons, we limited the decelaration maximum value (to $-3\ m / s^2$), it is physically impossible for the Fuzzy AICC vehicle to sufficiently brake when it goes too fast close to a target vehicle detected almost at the last moment.
Ergonomic studies have already started in order to find out if stronger decelerations, for example up to 1 G, could bother or even surprise the driver. Wih the present values, we can simply say that driving with the Fuzzy AICC system is pleasant.

The other regulation limits of the system are essentially due to the Lidar. Also, the situations involving an instantaneous loss of the target, such as when the target vehicle overtakes a truck preceding it, lead to an unwanted inertia in the system's reaction. On the other hand though, such inconvenience can be desirable since it seems necessary to maintain a certain latent period so that a very brief and temporary target loss does not interrupt the regulation process.

The case of dense and fast traffic leads to two types of potential problems, too. First of all, there is a risk of multiple echoes at the Lidar level : it then becomes problematical for the image processing algorithm to select the right target ; now, the quality of the regulation strategy provided by the fuzzy supervisor totally depends on the reliability of the data given by the Lidar (relative speed and relative distance).
The second type of problem we could be confronted with in fast and dense traffic scenarii is the risk of the instantaneous interposition of a car which would cut in right in front of the Fuzzy AICC vehicule ; the system would then have a tendency to suddenly brake, in order to find its order distance again, which would not necessarily be done by a driver who would anticipate the situation.

Apart from the solving of the previous problems, the prospect of furthering the research consists on the one hand in treating dense but slow traffic situations which correspond to traffic jams ("Stop and Go") ; that would enable to increase the driving comfort in such stressful situations.
On the other hand, it appears quite interesting to give the system a training ability which would authorize it to analyse the driver's driving style in order to imitate him ; the fuzzy controller calibration would then be done automatically and that would work towards facilitating the acceptability of such a system.

In a general way, the problem of assistance to the driver still remains : with a system that would act instead of him, would not the driver run the risk of falling into a state of hypovigilance since he has almost nothing to decide by himself ? Would not that end up in dulling his reflexes ? Only detailed behavior studies could find the answer, all the more so as the driver's reactions depend a great deal on his temperament.

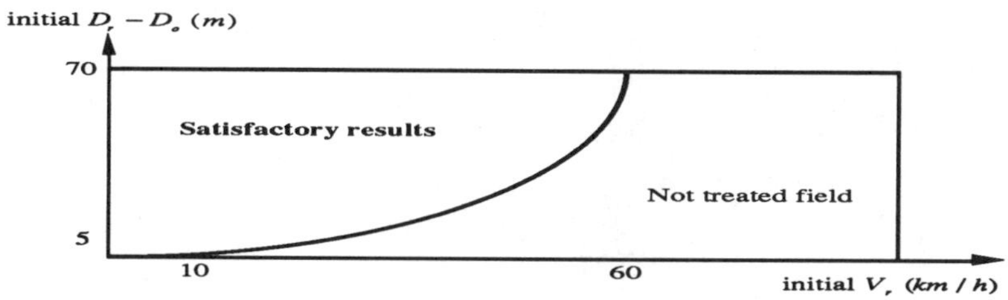

Figure 6 : Validity field of the Fuzzy AICC system

REFERENCES

[BRA79] BRAAE M., RUTHERFORD D.A.
Theoretical and linguistic aspects of the fuzzy logic controller. Automatica, 15, pp 15-30, 1979.

[LEE90] LEE C. C.
Fuzzy logic in control systems : Fuzzy Logic Controller, Part 1 & 2. IEEE vol. SMC 20, n°2, pp 404-435, March/April 1990.

[LEZ91] LEZY P.
AICC : Autonomous Intelligent Cruise Control. Présentation interne RENAULT, Direction de la Recherche, Département Systèmes Véhicules, réf. H353101, 3 octobre 1991.

[MAL91] MALVACHE N., WILLAEYS D.
La commande floue de processus. Rapport interne LAIH - Unité associée CNRS n°1118, Université de Valenciennes, 1991.

[MAM74] MAMDANI E.H.
Application of fuzzy algorithms for control of a simple dynamic plant. Proc. of IEE, vol. 121, n°12, pp 1585-1588, December 1974.

[MAM77] MAMDANI E.H.
Application of fuzzy set theory to control systems : a survey. In: Fuzzy Automata and Decision Processes (M.M. Gupta, G.N. Saridis, B. R. Gaines, eds.), North-Holland, New-york, pp 77-88, 1977.

[REN90a]
Choix de la loi de commande pour la régulation de vitesse et de distance. Rapport interne RENAULT, Direction de la Recherche, Département Systèmes Véhicules, 1990.

[REN90b]
Présentation de la bulle 3 du projet AICC. Rapport interne RENAULT, Direction de la Recherche, Département Systèmes Véhicules, 1990.

[SAN89] SANCHEZ E.
Applications industrielles de la commande floue au Japon. Interfaces n°80, pp 19-25, Juin 1989.

[SUG85a] SUGENO M.
An introductory survey of fuzzy control. Information Sciences, 36, pp 59-83, 1985.

[SUG85b] SUGENO M., NISHIDA M.
Fuzzy control of model car. Fuzzy Sets and Systems, 16, pp 103-113, 1985.

[VAR92] VARAIJA P.
Smart vehicles on smart highways. IEEE Conference on Decision and Control, Tucson, Arizona, USA, December 16th, 1992.

[ZAD65] ZADEH L.A.
Fuzzy sets. Information and Control, 8, pp 338-353, June 1965.

[ZAD88] ZADEH L.A.
Fuzzy logic. IEEE Computer, pp 83-93, 1988.

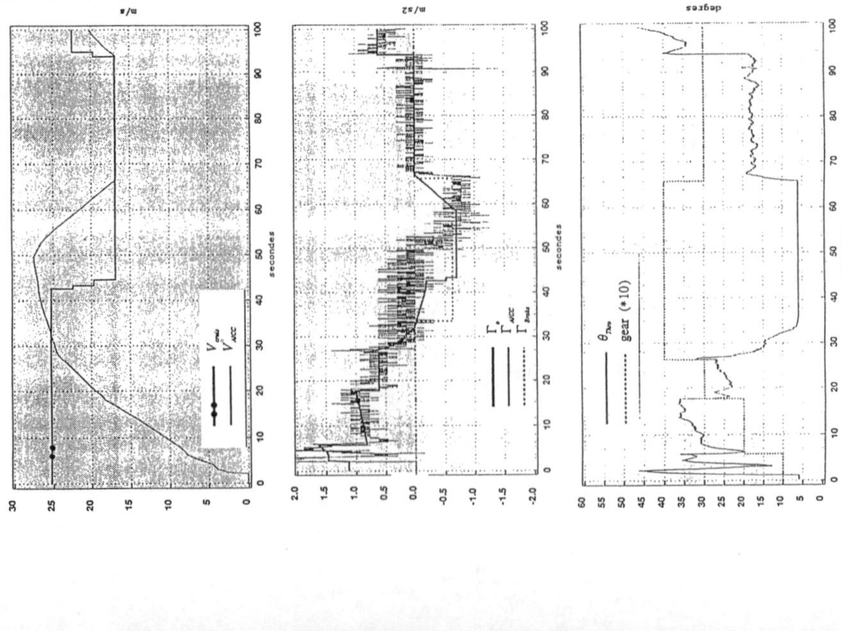

Figure 7 : Speed tracking in downhill - Speed tracking and regulation in climb. $V_{cruis} = 90\ km/h$ then $61\ km/h$.

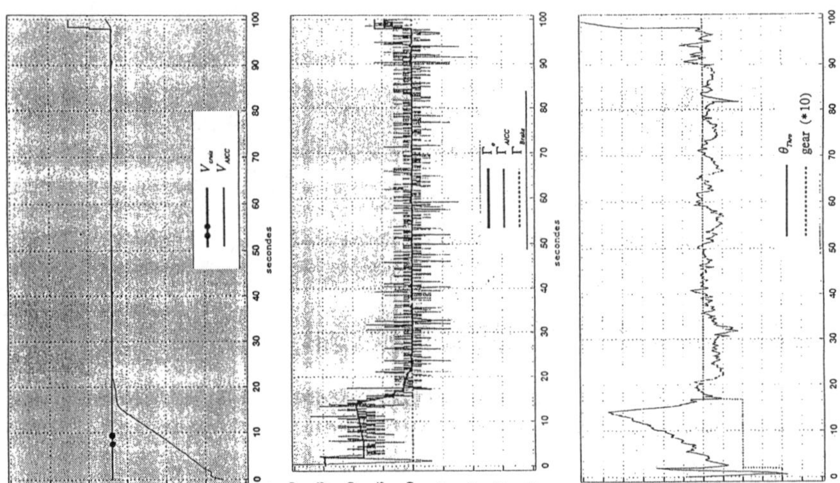

Figure 8 : Speed tracking and regulation in climb with turns. $V_{cruis} = 61\ km/h$.

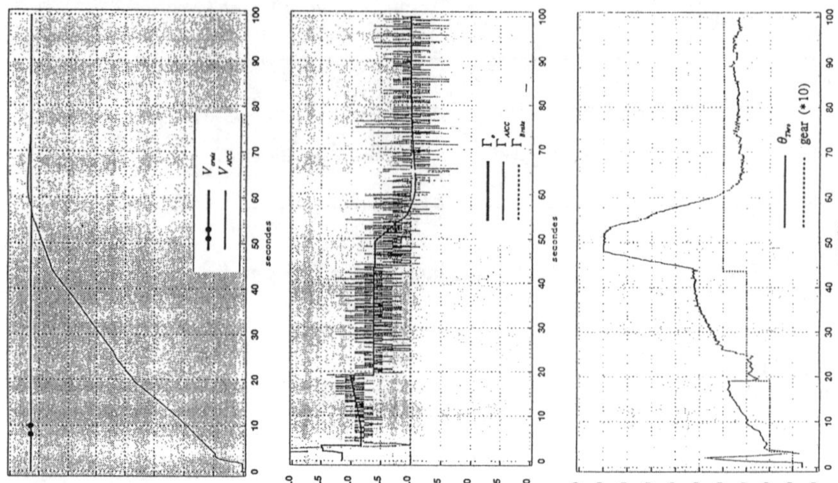

Figure 9 : Speed tracking and regulation on flat road. $V_{cruis} = 131\ km/h$.

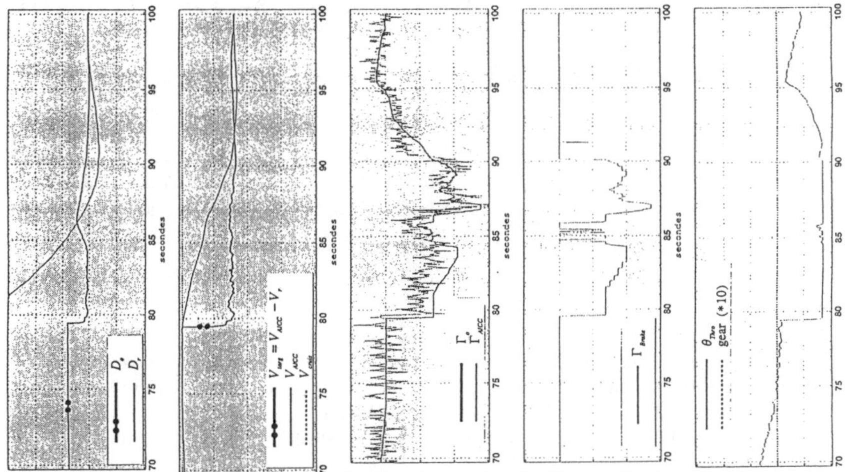

Figure 12 : Approach with initial $V_r = 54\ km/h$ and initial $D_r > 100\ m$. Distance regulation at $86\ km/h$.

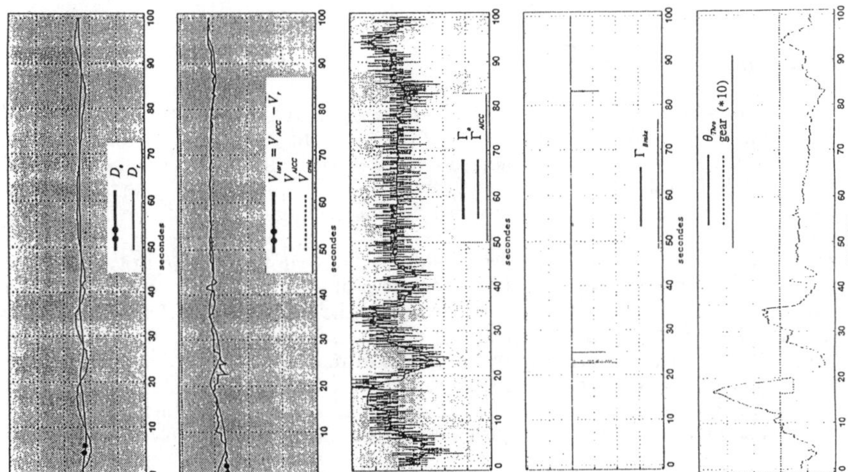

Figure 11 : Distance regulation in downhill. V_{targ} changes.

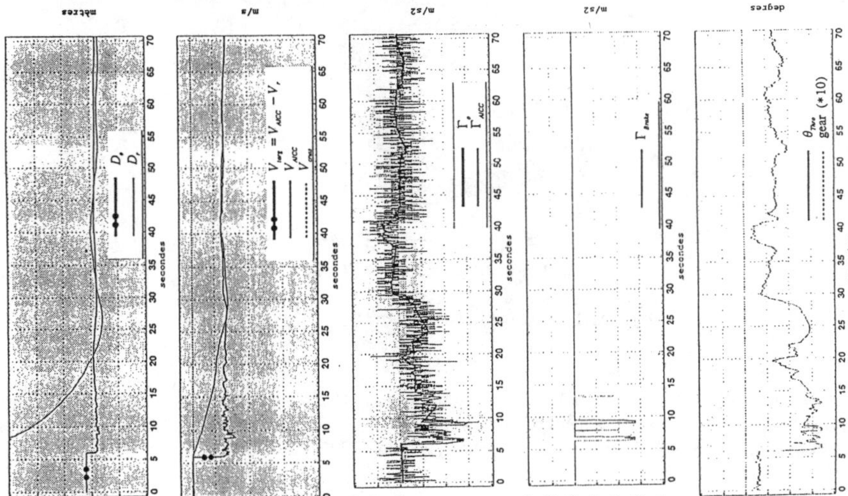

Figure 10 : Approach with initial $V_r = 35\ km/h$ and initial $D_r > 100\ m$. Distance regulation at $94\ km/h$.

APPLICATION OF FUZZY KNOWLEDGE-BASED SYSTEM TO CAR SEATS ADJUSTMENT

Sylvain Plommet*, Patrick Truchot**, Jean Marc JUDIC***, Patrick van Effenterre***

* Institut Polytechnique de Sévenans
Division CSP
90010 Belfort - France

** ENSAM
Laboratoire CPN
151, bd de l'hôpital
75013 Paris - France

*** BERTRAND FAURE FRANCE
Parc d'activités du Moulin de Massy
23, rue du Saule Trapu
91882 Massy Cedex France

ABSTRACT

A study was made to design an automatic system to find the required adjustments of a car seat according to the anthropometric features of the users. The main problems of car seating comfort were presented, and a fuzzy knowledge-based system was proposed as a solution. The construction of the knowledge base, the performance and the conditions of use of the system were described. A comparison with a program obtained through a conventional approach facilitated the validation of the system and showed the advantages of a fuzzy knowledge-based system for the Bertrand Faure industrial corporation.

I - Introduction

A car seat is one of the most important components of an automobile, because it both carries the image of the car and contributes to driving comfort. It is both a physical and sensory link between a vehicle and its user. From the first motorized vehicles with simple wooden benches, to the latest designs, the seat has evolved in parallel with the automotive industry in general. The research of a better posture comfort, and a greater complexity of the seating component are consequences of this evolution.

A car seat is made of three distinct sub-assemblies [1], which can be adjusted in various positions. The number of its functions increased largely during the past ten years. For exemple, longitudinal slide and backrest angle adjustment were generalized since the end of the 1970s, and other adjustments were progressively added, such as seat height, lower seat angle or lumbar support. These adjustments are now a standard feature of lower to mid-range vehicles (Renault 19, Citroën ZX, ...). Also, some upper range or experimental seats, such as MASSY 2000 [2], allow about ten different adjustments. Finally, these adjustments are often motorized to facilitate their use. Figure 1 presents the four most common seat adjustments proposed by the automotive industry around the World.

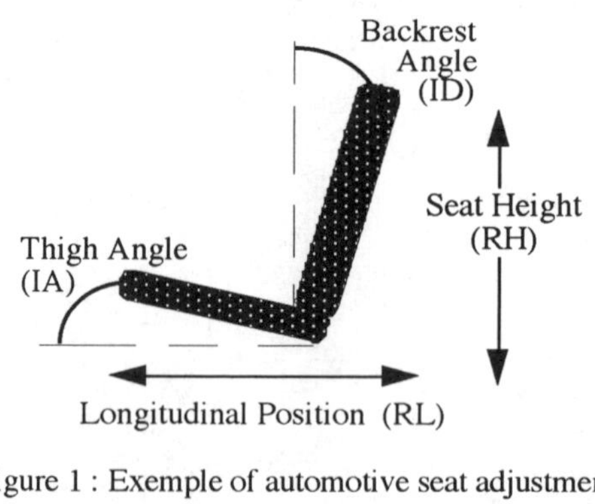

Figure 1 : Exemple of automotive seat adjustments

---

[1] The metal frame, with the primary function of providing rigidity to the whole seat, the foams, and the fabric cover.

[2] MASSY 2000 is an experimental seat allowing nine motorized adjustments. MASSY 2000 was presented by Bertrand Faure France during the Mondial de l'Automobile show, October 1992, Paris, France.

As for other mass consumption products, this evolution presents the problem of the man-machine interface. Notice that, according to a study made in the U.S. [1], about 90% of the VCR owners are unable to use all of the functionalitites available on their devices, especially the programming options.

In the 1980s, being the European leading producer of automotive seats, the Bertrand Faure France corporation initiated a research program concerning posture comfort to improve its products. As a part of this research, our study was oriented towards the application of knowledge-based systems to determine automotive seat adjustments according to the anthropometric characteristics of the user. The objective is to bring valuable answers to the problem of obtaining software that is adapted to computer supports used in the automotive industry.

II - The concept of posture comfort

Recent studies applied to automotive seating pointed out the multi-faceted aspects of posture comfort. Actually, its definition is based on data coming from various scientific domains.
To begin with, technical constraints concerning dimensions and joints of a seat [2] have to be considered. Then, ergonomy-related data have to be gathered. The activity of driving can be divided into elementary actions, either successive or simultaneous, that are translated in terms of collecting information, effecting motions, exerting forces, and operating at certain speeds.
Finally, physiological data about the driver must be obtained :

- The man-vehicle interface [2] consists of requirements describing the position of human body elements, with respect to vehicle elements (hands on the steering wheel, for exemple).
- A 2D model of the human skeleton [3] consists of 8 articulated elements (Figure 2), with the anthropometric characteristics of drivers.

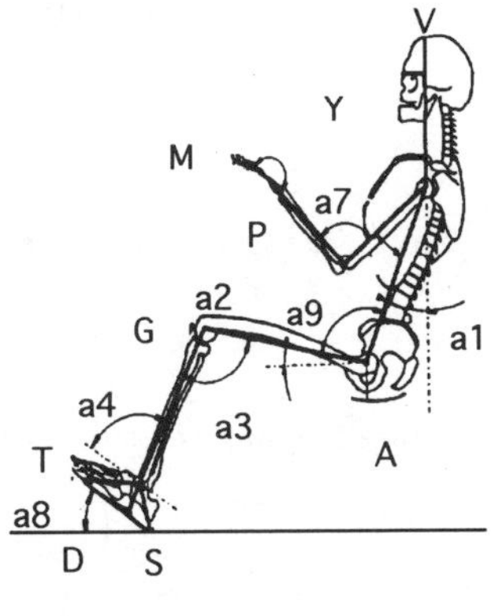

Figure 2 : Model of human skeleton using a system of articulated elements

- The notion of "angle of least discomfort" [4]. To be able to withstand the various constraints during an extended period of driving time, the human body must adapt. Concerning angles in the body joints, the need to adapt can be related to ranges. Out of the limits, pains may quickly appear in the muscle and ligament (Figure 3) areas.

Angles Definition	Ref.	Mini	Maxi
Torso Axis - Vertical	a1	20°	30°
Torso Axis - Thigh Axis	a2	95°	120°
Thigh Axis - Lower Leg Axis	a3	95°	135°
Leg Axis - Former Arm Axis	a4	78°	105°
Torso Axis - Upper Arm Axis	a5	0°	50°
Upper Arm Axis - Former Arm Axis	a6	80°	170°
Former Arm Axis - Hand Axis	a7	170°	190°
Former Arm Axis - Floor Pan	a8	40°	70°
Thigh Axis - Horizontal	a9	12°	25°

Figure 3 : Angles of least discomfort

III - Modeling the automotive man/seat system

More research is needed to built a coherent and organized set of data. However, the work done by the Bertrand Faure France corporation [5] allowed the construction of a model including an automotive seat and its passenger, within the inevitably limited environment consisting of a car interior. Based on the model, a computer program, called Ergosoft, was written to determine seat adjustments according to user anthropometric characteristics. To be validated, the model was applied to the MASSY 2000 experimental seat [3].
Ergosoft was written to help the Bertrand Faure France design team build new automotive seats. Powerful computer tools are required (Intel 386-type microprocessor or higher) to perform Ergosoft's calculations, due to the complexity of the mathematical model.

To better satisfy the needs of the consumers, the Bertrand Faure France corporation plans to design an on-board program that could determine seat adjustments, like Ergosoft. Thus ,the main problem in this study is to obtain a model of the man-seat system with computational tools that could be integrated into an automotive seat within industrial cost limits. Our objective is to show the advantages of fuzzy knowledge-based systems, and of a linguistics-based approach, for this kind of problem.

IV - Design of a fuzzy knowledge-based system

IV.1 - Description of the command system - Design constraints

For the preliminary study of technical feasibility, only the four most common seat adjustments offered by the automotive industry around the World were considered :

- RL, the longitudinal (front-to-back) position,
- ID, the backrest angle,
- RH, the seat height position,
- IA, the thigh angle.

Inputs to the system are, as for Ergosoft, the anthropometric characteristics of drivers :

- S, the driver height,
- T, the driver chest size,
- P, the driver weight.

Figure 4 shows the possible location of the control panel on an automotive seat.

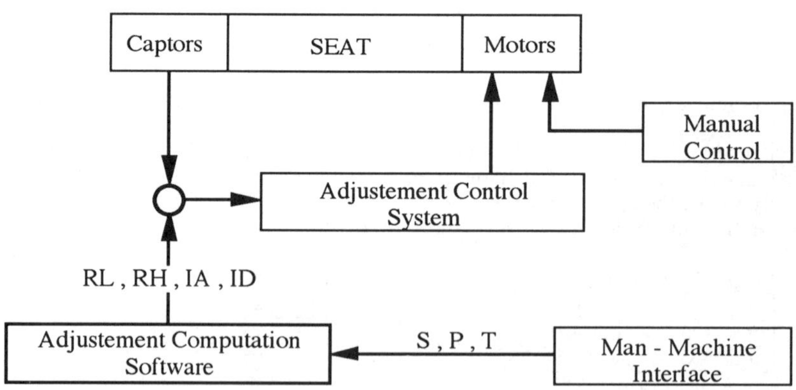

Figure 4 : Sample location of a control panel within a vehicle

The control system uses closed-loop commands, with the capability to adapt through open loops. The adjustment computation software generates signals that are transmitted to the computer controlling

[3] MASSY 2000 is an experimental seat offering nine completely motorized adjustments. MASSY 2000 and Ergosoft were presented during the Mondial de l'Automobile show, October 1992, Paris, France.

seat motions. The man-machine interface allows the user to enter height, chest size and weight values as program inputs. The driver can also modifiy the seat position through manual controls.

The selection of hardware for the control system is highly determined by cost constraints, whichever adjustment software approach is used. Our study is focused on the adjustment software, with the following design constraints :

Expected performances :

- Adjustment variables are functions of the anthropometric characteristics of seat occupants. The monotony of these functions must be respected (Figure 5).

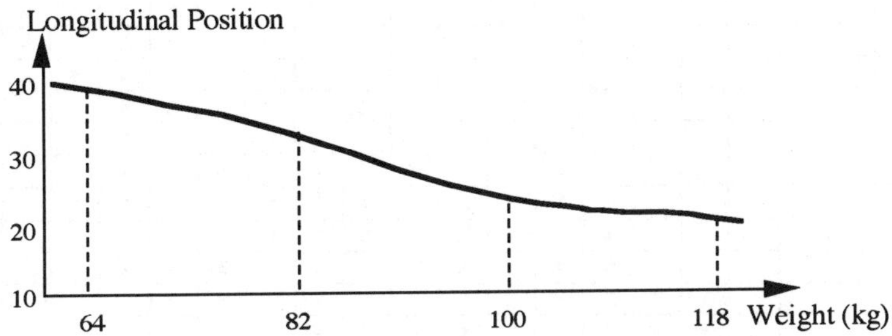

Figure 5 : Longitudinal position versus user weight

- A lower level of accuracy can be tolerated, because Ergosoft does not account for seat wear, or driving habits, for exemple. However, this difference of accuracy should not be perceived by the user, and its value depends upon the type of seat and the adjustment considered.

Technical constraints :

- The determination of adjustments must be fast. Computations times must be acceptable for the users (a value of about 1 second was accepted).
- The software must be compatible with the chosen micro-control device [4].

IV.2 - Building the knowledge base

The lack of experts about seat adjustment determination was the largest difficulty in this study. Each individual has an experience about driving self-propelled vehicles, but cannot extrapolate this knowledge to other people. The only useful knowledge available to us comes from Ergosoft. Each set of numerical values (S,T,P) is related to another set of values (RH, RL, IA and ID). Fuzzy logic rules like IF ... THEN ... can directly represent this knowledge as :

IF		"The driver height, S, is ..."
	AND	"The chest size, T, of the driver is ..."
	AND	"The driver weight P is ..."
THEN		"The longitudinal location, RL, of the seat is ..."
	AND	"The backrest angle, ID, is ..."
	AND	"The seat height, RH, is ..."
	AND	"The thigh angle, IA, is ..."

IV.3 - Quantifying the knowledge base

The anthropometric characteristics of all the potential car seat users (male and female) are within the following ranges :

Height: 1360 to 2020 mm,
Chest size: 740 to 1050 mm,
Weight: 34 to 134 kg.

---

4 Motorola MC68HC11

By combining these three variables, 500 000 theoretical bodies [5] can be defined. Actually, some of these theoretical bodies do not match any real person. For exemple, the probability to find a man with a height of 1.50 m, a chest size of 1 m, and a weight of 90 kg, is very small.

A better representation of the population can be obtained by selecting individuals through a frequency diagram including both the height and the chest size (Figure 6).

CHEST SIZE (cm)

HEIGHT (mm)	148	150	152	154	156	158	160	162	164	166	168	170	172	174	176	178	180	182	184	186	188	190	192
1000																			X		X	X	X
990																	X	X	X	X	X	X	X
980															X	X	X	X	X	X		X	
970															X	X	X	X	X	X	X		X
960													X	X	X	X	X	X	X	X			
950													X	X	X	X	X	X	X	X	X		
940												X	X	X	X	X	X	X	X		X		
930									X	X	X	X	X	X	X	X		X	X				
920								X	X	X	X	X	X	X	X		X	X					
910									X	X	X	X	X	X	X	X	X	X					
900						X	X	X	X	X	X	X	X	X	X	X							
890					X	X	X	X	X	X	X	X	X	X	X								
880					X	X	X	X	X	X	X	X	X	X	X								
870			X	X	X	X	X	X	X	X	X	X	X										
860	X	X	X			X	X	X	X	X	X	X			X								
850	X	X	X	X	X	X	X	X		X	X												
840	X	X	X	X	X	X	X	X	X	X													
830	X	X	X	X	X	X	X	X	X														
820	X	X	X	X	X	X																	
810		X	X	X	X	X	X	X															
795	X		X																				

Figure 6 : Frequency diagram using both driver height and chest size.

The approach is to select "reference drivers" that can represent population classes through the 4 seat adjustments associated to each of them. A fuzzy logic rule allows the attribution of a range of ajustments values to each population class.

The best solution is to determine a rule basis for each ajustment. This method allows the selection of rules providing the best results, with respect to the determination of ajustments. However, this solution does not satisfy the cost requirements. Thus, a common knowledge base is established for all the adjustments, and all variables have the same level of significance. But, the criteria to facilitate the selection of "good" rules are then unavailable. Therefore, reference drivers are chosen to provide a homogeneous distribution of the target population (each reference driver is represented by a square box in Figure 6). Section V shows that, later on, the performance analysis of this system provided hints to optimize the knowledge base.

Because of their uncommon anthropometric characteristics, people who are not representative of the population are not treated differently: The pre-adjustments for a person having a height of 166 cm and a chest size of 990 mm, are the same as for a person with the same height and a chest size of 960 mm.

A similar study combines driver height and weight in a frequency distribution diagram to refine the selection.

Computer simulations show that the best trade-off between expected performance and technical constraints requires a knowledge base consisting of 113 rules (or 113 reference drivers). Eighteen fuzzy sets (six for the height, seven for the chest size, and five for the weight) are required to describe the population groups associated to the 113 reference drivers. The four adjustment variables

[5] Increment step sizes were 5 mm for height and chest size, and 2 kg for the weight.

are described by sixteen fuzzy sets each. The numerical values related to the adjustment variables are seat positions (225 possible positions for each of the variables).

Computer simulations also guide the choice of triangular and trapeze-shaped fuzzy sub-sets. Figure 7 presents the fuzzy sub-sets representing the population according to weight groups (P variable).

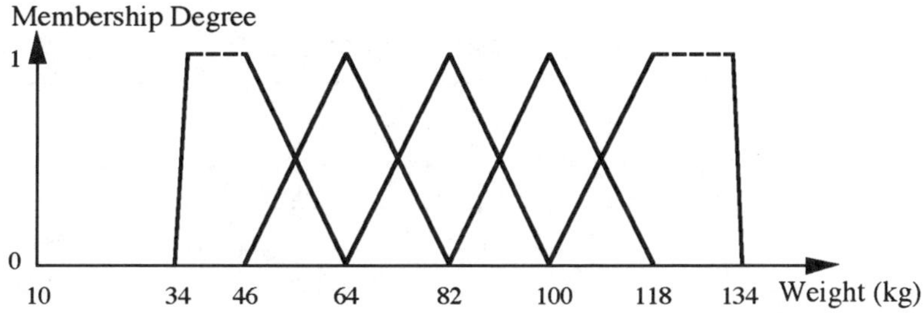

Figure 7 : Fuzzy sub-sets for the weight variable, P.

IV.4 - Selecting a fuzzy inference method

The inference method must satisfy the requirement for software speed of execution. Because the inference engine has to process 113 rules to determine the adjustments of the MASSY 2000 seat, an inference method using simple calculations is preferable. The general control method is retained, because it offers this feature. For the rule IF "X is A" THEN "Y is B", the general control method represents the fact "Y is B" by the function defined as :

$$\vartheta_{B'}: [0,1] \longrightarrow [0,1]$$
$$v \qquad \vartheta_{B'}(v) = \min(\vartheta_A(u_0), \vartheta_B(v))$$

V - Results

A performance comparison between the fuzzy knowledge-based system and Ergosoft, the reference software, allows the analysis of the fuzzy system's behavior, and the identification of opportunities for improvement. The results provided by the fuzzy knowledge-based system and Ergosoft are in general agreement. The discrepancies are mainly due to differences in hardware configurations (MC68HC11 micro-control for the new software, and 486 DX 33 PC for Ergosoft). Figures 8 and 9 present sample results.

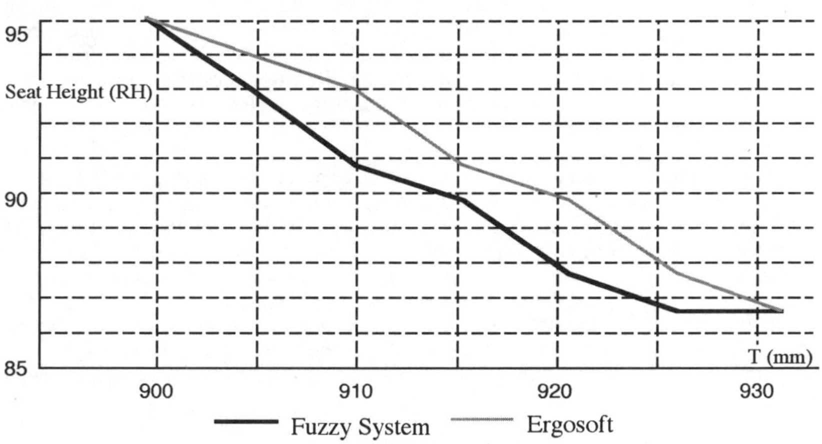

Figure 8 : Computed RH values versus actual chest size
(driver height: 174 cm, weight: 82 kg)

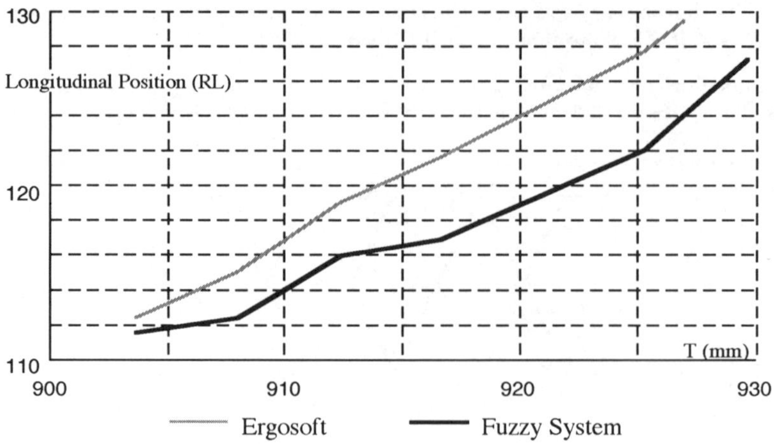

Figure 9 : Computed RL values versus actual chest size
(driver height: 174 cm, weight: 82 kg)

Differences in IA, ID and RH adjustments according to the software used are too small to be detected by users, who have higher thresholds. Longitudinal position has larger discrepancies, reaching up to 10 mm. Such a difference can be detected by users. These results can be explained in two ways. First, longitudinal position has a larger range of values than the other three variables. This is also amplified by the MASSY 2000 experimental seat, which has a much larger stroke than commercial seats. Second, users prove to be more sensitive to the value of this variable. Further developments of the software will have to take this problem into account.

VI. Summary and conclusion

This study shows that fuzzy knowledge-based systems are a promising method to take user physical characteristics into account during the design phase of a control system.

Comparisons of performance and of computation results show good agreement between a fuzzy knowledge-based system and the Ergosoft program. Also, the knowledge base can be modified to make the determination of the longitudinal position (RL) adjustment more accurate.

The approach used for this study is satisfactory for four of the variables. However, an MC68HC11 micro-control would not satisfactorily perform the calculations of a linguistics-based model for the 9 adjustment variables of the MASSY 2000 experimental seat. In the first section, fuzzy knowledge-based systems appear most powerful in single-input single-output control applications. This study shows that their use can be extended to a more complex process. However, technical constraints limit the possibilities of more developments.

Expert knowledge could be another area for research to improve this fuzzy knowledge-based system. Current knowledge is rather poor, because it is limited to numerical values of the (S, T, P) and (RL, RH, IA, ID) sets. Refining this knowledge is possible, but the problem statement would have to be completely revised. In this study, the main hypothesis is that the driver is not able to determine the appropriate adjustments, due to the complexity of the seat. However, the users master some of the adjustments, such as the longitudinal position and the backrest angle. Consequently, other approaches can be considered : the driver can adjust the seat for the longitudinal position, RL, and the backrest angle, IA, the calculation software taking care of the other adjustments.
Finally, more research is needed to evaluate the potential for actual use of the fuzzy knowledge-based system, before planning its evolution.

References

1 - "Les utilisateurs devront - ils devenir aussi intelligents que les produits qui leur sont proposés ? Le cas des télécommandes audiovisuelles" - Michel NAEL - 1989 - Actes du XXV ème congrès de la Société des Ergonomes de Langue Française (S.E.L.F.), Lyon

2 - "Le Confort, Elément Essentiel de la Qualité du Siège Automobile" - J. M. JUDIC - Acte du Congrès International de la Société des Ingénieurs de l'Automobile, "Les composants de l'Habitacle Automobile", 4&5 Décembre 1991, Paris

3 - "Free Body Diagrams as an Approach to the Mechanics of Human Posture and Motion" DEMPSTER, W. T. - 1961 - in, EVANS G., Biomechanical studies of the Musculokeletal System, Thomas, Springfield, 81-135

4 - "Le Siège du Conducteur. Son Adaptation aux Exigences Fonctionnelles et Anthropométriques" - R. REBIFFE - 1969 - in, E. GRANDJEAN, Posture Assise, Taylor and Francis, London, 139-147

5 - "More Objectives Tools for the Integration of Postural Comfort in Automotive Seat Design" - J. M. JUDIC, J. A. COOPER, P. TRUCHOT, P. VAN EFFENTERRE, R. DUCHAMP - Proceedings of S.A.E; International Congress and Exposition, March 1-5, Detroit, Michigan, U.S.A.

6 - "Application of Fuzzy Control for ISIS Vehicule Braking" - J. P. AURRAND-LIONS, L. FOURNIER, P. JARRI, M. de SAINT BLANCARD, E. SANCHEZ - Proceedings of IFSA'91, Brussels, Belgium, 7-12 July 1991

7 - "Application of Fuzzy Control in a Warm Water Plant" - W. J. M. KICKERT, H. R. VAN NAUTA LEMKE - Automatica, Vol 12, 301-308, 1976

8 - "Application of Fuzzy Algorithms for Control of Simple Dynamic Plant" - E. H. MAMDANI, B. E., M. Sc., Ph. D. - Proceedings of the Institution of Electrical Engineers, Vol 121, N° 12, December 1974

9 - "La Représentation et l'Utilisation de Connaissances Imprécises pour l'Aide à la Conduite de Procédés" - O. BOBROWICZ - Thèse de Doctorat, Janvier 1991, Université de Franche-Comté

A MULTIPLE ATTRIBUTE FUZZY DECISION SUPPORT SYSTEM: TWO APPLICATIONS

Rita Almeida RIBEIRO* James F. BALDWIN**

* Universidade Nova Lisboa
FCT, Dept. Informática
2825 Monte Caparica, Portugal

** University of Bristol
Department of Engineering Mathematics
Bristol BS8 1TR, United Kingdom

ABSTRACT

This paper presents two applications of a multiple attribute fuzzy decision support system (MAF-DSS) addressing different contexts. One determines the extent to which a person is a hero in ancient mythology, and the second is a military vehicle pattern recognition problem. The aim is to use the applications to demonstrate the generality, flexibility and main properties of the fuzzy multiple attribute decision support tool.

1. Introduction

The innovative MAF-DSS is a hybrid system incorporating theories, methods and tools from the domains of knowledge-base systems, case-based reasoning, fuzzy decision support, fuzzy sets and logic, support logic programming and evidential reasoning. It also incorporates a "pseudo" object-oriented representation.

The first application defines a hero in ancient mythology and tests the extent. This application is an unusual example of a multiple attribute problem, derived from Lord Raglan's rating of classical heroes [10], on twenty two attributes. The purpose was to justify that heroes of tradition stories are accounts of myths and not historical facts. After fuzzifying certain attributes of the proposed pattern the results should prove more accurate and increase the knowledge gained about the mythological heroes. Further, the natural language query module proves useful in deducing which of the defined attributes represent the core of the heroes of myth definition and in providing other interesting information about the heroes.

The second application addresses the recognition of military vehicles. Such vehicles usually have a multi-sensor for target recognition and this seems a problem well-suited for recognising potential enemy tanks. Methods for modelling uncertainty in computer vision systems have already received considerable attention (see for example, [10, 11, 14]), the novelty presented by the MAF approach is the use of 'blind' fuzzification. Blind fuzzification means that there is no need to define characteristics (features) to create an abstract model of the real object. A further important aspect of the MAF system is the assignment of importance to the attributes and objects. This ability allows the tuning of the system until the supports are the highest possible for all alternatives.

2. Multiple Attribute Fuzzy Decision Support System Overview

Multiple-attribute problems, a subset of decision making problems, usually refer to single goals, the selection of the best alternative, and multiple constraints or attributes [8]. Decision making within a fuzzy environment [5] extends multiple attribute problems to accept fuzzy knowledge representation and fuzzy inference methods.

The MAF-DSS is a general context-independent system for fuzzy multiple attribute type problems. It incorporates generalised knowledge from past experiences or cases using a case-base perspective and fuzzy knowledge and inference. Its theoretical basis derives from (a) the knowledge engineering steps in its development, (b) a representational scheme based on concepts from the object-oriented paradigm [9] to fuzzy linguistic representation [13]; (c) use of the case-based paradigm [6] coupled with support logic programming and evidential reasoning [1, 2]; (d) use of fuzzy sets and logic theories to select the best alternative [5, 12] and (e) a meaning representation language for natural languages queries [15]. The main features of the MAF system are: (1) it accepts and manipulates both types of goals of fuzzy multiple-attribute problems; (2) it is interactive and user-friendly in that all dialogue with the user is guided by scroll-bar menus and dialogue-boxes; (3) it is flexible in handling attributes and objects with regard to insertions and modifications of values. Further, it is able to solve numerous different fuzzy multiple attribute problems; (4) it incorporates simulation facilities, enabling tuning of the system and verification of the variables sensitivity; (5) it contains a 'learning ability' since the decision maker can learn about the problem by using the natural language query type and/or the simulation ability; (6) it provides various types of inferencing, since the user may employ more than one type of reasoning.

The components of the knowledge representation kernel are classes, attributes, objects, universes, objectives and cases. Classes represent a set of alternatives of the problem, e.g. ((class cars (volvo, fiat))). An attribute is a characteristic or feature common to a collection of objects which represent the alternatives, for example the colour of a car. Objects are elements of attributes that may be crisp or fuzzy and, if fuzzy, discrete or continuous. Fuzzy attributes and respective objects are associated with a specified universe of discourse which defines the problem space, e.g. ((attribute fuzzy c height (h1 h2 h3)). An objective is a relation, used for classification or selection of a class (set of alternatives) and a set of attributes (constraints) of the alternatives, e.g. ((objective buycar (cars) (price speed))). Cases depict instantiations of alternatives and their specific values for each attribute included in the abstract representation of objective, e.g. ((case buycar (volvo) (7000, 120))). Qualifiers and quantifiers provide a sentence classification such as 'most Portuguese are short' and are represented by fuzzy objects. Thus, there exists a one-to-many relation between objective and cases. In addition, the MAF-DSS is implemented with the FRIL language [4]. Hence, all examples, applications data and results are presented in FRIL syntax. The conceptual model of the MAF-DSS is depicted in Figure 1.

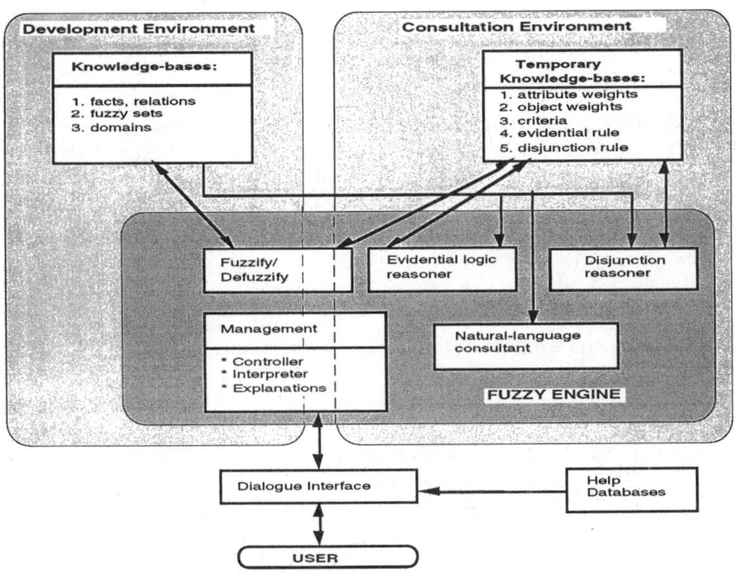

Figure 1. Conceptual model of MAF-DSS

After presenting the main entities of the MAF-DSS a brief introduction to the mathematical formalisation for the consultation environment is given. This formalisation corresponds to the options of 'Consult' menu 'Selection', 'Classification' and 'General Queries'. The fuzzy multiple-attribute problem formulation is:

Assume there is a set of decision alternatives $\mathbf{A} = \{\mathbf{A}_1, \mathbf{A}_2, \ldots\ldots, \mathbf{A}_n\}$ and a set of crisp and / or fuzzy criteria (attributes) $\mathbf{C} = \{\mathbf{C}_1, \mathbf{C}_2, \ldots\ldots\ldots, \mathbf{C}_m\}$. Let $\mathbf{R}_{ij}$ be the numerical rating (membership degree) of each $\mathbf{A}_i$, assessed by criteria j. Then the general decision problem is:

$$\mathbf{D} = \max_i \left\{ \sum_{j=1}^{m} \mathbf{R}_{ij} : \mathbf{A}_i \right\} \text{ where for } \mathbf{D} = \mathbf{A}_k \text{ chose } \mathbf{A}_k \text{ such that } \sum_j \mathbf{R}_{kj} \geq \sum_j \mathbf{R}_{ij} \text{ for } \forall \mathbf{i}.$$

The MAF-DSS module of the natural language general queries is an adapted implementation of query types I, III and IV of Zadeh's PRUF (Possibilistic Relational Universal Fuzzy) language [15]. PRUF is a natural language that handles the truth of propositions with linguistic variables, fuzzy quantifiers and qualifiers. In the MAF system propositions are expressions which involve classes or objects, attributes, connectives, qualifiers, quantifiers, e.g. the car volvo is quite cheap. The general formalisations used in the MAF-DSS are (with FRIL multiplication model):

Q1- Corresponds to PRUF Type I: "X is mF" $\Rightarrow \mathbf{X_i}$ is / has $\prod_k \mathbf{Y_k}(\mathbf{a_i})$

where $\mathbf{Y_k} = \mathbf{M_j} * \mathbf{A_k}$ is a composed fuzzy set; $\mathbf{X_i}$ is an object of a class; $\mathbf{M_j}$ is a qualifier; $\mathbf{A_k}$ is a fuzzy object of an attribute $\mathbf{k}$ and $\mathbf{a_i}$ is the value for object $\mathbf{i}$ of attribute $\mathbf{k}$.
E.g. John is quite tall.

Q2- Corresponds to PRUF type III: "QX is mF" $\Rightarrow \mathbf{Q_n}\left(\frac{1}{N}\sum_{i=1}^{n} \mathbf{Y_k}(\mathbf{a_i})\right)$

where $\mathbf{Q_n}$ is a quantifier; $\mathbf{Y_k}$ is a composed fuzzy set; $\mathbf{N}$ is the total number of objects $\mathbf{i}$ that share attributes $\mathbf{k}$. E.g. most cars are slow and very expensive

Q3- Corresponds to type IV: "X is F is m true" $\Rightarrow \mathbf{M_n}\left(\mathbf{X_i} \text{ is / has } \prod_k \mathbf{Y_k}(\mathbf{a_i})\right)$

where $\mathbf{M_n}$ is a qualifier over the truth of the sentence represented by the interval [0,1] and the remaining parameters are identical to Q1. E.g. John is quite tall is very true.

Quantifiers are functions that act as a filter for the proposition by giving a relative measure, e.g. '*most* cars are expensive'. Qualifiers are also functions, but only affect a single object, e.g. 'volvo is *very* expensive' (see for example [15]).

The second group in the consultation environment uses evidential reasoning [3] for its inference mechanism. It is divided into two sub-modules, one that selects the best alternative using criteria, operators, importance and the evidential rule, while the other module, selection, ranks the set of alternatives using only weights (importance) and evidential rules. The best alternative is the one with highest support. Mathematically this group formalisation is :

$$\mathbf{S}(\mathbf{A_i}) = \mathbf{F}\left(\sum_j \mathbf{w_j} * \mu(\mathbf{x_{ij}})\right)$$

where $\sum_j \mathbf{w_j} = 1$; $\mathbf{w_j}$ are the weights of the attributes or objects; $\mu(\mathbf{x_{ij}})$ is the membership degree of attribute or object $\mathbf{x_{ij}}$; $\mathbf{A_i}$ is alternative i and $\mathbf{F}$ is a function $\mathbf{F}:[0,1]\times[0,1]$.

The main dialogue interface of the MAF-DSS is performed using the following two menus:

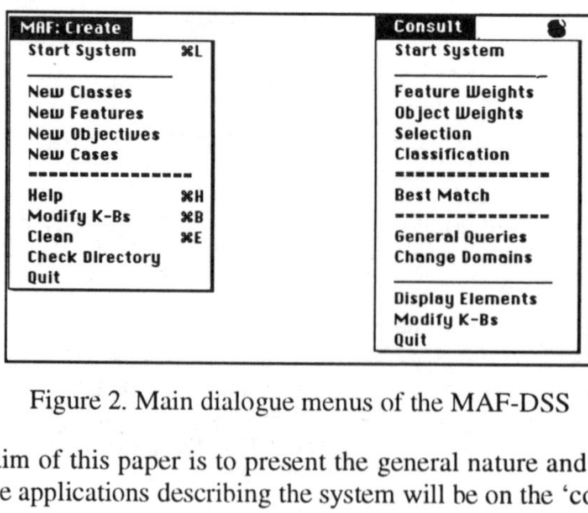

Figure 2. Main dialogue menus of the MAF-DSS

Considering that the aim of this paper is to present the general nature and properties of the MAF system, the focus of the applications describing the system will be on the 'consult menu'.

3. Heroes of Ancient Mythology

Analysing the life stories of twelve heroes of Greece, Lord Raglan [10] tabulated common incidents of their lives into a common pattern. These heroes stories, "were accounts not of fact but of ritual – that is, myths" [10] (pp. 173). After defining the pattern with a dozen heroes, Lord Raglan applied the pattern to the heroes and extended its use to heroes outside the classical area. This proved quite successful as a classification scheme for a mythological hero. Lord Raglan determined twenty two types of incidents for his common pattern in order to classify the heroes. From the application of the pattern to the heroes, each scores the number of incidents which match in their life stories. Lord Raglan's pattern is shown in table 1.

	PATTERN / ATTRIBUTES (transcribed): "
(1)	The hero's mother is a royal virgin;
(2)	His father is a king, and
(3)	Often a near relative of his mother, but
(4)	The circumstances of his conception are unusual, and
(5)	He is also reputed to be the son of a god.
(6)	At birth an attempt is made, usually by is father or his maternal grandfather, to kill him, but
(7)	He is spirited away, and
(8)	Reared by foster parents in a far country.
(9)	We are told nothing of his childhood, but
(10)	On reaching manhood he returns or goes to his future kingdom.
(11)	After a victory over a king and/or a giant, dragon, or wild beast,
(12)	He marries a princess, often the daughter of his predecessor, and
(13)	Becomes king.
(14)	For a time he reigns uneventfully, and
(15)	Prescribes laws, but
(16)	Later he loses favour with the gods and/or his subjects, and
(17)	His driven from the throne and city, after which
(18)	He meets with a mysterious death,
(19)	Often at the top of a hill.
(20)	His children, if any, do not succeed him.
(21)	His body is not buried, but nevertheless
(22)	He has one or more holy sepulchres. "

Table 1. Heroes of myth attributes. Source: [10] (pp 174)

The pattern was applied to twenty one heroes, ranging from those in ancient Greece to the Old Testament, Java, Upper Nile, Northern Europe and Celtic. They all proved to belong, to a greater or lesser degree, to the classification of heroes of mythology. All heroes score at least eleven points, except one who scored nine. After showing how his pattern could successfully define heroes of tradition and myth, Lord Raglan applied the same classification to historical heroes. Of the latter heroes, the maximum fit obtained for the pattern was less than seven points. Therefore he states that there are fundamental differences between the heroes of myth and the heroes of history. There is a threshold of more or less 10 points, above which a potential hero can be classified as a hero of tradition and myth. To simplify the understanding of how Lord Raglan describes the application of his pattern to twenty one cases, one of them, the hero Apollo, is quoted here [10] (pp 175):

> " His mother, Leto is (1) a royal virgin, and his father is Zeus (5), who is (3) her first cousin. At birth he is (6) in danger from Hera, but (7) his mother escapes with him, and (8) he is reared at Delos. We hear (9) nothing of his childhood, but on reaching manhood he (10) goes to Delphi, where he (11) kills the Python, becomes (13) king, and (15) prescribes the laws of music, etc. We can take him no further, but he has scored eleven points."

In the MAF-DSS twenty one attributes, instead of the original twenty two, were used. Seven categories were fuzzified as discrete fuzzy sets. This fuzzification is due to the semantic ambiguities in some of Lord Raglan classifications, such as considering that a hero's mother to be a royal virgin is 'similar' to being the son of a princess. Since these seven attributes deal with different semantic alternatives, and not with a numeric continuous scale, the attributes were fuzzified as fuzzy discrete types. All other categories were considered crisp, with one or more members in the set, as defined in the original pattern. A sample of the heroes' knowledge representation for the 21 heroes, in FRIL format, is then as follows:

```
((objective   is_hero (heroes) (mother_royal father_royal ........... body_not_buried holy_sepulchre))
((class       heroes     (oedipus theseus romulus heracles perseus jason bellerophon pelops asclepios
dionysos apollo zeus joseph moses elijah watu-gunung nyikang siegfried llew-llawgyffes arthur robin-
hood) ))
((attribute   fuzzy d mother_royal        (princess royal_virgin demigodess royal_parents)))
((attribute   fuzzy d father_royal        (prince  king royal_family holy_man )))
......................
((attribute   crisp d body_not_buried     (uncertain not_buried)))
((attribute   fuzzy d holy_sepulchre      (one_sepulchre few_sepulchre many_sepulchre)))
% FUZZY SET DEFINITIONS:
(fuz_mother_royal        {princess:0.7 royal_virgin:1 demigodess:1 royal_parents:0.4})
(fuz_father_royal        {prince:0.7 king:1 royal_family:0.4 holy_man:0.7})
(fuz_parents_relatives   {cousins:1 uncle:1 brothers:1 ancestors:0.5 near_relatives:0.3})
(fuz_son_of_god          {yes_songod:1 might_songod:0.7 })
(fuz_attempt_murder      {father:1 grandfather:0.7 brother:0.7 relative:0.5 other_murderer:0.2}
(fuz_kill_victory        {kill_king:1 giant:1 dragon:1 wild_beast:1 kill_other:0.5})
(fuz_holy_sepulchre      {one_sepulchre: 0.6 few_sepulchre: 0.8 many_sepulchre: 1})
NOTES: 1)Discrete fuzzy sets have prefix 'fuz_'; 2) The order of attributes is equivalent to the incidents-table 1.
```

As stated the original set of alternatives numbered twenty one. From the complete set some were excluded because they were considered quite obscure. The exclusions were: the Javanese hero, Watu Gunung; the Upper Nile hero, Nyikang; and the Northern Europe hero Siegfried. Hence, there are eighteen heroes, called cases, tested in the MAF system. The set includes classical heroes (originally twelve), heroes of the Old Testament (originally three) and Celtic heroes (originally three). A sample of two life stories, in FRIL format is:

```
((case is_hero (oedipus) (princess king ancestors unusual_birth yes_songod father hidden_at_birth
foster_parents unknown_childhood birthplace kill_king married_princess became_king reigns_well subjects
expelled_throne mysterious_death dies_high_place no_successors uncertain many_sepulchre)))

((case is_hero (apollo) (royal_virgin no cousins no yes_songod other_murderer hidden_at_birth foster_parents
unknown_childhood future_kingdom wild_beast no became_king reigns_well no no no no no no no)))
```

Table 2. Heroes Cases (2 samples)

The complete results obtained with the MAF system, using the option 'Classification' (which uses the evidential logic rule) of the consult menu, are:

RESULTS with EQUAL WEIGHTS for attributes and objects :			
((final_support is_hero for oedipus))	: (0.687 1)	((final_support is_hero for robin-hood))	: (0.502 1)
((final_support is_hero for theseus))	: (0.63 1)	((final_support is_hero for jason))	: (0.497 1)
((final_support is_hero for arthur))	: (0.621 1)	((final_support is_hero for bellerophon))	: (0.468 1)
((final_support is_hero for romulus))	: (0.592 1)	((final_support is_hero for asclepios))	: (0.4271)
((final_support is_hero for moses))	: (0.592 1)	((final_support is_hero for zeus))	: (0.389 1)
((final_support is_hero for dionysos))	: (0.58 1)	((final_support is_hero for pelops))	: (0.344 1)
((final_support is_hero for llew-llawgyffes))	: (0.56 1)	((final_support is_hero for joseph))	: (0.338 1)
((final_support is_hero for heracles))	: (0.544 1)	((final_support is_hero for apollo))	: (0.326 1)
((final_support is_hero for perseus))	: (0.521 1)	((final_support is_hero for elijah))	: (0.287 1)

Almost all the results are compatible with Raglan's ratings. For example, Oedipus scores 21 points and in the MAF system he obtains a support of 68.7% (the highest score in both results); Moses scores 20 points and the MAF-DSS support is 63% (the second highest in both); Apollo scores 11 points while the MAF-DSS support is 32.6% (the second worst in both results); and Elijah scores nine points while with the MAF-DSS his support is 28.7% (the worst in both results). The cases with different ranking in both approaches are: (a) Moses has a second place from Lord Raglan (scored 20 points) and only fourth place in the MAF-DSS because here all the first three attributes received small memberships. In the MAF-DSS, instead of assuming semantic equivalencies for different incidents, all incidents are considered but with smaller degrees of belonging (fuzzification process); (b) Perseus has a worse place than Heracles and Llew-llawgyffes in the MAF-DSS version by the same reasons as before; (c) Robin-Hood on the contrary, is better classified in the MAF-DSS even with 9 attributes that do not match, because the majority of the no matches have a small weight. Robin-Hood compared, for instance with Jason, who only has 5 no matches but all in attributes with high weights, such as attributes 4, 14, 16 and 19, is better classified even though he has 9 attributes that do not match. This is a disadvantage of the evidential inference of the MAF-DSS. It does not differentiate between having small memberships but belonging to the attribute, and not matching at all because it is a linear weighted average algorithm. All the other heroes have more or less the same ranking with both approaches. Both Apollo and Elijah are the 'lesser heroes' found in both approaches. According to the implicit threshold (10 points), proposed by Lord Raglan to be a full hero of myth, Elijah can be said to be only a 'partial' hero.

One important point must be discussed. The supports are not very high because the evidential logic rule (the inference scheme used) is a weighted average algorithm and therefore all elements of each attribute, intermediate results or body supports, are assigned importance. This means that for problems where only one object (element) of the set that constitutes an attribute, matches, the other elements (zero membership) are assigned importance that are wasted. Each intermediate result (per attribute) is reduced since the weighted average depends on the number of elements in the set. For example, an attribute with two elements having equal weights, 50% each, then the element that matches will be multiplied by 50% and therefore the attribute result could not exceed 50%. The final combination of attributes reflects this problem. Therefore, to guide the results explanation the heuristic followed is that a support of 50% is considered quite good. Henceforth, this heuristic will be considered when explaining results obtained with the evidential logic rule.

Raglan's score system fails to differentiate, for instance, Heracles and Llew-llawgyffes, since it gives both 16 points. In the MAF-DSS the results are 54.4% and 56.6% respectively. The classification obtained shows, for example, that Heracles does not matches attributes 7, 8 and 9, corresponding to have been spirited away, reared by foster parents and to ignore details of childhood. All these attribute comparisons are available in the intermediate results provided by the MAF-DSS, on demand. They are a useful capability of the MAF-DSS.

The main advantages of using the option 'Classification' and therefore evidential logic reasoning are the fuzzification of attributes and the intermediate results supports comparison. Fuzzification distinguishes elements of attributes that 'fit' an attribute better than others. The advantages allow finer results and differentiations between alternatives.

Another important aspect, and a further advantage of the MAF-DSS approach, is the possibility of performing natural language queries. This 'pseudo' natural language facility is useful for the heroes application, since it can provide interesting information about the heroes' characteristics. Observe the following queries carried out on the knowledge base, with the option 'General Queries' of the consult menu:

NATURAL LANGUAGE QUERIES :

some heroes have mothers royal_virgin ?	**most heroes have been hidden_at_birth?**
Relative sigma_count is 0.444444	Relative sigma_count is 0.6666667
THE SUPPORT IS: (0.37037)	THE SUPPORT IS: (0.266667)
about_half heroes have mothers royal_virgin?	**many heroes have been hidden_at_birth ?**
Relative sigma_count is 0.444444	Relative sigma_count is 0.666667
THE SUPPORT IS: (0.444444)	Test_score is (0.833333)
few heroes are sons of princess ?	**most heroes had a mysterious_death ?**
Relative sigma_count is 0.155556	Relative sigma_count is 0.777778
THE SUPPORT IS: (0.472222)	Test_score is (0.711111)
few heroes have fathers holy_man ?	**oedipus is son of princess ?**
Relative sigma_count is 0.0388889	Composed query is : ((match mother_royal princess))
THE SUPPORT IS: (1)	THE SUPPORT FOR QUERY IS: (0.7)
all heroes have became_king ?	**perseus has killed a dragon is very true ?**
Relative sigma_count is 1	Composed query is : ((match dragon dragon))
THE SUPPORT IS: (1)	THE SUPPORT FOR BEING very true IS : (1)

Quantifiers and qualifiers used:
(few [0.05:1 0.25:0]); (many [0.5:0 0.7:1 0.76:1 0.94:0]); (about_half [0.4:0 0.5:1 0.6:0]); (most [0.6:0 0.85:1]); (all [0.95:0 1:1); (very [0.68:0 0.88:1]); (some [0.1:0 0.25:1 0.35:1 0.5:0])

As can be observed all heroes have became a king (support 100%) which certainly stresses the importance of this characteristic to determine a hero of mythology. Another important characteristic is that few heroes married princesses, since the support is 27%. It seems that to have been hidden at birth is not very important since support is 0.26%. For this last item when the query uses 'many' instead of 'most' the support is 83%. The inference to be drawn is that 'most' of the heroes have not been hidden at birth but certainly 'many' have. Another important aspect is that few heroes have fathers who were holy men. Individual queries are: Oediopus is son of a princess has a support of 70% and "is very true" that Perseus killed a dragon. The other queries are considered self-explanatory.

The quantity of information learned with this type of natural language queries is obvious. Many conclusions and knowledge about the importance of objects (elements) were highlighted. The extra information about the importance of elements could be 'fed' back into the system in the 'Objects weights' option, in order to change the equal weights previously used. This would allow a tuning of the system leading to increased final weights since most common incidents will have higher weights. Here, changes in the weights will not be simulated because the results already obtained are considered expressive enough.

In conclusion, the MAF-DSS system proves useful for dealing with problems with multiple attributes. It is also useful when ambiguous or vague semantic definitions need to be used. It provides finer results and therefore better differentiations between alternatives. More information and knowledge about the attributes and respective objects is available. The natural language queries also provide a means to make deductions about which qualities are more common and therefore proposes a ranking for the attributes. This ranking could improve the classification of heroes of tradition and mythology and provide a better framework for defining them.

4. Military Vehicle Recognition

This application deals with the recognition of main battle tanks, light tanks and reconnaissance vehicles. The aim here is to demonstrate that the MAF-DSS can be used successfully for pattern recognition. The MAF system is sufficiently general to abstract and fuzzify features from any role model (image). Procedurally, for the military vehicle images, the user: (1) defines the number of lines (features/attributes) and columns (objects) the image matrix should be divided into, (2) provides the number of pixels (zero squares in this case) and respective deviation factor, per object, for the alternative considered a role model, (3) and provides the number of pixels, per object, for each image to be recognised.

The selection of the vehicles was arbitrary and the only constraints were to have different types and shapes over the same context and to have been built by different countries. The military vehicles used were scanned from "The Concise Illustrated Book of Tanks" [7]. The images depict modern military vehicles. The cases consist of seven main battle tanks, one light tank, two reconnaissance vehicles and four noisy images of the first four main battle tanks. The choice was based on the desire to discover which type of recognition the MAF-DSS is better suited to: noisy images or different shapes. Table 3 describes the vehicles used.

Name	Nationality	Comments
1- Challenger	U. K.	Primary British army fighting vehicle. Used as *role model.*
2- AMX-32	French	In service in nine countries.
3- Centurion MK 10	U. K.	In service in ten countries.
4- Leopard 1	West German	Most successful mass production tank since World War II.
5- Stridsvagn 103	Swedish	Offers the smallest target area to any opponent.
6- T-34/85	Soviet Union	Premier mass-production tank. In service in a number of small countries (Africa, Middle East, Asia)
7- TAM	W.German/Argent.	One of the fastest. Mainly used in Argentina.
8- M4A1 Sherman	United States	Most famous tank of World War II. Light vehicle.
9-EngesaEE9 Cascavel	Brazilian	Widespread service around Africa and South America. Reconnaissance.
10- Spahpanzer Luchs	West German	Fast and fully amphibious. Reconnaissance vehicle.
11,12,13,14		Identical to numbers 1, 2, 3 and 4 respectively but considering objects (noise) in front of tanks.

Table 3. Military Vehicles Description

The scanned images of the tanks, shown in the next figure, suffer some operations to facilitate its visibility and use in the MAF-DSS. First the image was inverted, the tank white and the frame black. Second the tank was 'filled' with squares, representing pseudo-pixels. Each image was considered a matrix of 65 X 28. Even though fuzzy sets are quite well suited to represent grey scales in images, here the original scanned images were all black, therefore this ability was not tested. Figure 3. depicts the scanned images of the tanks used in the MAF-DSS, with a scale reduction of 60%.

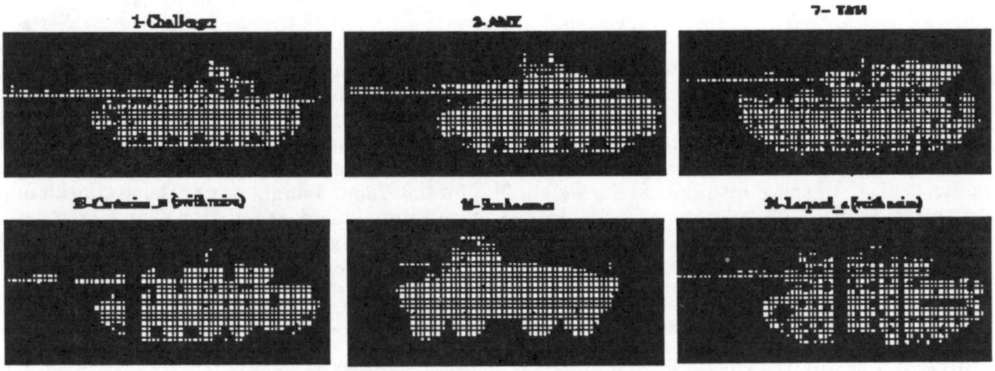

Figure 3. Sample of six scanned military vehicles

In order to define the knowledge base for the image the attributes, objects and parameters of the objects must be defined. The attributes in the MAF-DSS correspond to five columns in image terms. Objects correspond to three lines of nine pixels in the image. The attribute Xi, of X = {1, 2, ... n} is called lineX, and object XY, with Y={1,.. ,m}, of attribute X is called cell XY. For example attribute line1 has cell11, cell12 and cell13 for objects. The tanks application considered cells of size of about 121 pixels, since (65 X 28)/15 = 121.3.

As stated, the MAF system needs a role model, corresponding to the best image desired or known. Here the main battle tank 'Challenger' was used as the role model to be generalised. In order to fuzzify (generalise) the cells/objects of the role model, a factor of deviation should be provided. For objects/cells X2 and X3 10% of the cell size was used, as a deviation factor, and for objects X1 5%. This latter choice was due to the observation that tanks have a somewhat triangular shape. Hence, the tanks knowledge base, including two tanks' case and in FRIL notation, is:

```
                         Knowledge-base of tanks (with 2 sample cases):
((objective   classify_tanks (tanks)    (line1 line2 line3 line4 line5) ))
((attribute   fuzzy c line1   (cell11 cell12 cell13)))
((attribute   fuzzy c line2   (cell21 cell22 cell23)))
((attribute   fuzzy c line3   (cell31 cell32 cell33)))
((attribute   fuzzy c line4   (cell41 cell42 cell43)))
((attribute   fuzzy c line5   (cell51 cell52 cell53)))
(cell21       [11:0 17:1 23:0])
(cell22       [102:0 114:1 126:0])
(cell23       [33:0 45:1 57:0])
((case        (amx)             ((0 54 33) (10 116 70) (14 97 64) (0 46 30) (0 20 0)) ))
((case        (challenger)      ((0 52 21) (17 114 45) (1 101 46) (0 52 23) (0 17 0)) ))
```

In order to obtain the classification for the military vehicles the option 'Classification' in the consult menu was used. This option requires the importance for the attributes and objects to be defined. Considering equally alike weights for the attributes and objects the results are:

%1- RESULTS with equal weights :			
((final_sup tanks for cascavel))	: (0.266)	((final_sup tanks for leopard))	: (0.455)
((final_sup tanks for challenger))	: (0.999)	((final_sup tanks for spahpanzer))	: (0.25)
((final_sup tanks for tam))	: (0.292)	((final_sup tanks for t34_85))	: (0.2)
((final_sup tanks for sherman))	: (0.226)	((final_sup tanks for challenger_a))	: (0.622)
((final_sup tanks for amx))	: (0.543)	((final_sup tanks for centurion))	: (0.549)
((final_sup tanks for leopard_a))	: (0.525)	((final_sup tanks for centurion_a))	: (0.405)
((final_sup tanks for amx_a))	: (0.478)	((final_sup tanks for stridsvgn))	: (0.456)

Observing the results the Cascavel, Tam, Sherman, Spahpanzer and T34-85 have supports of less than 30% which show that the images are quite different from the role model, the Challenger. The results for the noisy images, the ones that have the same name as their counterparts with an added suffix '_a', are quite good, with only one, the Centurion_a being under 50%. Therefore, it can be deduced that it is easier, for the MAF-DSS, to recognise noisy images than different shapes.

Analysing the intermediate results, lines 1 and 5 always have strong support, line 2 and 4 normal support and line 3 has almost always extremely small support. Semantically, the supports show that in order to be recognised as a tank, attribute line1 and line5 are more important than lines 2, 4 or 3. Using the extracted information, a new simulation was performed, with weight 5 for line 1 and 5; 3 for lines 2 and 4 and weight 1 for line 3. The results obtained show that all the supports improved and only three tanks are below the 30% support. Observing again the intermediate results (not shown because of space restrictions) the weight of 3 for line 2 and 4 might be excessive considering that many tanks have supports of zero. Hence, performing a new simulation using extremely important (5) for line 1 and 5; not important (2) for line 2 and 4; and extremely unimportant (1) for line 3, the results are:

%3-RESULTS with line1,5 = 5 line2,4=2 and line3=1			
((final_sup classify_tanks for cascavel))	: (0.415)	((final_sup tanks for leopard))	: (0.593)
((final_sup classify_tanks for challenger))	: (0.999)	((final_sup tanks for spahpanzer))	: (0.278)
((final_sup classify_tanks for tam))	: (0.399)	((final_sup tanks for t34_85))	: (0.333)
((final_sup classify_tanks for sherman))	: (0.321)	((final_sup tanks for challenger_a))	: (0.69)
((final_sup classify_tanks for amx))	: (0.662)	((final_sup tanks for centurion))	: (0.708)
((final_sup classify_tanks for leopard_a))	: (0.642)	((final_sup tanks for centurion_a))	: (0.578)
((final_sup classify_tanks for amx_a))	: (0.62)	((final_sup tanks for stridsvgn))	: (0.522)

The supports obtained improve again and show a good degree of recognition. All but one tank has a support of more than 30%, and six have a support of more than 60% Considering that a linear combination of attributes and not the selection of the best of the attributes is used, these supports are impressive. Hence, it can be claimed that the MAF-DSS deals much better with fuzzy/noisy images than with different shapes, but is useful for fuzzy images recognition. The simulation capability of the MAF-DSS proves very useful for fine tuning the importances of the attributes within a certain context, thus providing more accurate results.

A second set of tests could be performed using six attributes instead of five. Increasing the number of attributes and/or objects is another simulation alternative for any image recognition problem. The rationale is to perform simulations until the results obtained are considered satisfactory, by the decision maker, in terms of supports obtained. This represents the tuning of the system to increase accuracy.

Too much specialisation acts as a counterpart to fuzzification because the borders of the fuzzy sets are reduced, which will affect the fuzzy recognition process. A proper balance must be decided upon by the decision maker. Many other choices regarding the assignment of weights to attributes, are provided by the MAF-DSS. They range from equally alike, ranking, numeric scale, manual and even automatic for objects. These were not tested because the results obtained are considered enough to emphasise the two most important observations. First, that the MAF-DSS reacts better to noisy images than to shapes and second that its simulation ability allows a tune-up of recognition process efficiency and capacity. Furthermore, the results emphasise that even with very scarce information, no features (distinctive features), objects or context definition are provided, the MAF-DSS can handle the recognition of any image. The only requirement is to 'arrange' the image matrix into lines and rows arbitrarily, though even this can be changed if required.

Concluding, fuzzy image recognition using multiple attributes and a generalised role model, proves a simple and quite effective method. Any image can be considered as a multiple attribute problem, lines and columns being the attributes and objects. This process allows a 'blind' recognition capable of dealing with any image for which the features are difficult to define. The use of a generalised (fuzzified) role model helps the pattern matching process. The matching is done automatically by comparison with a role model. Unfortunately the MAF-DSS can only handle one role model at a time, a problem to be addressed in the future. Any differences are handled by the fuzziness ability and 'similar enough' alternatives are detected. The latter demonstrates the suitability of fuzzy sets for handling grey scales. The MAF-DSS proves to be better suited to differentiating between noisy images than differentiate shapes. The demonstrated results are proof. In addition, in a real military context, this method could be used successfully in an automatic target recognizer. Mainly, it seems appropriate for night image recognition, where the detected images are quite noisy.

5. Conclusions

The two applications tested belong to quite different contexts and deal with quite different types of problems. The common factor is that they can be viewed and addressed as fuzzy multiple attribute problem, with objectives so disparate as ranking, rating, selection or simply answering natural language queries. The MAF-DSS covers all these objectives.

The first application results show that the MAF-DSS can be an efficient tool when dealing with vague or uncertain information, certainly providing better quality results. This use of the natural language query facility shows that more and deeper information about specific attributes and alternatives is available. For the second application, military vehicle recognition, the results show that there are alternative means for computer vision recognition, not dependent on abstracting an object to define a model by its features. Further, the problem of noisy images could imply that the MAF-DSS is a good approach. Of course, since this type of tool belongs to a higher level of computer recognition it could be combined with other methods and algorithms. It is not intended to be a full substitute.

Summing up, the MAF-DSS is a general and flexible system that can perform ranking, rating, selection and answer natural language queries, for fuzzy multiple attribute problems, viewed from an object-oriented perspective.

References

[1] Baldwin, J. F. (1986) Support Logic Programming. Int. J. of Intelligent Systems 1, 73-104.

[2] Baldwin, J. F. (1987) Evidential Support logic Programming. Fuzzy Sets and Systems 24, 1-26.

[3] Baldwin, J. F. (1994) FRIL - Fuzzy and Evidential Reasoning in AI. Research Studies Press (To appear).

[4] Baldwin, J. F., T. Martin and B. Pilsworth (1988) FRIL Manual. FRIL Systems Ltd,Bristol Business Centre. Bristol, BS8 1QX, UK.

[5] Bellman, R. E. and L. A. Zadeh (1970) Decision-Making in a Fuzzy Environment. Management Science. Vol.17 (4), 141-164.

[6] Kolodner, J. and W. Mark (1992) Case-Based Reasoning. IEEE Expert Vol. 7(5), 5-6.

[7] LLoyd, M. (1993) The Concise Illustrated Book of Tanks. Grange Books.

[8] MacCrimmon, K. R. (1973) An Overview of Multiple Objective Decision Making. In: Multiple Criteria Decision Making, Cochrane and Zeleny (Ed), University of South Caroline Press.

[9] McCabe, F. G. (1992) Logic and Objects. Series in Computer Science. Prentice Hall.

[10] Raglan, F. R. (1975) The Hero - A Study in Tradition, Myth and Drama. Greenwood press.

[11] Ribeiro, R. and J. F. Baldwin (1991) Partial Matching in Fuzzy Image Recognition. Proc. Developing and Managing Expert Systems Programs, IEEE Computer Society.

[12] Yager, R. R. (1977) Multiple Objective Decision-making Using Fuzzy Sets. Int. J. Man-machine Studies 9, 375-382.

[13] Zadeh, L. A. (1975) Fuzzy Logic and Approximate Reasoning. Synthese 30, 407-428.

[14] Zadeh, L. A. (1977) Fuzzy Sets and Their Application to Pattern Classification and Clustering Analysis. Classification and Clustering. Academic Press, NY, 251-300.

[15] Zadeh, L. A. (1978) PRUF-a Meaning Representation Language for Natural Languages. Intern. J. of Man-Machine Studies 10, 395-460.

SPECIAL INVITED PAPER

VIEWNET ARCHITECTURES FOR INVARIANT 3-D OBJECT LEARNING AND RECOGNITION FROM MULTIPLE 2-D VIEWS

Stephen Grossberg[1] and Gary Bradski[2]

[1] Department of Cognitive and Neural Systems
and
Center for Adaptive Systems
Boston University
Boston, MA 02215[3]

[2] First Union Bank, NC-0600
One First Union Center, TW8
Charlotte, NC 28288-0601

ABSTRACT

The recognition of 3-D objects from sequences of their 2-D views is modeled by a family of self-organizing neural architectures, called VIEWNET, that use View Information Encoded With NETworks. VIEWNET incorporates a preprocessor that generates a compressed but 2-D invariant representation of an image, a supervised incremental learning system (Fuzzy ARTMAP) that classifies the preprocessed representations into 2-D view categories whose outputs are combined into 3-D invariant object categories, and a working memory that makes a 3-D object prediction by accumulating evidence over time from 3-D object category nodes as multiple 2-D views are experienced. VIEWNET was benchmarked on an MIT Lincoln Laboratory database of 128x128 2-D views of aircraft, including small frontal views, with and without additive noise. A recognition rate of up to 90% is achieved with one 2-D view and of up to 98.5% correct with three 2-D views. The properties of 2-D view and 3-D object category nodes are compared with those of cells in monkey inferotemporal cortex.

1. Introduction: Converting Variant and Noisy Data into Invariant Recognition Codes

This chapter describes a class of self-organizing fuzzy neural architectures, generically called VIEWNET, wherein spatially and temporally variant, noisy, and imcomplete data may be transformed into invariant recognition events. VIEWNET illustrates how a hierarchy of processing stages may compute increasingly stable and invariant representations of fluctuating and uncertain data. No one processing stage, in itself, accomplishes the transformation of these data into an invariant code.

The simplest type of VIEWNET architecture, called VIEWNET 1, is capable of learning invariant representations of 3-D objects from sequences of their 2-D views (Bradski and Grossberg, 1993, 1995). In fact, the acronym VIEWNET stands for View Information Encoded With NETworks. More generally, VIEWNET architectures illustrate a computational strategy that may be generalized to accomplish more complex tasks of scene understanding.

[1]Supported in part by the Advanced Research Projects Agency (AFOSR 90-0083 and ONR N00014-92-J-4015) and the Office of Naval Research (ONR N00014-95-1-0409 and ONR N00014-95-1-0657).

[2]Supported in part by the National Science Foundation (NSF IRI-90-24877) and the Office of Naval Research (ONR N00014-92-J-1309).

[3]Acknowledgments: The authors wish to thank Diana Meyers for her assistance in preparing the manuscript.

The VIEWNET 1 architecture incorporates a preprocessor that generates a compressed but 2-D invariant representation of an image, a supervised incremental learning system that classifies the preprocessed representations into 2-D view categories whose outputs are combined into 3-D invariant object categories, and a working memory that makes a 3-D object prediction by accumulating evidence from 3-D object category nodes as multiple 2-D views are experienced.

The VIEWNET 1 preprocessor includes the CORT-X 2 filter, which discounts the illuminant, regularizes and completes figural boundaries, and suppresses image noise. This boundary segmentation is rendered invariant under 2-D translation, rotation, and dilation by use of a log-polar transform. The invariant spectra undergo Gaussian coarse coding to further reduce noise, compensate for small shifts in log-polar processing, partially overcome 3-D foreshortening effects, and increase generalization.

These compressed codes are input into the classifier, a supervised learning system based on the Fuzzy ARTMAP algorithm. Fuzzy ARTMAP learns compressed 2-D view categories that are invariant under 2-D image translation, rotation, and dilation as well as 3-D image transformations that do not cause a predictive error. Evidence from sequences of 2-D view categories converges at 3-D object nodes that generate a response which is invariant under changes of 2-D view. The properties of these 2-D view and 3-D object category nodes may be compared with those of analogous cells in monkey inferotemporal cortex.

These 3-D object nodes input to a working memory that accumulates evidence over time to improve object recognition. In the simplest working memory, each occurrence (nonoccurrence) of a 2-D view category increases (decreases) the corresponding node's activity in working memory. The maximally active node is used to predict the 3-D object.

Recognition has been studied with noisy and clean images using both slow and fast learning. Slow learning at the Fuzzy ARTMAP map field is adapted to learn the conditional probability of the 3-D object given the selected 2-D view category. VIEWNET 1 was originally benchmarked on an MIT Lincoln Laboratory database of 128x128 2-D views of aircraft with and without additive noise. A recognition rate of up to 90% was achieved with one 2-D view and of up to 98.5% correct with three 2-D views.

2. The Seibert-Waxman Architecture

Seibert and Waxman (1990a, 1990b, 1991, 1992) pioneered the development of neural network architectures that self-organize representations of 3-D objects from 2-D view sequences. Since the work on VIEWNET was inspired by their model and uses the same data base to benchmark its results, their model is first reviewed in some detail.

Seibert and Waxman relied on building a neural "cross-correlation matrix" as shown in Figure 1. This network provides a realization of the *aspect graph* concept of Koenerink and van Doorn (1979). An aspect graph represents 2-D views of a 3-D object by nodes of a graph, and legal transitions between nodes by the arcs between them. The Seibert-Waxman cross-correlation matrix was used to learn both 2-D views and 2-D view transitions and to associate the 2-D views and view transitions with the 3-D objects that produced them. To test their architecture, the 2-D views were taken from 3-D airplane models painted black and digitized against a light background for ease of segmentation.

Sequences of images were obtained by rotating the airplanes 360° for each of the many angles of inclination that the video camera was set at, from zero degrees (horizontal) to 72 degrees (looking down at the spinning jet). These images were then binarized using a high threshold to remove noise. Points of high curvature and the object centroid were found using a reaction-diffusion process. A log-polar transform around the object centroid was used to

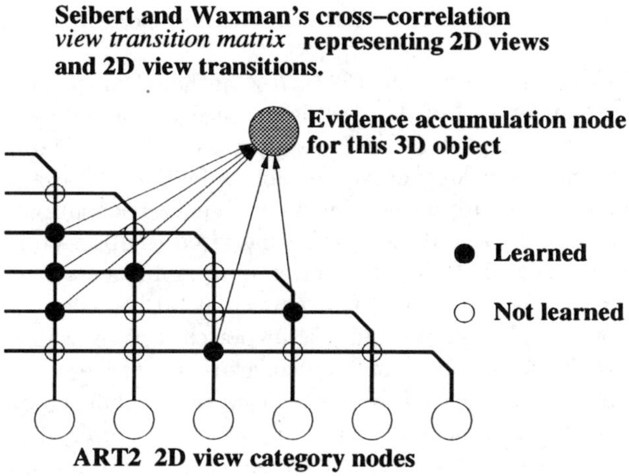

Figure 1: View Transition Matrix from the architecture of Seibert and Waxman computes cross-correlations between its input 2-D view categories. The matrix represents the correlation between the present categorical 2-D view and the decaying activation of the previous view. As preprocessed images enter ART 2, the resulting categorical views excite learned weights in this cross-correlation "aspect graph" devoted to each 3-D object. An evidence node integrates activation from the learned connections in the matrix. Another network (not shown) chooses the evidence node with the highest accumulated activation as the the 3-D object most likely being viewed by the architecture.

remove 2-D rotation and scale variations. The output of this filter was (approximately) invariant under 2-D translations, rotations, and dilations of the image. In order to compress this invariant spectrum and reduce 3-D foreshortening effects, the result was coarse coded (compressed to 5x5 pixels from 128x128) using Gaussian filters.

The coarse coded vectors (25 data points) were fed into an ART 2 (Carpenter and Grossberg, 1987) network for unsupervised learning and categorization. These "categorical" 2-D views further compressed the 2-D representation so that a new 2-D view category was chosen only if significant changes occurred in the 2-D appearance of the object that were not invariant under translation, rotation, dilation, or modest foreshortening. How much change was tolerated was controlled by the ART 2 vigilance parameter. These 2-D view categories were then fed into a series of cross-correlation matrices, or view graphs, one for each possible 3-D object, so that views and view transitions could be learned by a 3-D object categorization layer. The 3-D categorization layer incorporated "evidence accumulation" nodes which integrate activations that they receive from learned connections to the correlation matrix. Decay terms in these integrator nodes determine how long they stay active without input support.

Seibert and Waxman's approach of automatically generating aspect graphs directly from the imagery that the architecture experiences vastly simplifies earlier attempts which generated aspect graphs by constructing projections from mathematical descriptions of the 3-D objects. However, using view transition information comes at a cost: Given N 2-D views and M objects, the architecture must have the potential to encode on order of $N^2 \times M$ 2-D view transitions and the corresponding adaptive weights. Another potential problem is that an error in identifying a 2-D view may introduce a spurious 2-D view transition. Finally, unless one presumes a 2-D view frame capture rate fast enough to capture the highest speed movement that an object can make, view transitions may be skipped inadvertently by fast object motion.

As reported in Seibert and Waxman (1992), 75% of the 2-D airplane images were ambiguous to some degree. That is, 75% of the 2-D view categories formed by ART 2 gave evidence for more than one type of airplane. Two possible reasons for this level of ambiguity exist: (1) Image preprocessing using high curvature points followed by coarse coding may lump together object features that are needed for unambiguous recognition. (2) The 2-D views were categorized by ART 2 without using any supervised feedback to help correct category boundaries. Although most 2-D view categories were ambiguous, the transitions between them were used to unambiguously identify a particular 3-D object. Thus, view transitions are critically important in the Seibert and Waxman architecture, which may then incur the cost of needing up to $N^2 \times M$ view transition correlation matrices.

This analysis suggests that a tradeoff exists between the choice of preprocessor, learned categorizer, and evidence accumulation parts of the architecture. If the preprocessor and categorizer generate 2-D view categories that are too coarse or ambiguous, then the evidence accumulation network may have to be enhanced to overcome these limitations. VIEWNET explores this tradeoff by using a different preprocessor and categorizer that generate less ambiguous 2-D view categories. In fact, VIEWNET can categorize individual views on the Seibert-Waxman data base with high accuracy (up to 90%), in accord with the human experience that many objects can be identified with a single view, except when they are observed from an unfamiliar perspective or from a perspective that reduces the object's apparent dimension. A computationally less costly evidence accumulation, or working memory, network could then be used, at least on these data. The general problem is to design the optimally balanced preprocessor, categorizer, and working memory networks to handle the largest possible set of images.

3. Overview of VIEWNET 1

As diagramed in Figure 2, VIEWNET consists of three parts: an image preprocessor, a self-organizing recognition network that may operate in either unsupervised or supervised modes, and a working memory network to accumulate evidence over multiple views. It is assumed that the figure to be recognized has already been separated from its background. Neural networks for figure-ground separation that use computations consistent with those in the VIEWNET preprocessor were described in Grossberg (1994) and Grossberg and Wyse (1991, 1992). The image figure is then processed by a boundary segmentation network, called the CORT-X 2 filter (Carpenter, Grossberg, and Mehanian, 1989; Grossberg and Wyse, 1991, 1992). CORT-X 2 is a feedforward network that first compensates for variable illumination, extracts ratio contrasts, and normalizes image intensities. It then suppresses image noise while it completes and regularizes a boundary segmentation of the figure. Thus the maximal curvature point representation of Seibert and Waxman is replaced by an illumination-compensated, noise-suppressed boundary segmentation of the entire figure.

This boundary segmentation is then rendered invariant under 2-D rotation, translation, and scale by a centering, log-polar, centering operation as described in Schwartz (1977). As in Seibert and Waxman, the resulting spectra are then compressed by coarse coding to gain some insensitivity to pixel shift and 3-D deformation effects and to reduce memory requirements. Coarse coding was done by two methods in Bradski and Grossberg (1995) whose performance was compared: a many-to-few pixel simple spatial averaging, and Gaussian spatial averaging. The latter generated somewhat better results for reasons summarized below. The output of this preprocessor is a coarse-coded, invariant spectrum of an illumination-

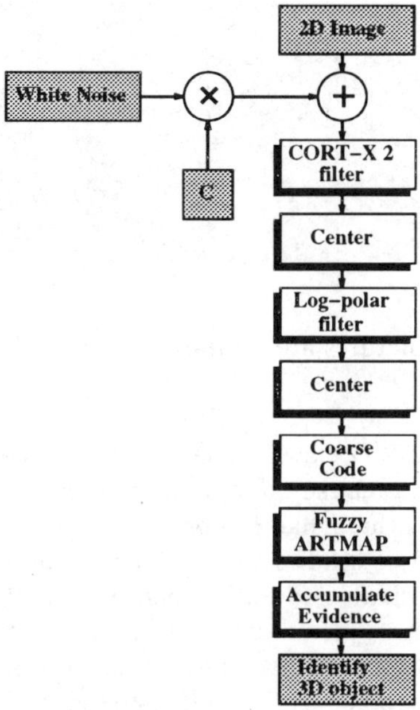

Figure 2: The image processing flow chart of the VIEWNET system.

compensated, noise-suppressed boundary segmentation. This representation provides the input vectors to the self-organizing neural network classifier.

Fuzzy ARTMAP (Carpenter, Grossberg, Markuzon, Reynolds and Rosen, 1992) was used to categorize the output spectra. This architecture is capable of fast, stable learning of recognition categories in response to nonstationary multidimensional data, and of learning to generate many-to-one output predictions from the recognition categories to output labels. Fuzzy ARTMAP runs under either unsupervised or supervised learning conditions. Under supervised conditions, erroneous predictions trigger further hypothesis testing, or memory search, in the input classifier. Fuzzy ARTMAP converts the vigilance parameter of unsupervised ART classifiers, such as ART 2, into an internally controlled parameter. When an erroneous prediction occurs, vigilance is increased just enough to trigger a new bout of hypothesis testing to discover a better category. Multiple 2-D views can be compressed into a single 2-D view category by this mechanism.

This control scheme is called *match tracking* because the vigilance parameter tracks the match value that encodes how well the selected category's prototype matches the input spectrum. Using match tracking, memory search discovers and learns recognition categories that conjointly maximize code compression and minimize predictive error. Fuzzy ARTMAP can hereby use supervised learning to rapidly fit the number, size, and shape of input categories to the statistical demands of the environment. This added power helps Fuzzy ARTMAP to learn 2-D view categories that tend to fit the data better than ART 2. It is this combination of boundary segmentation preprocessing combined with supervised ART learning that achieves correct prediction of up to 90% accuracy in response to a single airplane view.

In the VIEWNET 1 architecture, Fuzzy ARTMAP automatically combines the outputs of 2-D view categories at 3-D object category nodes that are invariant under changes of experienced 2-D views. This learned fusion of 2-D view categories into invariant 3-D object categories occurs at the Fuzzy ARTMAP *map field* that is defined below. A similar type of

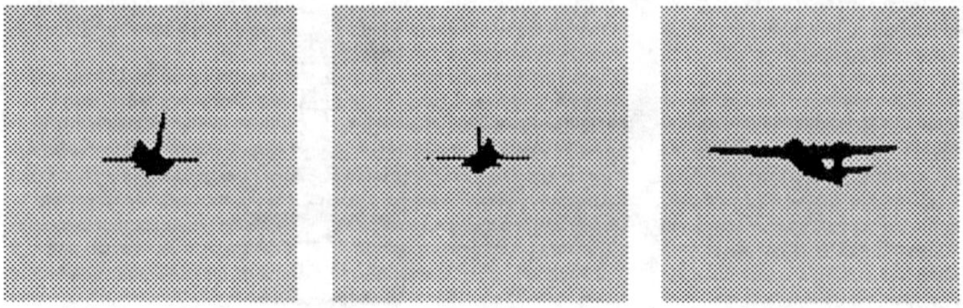

Figure 3: The two left-hand pictures are frontal views of an F-16. The right hand picture is of an HK-1.

hierarchical organization from 2-D view to 3-D object has been reported in neurophysiological studies of cell responses in monkey inferotemporal cortex, where some cells respond to individual 2-D views whereas others, like 3-D object nodes, respond to a wide range of views (Logothetis *et al.*, 1994). These studies were motivated by the regularization networks of Poggio and Girosi (1990) which also add up responses from 2-D views at 3-D object nodes. These networks do not, however, incrementally learn their categories in real-time and have not yet been incorporated into a self-organizing image processing architecture.

Given the high accuracy attained by individual 2-D view categories, the simplest possible working memory was used in VIEWNET 1 to illustrate the tradeoff between preprocessor, categorizer, and working memory. No view transitions were used. In fact, no temporal order information was used. Instead, the working memory simply updated its representation of each 3-D object category every time one of its 2-D views was experienced, and the 3-D object was predicted by voting for the winning 3-D category. Using this scheme, voting with two views achieves up to 94%, and with three views up to 98.5% accuracy. It was hereby shown that the simplest evidence accumulation from multiple views, without view transitions or even a representation of view temporal orders, can lead to high recognition on the Seibert-Waxman database if the preprocessor and classifier are designed as indicated.

The remainder of the paper describes these operations in more detail. Section 4 describes the airplane database. Section 5 describes the CORT-X 2 boundary segmentation network. Appendix describes its equations. Section 6 describes the operations to generate an invariant representation of the boundary segmentation. Section 7 describes the coarse coding algorithm. Section 8 describes the Fuzzy ARTMAP network. Appendix describes the equations. Section 9 describes computer simulation results in response to a single view, using fast or slow learning with or without image noise. The results are robust across all simulation conditions. Section 10 summarizes how multiple views may improve recognition scores. Section 11 provides concluding remarks.

4. Airplane Data Base

The image database which Bradski and Grossberg (1995) used to test VIEWNET 1 consists of multiple 2-D images of three jets airplanes: an F-16, F-18, and HK-1. Michael Seibert and Allen Waxman of MIT Lincoln Laboratory kindly provided access to this database. Video images were taken of three models of these airplanes. Each was painted black and suspended by string against a light background to aid in segmentation. The camera was mounted anywhere in an arc around the jets that started at 0.0 degrees above horizontal and went in increments of 4.5 degrees to a maximum of 72.0 degrees above horizontal. For each camera angle, the jets were spun and frames covering one full revolution (an average of 88 frames) were retained resulting in 1200 to 1400 images per object. The images themselves were 128x128 pixel gray scale. The images were then thresholded and binarized into a SUN raster format to form the "raw" database. For our processing, data was turned into a floating point format scaled between 0.0 and 1.0 and an additive noise process was introduced. The noise consisted of a 128×128 pixel images with each pixel taken from a uniform distribution between 0.0 and 1.0 scaled by a constant $C \geq 0.0$. That is, every pixel in the noise image is multiplied by C to create a "scaled" noise image. Thus, if say $C = 0.5$, the noise image would consist of pixels that varied randomly in amplitude with uniform distribution between 0.0 and 0.5. These scaled, 128×128 noise images were then added to the 128×128 jet images prior to preprocessing. Thus, both noise-free and noisy 2-D views covering a half-sphere surrounding the 3-D object were collected, keeping their spatial relationships intact.

Even numbered rotation images from each camera angle were taken as the training set with the odd numbered images forming the test set. The system was trained using random walks over the half-sphere of training images. Testing was done using random walks over the half-sphere of test images so that the paths taken and views seen were never the same between the training and test sets. The difficulty of the problem derives in part from the existence of small ambiguous frontal views in the data base, as in Figure 3.

5. CORT-X 2 Filter

The CORT-X 2 filter was used to preprocess the 128x128 airplane images. This is a feedforward network that detects, regularizes, and completes image boundaries from edge, texture and shading contrasts, while suppressing noise that does not have an underlying anisotropic structure. The CORT-X 2 filter is an enhancement of the original CORT-X filter (Carpenter, Grossberg, and Mehanian, 1989). It generates better boundary segmentations, deals better with noise, and may also be used for figure-ground separation. The figure-ground separation properties were not needed in this research because the data images were already separated from their backgrounds. These CORT-X filters are simplifications of the Boundary Contour System (or BCS) for boundary segmentation (Grossberg, 1987, 1994; Grossberg and Mingolla, 1985; Grossberg, Mingolla, and Todorović, 1989; Grossberg, Mingolla, and Williamson, 1994). The BCS includes internal feedback loops to generate coherent boundary completions even over image regions defined by sparse image contrasts. A full BCS system can be inserted into the VIEWNET architecture to handle a wider class of imagery.

The CORT-X 2 filter incorporates a number of features that are useful for processing noisy imagery, including ON and OFF cells and multiple size scales, that respond to images in complementary ways. The CORT-X 2 filter embodies a computational strategy for combining these complementary computations to achieve enhanced image processing. The processing stages of the CORT-X 2 filter are schematized in Figure 4 and summarized below. Equations and parameters are listed in Appendix A.

Step 1. Discount the Illuminant. The first processing stage compensates for variable illumination, and thereby extracts ratio contrasts form the image while normalizing the overall activity level in each image region. Both ON cells and OFF cells process the image in this way, using parallel shunting on-center/off-surround ("ON-C") and off-center/on-surround ("OFF-C") networks. Parameters are set so that the ON-C network has a zero baseline activity and the OFF-C network has a positive baseline activity. The OFF-C filter performs an image inversion because it has a positive baseline activity that is inhibited by

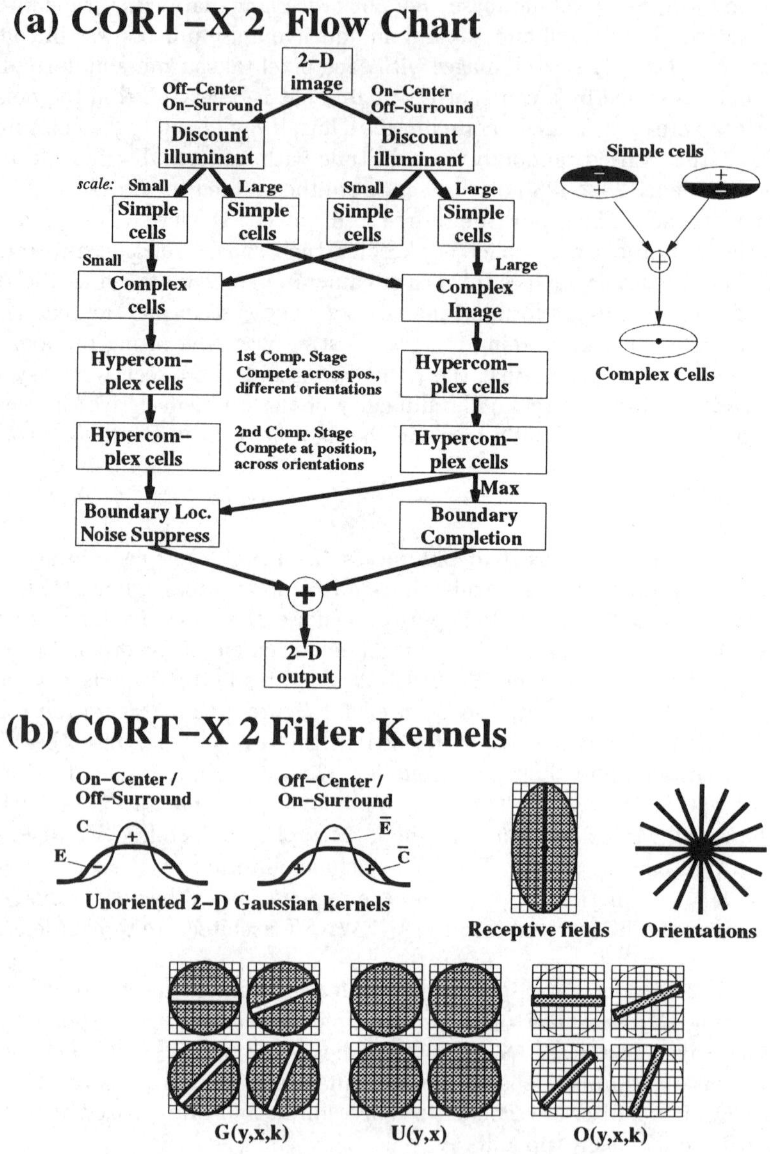

Figure 4: CORT-X 2 flow chart and filter kernels. The image is processed in parallel with small and large scale filters. Grey areas in the kernels are the active regions. All kernels are normalized to have an area equal to one.

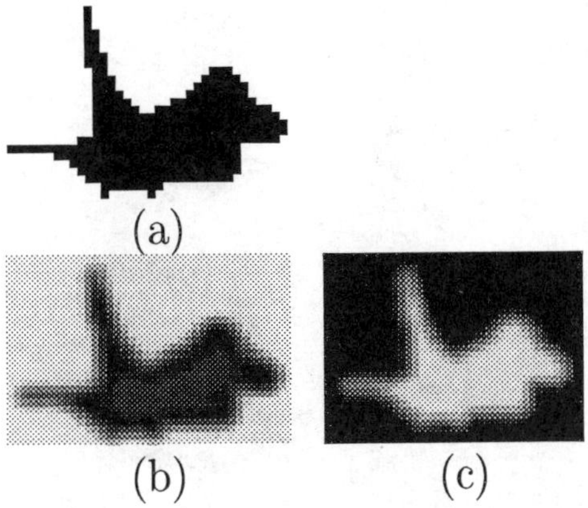

Figure 5: (a) The original F16 image. (b) CORT-X 2 ON-C output. (c) CORT-X 2 OFF-C output.

positive signal values in the image. Figure 5 shows a noise-free image as well as the ON-C and OFF-C outputs.

Along straight contrast boundaries in an image, both the ON-C and OFF-C networks enhance the contrast. On the other hand, the ON-C network has a stronger response to concave corners of activity in an image than the OFF-C network, while the converse is true at convex corners, as was noted by Grossberg and Todorović (1988). These complementary responses are joined at later processing stages used to build more complete boundary segmentations.

Step 2. Boundary Segmentation. The CORT-X 2 filter transforms the normalized ON-C and OFF-C images into a boundary segmentation using fast feedforward computations with oriented contrast-sensitive cells. Each processing stage possesses a full range of oriented cells as at each pixel in the image. Image processing is done in parallel using two sets of convolution kernels at two different size scales. As noted in Carpenter, Grossberg, and Mehanian (1989), larger scale oriented cells are better able to complete gaps in image boundaries and to suppress noise. On the other hand, smaller scale oriented cells do a better job of boundary localization than do larger scale cells. Interactions between the two size scales are designed to generate boundary outputs that combine good localization, boundary completion, and noise suppression properties.

The first stage, called the *simple cell layer*, consists of oriented contrast detectors that are sensitive to the orientation, amount, direction, and spatial scale of image contrast at a given image location. The orientation sensitivity results from using an elliptically shaped kernel, or input field, one for each of eight orientations spaced 45° apart that operate in parallel at each position in the image. Sensitivity to direction-of-contrast results from a kernel in which one half is excitatory and the other half inhibitory. At each orientation, a pair of detectors sensitive to opposite directions-of-contrast processes the image. The net activity of each detector is rectified, giving rise to a half-wave rectified output signal. Figure 7 shows the results of processing the ON-C (Figure 6a) and OFF-C (Figure 6b) image with the small spatial scale (6x3 pixels) simple cell layer. The line lengths in the figure indicate the magnitude of the simple cell responses at each orientation and position.

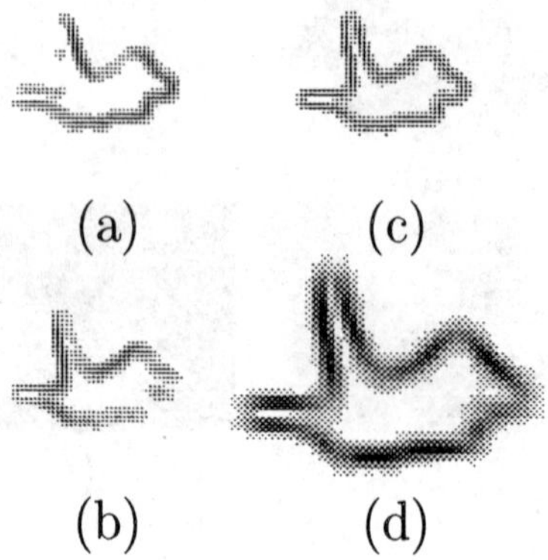

Figure 6: CORT-X 2 processing. (a) Output resulting from the ON-C network, using left sided elliptical filters (simple cell output) with a small spatial scale (6 x 3 pixels). A "left sided" filter refers to filters that respond to a vertical left-to-right, high-to-low contrast transition area in the image when the filter is in vertical orientation. A "right sided" filter is the opposite. Lines in the figure are proportional to the magnitude of the response at each orientation at each position. (b) Output from the OFF-C network using small left sided elliptical filters. (c) Hypercomplex cell output for the small scale. (d) The final CORT-X 2 output.

The next processing stage generates a cell type whose output is insensitive to direction-of-contrast, or contrast polarity. Such an operation enables image boundaries to bridge textured and shaded image regions where contrast polarity reverses. This *complex cell layer* combines outputs from the simple cells at each position as shown in Figure 5. Complex cells perform a full-wave rectification of the image that is sensitive to the orientation, amount, and spatial scale of the contrast in the image, but not to its direction-of-contrast. To achieve this, complex cells sum up the half-wave rectified outputs of like-oriented and scaled simple cells of both directions-of-contrast at each position from both the ON-C and OFF-C networks. This is done in two parallel circuits at both the small and large spatial scales (see Figure 4).

Complex cells excite hypercomplex cells in the next layer at their position and orientation while inhibiting hypercomplex cells at nearby locations that are not colinear with the complex cell's orientation. This positional competition is called the *first competitive stage.* It positionally sharpens the location of the segmentation, especially in response to textured and shaded images. Figure 6c shows the output of the hypercomplex cell layer for the small scale.

The next layer, called the *second competitive stage*, chooses the hypercomplex cell whose orientation is maximally activated to represent the activity at each position. These orientationally favored hypercomplex cells are often called higher-order hypercomplex cells in the full BCS. Figure 7a displays the small scale output of these higher-order cells.

The final stages of CORT-X 2 involve cooperative interactions between the large and small scale filters to join together the better boundary completion and noise suppression properties of the larger scale cells and the better localization properties of the smaller scale cells. One type of cooperative interaction is used to complete boundaries across boundary gaps caused by image noise. In the full BCS, boundary completion bridges gaps between

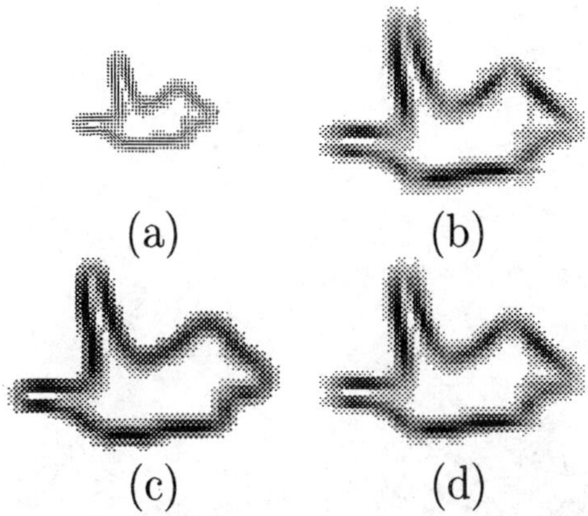

Figure 7: CORT-X 2 final stage outputs. (a) The maximal orientations of the hypercomplex cells (second competitive stage). (b) The long-range boundary completion output. (c) The multiple scale interaction output. (d) The final CORT-X 2 output from the additive combination of the top right and bottom left outputs.

distant image contrasts using a feedback loop that includes another cell type, called the bipole cell. Lacking a feedback loop, the CORT-X 2 filter uses a simplified interaction that captures the main heuristic for completing boundaries across small image gaps. In particular, cooperative interactions among the higher-order hypercomplex cells activate an inactive cell if enough cells that share the inactive cell's orientation are active on both sides of its oriented axis.

Another type of cooperative interaction combines large and small scales in such a way that the better localization properties of the smaller scale filters have an effect only within regions where the larger scales have located a boundary. The output of the boundary-completing cooperative cells is shown in Figure 7b. Figure 7c displays the multiple scale cooperative interaction, and Figure 7d shows the final CORT-X 2 output consisting of the sum of output of the boundary-completion and multiple scale localization interactions. Figure 8 shows the results of processing images with two levels of additive noise: $C = 0.5$ or 50% noise (a), and $C = 1.0$, or 100% noise (b). Because of the relative simplicity of the images being processed, only amplitude information was used. The CORT-X filter also computes potentially useful information about the relative orientation of different object parts, as in Figure 7a. The boundary completion capabilities of the CORT-X filter are also not greatly taxed by these images.

6. Translation, Rotation and Scale Invariance

The next processing stages generate a representation, using standard methods, that is invariant under 2-D translation, rotation, and dilation. First, the 2-D boundary segmentation is centered by dividing its 1st moments by its 0th moment to find the figure centroid, subtracting off the center of the image and then shifting the figure by this amount. A log-polar transform is then taken with respect to the center of the image. Each point (x, y) is represented as $re^{i\theta}$. Taking the logarithm yields coordinates of log radial magnitude and angle. As is well known (Schwartz, 1977), figural sizes and rotations are converted into figural

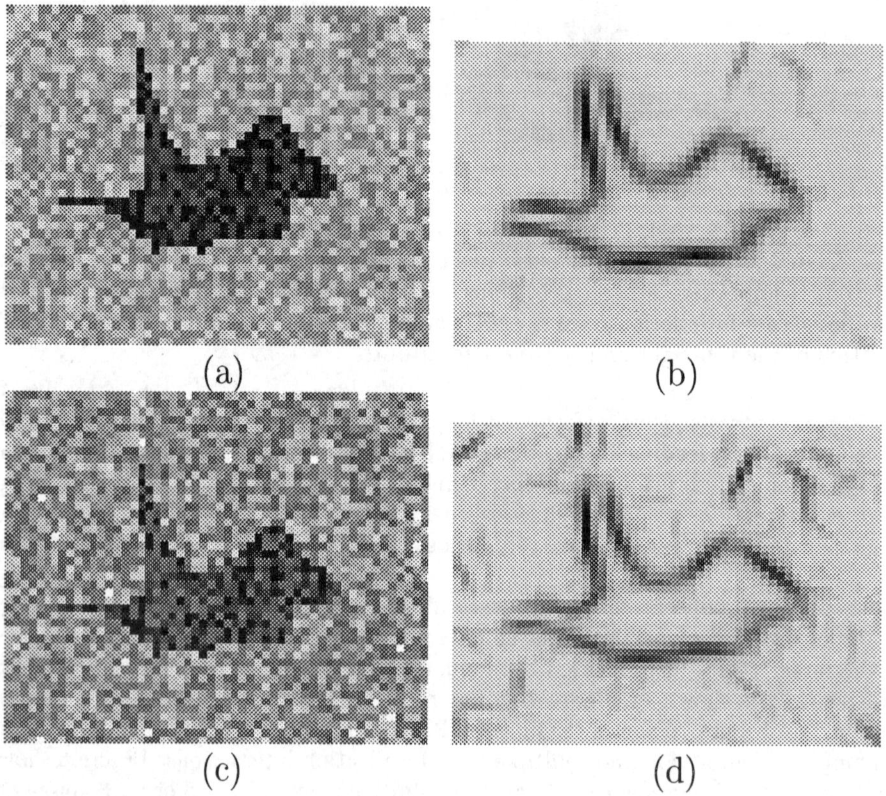

Figure 8: Results of processing noisy images with CORT-X 2. Uniform random noise was added to every pixel in the original image. The original image (left column) had pixels with activity levels between 0.0 and 1.0. Uniform random noise with pixel values ranging between 0.0 and 1.0 was scaled by C and added to the clean image prior to processing by CORT-X 2 with results shown in the right column. At top (a,b), random noise between 0.0 and 0.5 ($C = 0.5$ or 50% noise) was added, on the bottom (c,d), noise between 0.0 and 1.0 ($C = 1.0$ or 100% noise) was added. For example, 100% noise refers to the fact that the magnitude of any given noise pixel can be as large as the largest magnitude pixel in the noise-free image.

shifts under log-polar transformation. Using these shift parameters to center the log-polar transformed image leads to a figural representation that is invariant under 2-D changes in position, size and rotation. Figure 9 shows the results of processing two F-18 images which are identical except for scaling and rotation. The images become very similar in the centered log-polar domain.

7. Coarse Coding

Coarse coding reduces memory requirements, as it compensates for modest 3-D foreshortening effects and inaccuracies of figural alignment in the invariant filter. Coarse coding by averaging in the space domain is equivalent to low pass filtering in the frequency domain. Neighboring pixel features are hereby blurred, compensating for slight alignment variations. In addition, 2-D images of 3-D objects suffer from 3-D perspective distortions that cannot be corrected by log polar transforms. These are viewpoint-specific foreshortening effects and self-occlusions. The blurring associated with coarse coding can help to increase generalization by causing foreshortened and non-foreshortened images to map to nearly the same image. On the other hand, too much blurring can obscure critical input features and thereby harm recognition performance. Our analysis suggests how to balance these effects to maximize the benefits of coarse coding.

Data reduction is an important practical issue in its own right in many realistic pattern recognition problems. Seibert and Waxman's database contains 4300 images of 3 distinct objects. Each image is made up of 128x128 floating point pixels of 4 bytes each for a total of $128 \times 128 \times 4 \times 4300 = 281,804,800$ bytes. Yet, as reported below, object identification performance based on single 2-D views did well even when images were reduced to just 4x4 pixels, yielding a database of just $4 \times 4 \times 4 \times 4300 = 275,200$ bytes. This reduction affords an enormous saving in both computation and memory storage.

Coarse coding of the 2-D images used a spatial averaging method that preserves sufficient information for accurate recognition. This method was selected as follows. Spatial averaging consists of convolving the original image I with a function Ψ and then sampling the resultant image with delta functions spaced every T pixels: $\delta(x - nT, y - kT)$. For simplicity, in 1-D this is

$$(I * \Psi) \cdot \sum_{n=-\infty}^{\infty} \delta(x - nT). \tag{1}$$

If the Fourier transform of I is $\hat{I}$, and that of Ψ is $\hat{\Psi}$, then the Fourier transform of equation (1) is

$$(\hat{I} \cdot \hat{\Psi}) * \frac{2\pi}{T} \sum_{k=-\infty}^{\infty} \delta(\Omega - k\Omega_s), \tag{2}$$

where $\Omega_s = 2\pi/T$, and T is the sampling period in pixels. If Ω_N is the highest frequency in the image, then for the image to be uniquely determined by its samples, we must have by the Nyquist sampling theorem that

$$\Omega_s = \frac{2\pi}{T} > 2\Omega_N. \tag{3}$$

Two simple spatial averaging functions Ψ are: (I) uniform averaging of the input image so that all pixels in a window of some width are summed and divided by the number of pixels in the window; (II) Gaussian averaging of the input image so that a normalized Gaussian

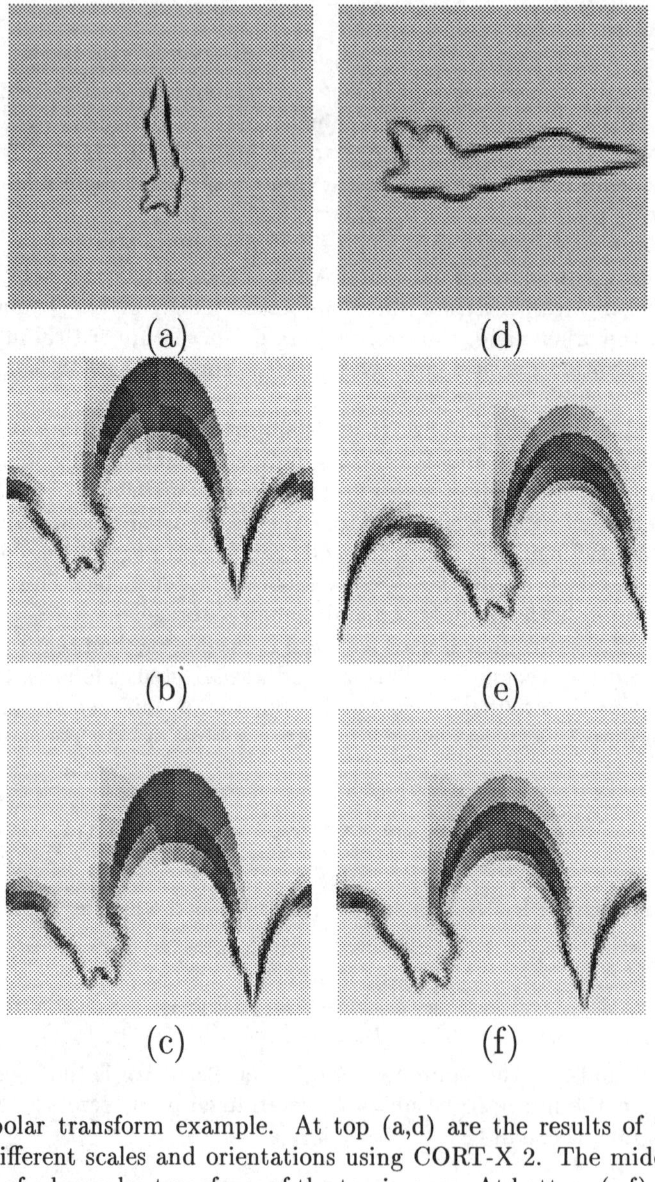

Figure 9: Log-polar transform example. At top (a,d) are the results of processing F-18 images at two different scales and orientations using CORT-X 2. The middle images (b,e) show the results of a log-polar transform of the top images. At bottom (c,f) are the centered log-polar images that have been made more identical in the sense that, overall, many more (x, y) pixel locations obtain similar values between the two images after processing. In the log-polar images, the ordinate is the log-radial magnitude and the abscissa is the periodic angle axis.

weighted sum of all pixels is taken over a window of some width. Both approaches were investigated by Bradski and Grossberg (1995), who noted that method (I) is a rectangular filter in space, and thus a sinc function in frequency. The side lobes of the sinc function can introduce high frequency aliasing ("ringing") in the resultant image, which can be reduced by using a function that provides a better low pass filter, such as the Gaussian filter method (II). The Gaussian filter has the further advantage of being an eigenfunction of a Fourier transform, since the Fourier transform of a Gaussian is a Gaussian with reciprocal variance, thereby simplifying calculation.

To optimize Gaussian-based spatial averaging, we must determine how to best set the standard deviation σ of the Gaussians. Let us define two standard deviations away from the Gaussian midpoint to be essentially zero. The cutoff frequency of such a low pass filter is then $\pi/2\sigma$. Equation (3), then implies

$$\frac{2\pi}{T} > \frac{2\pi}{2\sigma}. \tag{4}$$

which yields at equality

$$\sigma = \frac{T}{2}. \tag{5}$$

By (5), the standard deviation should be set to half the Gaussian center-to-center sampling period so that the zero point of each Gaussian just touches the center of the next Gaussian as in Figure 10c. Figure 10d-f shows the results of coarse coding the image in Figure 10b in this way to reduce the original image of 128 × 128 pixels down to 16 × 16, 8 × 8, and 4 × 4 pixels, respectively.

8. Learning Recognition Categories Using Fuzzy ARTMAP

The Fuzzy ARTMAP architecture was used to learn 2-D view categories from the coarse-coded invariant spectrum of the noise-suppressed boundary segmentations. Fuzzy ARTMAP was chosen because it can achieve stable fast incremental learning of categories in response to unlimited amounts of nonstationary input data, can run in both unsupervised and supervised modes, and in its supervised mode can fit the size, number, and shape of its categories to the input statistics. Fuzzy ARTMAP parameters were chosen to enable the network to learn the conditional probability of the true 3-D object given the selected 2-D view category at the map field, which is now defined.

We utilized the simplified version of the Fuzzy ARTMAP network of Carpenter, Grossberg, Markuzon, Reynolds, and Rosen, (1992) that was employed in Carpenter, Grossberg, and Iizuka (1992). This circuit consists of a Fuzzy ART module (Carpenter, Grossberg, and Rosen, 1991) ART_a that learns 2-D view categories and a field of 3-D object category output nodes F^b. The 2-D view and 3-D object category nodes are linked together by an associative memory F^{ab} that is called the *Map Field* (Figure 11). In supervised learning mode, Fuzzy ARTMAP receives a sequence of input pairs $(\mathbf{a}_p, \mathbf{b}_p)$ where $\mathbf{b}_p$ is the correct 3-D object class given the analog 2-D view input pattern $\mathbf{a}_p$. The ART_a module classifies analog input vectors $\mathbf{a}_p$ into categories and the Map Field makes associations from the ART_a categories to the outputs $\mathbf{b}_p$ in F^b. This simplified architecture thus does not include an ART_b module for independently clustering the outputs $\mathbf{b}_p$ into their own categories. In the present application, the ART_a categories are 2-D view categories and their associations converge at the map field upon invariant 3-D object categories. In the present application, the ART_a categories are 2-D view categories and their associations converge at the map field upon invariant 3-D

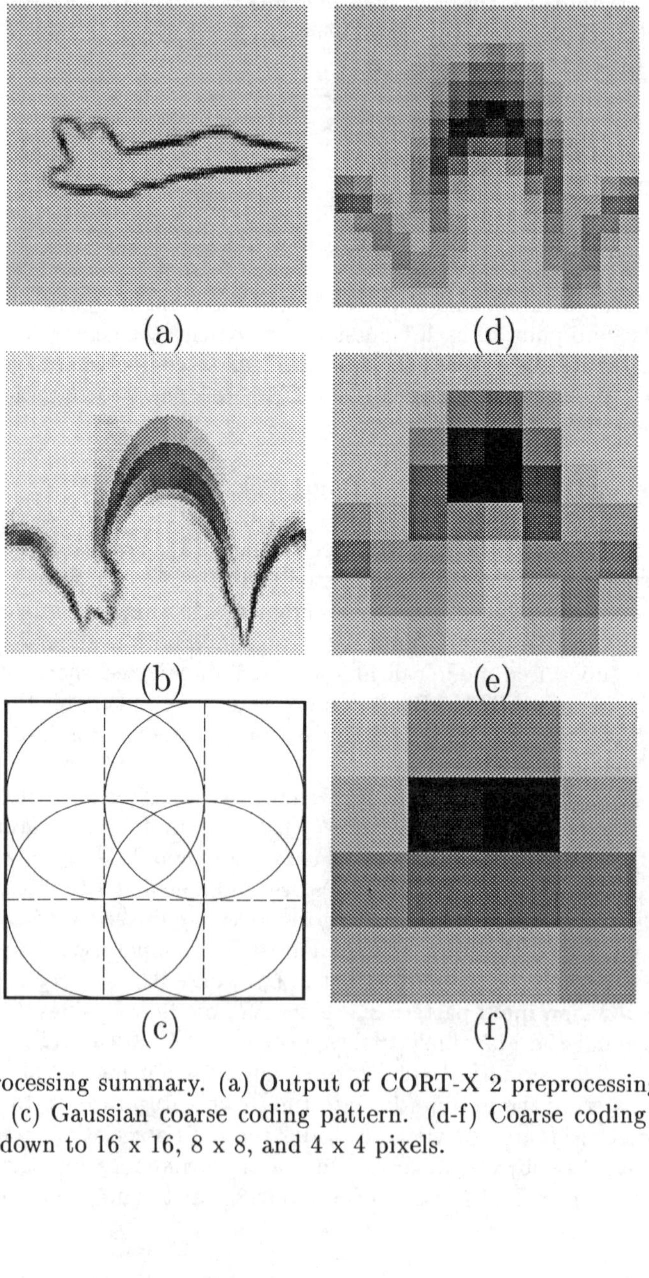

Figure 10: Preprocessing summary. (a) Output of CORT-X 2 preprocessing. (b) Centered log-polar image. (c) Gaussian coarse coding pattern. (d-f) Coarse coding reduction from 128 x 128 pixels down to 16 x 16, 8 x 8, and 4 x 4 pixels.

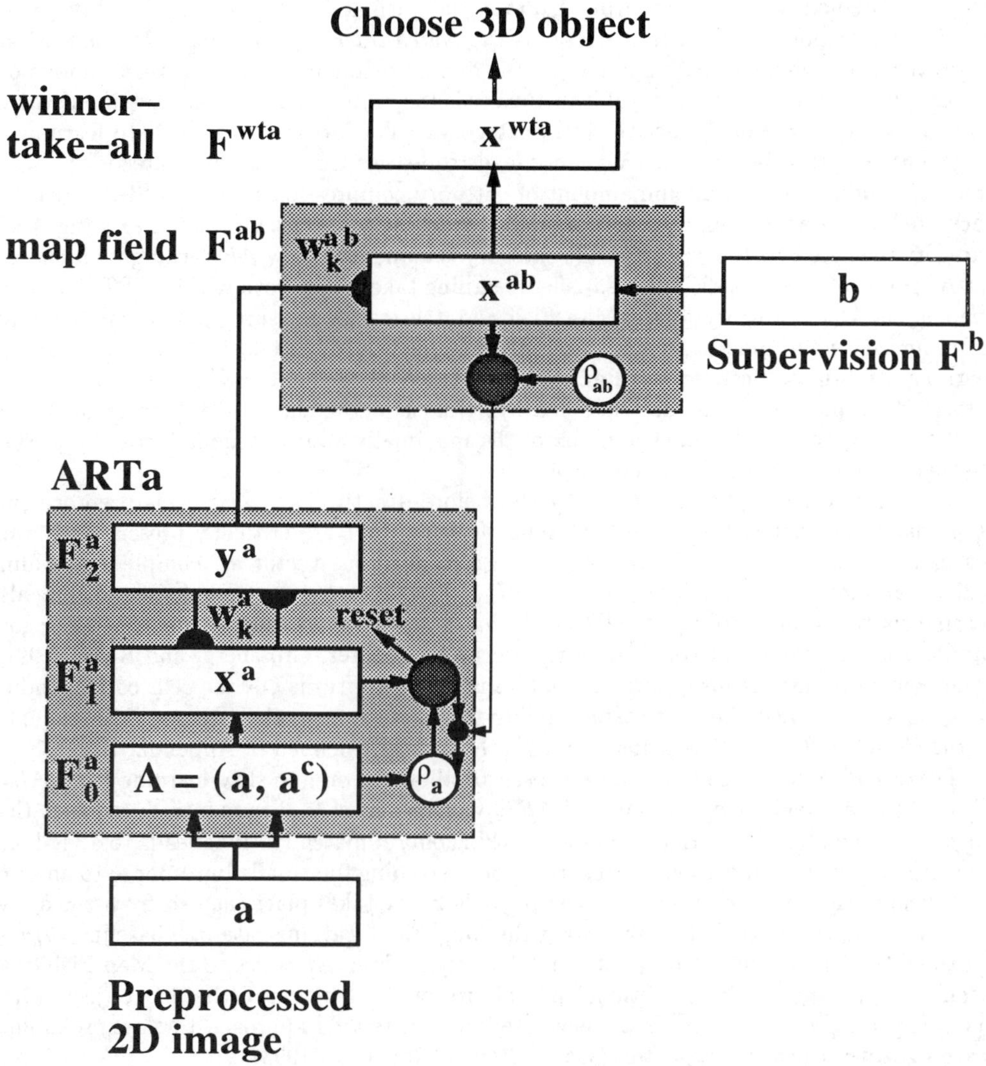

Figure 11: Fuzzy ARTMAP architecture. Each preprocessed 2-D input vector **a** is fed sequentially to the network as it becomes available. The inputs are complement coded which transforms the M-vector **a** into the $2M$-vector $\mathbf{A} = (\mathbf{a}, \underline{f}\ a_j^c)$ at field F_0^a which is then fed into the input field F_1^a. A category node k is chosen at F_2^a which reads out its prediction to the Map Field via weights $\mathbf{w}_k^{ab}$. If the prediction is disconfirmed, a match tracking process is invoked in ART_a. Match tracking raises the ART_a vigilance ρ_a to just above the match ratio $|\mathbf{x}^a|/|\mathbf{A}|$. This triggers an ART_a search which activates either a different existing category, or a previously uncommitted category node at F_2^a. After the search process concludes, F^{wta} chooses the maximally activated node in F^{ab} as the 3-D object being viewed.

object categories.

Under supervised supervised learning conditions, if $\mathbf{a}_p$ is categorized into an ART_a category that predicts an incorrect output $\mathbf{b}_p$, then the mismatch between actual and predicted $\mathbf{b}_p$ causes a memory search within ART_a via the *match tracking* mechanism. Match tracking raises the ART_a vigilance parameter ρ_a by the minimum amount that will trigger a memory search. In particular, ρ_a grows until it just exceeds the match value between the input vector $\mathbf{a}_p$ and the prototype of the active ART_a category. Since low vigilance leads to learning of large, coarse categories and high vigilance leads to learning of small, fine categories, match tracking sacrifices the minimum amount of category compression that is needed to correct each predictive error. Memory search by match tracking continues until a pre-existing ART_a category that predicts the correct ART_b category is found, or a new ART_a category is chosen.

After one of these conditions is satisfied, learning takes place both within ART_a and from the chosen ART_a category to the Map Field. Match tracking assures that predictive error is minimized while maintaining maximum generalization during fast or slow incremental learning conditions. Between learning trials, vigilance relaxes to its baseline vigilance $\bar{\rho}_a$. In test mode, input vectors $\mathbf{a}_p$ are classified by ART_a and the chosen category reads out its prediction to the Map Field. The index of the maximally activated node in the Map Field is taken to represent the predicted output class.

The input vector $\mathbf{a}_p$ is *complement coded* before it activates ART_a. This preprocessing step enables the network to code both input features that are critically present and input features that are critically absent. In response to an input vector $\mathbf{a}_p$, complement coding delivers an input vector $\mathbf{A}_p = (\mathbf{a}_p, \mathbf{a}_p^c)$ to ART_a, where $\mathbf{a}_p^c = \mathbf{1} - \mathbf{a}_p$. Complement coding also normalizes the total input $\mathbf{A}_p$ to ART_a such that $\| \mathbf{A}_p \|_1 = 1$. It thereby prevents a category proliferation problem that could otherwise occur (Carpenter, Grossberg, and Rosen, 1991). Complement coding means intuitively that an input vector turns ON the cells corresponding to $\mathbf{a}_p$ as it turns OFF the cells corresponding to $\mathbf{a}_p^c$, much as in the ON and OFF channels of the CORT-X 2 filter. The algorithm is mathematically defined in Appendix B.

Fuzzy ARTMAP parameters were chosen to allow for on-line slow learning from ART_a F_2^a to the Map Field nodes. A maximal ART_a vigilance level, $\bar{\rho}_{max}$ was introduced such that an error at the Map Field triggers match tracking only if match tracking leads to a vigilance $\rho_a \leq \bar{\rho}_{max}$. This bound prevents categories from becoming too small. In response to an error that would otherwise cause ρ_a to exceed $\bar{\rho}_{max}$, learning takes place instead from the active node in F_2^a to the Map Field. By setting the Map Field learning rate β_{ab}, baseline ($\bar{\rho}$) and maximal ($\bar{\rho}_{max}$) vigilance levels appropriately, weights from F_2^a nodes to the Map Field may begin to approximate the conditional probability of the true class (the 3-D object) given the selected F_2^a category (the 2-D view category). A related approach to slow probability learning is described in Carpenter, Grossberg, and Reynolds (1993).

9. Illustrative Simulation Results

A computer simulation was run on the airplane database using the CORT-X 2 parameters shown in Table 1. The database was processed twice by CORT-X 2, once with a large pair of large and small oriented filters and once with a smaller pair of large and small filters. The large oriented filter pairs consisted of elliptical receptive fields with axes 16×8 and 10×5 pixels. The small oriented filter pair consisted of oriented ellipses with axes 10×5 and 6×3 pixels. This was done so that recognition results could be compared when images were processed at different scales. Coarse coding was done with both simple spatial averaging and Gaussian averaging, reducing the image down to 16×16, 8×8, and 4×4 pixels from an original size of 128×128. The window for simple spatial averaging was square with a

Parameter	*Description*
$\kappa_{onC} = 7.0$	On-center magnitude
$\alpha_{onC} = 1.3$	On-center standard deviation
$\kappa_{offS} = 3.333$	Off-surround magnitude
$\alpha_{offS} = 1.875$	Off-surround standard deviation
$\overline{D} = B = 1.0$	Shunting values
$\overline{B} = D = 0.5$	Shunting values
$S = 0.2$	Spontaneous activity level
$A = 134$	Shunting decay
$\alpha_1 = \alpha_2 = 1.1$	Threshold contrast parameters
$\beta_1 = \beta_2 = \beta = .003$	Threshold noise parameters
$F = 0.5$	Complex cell scaling constant
$\epsilon = 0.1$	Hypercomplex cell divisive offset
$\mu = 5.0$	Hypercomplex cell convolution scaling
$\tau = 0.004$	Hypercomplex cell threshold
$\delta = 0.001$	Long range cooperation threshold
$\pi/8$	Oriented kernel orientation spacing
$(a_2, b_2)_{large} = (16, 8)$	Large set, large ellipse axis
$(a_1, b_1)_{large} = (10, 5)$	Large set, small ellipse axis
$(a_2, b_2)_{small} = (10, 5)$	Small set, large ellipse axis
$(a_1, b_1)_{small} = (6, 3)$	Small set, small ellipse axis
$G_1 = 2a_1/3$	Hypercomplex small kernel diameter
$G_2 = 2a_2/3$	Hypercomplex large kernel diameter
$U = 2a_2/5$	Multiple scale interaction kernel diameter
$O = 3a_2/5$	Long-range cooperation kernel length

Table 1: The parameter set used for CORT-X 2 in the simulations.

Parameter	*Description*
$\alpha = 0.6$	Fuzzy ART search order
$\beta_a = 1.0$	Fuzzy ART learning rate
$\beta_{ab} = 1.0$	Map Field learning rate
$\bar{\rho} = 0.1$	Baseline Fuzzy ART vigilance ρ_a
$\bar{\rho}_{max} = 1.0$	Maximum ART_a vigilance
$\rho_{ab} = 1.0$	Map Field vigilance

Table 2: The Fuzzy ARTMAP parameter set used for the simulations.

	% noise surviving CORT-X 2 filtering and Coarse Coding:	
	Large CORT-X 2 filters (16x8, 10x5)	*Small CORT-X 2 filters (10x5, 6x3)*
	1.79	2.42
	After Gaussian coarse coding from 128x128 down to:	
16x16	0.33	0.34
8x8	0.23	0.29
4x4	0.19	0.26
	After spatial average coarse coding from 128x128 down to:	
16x16	0.40	0.40
8x8	0.28	0.30
4x4	0.21	0.28

Table 3: Percent of additive white noise surviving processing by CORT-X 2 and coarse coding.

width twice the sampling period T; that is, a window centered at one pixel extended until it just touched its neighboring pixels. The standard deviation for Gaussian averaging was set to $T/2$, as discussed in Section 7. Training and testing sets were assembled as discussed in Section 4 with even numbered images forming the training set and odd numbered images forming the testing set. Except where explicitly mentioned, the simulations were run with the parameters shown in Tables 1 and 2.

The data were presented to the network in two different ways: (1) 2-D views were presented in the "natural" order in which they would appear if viewing the actual object in motion; (2) 2-D views were presented in random order. These two methods of data presentation were used to test whether presenting views in natural order helps recognition scores. Training in natural order consisted of 160 runs of from 1 to 50 views over each object. Training in random order consisted of a series of 40 runs of 100 training set views over each object. Recognition scores are taken as an average of fifteen separate training-testing cycles.

Fast Learning With and Without Noise

No clear advantage results from ordered presentation as compared to unordered presentation using noise-free data ($C = 0$) and fast learning, as shown by the results in Table 4. It can be seen that the smaller CORT-X 2 filter set resulted in better recognition performance overall and did better given more detail (less coarse coding).

The system was also tested with noisy data using additive white noise scaled by $C = 1.0$; that is, each pixel could have noise added to it less than or equal to the value of the maximum activity of pixels ($= 1.0$) in the original image. Table 3 shows what percent of the additive noise survives processing by CORT-X 2 alone, and by CORT-X 2 and coarse coding together,

CORT-X 2 filter set	*Data presentation*	*Coarse code using spatial avg / Gaussian avg* 4x4	8x8	16x16
Small	Ordered	81.0/83.1	84.4/86.4	86.7/90.5
Small	Unordered	80.3/83.9	84.9/86.5	86.8/89.3
Large	Ordered	76.8/78.7	79.0/81.6	79.1/80.1
Large	Unordered	77.4/79.7	80.5/81.5	77.1/80.5

Table 4: Recognition results on a noise free database ($C = 0$). CORT-X 2 filter sizes refers to the size of the oriented receptive field filters. CORT-X 2 was run twice using a larger and a smaller set of its large and small elliptical oriented filters. In the table, "Large" refers to the run with the larger set of oriented ellipses with axes 16x8 and 10x5 pixels; "Small" refers to the run with the smaller set of oriented ellipses with axes 10x5 and 6x3 pixels. Views were presented either in natural order or in random order. Data was coarse coded from 128x128 down to 4x4, 8x8, or 16x16 using simple spatial averaging or Gaussian averaging. Recognition scores refer to the percent of 2-D views correctly associated with a 3-D object.

CORT-X 2 filter set	*Data presentation*	*Coarse code using spatial avg / Gaussian avg* 4x4	8x8	16x16
Small	Ordered	80.1/83.3	84.5/85.9	84.2/89.1
Small	Unordered	79.4/83.2	83.9/86.4	84.3/88.0
Large	Ordered	76.6/79.4	79.3/80.8	75.8/79.3
Large	Unordered	76.0/79.7	78.4/80.7	75.5/79.0

Table 5: Recognition results on noisy data ($C = 1$) with fast learning ($\beta_{ab} = 1.0$). These results differ little from the noise-free results in Table 4 (no noise condition) with the exception of some consistent reduction in scores for the 16x16 coarse coding.

for two different filter sets and three different coarse codings. The percent noise surviving these transformations was measured by the following formula:

$$\max_{\forall(x,y)} \left[\frac{\Psi(\mathbf{I}+\mathbf{N}) - \Psi(\mathbf{I})}{C} \right] \times 100, \tag{6}$$

where $\mathbf{I}$ is the image, $\mathbf{N}$ is the noise image, Ψ is the CORT-X 2 filter, $C > 0$ is the noise scaling parameter and (x, y) is the pixel index in the images. Table 3 represents the average results from ten measurements using equation (6).

It can be seen that the combination of CORT-X 2 filtering followed by coarse coding is effective in reducing additive noise. Using a fast learning paradigm, the recognition results shown in Table 5 were similar to those for the noise-free case in Table 4, except for some minor falling off of recognition scores at the lowest level of coarse coding (the 16×16 case). Less coarse coding has the same effect on noise as raising the cutoff frequency of a low pass filter. Thus, as seen in Table 3, more noise gets through with less coarse coding, yielding slightly lower recognition performance.

Table 6 shows the number of nodes created by the network after training for the no noise (left entry) and noise (right entry) results reported above. Noise causes a small increase in the number of categories formed on average as the network attempts to correct a greater number of noise-induced errors during supervised training.

CORT-X 2 filter set	*Data presentation*	*Coarse code using spatial avg / Gaussian avg* 4x4	8x8	16x16
Small	Ordered	[172, 184]/[165, 169]	[77, 73]/[70, 73]	[34, 33]/[33, 35]
Small	Unordered	[191, 198]/[175, 179]	[76, 77]/[73, 76]	[34, 35]/[35, 36]
Large	Ordered	[168, 179]/[160, 162]	[71, 68]/[67, 71]	[31,33]/[30, 31]
Large	Unordered	[183, 192]/[169, 174]	[73, 75]/[69, 72]	[32, 32]/[33, 32]

Table 6: Average number of ART_a categories formed during training for the simulations of Table 4 (no noise) and Table 5 (noise). The format in the table is as follows: [spatial avg.]/[Gaussian avg.] = [No noise, Noise]/[No noise, Noise].

CORT-X 2 filter set	*Data presentation*	*Coarse code using spatial avg / Gaussian avg* 4x4	8x8	16x16
Small	Ordered	79.9/83.1	84.0/85.6	84.7/89.9
Small	Unordered	78.8/83.3	83.2/85.7	84.9/89.1
Large	Ordered	76.3/78.2	78.5/81.5	77.0/78.8
Large	Unordered	77.4/80.2	79.6/80.41	75.8/79.2

Table 7: Recognition results on noisy data ($C = 1$) with slow learning to the Map Field ($\beta_{ab} = 0.2$, $\bar{\rho}_{max} = 0.95$). Due to the low levels of noise surviving preprocessing, the recognition results here are not substantially different than those found using fast learning in noise in Table 5 except where noise was highest as in the 16x16 coarse coding. As noise increases, slow learning becomes more important for maintaining good recognition scores.

Slow Learning Simulation With Noise

The network was also run on the noisy data using slow learning to the Map Field ($\beta_{ab} = 0.2$ in equation (42)). Fast learning was still used within the ART_a module itself however ($\beta_a = 1.0$ in equation (41)). In addition, a maximum vigilance level in ART_a ($\bar{\rho}_{max} = 0.95$) was set so that when match tracking due to error feedback attempts to create an ART_a category smaller than a given size, no new category forms and learning takes place for the current category instead. With noisy inputs, rather than continually making new categories to correct for the noise, ART_a categories below a set prescribed size begin to learn the conditional probability of the true class (the 3-D object) given the selected ART_a category (the categorical 2-D view).

For $\bar{\rho}_{max} = 1.0$, the results for slow learning and fast learning to the Map Field are equivalent. They are equivalent because, with Map Field vigilance set to $\rho_{ab} = 1.0$ as in Table 2, the slightest mismatch at the Map Field will invoke match tracking and a new category will be created. The main difference between slow and fast learning using $\bar{\rho}_{max} = 1.0$ is that ART_a categories may learn their associations to nodes in the Map Field at different rates. The weights from an F_2^a node in ART_a to the correct node in the Map Field will always have a value of 1.0, however, since any error is corrected by forming a new category. Weights to the other nodes in the Map Field will be less than 1.0 (slow learning) or equal to 0.0 (fast learning). Recognition results on the test set are not hereby affected, since a winner-take-all field chooses the maximum activation in the Map Field as the recognition code via equation (43) in Appendix .

To derive benefit from slow learning, in the case $\rho_{ab} = 1.0$, we set $\bar{\rho}_{max} = 0.95$. For

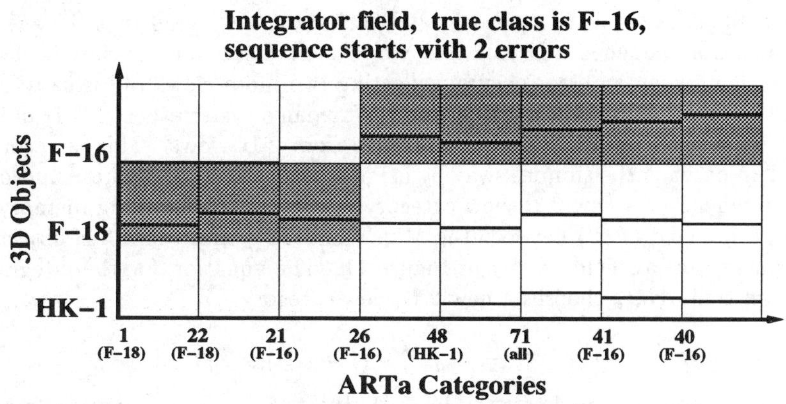

Figure 12: Simulation of the integration field. The ordinate axis contains the integrator nodes representing the 3-D objects: F-16, F-18, and HK-1. The abscissa represents ART_a F_2^a categories chosen by the preprocessed 2-D views being presented to VIEWNET. Black horizontal lines denote activity of the corresponding integrator node through time. Gray shading in the figure indicates the integrator node with maximal activation which represents the VIEWNET decision as to which object is being presented. Categories 1 and 22 erroneously code for the F-18 jet here. Category 48 erroneously codes for the HK-1. Category 71 has not been chosen before and so selects all objects simultaneously. The rest of the categories code correctly for the F-16. The integration step size was set to $\beta_{int} = 0.2$.

$\rho_{ab} = 1.0$, we may then compare the results of fast learning to the Map Field using $\bar{\rho}_{max} = 1.0$ with the results of slow learning to the Map Field using $\bar{\rho}_{max} = 0.95$. Table 7 records the results using slow learning in large amplitude noise ($C = 1$). Where noise levels after preprocessing were very small, the results were approximately the same as in the fast learning case shown in Table 5. Slow learning begins to help when the noise level increases, as with the 16 × 16 coarse coding. Table 8 records the average number of categories formed for the noisy data case using fast learning and slow learning. Slow learning with $\bar{\rho}_{max} = 0.95$, caused approximately 10% fewer categories to be formed than with $\bar{\rho}_{max} = 1.0$, since noise-induced errors do not always cause the formation of a new category in the former case.

CORT-X 2 filter set	*Data presentation*	*Coarse code using spatial avg / Gaussian avg*		
		4x4	*8x8*	*16x16*
Small	Ordered	[184, 165]/[169, 150]	[73, 67]/[73, 66]	[33, 30]/[35, 32]
Small	Unordered	[198, 180]/[179, 163]	[77, 69]/[76, 70]	[35, 32]/[36, 33]
Large	Ordered	[179, 160]/[162, 147]	[68, 61]/[71, 66]	[33 , 30]/[31, 29]
Large	Unordered	[192, 175]/[174, 160]	[75, 69]/[72, 67]	[32, 30]/[32, 30]

Table 8: Average number of nodes formed during training for the simulations of Tables 5 (noise with fast learning) and 7 (noise with slow learning). The format in the table is as follows: [spatial avg.]/[Gaussian avg.] = [fast learning, slow learning]/[fast learning, slow learning]. It can be seen that slow learning reduced the number of nodes formed by approximately 10%.

For a recognition system that can gather information from successive 2-D views, a key question is: Given that an error occurs, how many successive errors will follow on average?

For the airplane data set as processed by VIEWNET, it was found that the average overall length of an error sequence was 1.31 2-D views with a standard deviation of 0.57 views. On average then, when an error occurs, collecting two more views will usually be sufficient to correct the error. Thus, as in Seibert and Waxman's system, better 3-D object predictions may be derived by accumulating evidence from 2-D views. This is accomplished in VIEWNET in perhaps the simplest way by using a working memory whose unordered states are updated whenever a new 2-D view category is chosen. The working memory is realized as an integration field (F^{int}) between the Map Field (F^{ab}) that codes 3-D object categories and the winner-take-all field (F^{wta}) in Figure 11. The equation for the integrator field is updated each time ART_a chooses a new 2-D view category:

$$(x_k^{int})^{new} = \beta_{int} x_k^{ab} + (1 - \beta_{int})(x_k^{int})^{old}, \tag{7}$$

where x_k^{int} is an integrator node for the kth 3-D object, β_{int} is the integration rate each time the equation is stepped, and x_k^{ab} is the kth Map Field 3-D object category. The integration node with the largest activation represents the prediction of the 3-D object being viewed by the system. This maximum activation in the integration field is chosen by the winner-take-all field (F^{wta}) as the network's prediction of the 3-D object.

10. Evidence Accumulation by Voting and View Transitions

Figure 12 shows a simulation of the integrator field where equation (7) is stepped once each time ART_a chooses a category. Three integrator nodes are shown along the ordinate, one for each of the airplanes. Grey shading in the figure shows which 3-D object is "winning". In this simulation, the 3-D object being viewed is an F-16. The sequence of ART_a F_2^a categories that occurred is shown along the abscissa. The first two categories, 1 and 22 were in error since they were associated with the F-18 node in the Map Field. Categories 21 and 26 code correctly for the F-16 followed by category 48 which codes for the HK-1. Next, category 71 is selected. It is an uncommitted category that has never been activated before. By default, it gives equal activation to all integrator nodes. The remaining categories code correctly for the F-16.

Implementing evidence accumulation in this way is similar to voting for the 3-D object over a sequence of 2-D view inputs, but with recent views being given more weight. For $1 \geq \beta_{int} \geq 0$, the closer β_{int} is to 1, the more weight is given to recent votes. To measure performance on the test set with voting, the integrator field was allowed to collect first two (for the two votes score), or three (for the three votes score) activations before VIEWNET's recognition decision was recorded. The integrator field was then cleared and two or three more activations were again collected before the next decision was made. This process was repeated until 1000 views had been seen in the test set for each object at which time the percent correct recognition score was computed. Figure 13 shows the average recognition scores for voting with $\beta_{int} = 0.2$ over one, two, and three views under CORT-X 2 preprocessing with large and small scale filter sets and coarse coding to 4×4, 8×8 and 16×16 pixels using both Gaussian and spatial averaging. Voting over 3 frames improves recognition results by an average of ten percent with the best results being 98.5% correct for small scale filtered, 16×16 Gaussian coarse-coded data. The black dots and squares in the figure show recognition results from using 2-D view transition information. Note that view transitions

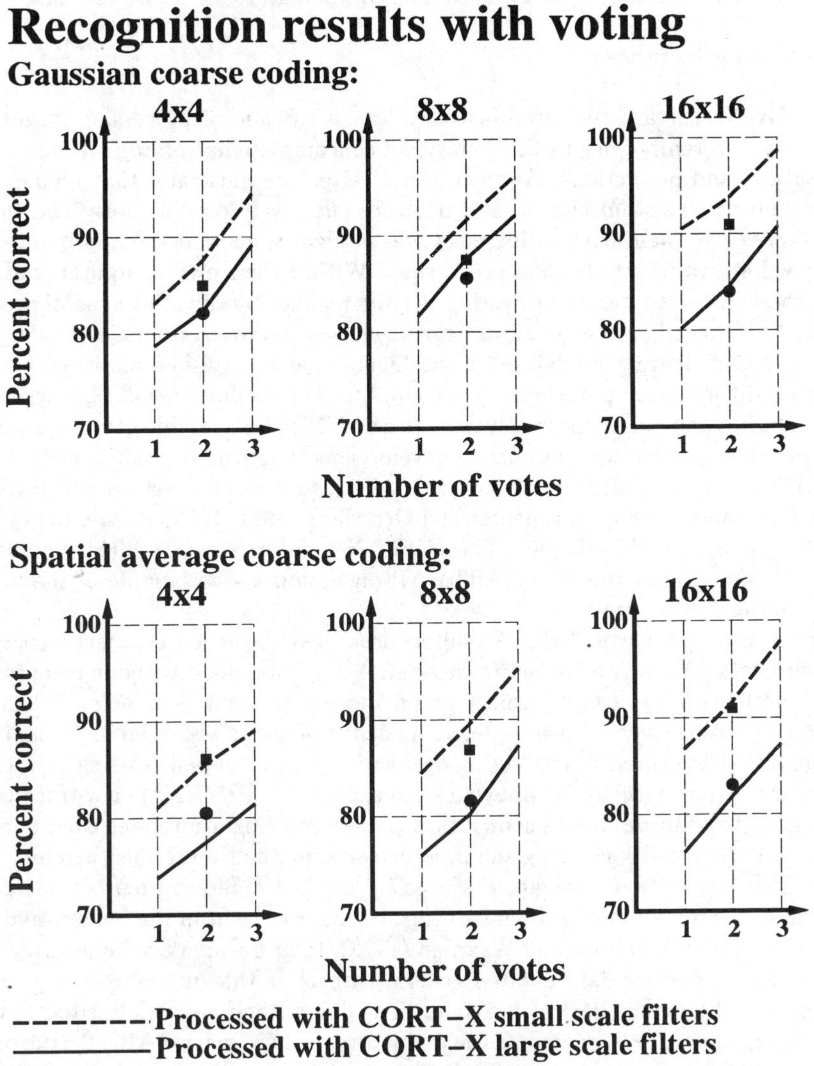

Figure 13: Recognition results for voting with an integration rate of $\beta_{int} = 0.2$. The graphs show the recognition results after gathering evidence over one, two and three 2-D views for data preprocessed using large (solid line) and small (dotted line) scale CORT-X 2 filters. Results from both Gaussian and spatial averaging coarse coding methods are shown where the images were reduced from 128×128 down to 4×4, 8×8 and 16×16 pixels. The circles and squares represent recognition scores resulting from using view transitions as discussed in Section 8.1. The black circles represent the recognition scores using view transitions for preprocessing with the large scale CORT-X 2 filters, the black squares represent recognition scores using view transitions for preprocessing with small scale CORT-X 2 filters.

did not improve performance; see Bradski and Grossberg (1995) for further details.

11. Concluding Remarks

The VIEWNET 1 architecture illustrates how a cascade of processing stages can convert spatially and temporally variant, noisy, and incomplete data into invariant 3-D object representations and predictions. As such, the architecture illustrates that no one computational method or algorithm can, in and of itself, effectively process noisy and fuzzy data. Instead, there exist qualitatively different types of algorithms that are appropriate to apply at prescribed stages of information processing. Within their own computational domains, however, these distinct types of processing may be realized by the most general processing of that type. I like to call such algorithms *general purpose solutions of modal problems* in order to emphasize that distinct modalities of intelligence, such as vision, audition, and recognition, require different computational constraints for their solution, but that each modality seems to admit a general-purpose solution. In the VIEWNET architecture, simplified algorithms from theories that are working to develop general-purpose vision architectures, such as FACADE theory (Grossberg, 1987b, 1988, 1994), and recognition architectures, such as Adaptive Resonance Theory (Carpenter and Grossberg, 1991, 1993, 1994; Grossberg, 1987a, 1988, 1995), illustrate this theme. The CORT-X 2 filter is a simplified module inspired by FACADE theory and the Fuzzy ARTMAP algorithm is one module of the larger ART cognitive and neural theory.

Using the smaller scale of CORT-X 2 filters described above, a 3-D object recognition rate of approximately 90% was achieved from single 2-D views alone without recourse to more elaborate methods of generating aspect graph models of the 3-D objects. When evidence integration or voting over a an unordered sequence of views was added, recognition rates reached 98.5% within three views. Voting over two views did as well as using view transitions on this database, but without the drawback of needing to learn $O(N^2)$ view transitions given N, 2-D views. In addition, it was shown that the above recognition rates can be maintained even in high noise conditions using the preprocessing methods described herein.

These high recognition rates are achieved by using a different preprocessor and supervised learning to create more optimal category boundaries than in the Seibert and Waxman studies. As reported in Seibert and Waxman (1992), their unsupervised clustering of coarse-coded maximal curvature data created general categories that unambiguously selected for the correct 3-D object only 25% of the time. In so doing, their network created 41 categories during training. In order to overcome the ambiguity of their general ART 2 categories, Seibert and Waxman used explicitly coded 2-D view category transitions to help identify the 3-D objects.

Using this approach, the network must be able to represent possible cross-correlations between every categorical 2-D view in the view transition matrices, one for each object, even if no correlations are eventually found between some of the categories. This is because a view transition matrix represents a definite network structure that explicitly codes the particular sequence of view transitions that ends up coding a prescribed 3-D object. Thus, such an algorithm is committed to represent all possible correlations between each of the 41 2-D view categories. The total number of correlations is then $(41^2 - 41)/2 = 820$, since transitions and their reverse are equivalent and there are no self-transitions, this is done for each object, then a total of $820 \times 3 = 2460$ correlation matrices would be needed. Add to this the 41 ART 2 categories and up to 2501 activations could be needed to recognize the 3 airplanes. As seen in Table 6, VIEWNET needs only 30 some odd nodes to categorize this database of 4000 examples since, by using unordered voting, it avoids the $O(N^2)$ penalty for using view transition information given N 2-D views.

Appendix A: CORT-X 2 Equations

The equations for the CORT-X 2 filter as described in Section are discussed below. Figure 4a shows the model flow chart and Figure 4b shows the kernels used in the CORT-X 2 filter algorithm. Table 1 summarizes the parameters used in the simulations.

Step 1. Discounting the Illuminant

The 1D cross-sections of the on and off center kernels are shown in Figure 4a.

ON-C and OFF-C Network

The activation x_{ij} at node v_{ij} at position (i,j) obeys the shunting on-center off-surround equation:

$$\frac{d}{dt}x_{ij} = -Ax_{ij} + (B - x_{ij})C_{ij} - (x_{ij} + D)E_{ij}, \tag{8}$$

and $\bar{x}_{ij}$ obeys the off-center, on-surround equation:

$$\frac{d}{dt}\bar{x}_{ij} = -A(\bar{x}_{ij} - S) + (\overline{B} - \bar{x}_{ij})\overline{C}_{ij} - (\bar{x}_{ij} + \overline{D})\overline{E}_{ij} \tag{9}$$

where C_{ij}, $\overline{C}_{ij}$, E_{ij}, $\overline{E}_{ij}$ are discrete convolutions of the input with Gaussian kernels of the form:

$$K_{ij} = \sum_{p,q} I_{pq}K_{pqij} \tag{10}$$

with

$$K_{pqij} = \kappa \exp\left\{-\alpha^{-2} \log 2[(p-i)^2 + (q-j)^2]\right\}. \tag{11}$$

The on-center kernel of $\bar{x}_{ij}$ is the off-surround kernel of x_{ij}, and the off-surround kernel of $\bar{x}_{ij}$ is the on-center kernel of x_{ij}. Then $\overline{C}_{ij} = E_{ij}$, $\overline{E}_{ij} = C_{ij}$. Also in equations (8) and (9), $\overline{B} = D$ and $\overline{D} = B$. At equilibrium in the ON-C network,

$$x_{ij} = \frac{\sum_{(p,q)}(BC_{pqij} - DE_{pqij})I_{pq}}{A + \sum_{(p,q)}(C_{pqij} + E_{pqij})I_{pq}}, \tag{12}$$

and in the OFF-C network.

$$\overline{x}_{ij} = \frac{AS + \sum_{(p,q)}(DE_{pqij} - BC_{pqij})I_{pq}}{A + \sum_{(p,q)}(C_{pqij} + E_{pqij})I_{pq}}. \tag{13}$$

Step 2. CORT-X 2 Filter

Oriented receptive fields are elliptical as shown in Figure 4b so that

$$\frac{y^2}{a_s^2} + \frac{x^2}{b_s^2} = 1, \tag{14}$$

where a_s is the major axis and b_s is the minor axis with $a_s \geq b_s$. Two sizes of receptive fields were used, indexed by the subscript s with $1 =$ small and $2 =$ large scale. Orientations were chosen at angles spaced every $\pi/8$ degrees indexed below by the subscript k.

Simple Cells: The output of the pair of simple cells of scale s with activation variable $x = x_{ij}$ and receptive field orientation k is defined by

$$S_{sL}(i,j,k) = \max[L_s(x,k) - \alpha_s R_s(x,k) - \beta_s, 0] \tag{15}$$

and

$$S_{sR}(i,j,k) = \max[R_s(x,k) - \alpha_s L_s(x,k) - \beta_s, 0], \tag{16}$$

where $L_s(x,k)$ and $R_s(x,k)$ are the image inputs to the left and right oriented receptive fields

$$L_s(x,k) = \frac{\sum_{(p,q)\in l_s(i,j,k)} x_{pq} w_{pq}}{\sum_{(p,q)\in l_s(i,j,k)} w_{pq}} \tag{17}$$

and

$$R_s(x,k) = \frac{\sum_{(p,q)\in r_s(i,j,k)} x_{pq} w_{pq}}{\sum_{(p,q)\in r_s(i,j,k)} w_{pq}}, \tag{18}$$

and w_{pq} is a weighting factor proportional to the area of a cell covered by the receptive field. L and R in S_{sL} and S_{sR} indicate that each receptive field is sensitive to the opposite direction-of-contrast from its companion. The ON and OFF networks have separate sets of simple cells with the ON simple cells denoted by S^+_{sL} and S^+_{sR}, and the OFF simple cells denoted by S^-_{sL} and S^-_{sR}.

Complex Cells: The complex cell output $C_s(x,k)$ is defined by

$$C_s(i,j,k) = F[S^+_{sL}(i,j,k) + S^+_{sR}(i,j,k) + S^-_{sL}(i,j,k) + S^-_{sR}(i,j,k)]. \tag{19}$$

These cells are sensitive to the spatial scale s and amount-of-contrast x with orientation k, but are insensitive to direction-of-contrast.

Hypercomplex Cells (First Competitive Stage): The hypercomplex cells $D_s(i,j,k)$ receive input from the spatial competition among the complex cells

$$D_s(i,j,k) = \max\left[\frac{C_s(i,j,k)}{\epsilon + \mu \sum_m \sum_y C_s(p,q,m) G_s(p,q,i,j,k)} - \tau, 0\right]. \tag{20}$$

The circular oriented competition kernel $G_s(p,q,i,j,k)$ shown in Figure 4c is normalized such that

$$\sum_{p,q} G_s(p,q,i,j,k) = 1. \tag{21}$$

Partial cells at the kernel periphery are weighted in proportion to their area (taken to be one square unit). Grey areas in Figure 4c are inhibitory. Cells with centers within the one unit wide band through the middle of the kernel do not contribute to the inhibition. In our simulations, the small and large scale kernels were 2/3 the diameter of the small and large scale major axis's of the ellipses shown in Figure 14b, respectively.

Hypercomplex Cells (Second Competitive Stage): Hypercomplex cells $D_s(i,j)$ compute the competition among oriented activities $D_s(i,j,k)$ at each position. This process is simplified as a winner-take-all process

$$D_s(i,j) = D_2(i,j,K) = \max_k D_s(i,j,k), \tag{22}$$

where K denotes the orientation of the maximally activated cell.

Multiple Scale Interaction: The interaction between the small and large scales is defined by

$$B_{12}(i,j) = D_1(i,j) \sum_{p,q} D_2(p,q) U(p,q,i,j), \tag{23}$$

where the unoriented excitatory kernel $U(p,q,i,j)$ is circular as in Figure 5b and is normalized so that

$$\sum_{p,q} U(p,q,i,j) = 1. \tag{24}$$

In our simulations, $U(p,q,i,j)$ had a diameter 2/5 as large as the major axis of the large scale elliptical filter. Cells covered by the kernel contribute to the excitation to the extent that their area is covered by the kernel. The smaller kernel $D_1(i,j)$ in (23) localizes boundary segments and suppresses noise near the boundary, while the larger kernel $D_2(p,q)$ suppresses noise far from the boundary.

Boundary Completion: The large detectors $D_2(i,j)$, are capable of responding at locations where pixel signal strength has been reduced by noise. Such boundary signals may, however, be poorly localized. To overcome this tradeoff between boundary completion and localization, large-scale cells interact cooperatively as

$$B_2(i,j) = D_2(i,j) \max \left[\sum_{p,q} D_2(p,q,K) O(p,q,i,j,K) - \delta, 0 \right]. \tag{25}$$

Kernel $O(y,x,k)$ is defined by the one-unit-wide white strips in Figure 4e. Cells with centers lying within the one unit wide band contribute to the cooperative process. The kernel is normalized so that

$$\sum_{(p,q) \text{ in kernel}} O(p,q,i,j,k) = 1. \tag{26}$$

In the simulations, the length of the kernel is 3/5 as long as the major axis of the large-scale ellipse.

CORT-X 2 Output: The final output of the CORT-X 2 filter is the sum of the multiple scale interaction and the cooperative process:

$$B(i,j) = B_{12}(i,j) + B_2(i,j). \tag{27}$$

Appendix B: Fuzzy ARTMAP Equations

In the following, all architectural references are to Figure 11. Three parameters determine Fuzzy ART dynamics: a choice parameter $\alpha > 0$; a learning rate parameter $\beta_a \in [0,1]$; and a vigilance parameter $\rho_a \in [0,1]$.

Input preprocessing: Input **A** into a Fuzzy ART module is normalized by preprocessing the vector A as:

$$\mathbf{A} = \frac{\mathbf{a}}{|\mathbf{a}|} \tag{28}$$

where the norm operator, $| * |$ is defined as

$$|\mathbf{a}| \equiv \sum_{k=1}^{M} |a_k|. \tag{29}$$

Inputs are then **complement coded** by setting $\mathbf{A} = (\mathbf{a}, \mathbf{a}^c)$ where

$$a_k^c \equiv 1 - a_k. \tag{30}$$

Category choice: Category choice is determined by choosing the maximum choice function T_j^a,

$$T_J^a = max\{T_j^a : j = 1 \dots N\} \tag{31}$$

where N is the number of nodes in F_2^a, and T_j^a is defined by

$$T_j^a(\mathbf{A}) = \frac{|\mathbf{A} \wedge \mathbf{w}_j^a|}{\alpha + |\mathbf{w}_j^a|}, \tag{32}$$

where $\mathbf{w}_j^a$ is the template belonging to the jth F_2^a node, and the operator $\wedge$ is defined as the fuzzy AND operation.

$$(p \wedge q)_k \equiv \min(p_k, q_k) \tag{33}$$

for M-dimensional vectors **p** and **q** (Zadeh, 1965). If more than one T_j^a is maximal, the category j with the smallest index is chosen. When category J is chosen, the node representing that category has activation $y_J^a = 1$. $y_j^a = 0$ for $j \neq J$. The F_1^a activity vector $\mathbf{x}^a$ obeys

$$\mathbf{x}^a = \begin{cases} \mathbf{A} & \text{if } F_2^a \text{ is inactive} \\ \mathbf{A} \wedge \mathbf{w}_J^a & \text{if the } J^{th} F_2^a \text{ node is chosen.} \end{cases} \tag{34}$$

Match resonance and mismatch reset: A match and resonance are said to occur if

$$\frac{|\mathbf{A} \wedge \mathbf{w}_J^a|}{|\mathbf{A}|} \geq \rho_a. \tag{35}$$

Otherwise, a mismatch reset occurs. When a reset occurs, the currently active node J is set to zero for the duration of the current input presentation and a new maximal node is found until the chosen node J satisfies equation (35).

Map Field: The Map Field F^{ab} becomes active whenever one or both of ART_a or the supervised input b is active. Supervised inputs b_k are in 1-to-1 correspondence with nodes in the Map Field. If node J of F_2^a is active, then its weights $\mathbf{w}_J^{ab}$ activate F^{ab}. If a supervised input b_k becomes active, it directly activates F^{ab} node x_k^{ab} at an activation level of $b_k = 1$. The F^{ab} activity vector $\mathbf{x}^{ab}$ obeys

$$\mathbf{x}^{ab} = \begin{cases} \mathbf{b} \wedge \mathbf{w}_J^{ab} & \text{if the Jth } F_2^a \text{ node is active and } \mathbf{b} \text{ is active,} \\ \mathbf{w}_J^{ab} & \text{if the Jth } F_2^a \text{ node is active and } \mathbf{b} \text{ is inactive,} \\ \mathbf{b} & \text{if } F_2^a \text{ is inactive and } \mathbf{b} \text{ is active,} \\ 0 & \text{if } F_2^a \text{ and } \mathbf{b} \text{ are inactive.} \end{cases} \tag{36}$$

Match tracking: At the start of an input presentation to ART_a, the vigilance parameter is set to a baseline vigilance $\rho_a = \bar{\rho}$. Parameter ρ_{ab} is the Map Field vigilance parameter with $0 \leq \rho_{ab} \leq 1$. If

$$|\mathbf{x}^{ab}| < \rho_{ab}|\mathbf{b}|, \tag{37}$$

then subject to

$$\rho_a \leq \rho_{max}, \tag{38}$$

match tracking causes ρ_a to be increased so that

$$\rho_a = \frac{|\mathbf{A} \wedge \mathbf{w}_J^a|}{|\mathbf{A}|} + \epsilon \tag{39}$$

where ϵ is a small positive constant. If match tracking causes equation (38) to be violated, then match tracking is inhibited and learning takes place. If after match tracking, equation (38) is still satisfied, then a new search cycle in ART_a leads to a different F_2^a node J with

$$|\mathbf{A} \wedge \mathbf{w}_J^a| \geq \rho_a|\mathbf{A}| \quad \text{and} \quad |\mathbf{b} \wedge \mathbf{w}_J^{ab}| \geq \rho_{ab}|\mathbf{b}| \ , \tag{40}$$

or, if no such node exists, ART_a is shut down till it next gets an input.

Learning: Once search ends and a winning node y_J^a is chosen in ART_a, the Fuzzy ART weight vector $\mathbf{w}_J^a$ of the winning node is updated according to

$$(\mathbf{w}_J^a)^{new} = \beta_a(\mathbf{A} \wedge (\mathbf{w}_J^a)^{old}) + (1 - \beta_a)(\mathbf{w}_J^a)^{old}, \tag{41}$$

where $\mathbf{w}_J^a(0) = 1$. The weight vectors $\mathbf{w}_j^a$, $j \neq J$ of non-winning nodes are not updated.

The weight vectors from the Fuzzy ART modules to the Map Field are updated according to

$$(w_{jk}^{ab})^{new} = \begin{cases} \beta_{ab} h[x_k^{ab}] + (1 - \beta_{ab})(w_{jk}^{ab})^{old} & \text{if } j = J \\ (w_{jk}^{ab})^{old} & \text{if } j \neq J \\ 1 & \text{initially} \end{cases} \tag{42}$$

where $h[x_k^{ab}] = 1$ if $x_k^{ab} > 0$, else $h[x_k^{ab}] = 0$. When $\beta_{ab} = 1$, Fuzzy ARTMAP is said be in fast learning mode, when $0 \leq \beta_{ab} < 1$, Fuzzy ARTMAP is in slow learning mode.

Winner-take-all: A choice, or winner-take-all network F^{wta} sits above the Map Field F^{ab}. A winner-take-all field is needed above the map field when slow learning has been in effect from ART_a to the Map Field because it is then possible that a chosen F_2^a node may read out activation to more than one node in the Map Field during test mode.

The winner-take-all field is implemented algorithmically as

$$x_j^{wta} = \begin{cases} 1 & \text{if } j = \arg\max_k \, [x_k^{ab}], \\ 0 & \text{otherwise.} \end{cases} \tag{43}$$

Since nodes in F^{wta} are in one-to-one correspondence with nodes in the supervised field F^b, the winning node in F^{wta} represents VIEWNET's choice of the 3-D object given the single 2-D view that the network has experienced. When evidence accumulation is used, a working memory, as in Section 10 interpolates the Map Field and the winner-take-all 3-D object recognition field.

REFERENCES

Bradski, G., & Grossberg, S. (1995). Fast learning VIEW.NET architectures for recognizing 3-D objects from multiple 2-D views. *Neural Networks,, in press.* Special issue on automatic target recognition.

Carpenter, G., & Grossberg, S. (1987). ART 2: Self-organization of stable category recognition codes for analog input patterns. *Applied Optics*, *26*, 4919–4930.

Carpenter, G., & Grossberg, S. (Eds.). (1991). *Pattern recognition by self-organizing neural networks.* Cambridge, MA: MIT Press.

Carpenter, G., & Grossberg, S. (1993). Normal and amnesic learning, recognition, and memory by a neural model of cortico-hippocampal interactions. *Trends in Neurosciences*, *16*, 131–137.

Carpenter, G., & Grossberg, S. (1994). Integrating symbolic and neural processing in a self-organzing architecture for pattern recognition and prediction. In Honavar, H., & Uhr, L. (Eds.), *Artificial intelligence and neural networks: Steps towards principled predictions*, pp. 387–421. San Diego: Academic Press.

Carpenter, G., Grossberg, S., & Iizuka, K. (1992a). Comparative performance measures of fuzzy artmap, learned vector quantization, and back propagation for handwritten character recognition. In *Proceedings of the International Joint Conference on Neural Networks*, Vol. I, pp. 794–799. Piscataway, NJ: IEEE Service Center.

Carpenter, G., Grossberg, S., Markuzon, N., Reynolds, J., & Rosen, D. (1992b). Fuzzy ARTMAP: A neural network architecture for incremental supervised learning of analog multidimensional maps. *IEEE Transactions on Neural Networks*, *3*, 698–713.

Carpenter, G., Grossberg, S., & Mehanian, C. (1989). Invariant recognition of cluttered scenes by a self-organizing ART architecture: CORT-X boundary segmentation. *Neural Networks*, *2*, 169–181.

Carpenter, G., Grossberg, S., & Reynolds, J. (1995). Fuzzy ARTMAP, slow learning and probability estimation. *IEEE Transactions on Neural Networks, in press.* Boston, MA: Boston University Technical Report CAS/CNS-TR-93-014.

Carpenter, G., Grossberg, S., & Rosen, D. (1991). Fuzzy ART: Fast stable learning and categorization of analog patterns by an adaptive resonance system. *Neural Networks, 4*, 493–504.

Grossberg, S. (1987a). *The adaptive brain, Volumes 1 and 2.* Amsterdam: Elsevier/North-Holland.

Grossberg, S. (1987b). Cortical dynamics of three-dimensional form, color, and brightness perception. I: Monocular theory. *Perception and Psychophysics, 41*, 87–116.

Grossberg, S. (Ed.). (1988). *Neural networks and natural intelligence.* Cambridge, MA: MIT Press.

Grossberg, S. (1994). 3-D vision and figure-ground separation by visual cortex. *Perception and Psychophysics, 55*, 48–120.

Grossberg, S. (1995). Are there universal principles of brain computation?. *American Scientist*, in press. Tech Report: Boston, MA: Boston University CAS/CNS-TR-95-012.

Grossberg, S., & Mingolla, E. (1985). Neural dynamics of perceptual grouping: Boundary completion, illusory figures, and neon color spreading. *Psychological Review, 92*, 173–211.

Grossberg, S., Mingolla, E., & Todorović, D. (1989). A neural network architecture for preattentive vision. *IEEE Transactions on Biomedical Engineering, 36*, 79–102.

Grossberg, S., Mingolla, E., & Williamson, J. (1995). Synthetic aperture radar processing by a multiple scale neural system for boundary and surface representation. *Neural Networks, Special Issue on Automatic Target Recognition*, in press. Boston University Technical Report CAS/CNS-TR-94-001.

Grossberg, S., & Todorovic, D. (1988). Neural dynamics of 1-D and 2-D brightness perception: A unified model of classical and recent phenomena. *Perception and Psychophysics, 43*, 241–277.

Grossberg, S., & Wyse, L. (1991). A neural network architecture for figure-ground separation of connected scenic figures. *Neural Networks, 4*, 723–742.

Grossberg, S., & Wyse, L. (1992). A neural network architecture for figure-ground separation of connected scenic figures. In Pinter, R., & Nabet, B. (Eds.), *Nonlinear vision: Determination of neural receptive fields, function , and networks*, pp. 516–543. Boca Raton, FL: CRC Press, Inc.

Logothetis, N., Pauls, J., Buelthoff, H., & Poggio, T. (1994). View-dependent object recognition by monkeys. *Current Biology, 4*, 401.

Poggio, T., & Girosi, F. (1990). Regularization algorithms for learning that are equivalent to multilayer networks. *Science, 247*, 978–982.

Schwartz, E. (1977). Spatial mapping in primate sensory projection: Analytic structure and relevance to perception. *Biological Cybernetics, 25*, 181–194.

Seibert, M., & Waxman, A. (1990a). Learning aspect graph representations of 3-D objects in a neural network. In *Proceedings of the International Joint Conference on Neural Networks (IJCNN-90), Washington, D.C.*, Vol. 2, pp. 233–236.

Seibert, M., & Waxman, A. (1990b). Learning aspect graph representations from view sequences. In Touretzky, D. (Ed.), *Advances in neural information processing systems 2*, pp. 258–265. San Mateo, CA: Morgan Kaufmann Publishing.

Seibert, M., & Waxman, A. (1991). Learning and recognizing 3-D objects from multiple views in a neural system. In Wechsler, H. (Ed.), *Neural networks for perception.* New York, NY: Academic Press.

Seibert, M., & Waxman, A. (1992). Adaptive 3-D-object recognition from multiple views. *IEEE Transactions on Pattern Analysis and Machine Intelligence, 11*, 107–124.

Zadeh, L. (1965). Fuzzy sets. *Information Control, 8*, 338–353.